Corrosion Science and Technology

Corrosion Science and Technology

Editor: Theodore Schneider

|STATES|
ACADEMIC PRESS
www.statesacademicpress.com

Published by States Academic Press
109 South 5th Street,
Brooklyn, NY 11249, USA
www.statesacademicpress.com

Corrosion Science and Technology
Edited by Theodore Schneider

International Standard Book Number: 978-1-63989-128-3 (Hardback)

Cataloging-in-Publication Data

Corrosion science and technology / edited by Theodore Schneider.
p. cm.
Includes bibliographical references and index.
ISBN 978-1-63989-128-3
1. Corrosion and anti-corrosives. 2. Chemistry, Technical. 3. Metals--Surfaces.
4. Protective coatings. I. Schneider, Theodore.
TA418.74 .C67 2022
620.112 23--dc23

Contents

Preface

This book has been an outcome of determined endeavour from a group of educationists in the field. The primary objective was to involve a broad spectrum of professionals from diverse cultural background involved in the field for developing new researches. The book not only targets students but also scholars pursuing higher research for further enhancement of the theoretical and practical applications of the subject.

Corrosion refers to the damage and destruction of materials caused by chemical reactions with their environment. It leads to the conversion of a refined metal to a form that is more chemically-stable like sulfide, hydroxide and oxide. The field that is concerned with the control and prevention of corrosion is referred to as corrosion science. Many types of corrosion are studied under corrosion science and technology. Some of them are general corrosion, localized corrosion, marine corrosion, atmospheric corrosion, high temperature corrosion, microbiologically induced corrosion, etc. Some common methods used in protection from corrosion are cathodic protection, applying coatings, anodization, surface treatments, anodic protection, etc. This book outlines the applications of corrosion science and technology in detail. It elucidates new techniques and their applications in a multidisciplinary manner. Those in search of information to further their knowledge will be greatly assisted by this book.

It was an honour to edit such a profound book and also a challenging task to compile and examine all the relevant data for accuracy and originality. I wish to acknowledge the efforts of the contributors for submitting such brilliant and diverse chapters in the field and for endlessly working for the completion of the book. Last, but not the least; I thank my family for being a constant source of support in all my research endeavours.

Editor

The Analysis of the Influence of Various Factors on the Development of Stress Corrosion Defects in the Main Gas Pipeline Walls in the Conditions of the European Part of the Russian Federation

A. V. Afanasyev,[1] **A. A. Mel'nikov,**[1] **S. V. Konovalov**[ID][1,2] **and M. I. Vaskov**[3]

[1]*Samara National Research University, 34 Moskovskoye Shosse, Samara 443086, Russia*
[2]*Wuhan Textile University, 1 Fang Zhi Road, Wuhan 430073, China*
[3]*"Gazprom Transgaz Samara" LLC, 106a Novo-Sadovaya Str., Samara 443086, Russia*

Correspondence should be addressed to S. V. Konovalov; ksv@ssau.ru

Academic Editor: Jerzy A. Szpunar

This paper considers the factors influencing the formation and development of stress corrosion defects detected during the inspection and overhaul of the main gas pipeline section. The surveyed gas pipeline is made of large diameter steel pipes made by controlled rolling, produced by various companies, with the predominance of pipes produced by the Khartsyzsk Pipe Plant (KhPP). The correlation between the geometric parameters of defects is described, which makes it possible to estimate the depth of cracks by external parameters. Mechanical tests by cyclic loading of samples containing cracks, based on the site operation data for the last 11 years, showed no crack growth in the absence of a corrosive medium. Micro-X-ray spectral analysis of metal and corrosion products showed no trace of the influence of hydrogen sulphide and nonmetallic inclusions (sulphides) on the development process of SCC. According to the results of the research, the process of development of stress corrosion on the main gas pipelines located in the European part of the Russian Federation is described. The organization operating the gas pipeline is recommended to take into consideration the results of this work during drawing up their repair plan.

1. Introduction

To date, more than a third (36%) of accidents occur on the main gas pipelines (MG) of the Unified Gas Supply System (UGSS) belonging to «Gazprom» PJSC as a result of the development of stress corrosion defects or stress corrosion (hereinafter referred to as SCC). In the world practice of transporting natural gas through pipelines, this type of damage has the highest specific gravity among all other causes of accidents; therefore, during diagnostic examinations this type of defects is given the highest priority [1–3]. At the same time, along with the improvement of diagnostic tools, the number of newly detected CTC defects on UGS facilities is growing every year. If earlier, about 2,000 CWN defects were detected per year by methods of in-tube flaw detection (ILI) by magnetic shells; now with the use of electromagnetic-acoustic projectiles (EMA) this figure reaches 10,000 defects per year [4]. A large number of SCC defects are revealed by methods of nondestructive testing (NDT) in pits and during major overhauls. The absolute majority of defects detected by all inspection methods (almost 92%) have a measured depth of less than 10% of the wall thickness [5, 6]. According to experts, SCC crack of different depths was identified or is more likely to be detected in more than a million pipe sections during the next surveys [7–9]. This statement is consistent with the current trend to increase the number of small defects found.

The repair by replacing even a small part of these pipes will lead to a reduction in the total volume of UGSS pipeline overhaul due to the specific increase in the individual sections repair cost [10–12]. At the same time, the degree of defects' danger, the depth of which is less than 10–15% of the wall

FIGURE 1: Investigated pipeline section.

thickness, under condition that access of corrosive medium to them is restricted, is determined by many researchers as insignificant [13–16]. At the present time, there are methods of repairing the pipelines polymer insulation, which do not allow the cracks to develop and threaten the reliability of pipelines for a long time [17, 18]. However, methods for an accurate assessment of the SCC cracks depth are not developed, and the factors that influence the distribution of SCC defects in extended sections are not always considered systematically during major overhauls.

Thus, the development of systematic methods of combating negative manifestations of stress corrosion in the form of accidents (identification, assessment, and targeted removal of only those defects that may lead to an accident in the foreseeable future with account of repair of insulation on other defects) is an urgent task.

The purpose of this study is to determine the factors that affect the distribution of SCC defects and the rate of their development during the main gas pipelines operation period. The second goal is to determine the parameters by which the depth of cracks can be estimated.

2. Materials and Methods

2.1. Field Studies on the Gas Pipeline Extended Section. For detailed consideration of operational factors and sampling, a section of main gas pipeline located in the eastern part of the European part of the Russian Federation with a length of 25 km was selected (Figure 1).

The selection criteria for the site were the presence of pipes of various types and the possibility of subsequent objective control of their technical condition. It was planned to evaluate the cyclicity over a long period of operation and compare the data of the ILI and the rejection results obtained during upcoming major overhaul (100% NDT of the pipe surface). In addition, a comparison of the electrometric and physical profile of the section with a real distribution of SCC defects was carried out.

2.2. Estimation of Cracks Depth by Their External Manifestations. To create a mechanism for estimating the depth of SCC cracks using direct measurements of their geometric parameters, 15 samples were made from the rejected pipes, and 157 cracks were selected and described on these samples (Figure 2).

The surface of the samples was polished, the largest cracks in the colony were selected, and their length and width were measured. After that, the sample was cut and the depth of the cracks was directly measured on the cross section.

2.3. Cyclic Tests with a Load Similar to the Operational Load. To determine the effect of cyclic loading on the development of SCC defects under operating conditions with the excluded access of the corrosive medium to cracks, cyclic tests of 4 model samples were carried out under four-point loading conditions with accordance to the accepted industry method [19]. To justify the choice of cyclic test modes, the whole range of pressure fluctuations over a period of 11 years of operation was constructed based on the data from the operating modes log for selected pipeline section. The average value of the pressure (P_{av}) was defined as 5.84 MPa and the cycles with the greatest pressure deviations were identified. During the selected period, the pipeline section experienced 18 cycles of loading and unloading with the maximum deviation from the P_{av} in the range of 17–35%.

In the full spectrum of oscillations, a time interval (T) was determined, inside which the upper and lower half-periods were analyzed. In these half-cycles, the cycles with the greatest amplitude deviations from P_{av}, equal to 2÷11% of the working pressure (P_w), were identified. The parameters of the cyclic pressure change processes in each half-period (P_{cpc}) were calculated (Figure 3). Based on the analysis of operating conditions of the investigated section of the main gas pipeline, it was found that the total number of pressure drop cycles within one year could be divided into two parts: 30 pulsation cycles with an amplitude in the range of 2÷11% of P_w and 2 major cycles with an amplitude in the range of 17-35% of P_w.

As a result of the analysis the most conservative loading regime was chosen for the testing of model samples, which simulated the work of mail gas pipeline for 20 years of operation in real conditions. The load variation interval varied from 1.1 MPa to 7.4 MPa (the maximum working pressure allowed on the pipeline section), which is characterized by the asymmetry coefficient of the cycle (R = 0.15). The total number of cycles was calculated as the total number of all types of cycles affecting the gas pipeline for the year, multiplied by the planned interval of operation, equal to 20 years. Thus, the loading mode of the model samples was carried out in three stages with a change in the load every 640 cycles. The tests were carried out on a tensile machine using a four-point loading scheme (Figure 4).

To monitor the state of stress corrosion cracks, one control crack was selected on each sample, the photographic images of which were recorded with a metallographic microscope before the start of the tests, after the second stage, and at the end of the third stage of the test. The control was carried out relative to the initial state programmatically by measuring the number of pixels in the image along the line connecting the beginning and end of the crack.

2.4. Electron Microscopy. Cracks, cuts, and open corrosion cracks were studied by electron microscopy and micro-X-ray spectral analysis. In the course of the study, micro-X-ray spectral analysis of the elemental composition of corrosion

FIGURE 2: Samples with SCC cracks.

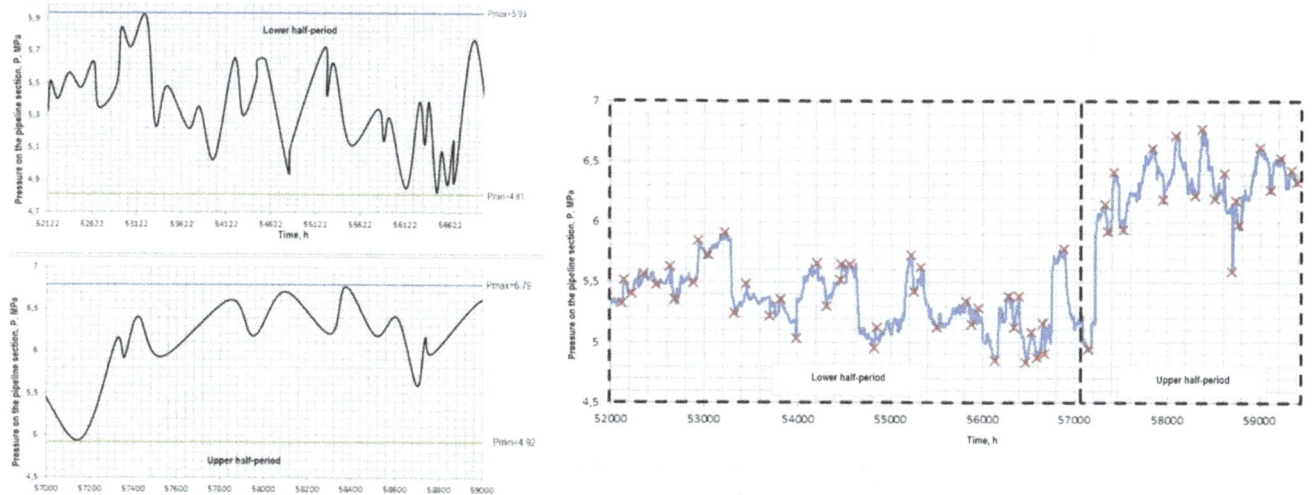

FIGURE 3: The upper and lower half-cycles of the oscillations in the chosen time interval (T).

products was carried out, and surface maps of the distribution of elements along the surface of destruction were made.

The fractographic and microscopic analysis of the crack surface was carried out on a TESCAN scanning electron microscope with VEGA software at magnifications up to 4000 times [19].

Spectral analysis to determine the quantitative chemical composition of the elements was carried out using the OXFORD INCAx-act energy dispersing attachment in accordance with ISO 15632: 2002.

Identification of the fracture behaviour by the type of surface relief was carried out in accordance with [20].

The general view of the TESCAN VEGA SBH Easy Probe scanning electron microscope and the OXFORD INCAx-act energy dispersive attachment is shown in Figure 5.

3. Results and Discussion

The density of the SCC defect distribution detected by the NDT during the overhaul was more than 97 times higher than expected by the results of the ILI. The average density of SCC defects on single-seam pipes (imported) is 50 times less than on double-seam pipes produced at Khartsyzsk Pipe Plant (KhPP). It was noted that stress corrosion damage localized mainly near the longitudinal weld. SCC defects of maximum

depth (up to 36% of the wall thickness) were also found on the KhPP pipes.

The density of the distribution of SCC defects correlates with the number of KhPP pipes in the pipeline layout. At the same time, the noted characteristics of the external environment, as the value of the electrical resistance of the soil and the height differences along the section profile, do not significantly affect the distribution of defects. (Figure 6)

During the overhaul, more than 13.5 km of pipes were rejected on the site, more than 11.5 km (more than 85%) of them due to SCC defects. In total, more than 58% of all the double-seam KhPP pipes inspected at the 25-kilometer section were rejected due to SCC defects. For single-seam pipes, including imported pipes, the rejection rate due to SCC was only 7%.

The majority of defects detected during NDT (more than 92%) have a depth of less than 10% of the wall thickness, which is below the detection threshold of magnetic in-tube flaw detectors. Therefore, undetected SCC defects can be present on other sections of the pipeline, and inspection with more sophisticated in-tube flaw detectors or NDT inspections in pits (for example, during the overhaul by reinsulation) will most likely reveal them.

During the research, it was established that there is a dependence between the main external parameters of SCC

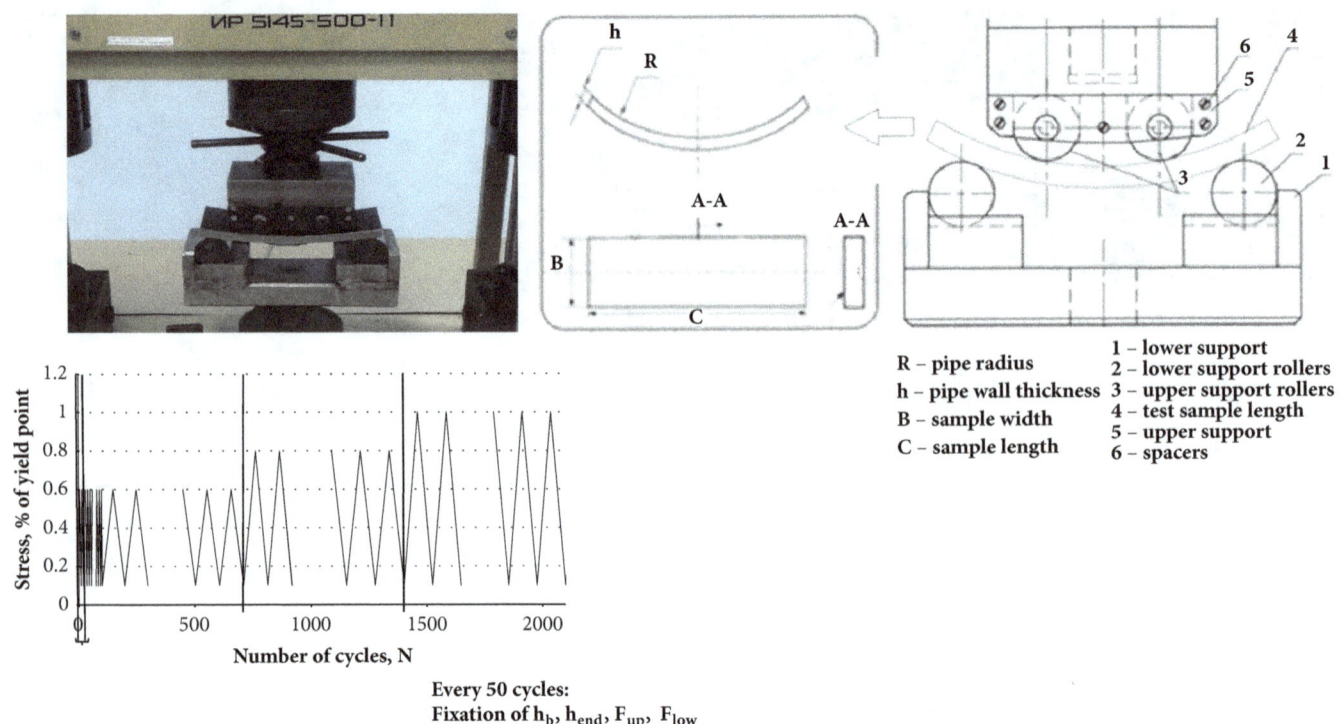

R – pipe radius
h – pipe wall thickness
B – sample width
C – sample length

1 – lower support
2 – lower support rollers
3 – upper support rollers
4 – test sample length
5 – upper support
6 – spacers

Every 50 cycles:
Fixation of h_b, h_{end}, F_{up}, F_{low}

Every 640 cycles:
1. Photographing the ends of the notches on metallographic microscope
2. Fixing the test time
3. Raising the amount of load stress per 0.2 of sample yield strength

FIGURE 4: Cyclic testing of samples according to the four-point loading scheme.

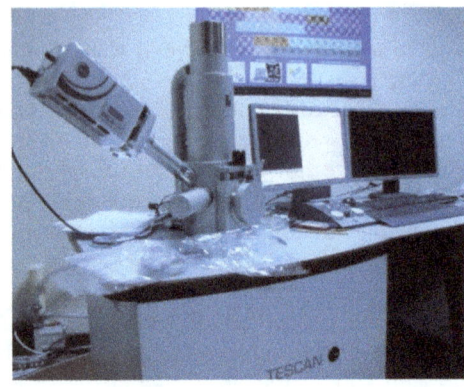

FIGURE 5: General view of the TESCAN VEGA SBH Easy Probe scanning electron microscope.

cracks. The crack length-to-width ratio was 10:1 and the width-to-depth ratio was 0.06:1. Thus, at a depth of 1-1.5 mm, the opening width is 0.06 mm; with a depth of 2-2.5 mm, the opening width is 0.12 mm (Figures 7(a) and 7(b)).

Measurement of the crack opening width was carried out using a measuring magnifier with an accuracy of ± 0.01 mm. In this case, the absolute limit error for the two measurements was ± 0.016 mm and the relative errors for each of the measurements were $\delta 1 = 28\%$ (section A) and $\delta 2 = 24\%$ (section B), respectively.

After carrying out the cyclic tests, which were controlled by the metallographic method (Figures 8 and 9), it was found that, in the absence of a corrosive environment, the stress corrosion crack retained its original state after all the test stages. During the tests, no formation of new cracks-branches and other changes in the morphology of the crack apex was recorded. Thus, during the three stages of testing cycles simulating the operation of the real pipeline section (load-unload cycles), the development of existing stress corrosion cracks has not occurred (Figure 9).

Ten cracks were opened out by excessive force in order to study the resulting fracture. Analysis of the fracture surface at small magnifications showed that the upper part of the crack is completely filled with oxides. The soil brine constantly permeated and oxidized the internal surface of the crack during the operation life of the pipeline. The oxides formed also wedged and deformed the inner surface. Micro-X-ray spectral analysis of the elemental composition of corrosion products also showed that the cracks are completely filled with oxide. The distribution of various elements along the fracture surface is represented on the distribution maps of the elements (Figure 10).

As it can be seen from the distribution maps of the elements, there are no obvious phase boundaries. Corrosion

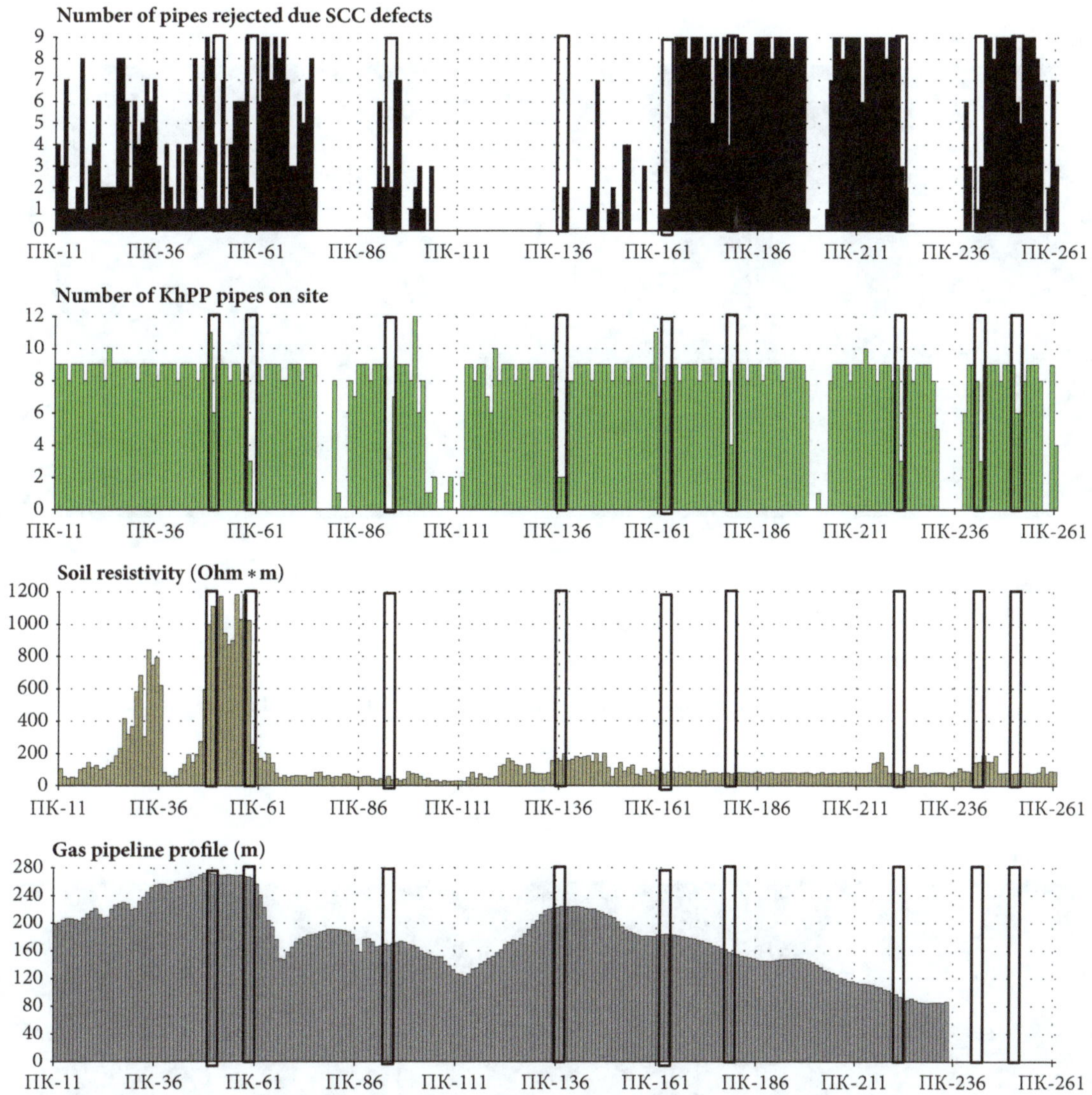

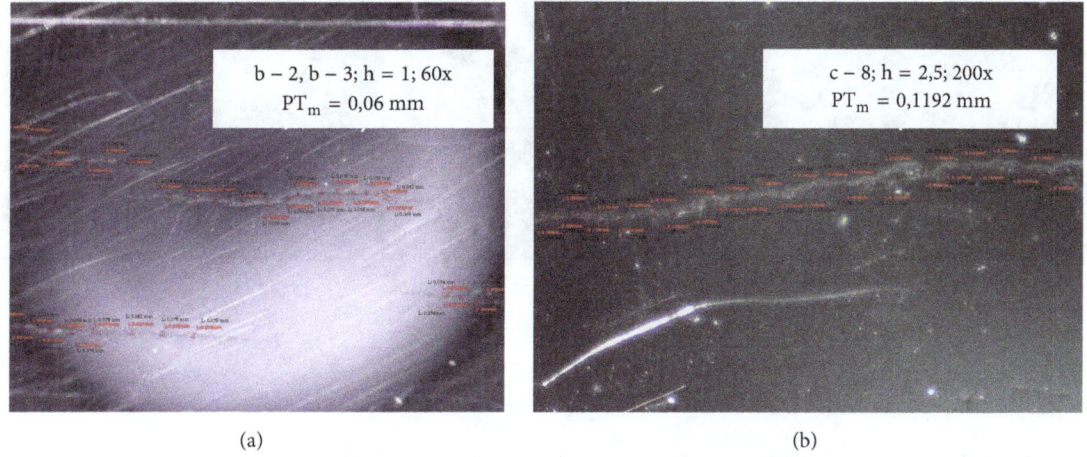

FIGURE 6: Number of pipes with SCC defects, number of KhPP pipes, soil characteristics, and section profile. The X-axis indicates the distance along the pipeline route. Pipeline sections with a small number of pipes produced by KhPP are marked in red.

FIGURE 7: Microscopic examination of cracked sections.

Base metal

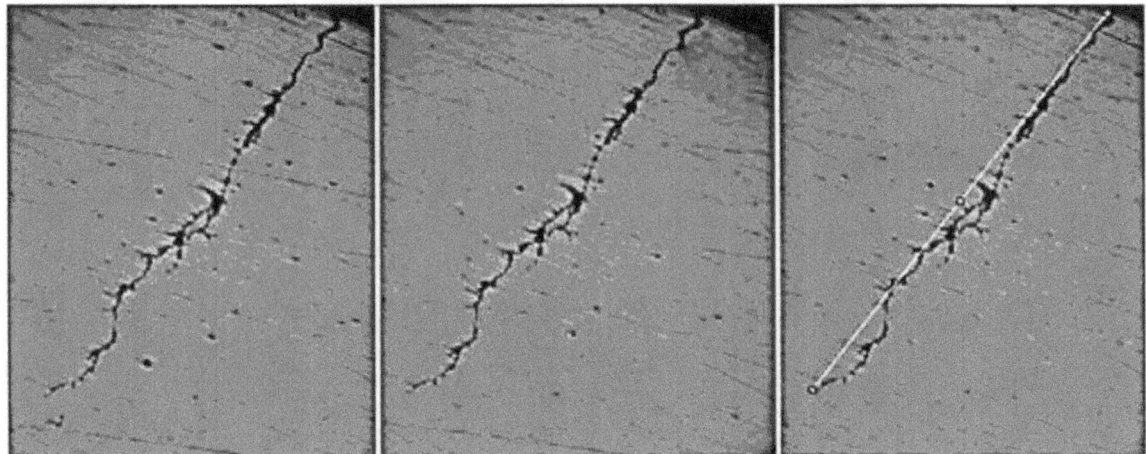

Results

	Mean	Angle	Length
1	166.487	-127.035	347.001
2	110.040	-127.569	344.426
3	113.190	-127.730	346.439

FIGURE 8: Results of crack depth stepwise measurement in the base metal of model sample.

weld metal

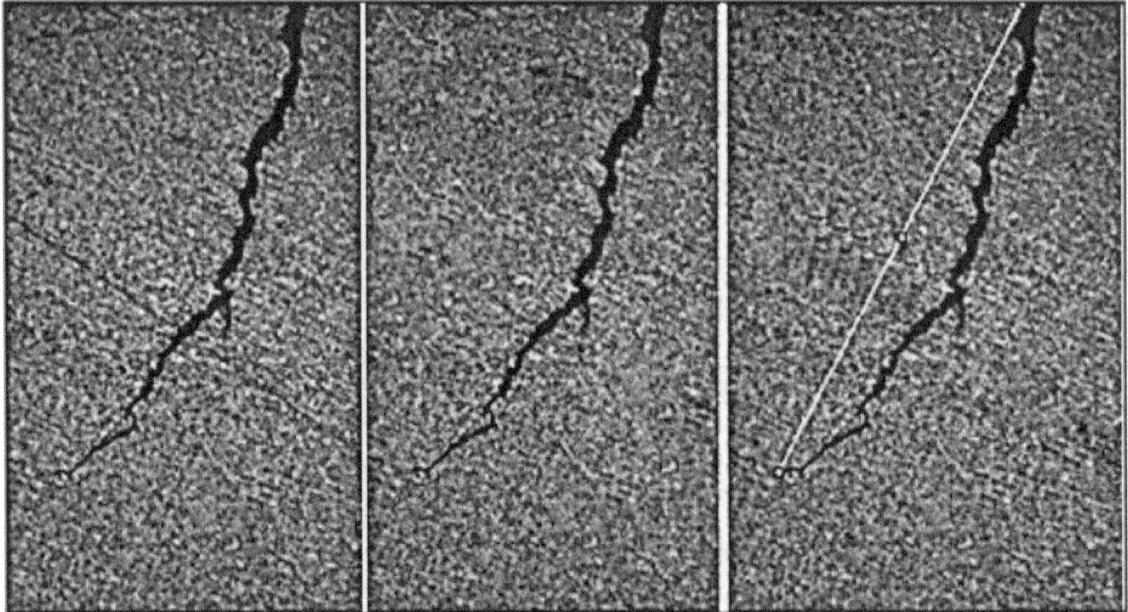

Results

	Mean	Angle	Length
1	106.151	-117.282	286.021
2	104.792	-117.196	284890
3	105.432	-117.375	285118

FIGURE 9: Results of crack depth stepwise measurement in the model sample with longitudinal weld section.

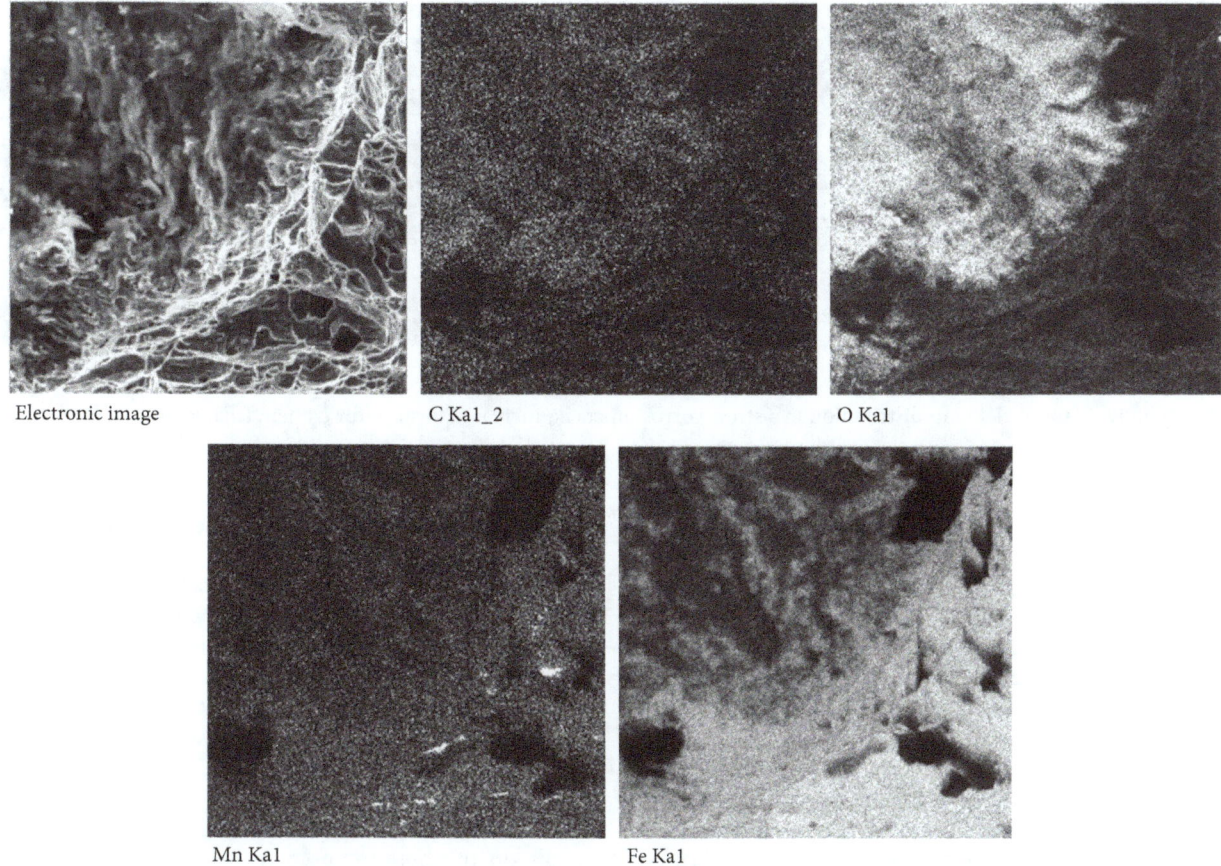

Electronic image

C Ka1_2

O Ka1

Mn Ka1

Fe Ka1

FIGURE 10: Elements distribution maps.

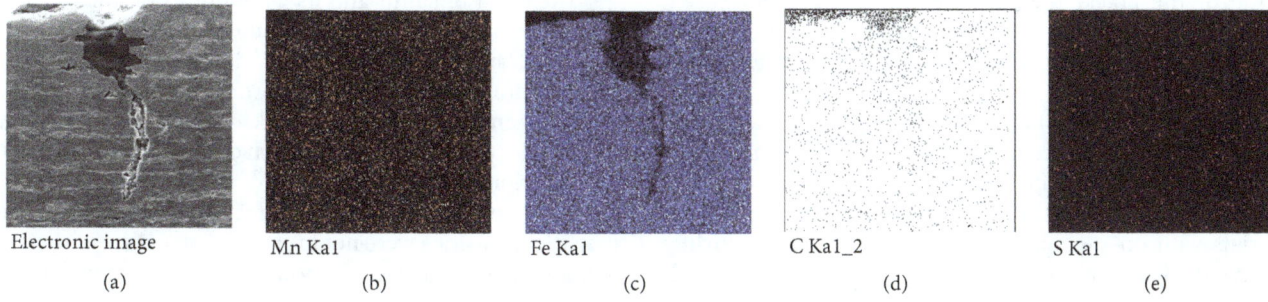

Electronic image   Mn Ka1   Fe Ka1   C Ka1_2   S Ka1

(a)   (b)   (c)   (d)   (e)

FIGURE 11: Maps of elements distribution in a stress corrosion crack: (a) electronic image of the crack; (b) distribution of manganese; (c) distribution of iron; (d) distribution of carbon; (e) distribution of sulfur.

products are evenly distributed inside the crack. The images in the iron, oxygen, and carbon spectra are homogeneous and equally bright, indicating either a monophasic break in the oxide or even mixture of several phases. The expected composition of corrosion products is either a homogeneous Fe_2O_3 or a mixture of Fe_2O_3 and $FeCO_3$.

To determine the effect of metallurgical impurities and additives on the development of cracks, the surface of the sections was also mapped in the spectra of sulfur and manganese. Due to limitations of the method, spectrometric data were not used to quantify the ratio of oxygen and carbon in corrosion products and in the base metal. During the

mapping process, the main goal was to identify points with a local increased content of individual chemical elements corresponding to nonmetallic inclusions, their subsequent detailed analysis in case of revealing their influence on the process of material destruction, and subsequent detailed analysis of such points in case of revealing their influence on the process of material destruction.

The maps show an image of cracks in the samples in the spectra of iron, oxygen, carbon, sulfur, and manganese (Figure 11). An increased sulfur content in corrosion products was not observed, as evidenced by the different brightness of the elements spectrum image in the transition from the base

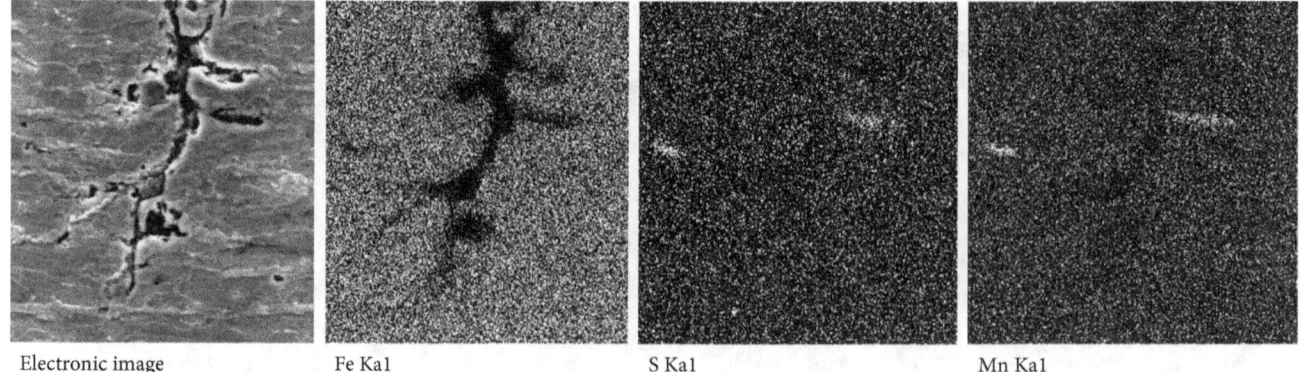

Electronic image   Fe Ka1   S Ka1   Mn Ka1

FIGURE 12: Maps of elements distribution in a stress corrosion crack. Increased local sulfur content (manganese sulphide).

metal to the crack. A reduced content of iron and manganese and a high content of oxygen and carbon in the products of corrosion were noted (Figure 11).

The sulfur content in the corrosion products does not exceed the sulfur content in the base metal. Images in the spectra of manganese and sulfur are homogeneous and equally bright both in the cross section of the base metal and in the section of the crack. In some samples, a lower sulfur content can be noted with the exception of some local points (Figure 12).

Thus, the coincidence of the increased concentration of manganese and sulfur in some sections of the map indicates the presence of sulphide nonmetallic inclusion. As seen in the electronic image, these nonmetallic inclusions are not the sources of development of destruction. In the remaining volume of the crack, the sulfur content does not exceed the sulfur content of the metal, and the total contamination of the material with sulfur does not go beyond technical specifications.

Based on the data obtained as a result of the research, it is possible to assume a scenario for the development of SCC defects in the main gas pipeline.

During the construction of the pipeline, film insulation materials with on-site application were used, which during the period of operation lost their necessary technological properties. This is reflected in the fact that, during the operation of the pipeline, corrugations and pockets appeared in some places on the insulation coating. Corrosion-active soil electrolyte penetrated damaged insulation areas.

Later, these corrosion pits continued their growth and became concentrators of mechanical stresses in the pipeline wall. The source of these stresses were constant pressure pulsations from operating gas pumping units and gas temperature fluctuations due to the nonstationary operation of the gas compressor station.

The first microdeformations appeared, leading to the appearance of sharp cracks. Further, the cracks developed irregularly, under the influence of operational loads. In the development of the defect, a continuous process of corrosion was of decisive importance, stimulated by a constant inflow of heat along with the transported medium. The soil brine continued to penetrate into the cracks, wash out corrosion

products, and deliver portions of the new electrolyte to the crack vertex.

Analysis of the effectiveness of various methods of non-destructive testing has shown that a colony of cracks can be identified by all available mass methods used in the oil and gas industry. However, in order to improve the efficiency of SCC defect detection, it is necessary to take into account and track factors that increase the predisposition of sites to this type of destruction.

The morphology of the detected defects corresponds to modern concepts of SCC defects. Defects are identified as cracks on the bottom of corrosive pits. They are grouped in colonies, oriented along the axis of the tube during their growth and branch along the section with the tendency to merge with one another.

The effect of sulfur and its compounds on the process of stress corrosion is not a significant factor, as there was no increased sulfur content in corrosion products. The phase composition of the corrosion products within the cracks is homogeneous across the crack cross section. The phase boundaries and the sections, which differ sharply in chemical composition, were not found.

Traces of the predominant effect of impurities and non-metallic inclusions were not found. The structure of the metal is homogeneous and corresponds to the structure of rolled metal.

It was found that there is a consistent pattern between the width of the opening and the length of the crack. A simple statistical generalization showed that the length-to-width ratio could be described as 10:1 and the width-to-depth ratio as 0.06:1. Since the majority of SCC cracks in the investigated samples are the same in origin, size, and morphology, it can be said that such a regularity exists in this particular case of SCC.

The appearance of cracks can be influenced by a variety of technological factors, laid down at various stages of the life cycle of the product. Such a factor can be the contamination of the pipe metal by harmful impurities and the presence of residual stresses in the pipe wall after and during manufacture at the stage of pipe production. In addition, these factors can include increased values of stresses in the pipeline wall, which arose due to flaws in construction design and installation work at the pipeline construction stage. This includes the nonprojected position of the pipe in the trench, unplanned soil movements, and welding stresses. The nonstationary operation mode of the pipeline also has a great influence, namely, the constant pulsation of the pressure and temperature of the pumped medium. Equally important is the corrosive activity of the ground in contact with the pipeline and the quality of the anticorrosive insulation at the stage of gas pipeline operation period.

Film insulation with on-site application does not provide long-term protection of the pipe body from SCC defects. After several years, due to soil movement and degradation of the coating, corrugations and pockets are formed in it, which collect soil moisture. Such conditions create a favorable environment for the development of corrosion defects. In the future, under the influence of these conditions and in the presence of an appropriate stress state, SCC defects can form. The insulation coatings used in newly constructed and repaired pipelines must provide long-term reliable protection of the pipe body against the corrosive effects of the environment in the operating conditions of main gas pipelines.

The main factor that can be detected and affects the distribution of stress corrosion, with other factors being equal, are the technological features of the pipe materials embedded in the product during its production stage. Presumably, this is the degree of local plastic deformation and the number of transitions during the formation of the pipe and sheet material and the thermal and high-speed welding of the longitudinal seam of two-seam pipes. To confirm this assumption, it is necessary to assess the residual stresses on the wall of the affected pipe and draw a map of them.

The presence of a corrosive medium is a prerequisite for the destruction of the pipeline wall due to the development of a SCC defect.

4. Confirmations

The presence of empirical dependence should be verified on a large statistical sample. Also, it is necessary to connect the external manifestations of SCC (the length of cracks and their opening) with the depth of stress corrosion damage. If such a relationship is confirmed in other cases of SCC, then it will be possible to describe the growth of a crack and the rate of its propagation in the general case with the sufficient accuracy. It will also be possible to build a mathematical model of the crack and calculate the residual life cycle of the pipeline using the finite element method.

Such an approach will make it safer to reuse the KhPP double-seam pipes during major overhaul, which is allowed only after a comprehensive evaluation of each pipe. In determining the degree of danger of each specific defect, a methodology for estimating the static strength of gas main pipeline with colonies of corrosion cracks is tested and used in "Gazprom transgaz Samara" LLC [21]. If, in addition to the nondestructive testing methods used in the survey, the external crack parameters are used and the calculation is performed using the largest depth obtained by different methods, then the assessment of the SCC crack danger will become more accurate.

The organization operating the gas pipeline is recommended to take into consideration the results of this work during drawing up their repair plan.

Conflicts of Interest

The authors declare that there are no conflicts of interest regarding the publication of this paper.

References

[1] S. V. Alimov, A. B. Arabej, and I. V. Ryaxovskix, *The concept of diagnosis and repair of main gas pipelines in regions with a high predisposition to stress corrosion Gazovaya promyshlennost*, vol. 724, 2015.

[2] Y. E. Cheng, *Stress Corrosion of Pipeline*, vol. 257, John & Sons Publishing, Hoboken, NJ, USA, 2013.

[3] F. King, "Stress corrosion cracking of carbon steel used fuel containers in a Canadian deep geological repository in sedimentary rock," report No. NWMO TR-2010-21, Toronto, Canada, 2010.

[4] A. I. Mixajlov, *Detection, identification and assessment of the depth of stress corrosion using combined magnetoacoustic in-tube flaw detectors// III Scientific and Practical Seminar "Increasing the Reliability of Main Gas Pipes Affected by Stress Corrosion Cracking", "Gazprom VNIIGAZ" LLC*, Moscow, 2017.

[5] I. Ryakhovskikh, R. Bogdanov, T. Esiev, and A. Marshakov, "Stress corrosion cracking of pipeline steel in near-neutral pH environment," in *Proceedings of Materials Science Technology*, Pittsburgh, PA, USA, 2014.

[6] Y. A. Perlovich, O. A. Krymskaya, M. G. Isaenkova et al., "OP Conference Series: Materials Science and Engineering 10, Development, Production and Application," in *Proceedings of the 10th International School-Conference on Materials for Extreme Environment: Development, Production and Application, MEEDPA 2015*, p. 012009, 2016.

[7] A. I. Zaitsev, I. G. Rodionova, O. N. Baklanova, K. A. Udod, T. S. Esiev, and I. V. Ryakhovskikh, "Structural factors governing main gas pipeline steel stress corrosion cracking resistance," *Metallurgist*, vol. 57, no. 7-8, pp. 695–706, 2013.

[8] V. Linton, E. Gamboa, and M. Law, "Strategies for the repair of stress-corrosion cracked gas transmission pipelines: Assessment of the potential for fatigue failure of dormant stress-corrosion cracks due to cyclic pressure service," *Journal of pipeline engineering*, vol. 6, no. 4, pp. 207–217, 2007.

[9] U. Marewski, *UKOPA/GP/009. Near neutral pH and high pH stress corrosion cracking: industry good practice guide*, U. Marewski and M. Steiner, Eds., UK onshore pipeline operators' association, Ambergate, Derbyshire, 2016.

[10] M. Gintten, "An integrated approach to the integrity management of stress corrosion cracking in pipelines: a case study," in *Proceedings of the Proc. of Rio Pipeline Conference Exposions*, M. Ginten, T. Penney, I. Richardson et al., Eds., 2013.

[11] A. B. Arabej, O. N. Melyoxin, I. V. Ryaxovskix et al., "Research into the possibility of long-term operation of pipes with low stress corrosion damage," *Vesti gazovoj nauki*, vol. 27, no. 3, pp. 4–11, 2016.

[12] R. V. Aginej, S. S. Gus'kov, V. V. Mussonov, R. A. Sadrtdinov, and V. A. Lapin, "Investigations of geometric parameters and peculiarities of the location of stress-corrosion damage on main gas pipelines," *Vesti gazovoj nauki*, vol. 27, no. 3, pp. 102–107, 2016.

[13] R. Bogdanov, A. Marshakov, V. Ignatenko, I. Ryakhovskikh, and D. Bachurina, "Effect of hydrogen peroxide on crack growth rate in X70 pipeline steel in weak acid solution," *Corrosion Engineering, Science and Technology*, vol. 52, no. 4, pp. 294–301, 2017.

[14] W. Chen, R. Kania, R. Worthingham, and G. V. Boven, "Transgranular crack growth in the pipeline steels exposed to near-neutral pH soil aqueous solutions: The role of hydrogen," *Acta Materialia*, vol. 57, no. 20, pp. 6200–6214, 2009.

[15] X. Chen, Q. Yuan, B. Madigan, and W. Xue, "Long-term corrosion behavior of martensitic steel welds in static molten Pb-17Li alloy at 550∘C," *Corrosion Science*, vol. 96, pp. 178–185, 2015.

[16] X. Chen, Z. Shen, X. Chen, Y. Lei, and Q. Huang, "Corrosion behavior of CLAM steel weldment in flowing liquid Pb-17Li at 480 ∘c," *Fusion Engineering and Design*, vol. 86, no. 12, pp. 2943–2948, 2011.

[17] A. B. Arabey, O. N. Melekhin, O. V. Burutin et al., "Studying the Possibility of Long-Term Operation of Pipes with Insignificant SCC," *3R*, no. 01-02, pp. 104–110, 2017.

[18] Y. Perlovich, O. Krymskaya, M. Isaenkova, N. Morozov, I. Ryakhovskikh, and T. Esiev, "Effect of layer-by-layer texture inhomogeneity on the stress corrosion of gas steel tubes," *Materials Science Forum*, vol. 879, pp. 1025–1030, 2017.

[19] C. J. B. Reed, *Electron Probe Analysis and Raster Microscopy (Translation)*, Moscow, Texnosfera, 2008.

[20] RD 50-672-88. Methodical instructions. Calculations and strength tests. Classification of metals fracture types// Moscow, VNIIGAZ, 2010– 56 p.

[21] V. A. Subbotin, I. V. Scherbo, S. A. Kholodkov, and M. G. Giorbelidze, "Estimation of static strength of sections of the linear pipeline portion with a colony of corrosion fine cracks," *Samara State Aerospace University Bulletin*, vol. 47, no. 5, pp. 151–157, 2015 (Russian).

Experimental Investigation on Corrosion Effect on Mechanical Properties of Buried Metal Pipes

Yingbo Hou,[1] Deqing Lei,[1] Shujin Li,[1] Wei Yang,[1] and Chun-Qing Li[2]

[1]School of Civil Engineering and Architecture, Wuhan University of Technology, Wuhan 430070, China
[2]School of Engineering, RMIT University, Melbourne, VIC 3001, Australia

Correspondence should be addressed to Chun-Qing Li; chunqing.li@rmit.edu.au

Academic Editor: Flavio Deflorian

Corrosion has been found to be the most predominant cause for failures of buried metal pipes. A review of published literature on pipe corrosion reveals that little research has been undertaken on the effect of corrosion on mechanical properties of pipe materials and almost no research has been conducted on corrosion effect on fracture toughness. The intention of this paper is to present a comprehensive test program designed to investigate the effect of corrosion on mechanical properties of metals in soil. Two types of metals, namely, cast iron and steel, are tested under corrosion in three different environments. A relationship between corrosion and deterioration of mechanical property of metals is developed. It is found in the paper that the more acidic the environment is, the more corrosion the metal undergoes and that the corrosion reduces both the tensile strength and fracture toughness of the metal. The results presented in the paper can contribute to the body of knowledge of corrosion behavior and its effect on mechanical properties of metals in soil environment, which in turn enable more accurate prediction of failures of buried metal pipes.

1. Introduction

Pipelines are essential infrastructure that play a significant role in a nation's economy, social well-being, and quality of life. Most of pipes are made of metals, for example, cast iron and steel, and located underground in soil. It is estimated that about 85% of water distribution pipes are cast iron and steel [1]. Due to their long-term service and exposure to aggressive environment in soil, aging and deterioration of metal pipes have resulted in an unexpected high rate of failures. For example, the failure rate of cast iron pipes can be as high as 39 bursts per 100 km per year in Canada [2] whilst the failure rate of water mains in Australia is 20 breaks per 100 km per year on average [3]. As is well appreciated, the consequence of pipe failures can be socially, economically, and environmentally catastrophic, resulting in massive disruption of daily life, considerable economic loss, widespread flooding, and subsequent environmental pollution and even casualties and so forth. Therefore, there is a well-justified need to thoroughly investigate the causes of pipe failures.

Experience and investigation of pipe failures suggest that corrosion of metals, both cast iron and steel, is the most predominant cause of pipe failures [4, 5]. Since corrosion is linked to almost all pipe failures, it has become a global problem for all stakeholders, in particular engineers and asset managers of buried metal pipes [6, 7]. As such, considerable research has been undertaken in the past few decades on corrosion of metal pipes, more perhaps for cast iron pipes, as represented notably by Doleac et al. [8], Dean Jr. and Grab [9], O'Day et al. [10], Randall-Smith et al. [11], Kirmeyer et al. [12], Camarinopoulos et al. [13], Sadiq et al. [14], Panossian et al. [15], and so on. Due to different environments, the mechanisms of corrosions are different for internal and external surfaces of the pipe. For internal corrosion, depending on the substance to be conveyed in the pipe, various factors, including microbial effects, can cause corrosion [16], whereas external corrosion is mainly due to corrosive chemicals in soil [17]. Pipe corrosion in soil is an interaction between the pipe materials and the soil environment [18]. There are several stimulating factors that lead to the pipe external corrosion in soil environment [5, 19]. Moisture, temperature, pH values, mineral salt content, sulfides, organics, precipitates, and so on are major factors that contribute to external corrosion of pipes in soil [20]. Metal corrosion in soils is determined primarily by a combined effect of these factors. It also depends on the physical and chemical characteristics of the soils.

A review of published literature on pipe corrosion, as cited above (and also see references), reveals that most of the current research focuses on corrosion mechanisms, corrosion progress, and corrosion rate from material perspective. Little research has been undertaken on the effect of corrosion on mechanical property change of pipe materials, and almost no research has been conducted on corrosion effect on fracture toughness of pipe materials. As is well known, it is the mechanical properties of the pipe materials that govern the behavior and eventual failure of the pipes. It is therefore imperative to thoroughly examine the effect of metal corrosion on its mechanical properties. The understanding and knowledge of corrosion induced deterioration of mechanical properties of metals can prevent future failures of metal pipes.

There are two main modes of pipe failure: by rupture due to the reduction of wall thickness of the pipes and by fracture due to the stress concentration at the tips of cracks, for example, corrosion pits or, in general, defects in the pipes [21]. The mechanical properties corresponding to these two failure modes are tensile strength and fracture toughness of the metal. A detailed examination of most published research in this area (see references) suggests that current research on corrosion induced pipe failures focuses more on loss of strength than toughness. An inspection of failures of trunk mains in service reveals that most cast iron water main failures are of fracture type; that is, the failure is caused by the growth of a crack and subsequent collapse of the pipe [22]. It is therefore essential to study the deterioration of both tensile strength and fracture toughness of the metals to enable more accurate prediction of pipe failures.

The intention of this paper is to experimentally investigate the effect of corrosion on mechanical properties of metals used as pipe material. A comprehensive test program is designed to observe, monitor, and evaluate corrosion behavior of metals and its effect on their mechanical properties in different environments. Two types of metals, namely, cast iron and steel, are tested under corrosion in three environments as represented by pH values. From the analysis of test results, a relationship between corrosion and deterioration of mechanical property of metals is developed. It is believed that tests on the effect of corrosion on mechanical properties of metals are one of few of the kind. The results produced from the tests can contribute to the body of knowledge of corrosion behavior and its effect on mechanical properties of metal in soil environment, which can equip engineers and asset managers in mitigating the risk of failures of metal pipes.

2. Design of Test Specimens

2.1. Specimen Materials. Cast iron and steel have been the most predominant pipeline material before the 1980s [21]. Among various types of cast irons and steel, grey cast iron and carbon steel are perhaps the most widely used pipe materials [12, 23]. Because of this, it is reasonable to select these two types of materials for corrosion investigation due to their wide use and also long service. Cast iron and steel have quite

TABLE 1: Chemical composition of test materials (wt.%).

Material	C	S	P	Mn	Si
Q235 steel	0.176	0.023	0.019	0.465	0.233
HT200 cast iron	3.2	0.12	0.015	0.9	1.6

different mechanical properties although they have been used for the same purpose of pipes. Cast iron is brittle material whilst steel is ductile. Cast iron has been widely used in pipeline industry due to its comparatively low cost but it has been replaced by steel in pipeline industry for its greater strength and ductility.

As is well known, the mechanical properties of metal are affected by its chemical composition, morphology, and microstructure which vary significantly. In this study, Q235 plain carbon steel and HT200 grey cast iron are selected as the testing materials due to their wide use in pipe industry in China and availability on market [1, 24]. The chemical composition of Q235 steel and HT200 cast iron is shown in Table 1.

2.2. Specimens for Tensile Strength. Specimens for tensile strength test were made according to ASTM E8M13 Standard Test Methods for Tension Testing of Metallic Materials [25]. The testing specimens are recommended single-edge bend [SE(B)] in the standard, of which the dimensions should comply with the following requirement:

$$L \geq 4D, \tag{1}$$

where L is the gauge length and D is the diameter of the middle part of the specimen within gauge length L. For the sake of comparison, the specimens of the two different materials for tensile strength test are intentionally made with the same dimensions. The specimen for tensile strength is shown in Figure 1(a).

2.3. Specimens for Fracture Toughness. Specimens for fracture toughness test were made according to ASTM E1820-13 Standard Test Method for Measurement of Fracture Toughness [26]. In this standard the key is to control the width of the specimen since it is the most important factor that affects the resulting fracture toughness. Depending on the width of the specimens, there can be two types of fractures: the plane stress fracture and plane strain fracture. For the plane stress fracture, the fracture toughness decreases with the increase of specimen width and stabilizes at a certain width for which plane strain fracture occurs. This width is determined according to ASTM E1820-13 as follows:

$$B \geq 2.5 \left(\frac{K_{IC}}{\sigma_y} \right)^2, \tag{2}$$

where B is the width of the specimen, K_{IC} is the fracture toughness, and σ_y is the yield strength of the material.

By (2), the width for selected cast iron specimen with the grade HT200 was calculated to be $B = 20$ mm. However, for the selected Q235 steel, the calculated width for the

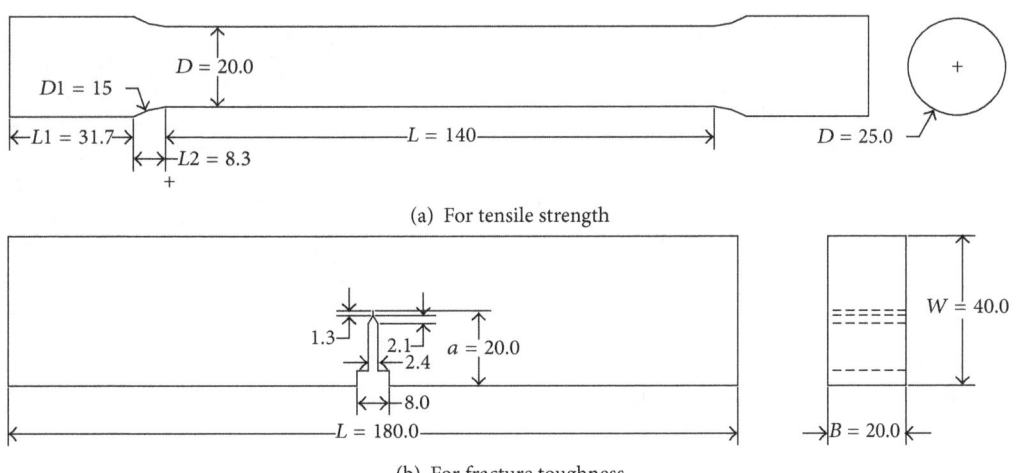

(a) For tensile strength

(b) For fracture toughness

FIGURE 1: Test specimen (unit: mm).

test specimen is as large as 1700 mm, which is too large to be practical for both corrosion and fracture tests. Since the primary purpose of this study is (1) to experimentally examine how corrosion affects the fracture toughness of the Q235 steel but not to determine its accurate value of fracture toughness and (2) to compare the corrosion effect on fracture toughness for different metals (i.e., steel and cast iron), it is justifiable to select a smaller but with same size as that for cast iron specimens for both corrosion and fracture toughness tests. This is because all test specimens should be under the same corrosion and fracture conditions and hence relative comparison of fracture toughness change over time and with each other is valid. Besides, the accurate value of fracture toughness of the Q235 steel has been determined with different methods as shown in, for example, Zhao et al. [27] and Dong et al. [28]. Small size specimens of steel for fracture toughness tests have also been used in other studies as shown in literature [29]. Therefore, the width of specimens for fracture toughness test for Q235 steel was selected the same as that for cast iron. The specimen for fracture toughness is shown in Figure 1(b).

2.4. Manufacture of Specimens. All test specimens were manufactured by specialist mechanical technicians. For tensile specimens the manufacture was straightforward. For fracture toughness, the specimens should theoretically be precracked by fatigue. Experience and literature survey have shown that it is impractical to obtain a reproducibly sharp, narrow machined notch that will simulate a natural crack well enough to provide a satisfactory fracture toughness test result [27]. The most effective alternative is to produce a precrack, a comparatively short fatigue crack, which is extended from a narrow notch. There are three forms of notches to start a fatigue crack (known as fatigue crack starter notch), which are straight through notch, chevron notch, and notch ending with drilled hole. Different forms of fatigue crack starter notches shall meet different dimension requirements. In this study, the straight through notch was employed as fatigue crack starter. For detailed specifications of specimen size, configuration, and preparation, refer to ASTM E1820 [26].

TABLE 2: Soluble chemical composition of soil sample (wt.%).

Chemical	CaO	MgO	K_2O	Na_2O
Content	0.92	1.54	2.17	0.60

TABLE 3: Chemical composition in simulated soil solutions (g/L).

Chemical	$CaCl_2 \cdot 2H_2O$	$MgSO_4 \cdot 7H_2O$	KCl	$NaHCO_3$
Content	0.036	0.190	0.069	0.540

3. Test Methodology

3.1. Simulation of Corrosive Soil. Pipe corrosion is electrochemical reaction between the pipe material and the corrosive agents in the ambient soil. In order to represent this reaction in the laboratory, it is necessary to simulate the working environment of the pipes. There are two methods to simulate the working environment; one is to bury the pipe in a box of real soil and the other is to immerse the pipe in a solution that contains main chemical elements extracted from the real soil (known as soil solution). Literature reviews suggest that most of current research employs soil solutions for pipe corrosion test in soil [30, 31]. Therefore, this study also employed the soil solution for corrosion test. One advantage of using soil solution is the ease to control the testing variables and also monitoring of corrosion behavior.

For convenience, soil in local land with pipes underneath was selected. The chemical composition of the selected soil was analyzed and is shown in Table 2. This composition was used to make soil solution. The chemical analysis of the soil indicates that the pH of the soil is 8.0. So the base solution used for corrosion test has pH of 8.0. The chemical composition of the soil solution used in the corrosion test is shown in Table 3 which was made based on the principle that the key chemical elements of soil sample and soil solution are the same [30, 31].

Since metal corrosion under natural soil conditions will take a long time to have any significant effect on its material properties and to achieve the research objective within the time period of the project, acceleration of corrosion appears to be necessary for almost all corrosion tests (e.g., [5, 32]). Thus, acceleration of corrosion was adopted in this test. A literature review suggests that pH value will accelerate the corrosion of metal exponentially [33]. For this reason, three values of pH were selected for the simulated soil solution so that the variation of pH effect on corrosion can be studied. Based on research experience and pretrial, pH of 3.0 would accelerate the corrosion sufficiently to have significant effect on the mechanical properties of the metal within the project period. With the pH of natural soil being 8.0, a middle value of pH of 5.5 was selected.

Different values of pH were achieved by adding sulfuric acid and maintained the same during the whole test period. It may be noted that the added sulfuric acid may react with the chemicals in the solution but this reaction would happen in exactly the same manner as with the soluble chemicals in natural soil [30, 31]. The point is that pH values of all solution were maintained the same and used as the measurement for the solution.

3.2. Test Variables.

As discussed in the instruction, of many factors that affect corrosion of metals in soil and its effect on mechanical property, the chemical compositions of soil and metal are the most influential. In this study, the chemical composition of the soil was represented by pH and that of metal by grade. Therefore, the pH values of the soil solution and the type of metal were selected as the main testing variables as well as their change with time. Three values of pH were selected for soil solution as discussed above, which are 8.0, 5.5, and 3.0. Two types of metal were selected for corrosion test and its effects on mechanical properties which are carbon steel and grey cast iron. To obtain the variation of corrosion and its effect on mechanical properties of metals over time, three points of time were selected which are 90, 180, and 270 days (or 3, 6, and 9 months), respectively, in addition to initial time, that is, before corrosion. Thus, there are four points in time in total. These times were selected based on the literature review and research experience to ensure the measurable corrosion and significant property change of mechanical properties of the specimens (e.g., [5, 32]).

For statistical studies, three duplicates were made for each specimen with the designated test variables. Therefore, the total number of test specimens is 3 (pH values) × 2 (two types of metal) × 2 (properties) × 4 (time points) × 3 (duplicates) = 144.

The measurement of the test includes (i) corrosion current; (ii) weight loss; (iii) tensile strength; and (iv) fracture toughness. Corrosion current was measured every day in the first week of the test and then weekly until the end of tests. Other three parameters were measured at initial point and three designated points of time, giving four measurements over time.

3.3. Test Setup and Procedure.

Immersion corrosion test was conducted according to ASTM G31-2012a *Standard Guide for Laboratory Immersion Corrosion Testing of Metals* [34] in room temperature with soil solutions of three pH values

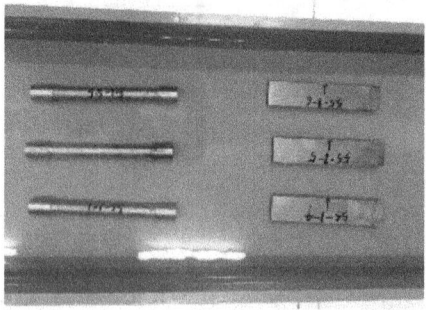

FIGURE 2: Specimens immersed in soil solution.

(3.0, 5.5, and 8.0) for a duration of three time periods of 90, 180, and 270 days. The specimens made of Q235 steel and HT200 cast iron were washed using 50% acetone. They were then dried and placed in the containers of designated soil solutions. In each container, 6 specimens, 2 for mechanical property (tensile and fracture) and 3 duplicates, were immersed in the solution as shown in Figure 2. In total, there are 18 containers representing different pH values (3.0, 5.5, and 8.0) and materials (cast iron and steel). During immersion time, pH values in different containers were measured using a pH meter and controlled by adding sulfuric acid. Corrosion tests in all 18 containers were run in parallel.

To monitor the corrosion behavior wires were welded on two specimens of each type (labeled 1# for tensile specimen and 4# for toughness specimen) in each container and corrosion currents of the specimens were measured using an ampere meter. At each point of three designated times, that is, 90, 180, and 270 days, specimens were taken out of the solution for measurement of weight loss, tensile strength, and fracture toughness for three pH values and both steel and cast iron. Weight loss of the specimens was measured according to ASTM G31-12a. Tensile strength was tested on WAW-1000 material testing system as shown in Figure 3(a). Fracture toughness was tested on MTS landmark testing system as shown in Figure 3(b). Both tensile and fracture toughness tests were carried out by laboratory technicians to ensure the quality of the test results.

4. Test Results and Analysis

4.1. Corrosion Current.

Corrosion current has long been used as a major indicator for corrosion behavior of metals [4, 12–15, 35]. In this study, corrosion currents were monitored over the whole test period and recorded using an ampere meter as shown in Figure 4. The results of corrosion currents are presented in Figures 5 and 6 for steel and cast iron specimens, respectively. It can be seen from the figures that the corrosion currents are in general very scattered. This is not unexpected due to the random occurrence and growth of corrosion. It may also be attributed to the accuracy in measuring the current due to aggressive environments.

Figures 5 and 6 indicate that although each point of measure corrosion current is scattered, the general trend of corrosion currents is clear, which is decreasing with the

(a) (b)

FIGURE 3: Test facilities: tensile (a) and fracture toughness (b) testing systems.

FIGURE 4: Monitoring of corrosion current.

exposure time. This means that current rate is high at the beginning of the corrosion and decreases over time. As it is well known, corrosion is an electrochemical process. The acidic environment can initiate the corrosion but the progress of corrosion needs the supply of oxygen which is not readily available to keep the high corrosion rate. These results are consistent with other results reported in the literature as well as research experience [5].

Figures 5 and 6 also show that corrosion currents are generally larger for more acidic solution, that is, smaller pH value, in particular at the beginning. This is again consistent with results published in literature. For example, a study by Panossian et al. [15] shows that the smaller pH is, the larger corrosion currents are, indicating that acid can induce more corrosion. Though corrosion currents in solution with smaller pH are comparatively larger, the decreasing rates of corrosion currents (i.e., the slope of the curve) are irregular, exhibiting the randomness of corrosion behavior. Corrosion currents in solutions with pH of 8.0 and 5.5 have the largest and the smallest decreasing rates, respectively. In general, the differences in variation rates of corrosion current (the slope of the curve) for different pH are 26.6% between 3.0 and 5.5 and −25.0% between 5.5 and 8.0.

The comparison of Figures 5 and 6 also shows that corrosion of cast iron is slightly faster than steel. As suggested by Dean Jr. and Grab [9], one of the reasons for this can be that a higher carbon content in metal can incur a larger corrosion rate.

4.2. Weight Loss. Before the test, all specimens were cleaned and weighed. After immersion, specimens were taken out

at three designated points of time, that is, 90, 180, and 270 days, respectively. Figure 7 shows the progress of corrosion activity in terms of change in color, rust accumulation, and distribution. Then they were dried, cleaned, and weighed again. Weight loss was calculated as reduction in weight of each specimen before and after immersion. Weight loss is normalized by surface area and expressed in $g \cdot m^{-2}$ to eliminate the influence of differences in shapes and exposure areas. Figure 8 shows the results of weight loss for steel and cast iron specimens, respectively, where each point is the average of three measurements of weight. The range of coefficients of variation of weight loss at each point is from 0.09 to 0.22 over the test period.

Figure 8 shows that the weight decreases with time almost linearly for both steel and cast iron specimens which is different from the results of corrosion current (which is nonlinear over time). The reason could be that the weight loss represents the cumulative effect of corrosion which is more gradual whilst the corrosion current represents the instantaneous rate of corrosion which is more fluctuated. It can be seen that weight loss is larger with smaller pH value, that is, pH = 3.0. This is consistent with the results of corrosion current. It can be also seen from Figure 8 that the trend of weight loss of both steel and cast iron specimens is almost the same although the weight loss of cast iron specimens due to corrosion is slightly larger. Again this is consistent with the results of corrosion current. As can be seen from the figure, there is not much difference in weight loss when pH values are between 5.5 and 8.0 in particular for cast iron.

4.3. Yield Strength Reduction. The main objective of this research is to investigate the effect of corrosion on mechanical properties of metals as represented by tensile strength and fracture toughness. For this purpose, specimens were taken out of the immersion after 90, 180, and 270 days of corrosion, respectively. Then they were cleaned and loaded to failure in tension on the testing machine in Figure 3(a). The results of tensile tests are shown in Figure 9, where each point represents an average of three testing results. The range of coefficients of variation of tensile strength reduction at each point is from 0.11 to 0.19 over the test period.

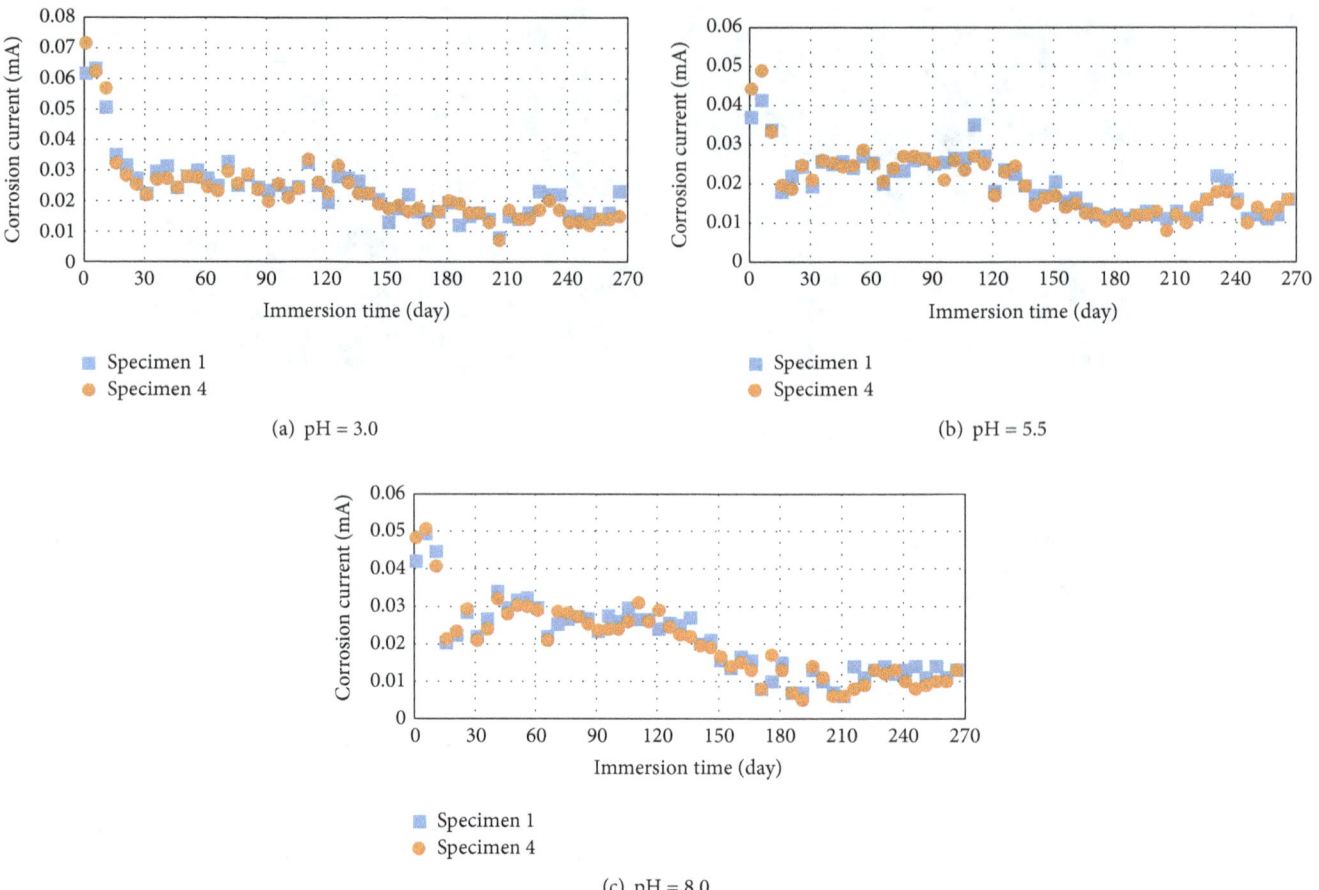

(a) pH = 3.0

(b) pH = 5.5

(c) pH = 8.0

FIGURE 5: Corrosion current in steel specimens in solutions of different pH values.

From Figure 9, it can be seen that the tensile strength decreases with time due to corrosion. This is the case for both steel and cast iron materials. The results of Figure 9 provide good evidence that corrosion does affect the mechanical property of metals. This is mainly due to the fact that corrosion penetrates the surface of the specimens, destroying the compactness of the specimen surface. It can be seen in Figure 7 that surfaces of all corroded specimens are rougher and more porous than intact metals, which makes it easier for corrosive agents or other elements, for example, O and Cl, to ingress into the metal. The ingress of corrosive agents and/or elements can alter the chemical composition of metal via chemical reactions of these agents and elements. It can also change the morphology or microstructure of the metal. As is known, chemical composition and morphology are main factors that determine the mechanical property of metals. As a result, the mechanical property of the metal changed.

Table 4 shows the corrosion induced deterioration of mechanical properties of steel and cast iron at three time periods of test. It can be seen that the reduction of tensile strength increases with the exposure time for both steel and cast iron. These results are in agreement with the results of weight loss which shows a linear increase with time (Figure 8)

as discussed above. Also seen from the table is the fact that the reduction of tensile strength of cast iron is larger than that of steel. This is again consistent with the results of both corrosion current and weight loss, indicating that high carbon content in metal may not only lead to more corrosion [9] but also have more effect on tensile strength. Table 4 also shows that in the first period of exposure the reduction of tensile strength is larger for more acidic environment (i.e., pH = 3.0) but later in the third period the reduction of tensile strength is larger for less acidic environment (i.e., pH = 8.0), indicating that high acidity may accelerate corrosion of metal but may not necessarily accelerate the corrosion effect on its mechanical properties which is determined by its chemical position, morphology, and microstructure as discussed previously.

4.4. *Fracture Toughness Reduction.* Fracture toughness of metals is determined by three-point bending test as shown in Figure 3(b). From this test, the fracture toughness can be calculated as follows:

$$K_{(i)} = \left[\frac{P_i S}{(BB_N)^{1/2} W^{3/2}} \right] f\left(\frac{a_i}{W} \right), \tag{3}$$

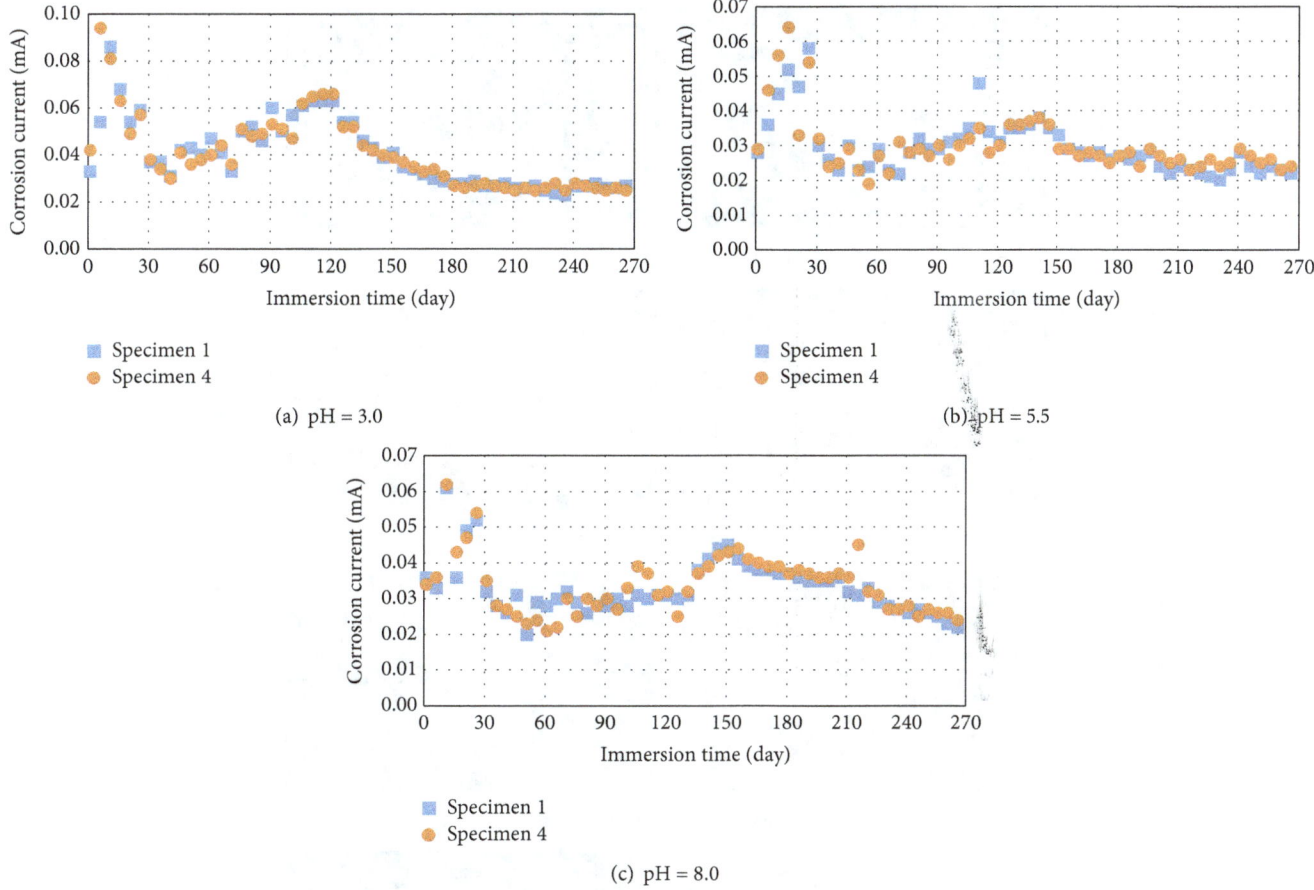

FIGURE 6: Corrosion currents in cast iron specimens in solutions of different pH values.

where

$$f\left(\frac{a_i}{W}\right) = \frac{3\left(a_i/W\right)^{1/2}\left[1.99 - \left(a_i/W\right)\left(1 - a_i/W\right) \times \left(2.15 - 3.93\left(a_i/W\right) + 2.7\left(a_i/W\right)^2\right)\right]}{2\left(1 + 2a_i/W\right)\left(1 - a_i/W\right)^{3/2}} \tag{4}$$

and $K_{(i)}$ is the calculated K for the bend specimen at load P_i, S is the support span, B and B_N are the width and net width of specimen, respectively, a_i is the current crack length, and W is the height of the specimen (see Figure 1(b)). In this study, the fracture toughness values, that is, K_{IC}, were determined directly from the testing machine as outputs, using the built-in program.

The results of fracture toughness reduction are shown in Figure 10, where again each point represents an average of three testing results. The range of coefficients of variation of fracture toughness reduction at each point is from 0.12 to 0.17 over the test period. It can be seen from the figure that the fracture toughness also decreases with time due to corrosion. This is true for both steel and cast iron materials. The results of Figure 10 again provide the evidence that corrosion does affect the mechanical property of metals for the same reason as explained for the tensile strength. In addition, it

can be seen in Figure 7 that the penetration of corrosion into the metal is not evenly distributed. In most cases, the locations that are damaged incur the most corrosion, forming localized corrosion pits. This is true especially when there is a precrack where the corrosion is the most severe, leading to the extension of the crack of the specimen. As it is known, the crack extension is the most affecting factor for the determination of fracture toughness [21].

In addition, from Table 4 it can be seen that the corrosion induced deterioration of fracture toughness is remarkably larger than that of tensile strength under the same conditions. This indicates that corrosion has larger effect on fracture toughness than tensile strength of the metal. The reason can be that, as explained above, corrosion pits extend the existing defects of the metal which reduce the fracture toughness. Also the reduction of fracture toughness for cast iron is almost twice that of steel. The results of all four measurements

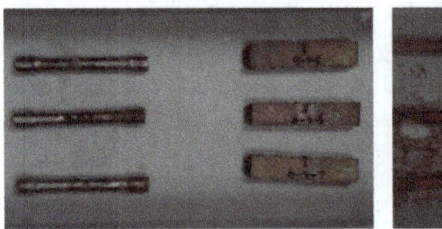

(a) For specimens at 6 days (steel left, cast iron right)

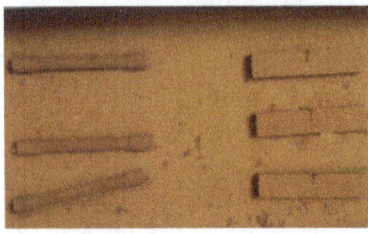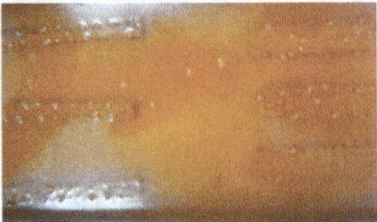

(b) For specimens at 30 days (steel left, cast iron right)

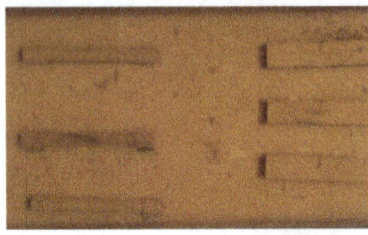

(c) For specimens at 90 days (steel left, cast iron right)

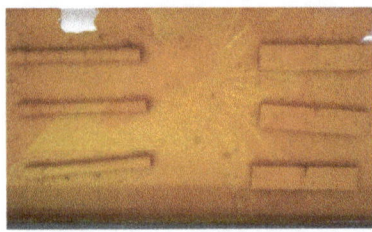

(d) For specimens at 180 days (steel left, cast iron right)

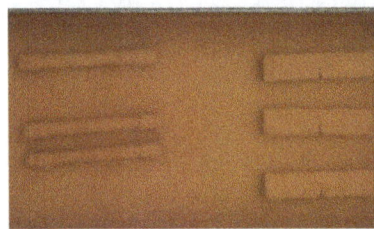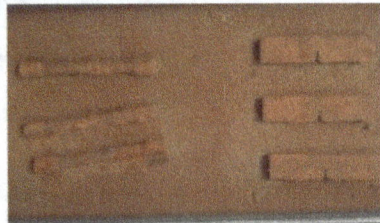

(e) For specimens at 270 days (steel left, cast iron right)

(f) Details of a very corroded specimen covered with rusts

FIGURE 7: Photos of corroded specimens at different exposure periods.

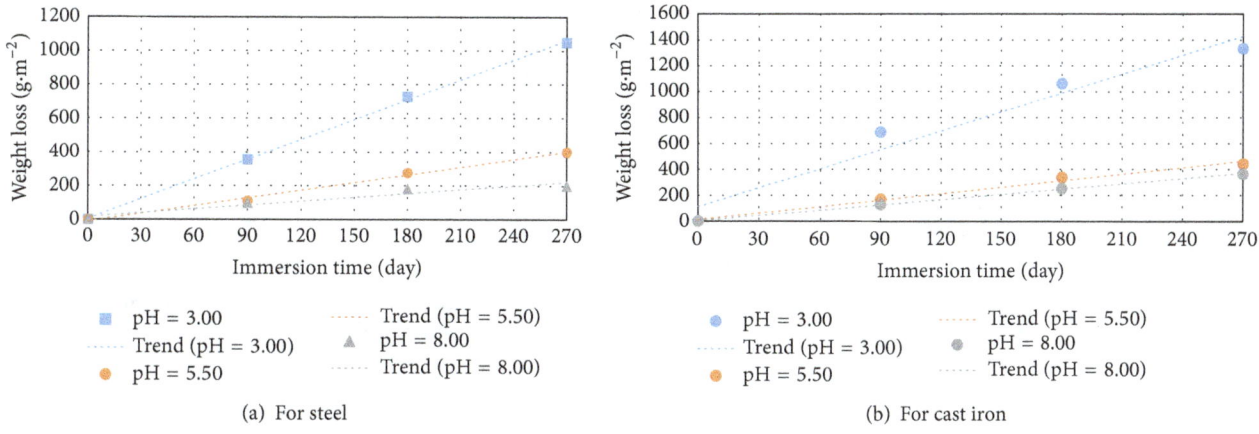

(a) For steel

(b) For cast iron

FIGURE 8: Weight loss of specimen.

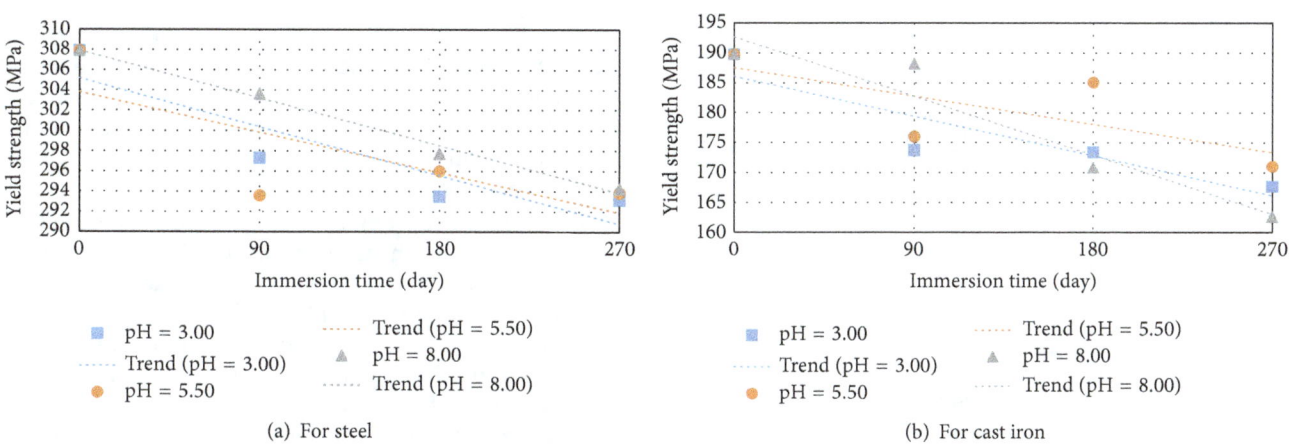

(a) For steel

(b) For cast iron

FIGURE 9: Tensile strength reduction of specimen.

(corrosion current, weight loss, tensile strength, and fracture toughness) suggest that high carbon content in metal can not only lead to more corrosion but also incur larger effect on mechanical properties.

5. Observation and Discussion

From the tests and test results further observation and discussion can be made. Photos of corroded specimens can provide some insight into the behavior of corrosion in different environments and states. From Figures 5 and 6, it can be seen that corrosion current is the highest at the onset of corrosion but from Figure 7(a) it can be seen that little visible change can be seen in terms of change of color (rust) of specimens although corrosion currents were at their peaks. This again indicates that the corrosion current measures instantaneous corrosion rate not the cumulative corrosion. At this stage, corrosive agents penetrated the oxidation film but little amount of corrosion was produced. This suggests that corrosion rate of the specimens is high but actual corrosion, in terms of products, that is, rusts, is not accumulated. The corrosion reaction can be expressed as follows (e.g., [35]):

(a) Anodic reaction

$$Fe \longrightarrow Fe^{2+} + 2e^{-} \qquad (5)$$

(b) Cathodic reaction

$$2H^{+} + 2e^{-} \longrightarrow H_2 \qquad (6)$$

After 6 days of immersion, the corrosion currents dropped sharply indicating that the corrosion rate is reduced but corrosion itself continues. This is demonstrated in Figure 7(a), where the solution turned into a brown color especially for the solution with pH = 5.5. The corrosion reactions at his stage can be expressed as follows [35]:

$$4Fe^{2+} + O_2 + 4H^+ = 4Fe^{3+} + 2H_2O \qquad (7)$$

$$2Fe + O_2 + 2H_2O = 2Fe(OH)_2 \qquad (8)$$

$$4Fe(OH)_2 + O_2 + 2H_2O = 4Fe(OH)_3 \qquad (9)$$

With the increase of corrosion, hydrogen bubbles showed up in particular in more acidic solution, for example, pH = 3.0, which can be explained by (6).

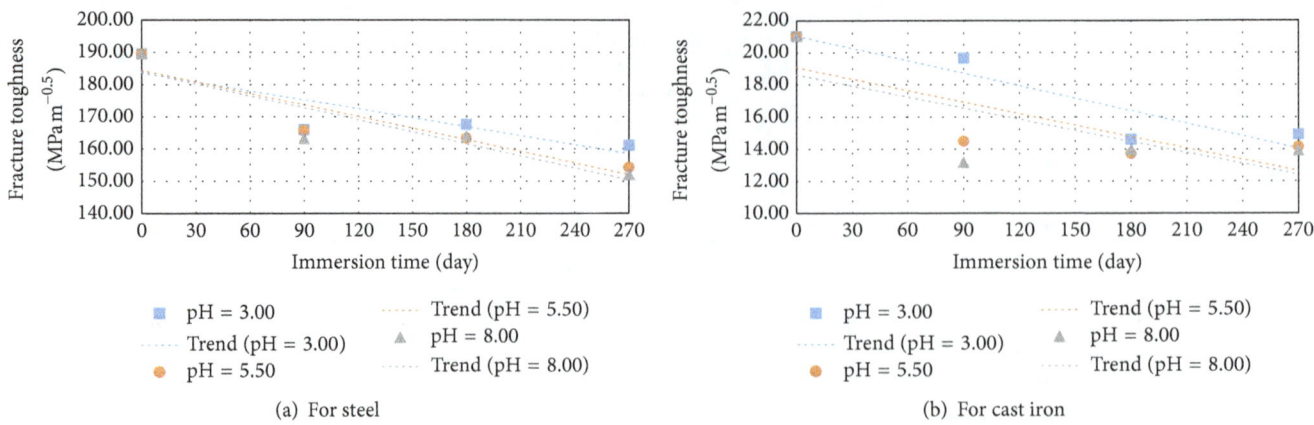

FIGURE 10: Fracture toughness reduction of specimen.

TABLE 4: Reduction of mechanical property (%).

(a) For steel

Time period (day)	pH	Tensile strength	Fracture toughness
90	3.0	3.43	9.29
	5.5	2.63	7.51
	8.0	1.39	6.35
180	3.0	4.66	11.51
	5.5	3.86	13.80
	8.0	3.31	13.51
270	3.0	4.77	14.93
	5.5	4.56	18.50
	8.0	4.43	19.77

(b) For cast iron

Time period (day)	pH	Tensile strength	Fracture toughness
90	3.0	8.45	6.56
	5.5	3.24	2.05
	8.0	0.84	1.38
180	3.0	8.62	20.36
	5.5	4.32	18.63
	8.0	9.98	33.67
270	3.0	11.64	29.05
	5.5	9.89	32.79
	8.0	14.31	33.99

It was also observed that corrosion in different conditions was of different forms. Corrosion pits were formed after surface oxidation film was penetrated, which was the case for all three scenarios of pH values. However, corrosion pits on specimens in more acidic solution, for example, pH = 3.0, were fewer and more evenly distributed than those on specimens in less acidic solutions, for example, larger pH values. This is shown in Figure 7(b) where, at 30 days, localized corrosion pits were more obvious and corrosion products began to become flakes. Deposits of corrosion products in two solutions with lower pH values (3.0 and 5.5) were in larger amount.

Figure 7(c) shows that specimens were further corroded and that corrosion products fell down to the bottom of the containers and covered up the surface of the specimens. The latter would prevent specimens from further corrosion so that the corrosion currents in specimens for all solutions reach the lowest values after 90 days.

It has also been observed from Figures 9 and 10 that the reduction of mechanical properties, both tensile strength and fracture toughness, does not follow the same trend as corrosion current and weight loss. That is, smaller value of pH, that is, more acidic solution, does not result in larger reduction of mechanical properties for both tensile strength and fracture toughness. This may be because whilst the pH can accelerate corrosion of metal, it may neither accelerate the reaction of corrosive agents with chemical elements of metal nor accelerate the change of microstructure of the metal. In other words, the effect of corrosion on mechanical properties of metals may not be in the same proportion as that of pH on corrosion, further indicating the randomness of both corrosion behavior and its effect on mechanical property of the metal. Also the reduction of mechanical properties is more scattered with respect to pH than corrosion current and weight loss. The precision of measurement can also contribute to the degree of the disperse. Obviously more experiments are needed to produce sufficient and quality data for developing models for corrosion induced deterioration of mechanical properties of metals.

For practical application of corrosion effect on mechanical properties of metal, it is desirable to develop a relationship between measurable parameters of corrosion, for example, weight loss, and the reduction of mechanical properties, for example, tensile strength and fracture toughness. This has been attempted in this study. Since the analysis of results presented in the previous section suggests that weight loss can be a better measure of corrosion than corrosion current, it is used in developing the relationship. Also ideally more data points (than four) can produce better correlation of this relation but time and resources are always the constraints. Literature and research experience (e.g., [32]) suggest that three data points are minimum. Figures 11 and 12 show the variation of tensile strength and fracture toughness with

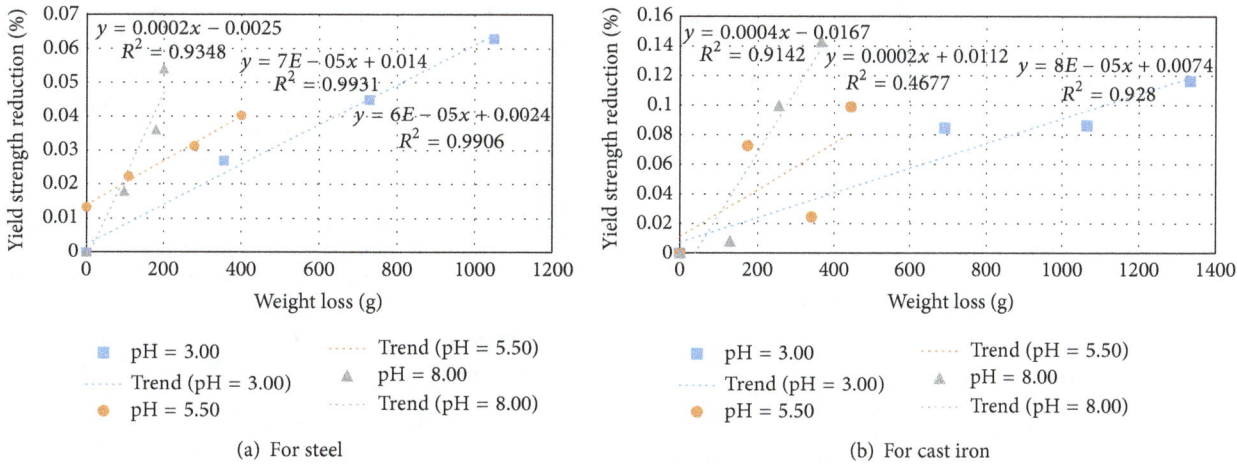

(a) For steel

(b) For cast iron

FIGURE 11: Tensile strength reduction with weight loss.

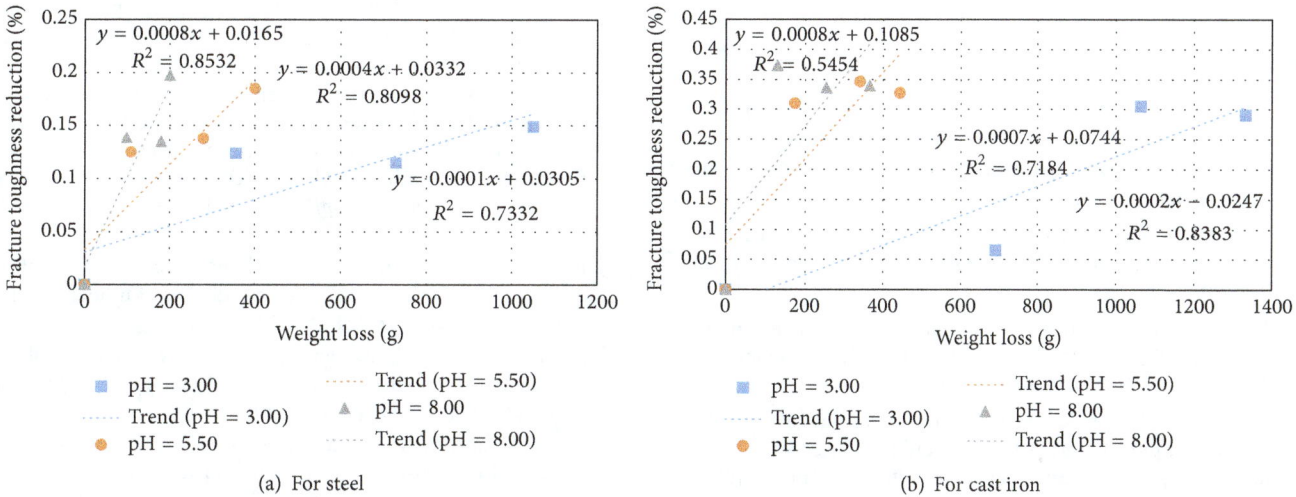

(a) For steel

(b) For cast iron

FIGURE 12: Fracture toughness reduction with weight loss.

weight loss for both steel and cast iron specimens under three tested environments. As can be seen from the figures, the reduction of both tensile strength and fracture toughness is by and large in linear relation with weight loss. Though the lower pH values contribute to greater weight losses, mechanical properties are more sensitive to weight loss in higher pH values, that is, steeper trend lines as shown in Figures 11 and 12. In practice, pH values of soil cannot be lower than 5. As such results of this study for pH lower than 5 can be closer to reality and hence can be of more practical use.

It needs to be noted that the test results presented in the paper are one step towards establishing understanding and knowledge on corrosion effect on mechanical properties of metals. The significance of these results lies more in their trend more qualitatively than quantitatively. It is acknowledged that more tests are necessary to produce larger pool of data for sensible quantitative analysis, based on which we develop theories and models for corrosion induced deterioration of mechanical properties of metals. This work is being continued by corresponding author's research team with

more test specimens under different testing variables and environments, such as real soil environment. More results will be submitted for publication once they are produced, processed, and analyzed.

6. Conclusion

A comprehensive test program has been presented in the paper to investigate the effect of corrosion on mechanical properties of buried metal pipes. The corrosion of two types of widely used metals, that is, cast iron and steel, and its effect on their mechanical properties have been observed, monitored, and evaluated in three different environments as represented by pH values. From the analysis of the test results, a relationship between corrosion and deterioration of mechanical property of metals has been developed. It has been found that the more acidic the environment is, the more corrosion of metal occurs and that grey cast iron corrodes more than carbon steel under the same environment. It has also been found that the corrosion reduces both the tensile strength and fracture toughness of the metal and the latter reduced more than the former. It can be concluded that the results presented in the paper can contribute to the body of knowledge of corrosion behavior and its effect on mechanical properties of metals in soil environment. This knowledge can enable more accurate prediction of failures of metal pipes.

Competing Interests

The authors declare that they have no competing interests.

Acknowledgments

Financial support from Natural Science Foundation of Hubei Province of China under 2015CFB510, Fundamental Research Funds by Wuhan University of Technology of China under 2015IVA012, and Australian Research Council under DP140101547 and LP150100413 is gratefully acknowledged.

References

[1] Y. Li, B. Jiang, H. Zhao, and C. Tao, "Experimental study on internal corrosion of grey cast iron pipeline in water distribution," *Journal of Chemical Information and Modeling*, vol. 53, article 160, 1989.

[2] J. M. Makar, "Preliminary analysis of failures in grey cast iron water pipes," *Engineering Failure Analysis*, vol. 7, no. 1, pp. 43–53, 2000.

[3] National Water Commission Australia, National performance report 2009-10, http://archive.nwc.gov.au/__data/assets/pdf_file/0015/11265/NPR_urban.pdf.

[4] B. Rajani, S. McDonald, and G. Felio, "Water mains break data on different pipe materials for 1992 and 1993," Report A-7019.1, National Research Council of Canada, Ottawa, Canada, 1995.

[5] H. Mohebbi and C. Q. Li, "Experimental investigation on corrosion of cast iron pipes," *International Journal of Corrosion*, vol. 2011, Article ID 506501, 17 pages, 2011.

[6] R. A. de Sena, I. N. Bastos, and G. M. Platt, "Theoretical and experimental aspects of the corrosivity of simulated soil solutions," *ISRN Chemical Engineering*, vol. 2012, Article ID 103715, 6 pages, 2012.

[7] C. Goulter, "An analysis of pipe breakage in urban water distribution networks," *Canadian Journal of Civil Engineering*, vol. 12, no. 2, pp. 286–293, 1985.

[8] M. L. Doleac, S. L. Lackey, and G. N. Bratton, "Prediction of time-to failure for buried cast iron pipe," in *Proceedings of the Annual Conference of the American Water Works Association (AWWA '80)*, pp. 31–38, Denver, Colo, USA, 1980.

[9] S. W. Dean Jr. and G. D. Grab, "Corrosion of carbon steel by concentrated sulfuric acid," *Materials Performance*, vol. 24, no. 6, pp. 21–25, 1985.

[10] D. K. O'Day, R. Weiss, S. Chiavari, and D. Blair, *Water Main Evaluation for Rehabilitation/Replacement*, American Water Works Association Research Foundation, Denver, Colo, USA, 1986.

[11] M. Randall-Smith, A. Russell, and R. Oliphant, *Guidance Manual for the Structural Condition Assessment of the Trunk Mains*, Water Research Centre, Swindon, UK, 1992.

[12] G. J. Kirmeyer, W. Richards, and C. D. Smith, *An Assessment of Water Distribution Systems and Associated Research Needs*, AWWA, 1994.

[13] L. Camarinopoulos, A. Chatzoulis, S. Frontistou-Yannas, and V. Kallidromitis, "Assessment of the time-dependent structural reliability of buried water mains," *Reliability Engineering and System Safety*, vol. 65, no. 1, pp. 41–53, 1999.

[14] R. Sadiq, B. Rajani, and Y. Kleiner, "Probabilistic risk analysis of corrosion associated failures in cast iron water mains," *Reliability Engineering and System Safety*, vol. 86, no. 1, pp. 1–10, 2004.

[15] Z. Panossian, N. L. de Almeida, R. M. F. de Sousa, G. de Souza Pimenta, and R. B. S. Marques, "Corrosion of carbon steel pipes and tanks by concentrated sulfuric acid: a review," *Corrosion Science*, vol. 58, pp. 1–11, 2012.

[16] L. A. Rossman, R. A. Brown, P. C. Singer, and J. R. Nuckols, "DBP formation kinetics in a simulated distribution system," *Water Research*, vol. 35, no. 14, pp. 3483–3489, 2001.

[17] V. Chacker and J. D. Palmer, Eds., *Effect of Soil Characteristic on Corrosion*, ASTM Special Technical Publication, Ann Arbor, Mich, USA, 1989.

[18] T. M. Liu, Y. H. Wu, S. X. Luo, and C. Sun, "Effect of soil compositions on the electrochemical corrosion behavior of carbon steel in simulated soil solution. Einfluss der Erdbodenzusammensetzung auf das elektrochemische Verhalten von Kohlenstoffstählen in simulierten Erdbodenlösungen," *Materialwissenschaft und Werkstofftechnik*, vol. 41, no. 4, pp. 228–233, 2010.

[19] B. Rajani and Y. Kleiner, "Comprehensive review of structural deterioration of water mains: physically based models," *Urban Water*, vol. 3, no. 3, pp. 151–164, 2001.

[20] J. Li, R. Tang, J.-F. Liu, and J.-S. Liu, "The analysis of soil corrosion factors in long-distance oil pipeline," *Equipment Manufacturing*, no. 9, pp. 31–33, 2012.

[21] C. Q. Li and S. T. Yang, "Stress intensity factors for high aspect ratio semi-elliptical internal surface cracks in pipes," *International Journal of Pressure Vessels and Piping*, vol. 96-97, pp. 13–23, 2012.

[22] P. Marshall, *The Residual Structural Properties of Cast Iron Pipes—Structural and Design Criteria for Linings for Water Mains*, Water Industry Research, London, UK, 2001.

[23] S. K. Singh and A. K. Mukherjee, "Kinetics of mild steel corrosion in aqueous acetic acid solutions," *Journal of Materials Science and Technology*, vol. 26, no. 3, pp. 264–269, 2010.

[24] Y. H. Wu, T. M. Liu, S. X. Luo, and C. Sun, "Corrosion characteristics of Q235 steel in simulated Yingtan soil solutions," *Materialwissenschaft und Werkstofftechnik*, vol. 41, no. 3, pp. 142–146, 2010.

[25] ASTM E8M13, *Standard Test Methods for Tension Testing of Metallic Materials*, ASTM International, 2013.

[26] ASTM-E1820, *Standard Test Method for Measurement of Fracture Toughness*, ASTM International, 2013.

[27] Z. Zhao, Y. Lu, and G. Sun, "Experimental measuring fracture toughness of Q235 steel by J integral," *Journal of Wuhan University of Technology*, vol. 24, no. 4, pp. 111–112, 2002.

[28] D. Dong, X. Zhu, and X. Mei, "Abaqus Q235 simulation on fracture toughness of material Q235 based on flexibility determination method of Abaqus," *Computer Aided Engineering*, vol. 21, no. 4, pp. 4–6, 2012.

[29] M. Wasim, C. Q. Li, D. J. Robert, and M. Mahmoodian, "Experimental investigation of factors influencing external corrosion

of buried pipes," in *Proceedings of the 4th International Conference on Sustainability Construction Materials and Technologies (SCMT '16)*, Las Vegas, Nev, USA, August 2016.

[30] M. Yan, J. Wang, E. Han, and W. Ke, "Local environment under simulated disbonded coating on steel pipelines in soil solution," *Corrosion Science*, vol. 50, no. 5, pp. 1331–1339, 2008.

[31] Z. Y. Liu, X. G. Li, C. W. Du, and Y. F. Cheng, "Local additional potential model for effect of strain rate on SCC of pipeline steel in an acidic soil solution," *Corrosion Science*, vol. 51, no. 12, pp. 2863–2871, 2009.

[32] C. Q. Li, "Initiation of chloride-induced reinforcement corrosion in concrete structural members—experimentation," *ACI Structural Journal*, vol. 98, no. 4, pp. 502–510, 2001.

[33] F. Long, W. Zheng, C. Chen, Z. Xu, and Q. Han, "Influence of temperature, CO_2 partial pressure, flow rate and pH value on uniform corrosion rate of X65 pipeline steel," *Corrosion and Protection*, vol. 26, no. 7, pp. 290–294, 2005.

[34] ASTM G31-2012a, *Standard Guide for Laboratory Immersion Corrosion Testing of Metals*, ASTM International, 2012.

[35] L. L. Shreir, R. A. Jarman, and G. T. Burstein, *Corrosion*, Butterworth Heinemann, 2010.

3

Detection of Reinforcement Corrosion in Reinforced Concrete Structures by Potential Mapping: Theory and Practice

Gino Ebell⊙, Andreas Burkert, and Jürgen Mietz⊙

Federal Institute for Materials Research and Testing (BAM), Berlin, Germany

Correspondence should be addressed to Gino Ebell; gino.ebell@bam.de

Guest Editor: Tao Cheng

Electrochemical potential mapping according to guideline B3 of DGZfP (German Society for Nondestructive Testing) is a recognized technique for the localization of corroding reinforcing steels. In reinforced concrete structures the measured potentials are not necessarily directly linked to the corrosion likelihood of the reinforcing steel. The measured values may be significantly affected, different from, e.g., stress measurement, by different influences on the potential formation at the phase boundary metal/concrete itself as well as the acquisition procedure. Due to the complexity of influencing factors there is a risk that the results are misinterpreted. Therefore, in a training concept firstly the theoretical basics of the test method should be imparted. Then, frequently occurring practical situations of various influencing factors will be made accessible to the participants by a model object specially designed for this purpose. The aim is to impart profound knowledge concerning the characteristics of potential mapping for detecting corrosion of reinforcing steel in order to apply this technique in practice as reliable and economical test method.

1. Introduction

In steel reinforced or prestressed concrete structures corrosion attack is generally impossible due to the high alkalinity of concrete environments. The reason for this protection is a passive layer on the steel surface which provides adequate corrosion resistance. The long-term durability of this protection against corrosion is connected with the stability of conditions necessary for the passive layer. Depending on concrete quality and constructional characteristics under certain environmental conditions the passivation effect of the concrete pore solution can be neutralized (carbonation, chloride ingress) leading to depassivation of the reinforcement surface. As a result corrosion reactions of the steel may occur. Due to the volume expansion of the generated iron oxide reaction products cracks or spalling of the concrete cover may develop.

In order to plan and perform necessary rehabilitation measures, early and mainly nondestructively determined information about the actual corrosion behavior of the steel reinforcement is of utmost importance. The potential mapping technique is an established and common method to assess the corrosion state of reinforcing steel in reinforced concrete structures [1–5]. Areas of active corroding reinforcement can be nondestructively localized. This method is usually applied for the detection of chloride-induced corrosion.

2. Measuring Technique

During the potential mapping of reinforced concrete structures the potential difference between the reinforcing steel embedded in concrete and an external reference electrode placed on the concrete surface is determined (Figure 1). As the reinforcement has to be locally contacted the method is quasi nondestructive.

The potential differences are obtained on a previously defined measuring grid. Different reference electrode types are available depending on the measurement task. Rod electrodes can be used for spot or individual measurements, while wheel-electrode systems are usually used for larger areas.

Figure 2 shows the so-called potential funnel which can be detected only in a 3D representation. Locally limited and pronounced potential shifts to more negative potential values

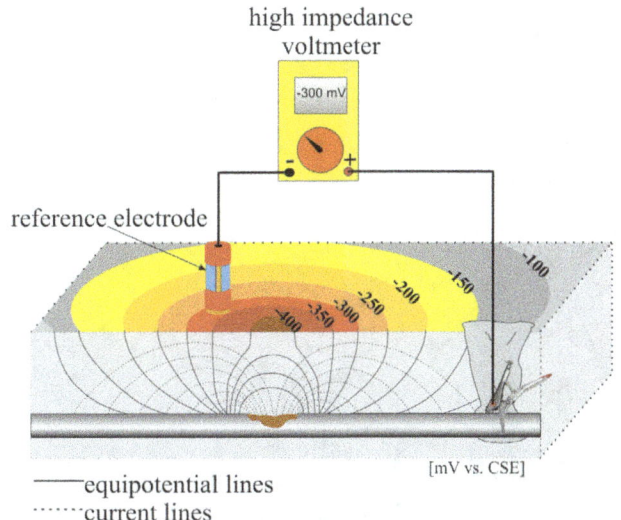

FIGURE 1: Principle of potential mapping.

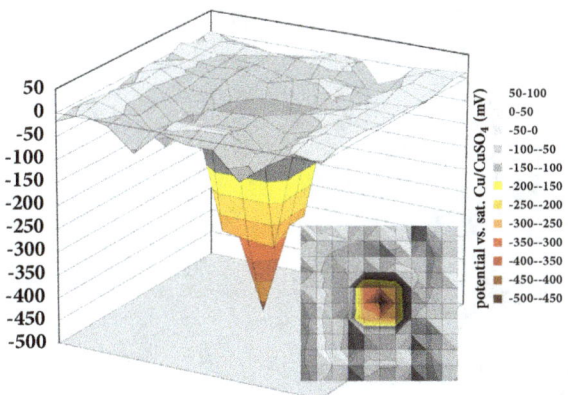

FIGURE 2: 3D representation of a potential funnel.

are often an indication of active reinforcement corrosion. The absolute potential values are of minor importance. According to [1, 3] the main focus is the detection of potential gradients forming such potential funnels. Contrary to this, in ASTM Standard C876 [4], the corrosion probability is designated to fixed potential values. The evaluation of potential mapping data only according to the threshold values given in [4] is not recommended as there is no direct correlation of corrosion probability and measured potential without considering the population.

The performing of the measurement and the requirements for the test personnel are described in detail in the DGZfP Specification B03 [1] and the SIA Guideline 2006 [3].

3. Evaluation and Interpretation

3.1. Additional Investigations. The interpretation of the potential mapping regarding the condition of corrosion of the steel reinforcement requires additional investigations taking into account possible influencing factors like varying concrete moisture or concrete cover.

The following additional requirements are indispensable for a reliable assessment of a potential mapping:

> complete measurement of the concrete cover
>
> checking the concrete surface for hollow locations/ delaminations
>
> complete investigation of the concrete surface regarding
>
>> exposed reinforcement
>>
>> remnants of sealing
>>
>> frost damage, concrete damage, and crack formations
>>
>> drainage installations
>
> sampling of bore dust samples to determine chloride profiles
>
> random determination of carbonation depth
>
> opening and assessment of the reinforcement at conspicuous areas

Optional additional investigations are, e.g.,

(i) determination of the electrolytic resistance of concrete

(ii) measuring concrete moisture by means of microwaves

3.2. Graphical Representation. The graphical representation of the measuring results should always be made with suitable boundary parameters which should enable a clear and effective visualization of the probability of reinforcement corrosion.

Figure 3 shows as an example a colour representation with a zoning of 50 mV for each assigned colour. In this case the dark red and dark orange colour indicate areas with high corrosion probability (potentials between -300 and -500 mV versus copper/copper-sulfate electrode).

For the graphical representation of potential mapping data the signal effect of the colours used should be regarded. In the given example (Figure 3) the most negative potential values (and hence the most critical ones) are represented by red and brown colours. The choice of other colours should consider that they provide a clear allocation between colour and probability of corrosion. Areas of high and low corrosion probability should be easily distinguished.

The statistical analysis (frequency distribution or cumulative frequency distribution) of potential values for the definition of potential limit values for the assignment of corrosive and noncorrosive areas requires a high level of expertise. Different moisture conditions in concrete can lead to a shift in the potential level by more than 100 mV leading to misinterpretations in the statistical analysis.

4. Final Appraisal and Random Opening

After graphical evaluation of the potential mapping data a random opening of the concrete cover to inspect the

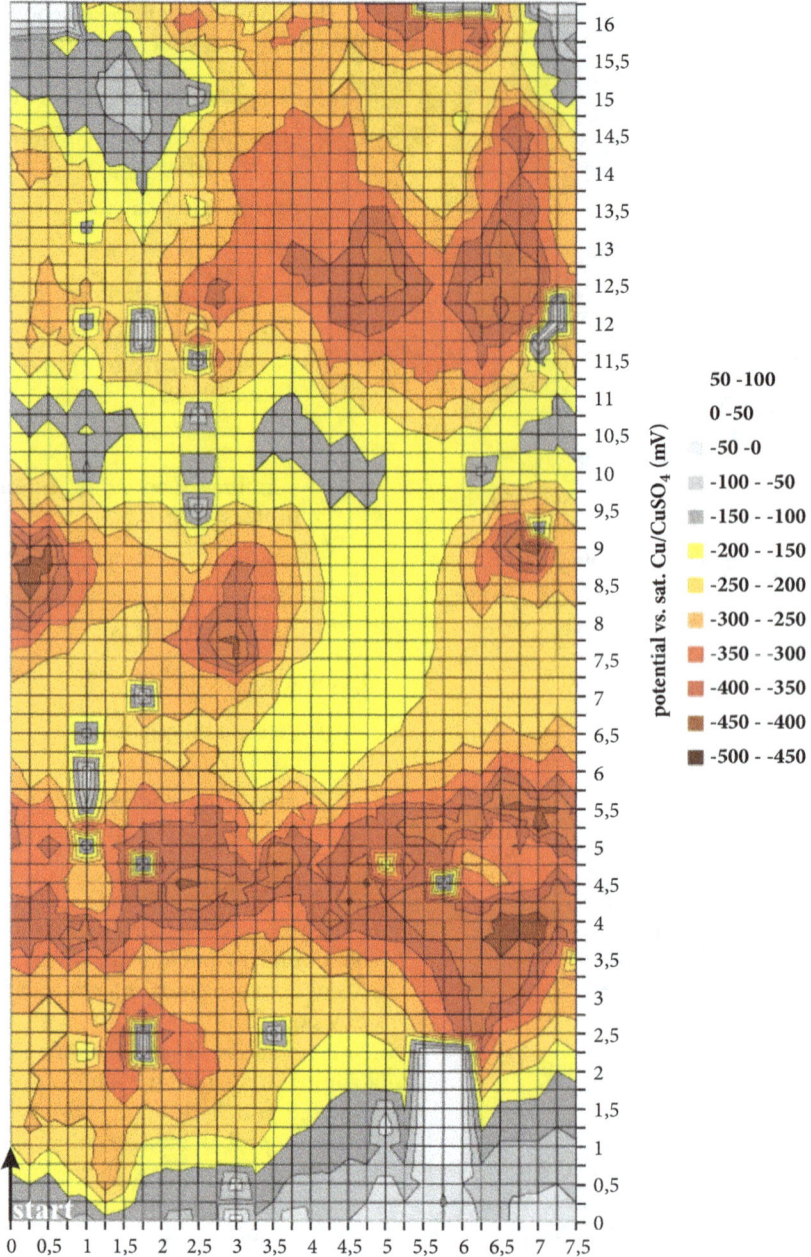

FIGURE 3: Graphical representation of measured potential field of a concrete slab.

FIGURE 4: Large-scale test specimen of BAM, Berlin.

reinforcement locally with respect to the actual degree of damage has to be performed. This should be done in areas with the most negative potentials. Also in areas with cracks or unclear potential distribution, local areas have to be opened and the reinforcement condition inspected. This procedure has to be conducted for each measuring surface. General conclusions from one measuring area to the whole structure are not possible. Only after validation of the measuring data with the corresponding local inspections a final assessment of the rehabilitation requirements can be made.

5. Test Personnel

Due to the high requirements for the test personnel the quali-fication is outlined in the DGZfP specification B03 as follows [1]: the process of potential mapping and the interpretation of the measurement results require sufficient expertise in the

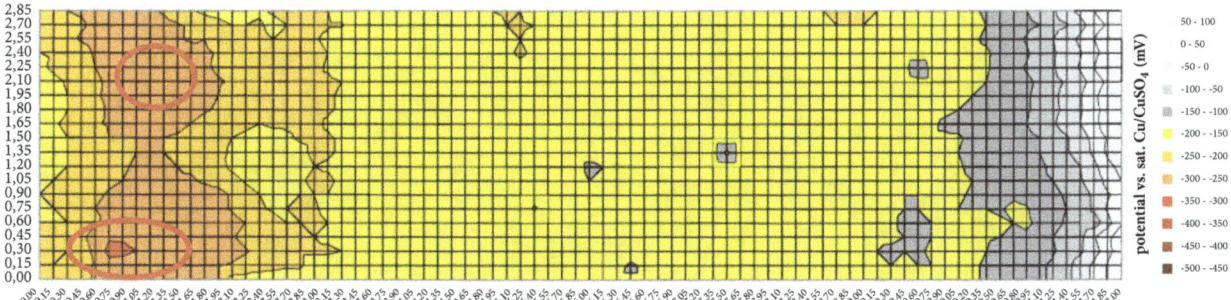

FIGURE 5: Potential mapping results of the large-scale specimen—two anodes.

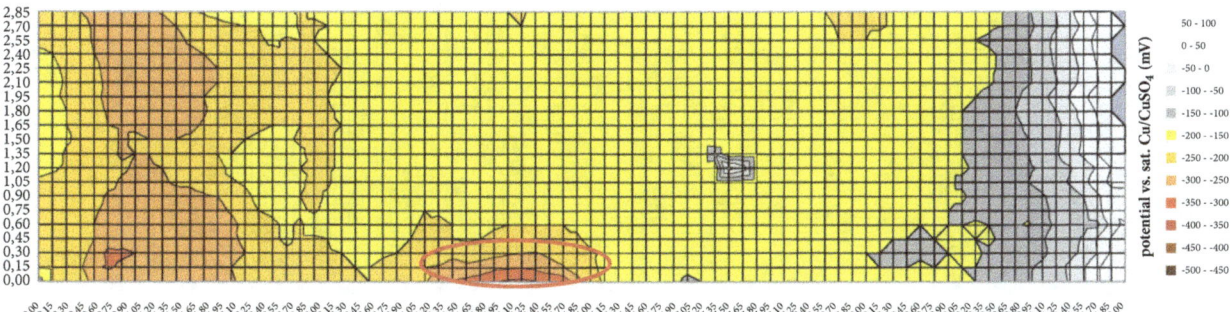

FIGURE 6: Potential mapping results of the large-scale specimen—two anodes and the galvanized stairway.

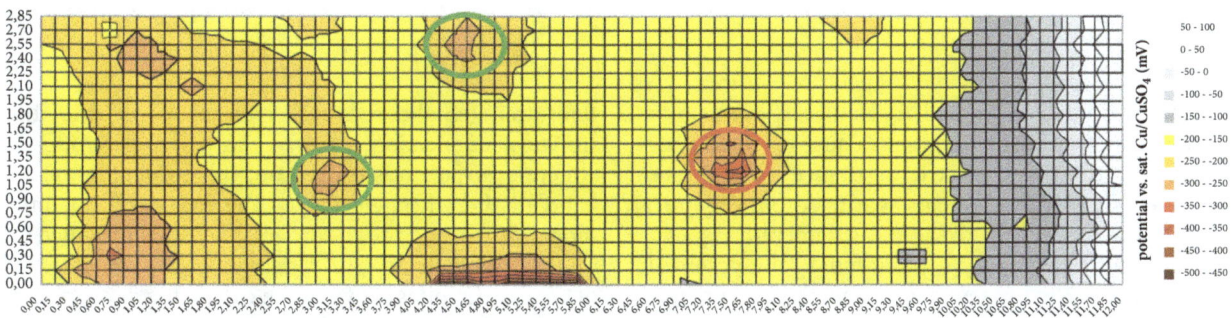

FIGURE 7: Potential mapping results of the large-scale specimen—four anodes, the galvanized stairway, and the connected floor drain.

area of both corrosion and corrosion protection, as well as concrete technology and at least basic knowledge in the field of structural engineering. The proof of this expertise may be given, e.g., via certifications or proof of formal qualifications of appropriate advanced training. The measurements themselves can also be conducted by test personnel without background knowledge in electrochemistry, if under the guidance of a responsible skilled inspector. In this case, the measurement collection process must be controlled by the inspector.

6. Results of the Practical Training Courses at the Large-Scale Test Specimen

Training courses for the acquisition of the proof of expertise for performing potential mapping measurements are carried out at three locations in Germany (Aachen, Berlin, and München). The course is subdivided into a theoretical and a practical training and lasts for 3 days including theoretical and practical examinations. The practical training takes place at a large-size test specimen including hidden corroding reinforcement as well as practice-relevant sources of error. Figure 4 shows the test specimen of the Federal Institute of Research and Testing (BAM), Berlin.

The specimen was constructed in 2011, the used concrete is based on a CEM III blast furnace slag cement mixture with a water/cement ratio of 0.45 and a maximum grain size of 16 mm. The lower and upper reinforcement-layer are electrically insulated by special constructed spacers. Different anodic areas as well as MnO_2 reference electrodes are embedded in the specimen. Additionally, influencing components like a galvanized stairway and a metallic floor drain are also integrated. They can be connected or disconnected to the reinforcement by a switch-control-center. In Figures 5–9 the measuring grid is always 15 by 15 cm; the used reference electrode was a saturated copper copper-sulfate electrode (sat. $Cu/CuSO_4$).

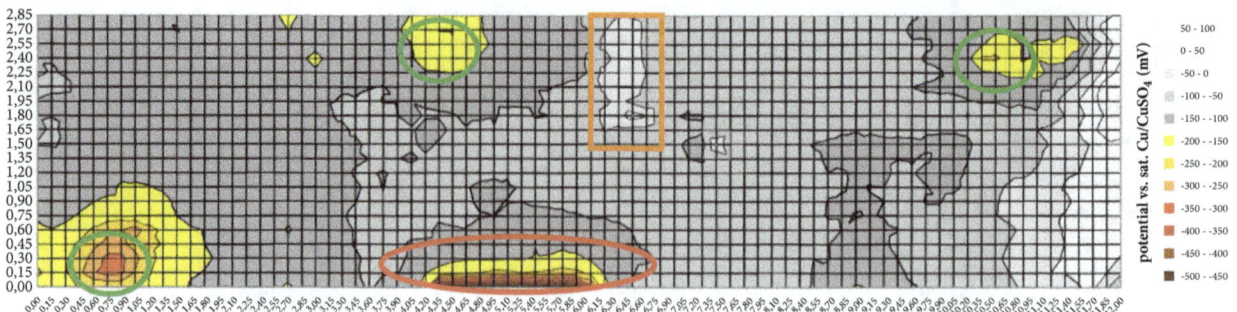

FIGURE 8: Potential mapping results of the large-scale specimen—three anodes, the galvanized stairway, and an epoxy coated area.

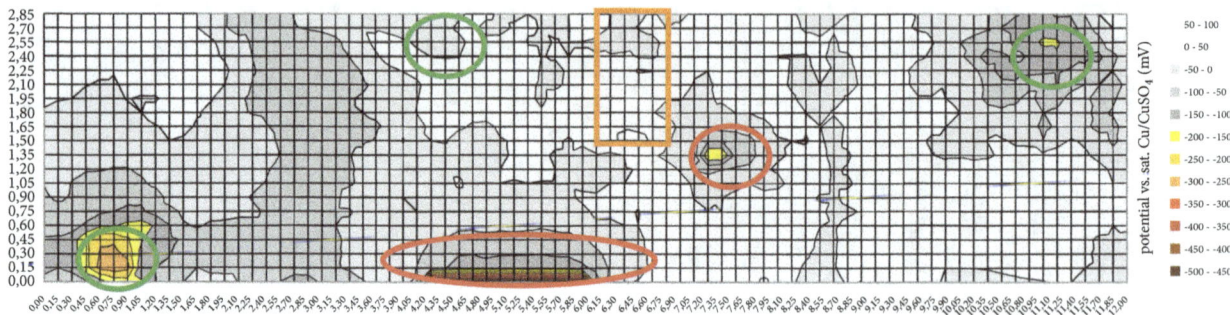

FIGURE 9: Potential mapping results of the large-scale specimen at a nonmoistened surface—three anodes, the galvanized stairway, connected floor drain, and an epoxy coated area.

At this test specimen as well as similar specimens in Aachen (ibac) and Munich (cbm) the training for measuring potentials, concrete cover, electric resistance of concrete, and other necessary measurements is carried out. Figure 5 shows one of the first results of the measurement-campaign from spring 2012 (specimen age 7 months). The potential measurement starts 30 minutes after moistening the concrete surface. During the measurement the concrete surface has to be moistened without any ponding. Two of the embedded anodes are connected (marked with red colour) none of the influencing metallic parts is connected.

Figure 6 shows results of a potential mapping from the same measurement-campaign as in Figure 5, in this case with two connected anodes (see Figure 5) and the galvanized stairway (marked with red colour).

Figure 7 shows similar results like Figure 6 but with a connected floor drain (marked with red colour) and two further anodes (marked with green colour).

Figure 8 shows a result of a measurement-campaign from 2017 (6-year-old specimen). In this measurement-campaign three anodes (marked with green colour) and the galvanized stairway (marked with red colour) are connected. In the year 2016 an epoxy based coating was partially applied (marked with orange colour). The influence of the coating is clearly visible by more positive potentials against the surrounding area. Furthermore, the potential of the passive reinforcement is about 50 to 100 mV more positive than 5 years before.

Figure 9 shows similar results like Figure 8, the main difference is the nonmoistened surface and the connected

floor drain (smaller red marker) in case of Figure 9. The electrolytical connection between the wheel-electrode and the concrete surface was carried out by moistening the sponge-band of the wheel-electrode. The measured potentials are more positive and the anodic areas are smaller, which is based on the higher electrolytic resistivity of the concrete. Just two of the three anodes are visible in case of the dry surface (green marked). The influence of the epoxy coated area is not visible.

Figures 5–9 clearly demonstrate the complex interrelations; the differences between the measured potentials from 2012 and 2017 are especially enormous. The measured potentials are not comparable—just the potential differences between anodic and cathodic areas are comparable. Based on these results it becomes clear that potential mapping results should be interpreted by the potential differences, so-called potential funnels, and not by measured absolute potential. Influences of metallic building parts like the galvanized stairway, when they are electrically connected to the embedded reinforcement, can be misinterpreted as an anodic area. In case of an electrically connected floor drain the interpretation of the measured potentials can be a bit more difficult. The negative potentials are definitely based on an active corroding system but it is difficult to distinguish between corroding floor drain or reinforcement in the surrounding area.

7. Conclusions

Years of experience with potential mapping have shown that clients often order potential mapping without considering the qualification of the test personnel. In a lot of cases they only get potential values without any interpretation and without information about boundary conditions regarding influencing factors. Rehabilitation measures are based on such measurements. And without knowing and considering the complex correlations between the different factors it can happen that areas will be rehabilitated without any real need to do so or—what is worse—that areas will not be rehabilitated although there is severe corrosion of the reinforcement.

Without any proof of sufficient expertise, the planning and assessment of potential mapping measurements could lead to serious misinterpretations and should be avoided.

Conflicts of Interest

The authors declare that they have no conflicts of interest.

References

[1] DGfZP: Specification B 03, *Electrochemical Half-Cell Potential Measurements for the Detection of Reinforcement Corrosion*, DGZfP, Berlin, Germany, 2014.

[2] S. Kessler, C. Gehlen, G. Ebell, and A. Burkert, "Potential mapping and its probability of detection," *Beton- und Stahlbetonbau*, vol. 106, no. 7, pp. 481–489, 2011.

[3] SIA and Merkblatt, *Planung, Durchführung und Interpretation der Potenzialmessung an Stahlbetonbauten*, Schweizerische Ingenieur- und Architektenverein, 2013.

[4] ASTM, *C 876-15 Standard Test Method for Corrosion Potentials of Uncoated Reinforcing Steel in Concrete*, ASTM, 2015.

[5] B. Elsener, C. Andrade, J. Gulikers, R. Polder, and M. Raupach, "RILEM TC 154 EMC: Electrochemical Techniques for Measuring Metallic CorrosionHalf cell potential measurements – Potential mapping on reinforced concrete structures," *Materials and Structures*, vol. 36, pp. 461–471, 2003.

4

Microbiologically Influenced Corrosion of Carbon Steel Exposed to Biodiesel

S. Malarvizhi[1] and Shyamala R. Krishnamurthy[2]

[1]Department of Science and Humanities, Faculty of Engineering, Avinashilingam Institute for Home Science and
 Higher Education for Women, Coimbatore 641 043, India
[2]Department of Chemistry, Avinashilingam Institute for Home Science and Higher Education for Women,
 Coimbatore 641 043, India

Correspondence should be addressed to S. Malarvizhi; s.malarvizhi@gmail.com

Academic Editor: Michael J. Schütze

Environmental concerns over worsening air pollution problems caused by emissions from vehicles and depletion of fossil fuels have forced us to seek fuels such as biodiesel which can supplement petrofuels. Biodiesels have the ability to retain water and provide a conducive environment for microbiologically influenced corrosion (MIC) which may cause difficulties during transportation, storage, and their use. This paper analyses the influence of bacteria on the corrosivity of biodiesel obtained from *Jatropha curcas* on carbon steel using mass loss method. Carbon steel showed the highest corrosion rates in B100 (100% biodiesel) both in the presence and in absence of bacteria. The surface analysis of the metal was carried out using SEM.

1. Introduction

The alarm over fossil fuel depletion, greenhouse gas emissions, and energy security has necessitated the development of alternative renewable energy sources for which biodiesels hold great promise [1]. Besides having a positive energy balance, use of biodiesel in conventional diesel engine results in substantial reduction of unburned hydrocarbons, carbon monoxide, and particulate matters. Nowadays all the countries in the world including those with surplus energy are banking upon biodiesel as an alternative energy source. India has also realized the enormous potential for biodiesel and increased its emphasis on biodiesel production. Under Indian conditions, two nonedible plants *Jatropha curcas* and *Pongamia pinnata* are widely used for the production of biodiesel. Of the two, jatropha has been identified as the most suitable one because it is a plant of tropics with high oil content and it can withstand extreme drought conditions [2, 3].

In spite of the numerous advantages that the biodiesel possesses it has its share of drawbacks. Corrosion is one

of the consequences of biodiesel compatibility issues [4]. Biodiesels may form sediments or cause corrosion when they come into contact with construction materials such as carbon steel, stainless steel, or aluminium which are used for making storage tanks, pipes, and pumping equipment [5]. Recent research has shown that biodiesel can accelerate corrosion of carbon steel used in manufacturing pipelines, storage tanks, and other components of fuel infrastructure. Hence there is an increasing need for research for material compatibility with biodiesel for its judicial application.

Biodiesels are fatty acid methyl esters produced from vegetable oils by a process known as transesterification. Biodiesel undergoes degradation through moisture absorption, oxidation, and attack by microorganisms during storage or use and becomes more corrosive. Both petroleum diesel and biodiesel are often contaminated with microorganisms. Water is one of the essential components for microbial growth to occur. Water enters into the fuel system during production, transportation, and storage. It is very hard to remove water from fuel systems especially when blended with biodiesel

TABLE 1: Elemental composition of carbon steel.

Element	Fe	C	Mn	Zn	P	Al	S	Si	Others
% composition	99.22	0.15	0.31	0.002	0.002	0.025	0.002	0.018	0.271

since biodiesels are more hygroscopic than petroleum diesel. While investigating the corrosion problems in petroleum product pipeline, Maruthamuthu et al. [6] reported 2–11% water contamination which also contained chloride ions. The corrosivity of two samples of biodiesels on mild steel was studied by Meenakshi et al. [7] in the presence of 3% NaCl solution to depict water contamination. Aktas et al. [8] reported that water present in the biodiesel contained ions like chloride and/or sulphide and caused pitting corrosion of metals. The water condenses and collects at the bottom of the fuel tanks or pipes and causes microbial growth leading to the formation of sediments, sludge, and slime resulting in fuel deterioration and corrosion which normally occurs under the resulting biomass [9]. Microbiologically influenced corrosion (MIC) is a serious problem in biodiesel handling facilities. Few works are available on this topic in the literature.

A pioneering work using wire beam electrodes technique carried out by Wang et al. [10] reported that corrosion of carbon steel occurred at the surface exposed to water while the cathodes were formed at water-biodiesel interfaces. Lee et al. [11] observed biofouling while evaluating MIC of metal and alloys in biodiesel, ultralow sulphur diesel, and their blends. Klofutar and Golob [12] reported that absence of water was one of the significant criteria for the prevention of microorganisms in fuels. Biodiesel blends showed an increased bacterial growth and activity compared to neat diesel when diesel-biodiesel blends were incubated with contaminated inoculation water collected from diesel storage tanks [13]. Hence knowledge about the nature of microbes that survive in biodiesel and the ingredients that help in their growth will help us control MIC and the present work aims to investigate the influence of bacteria isolated from the sediments of stored biodiesel obtained from *Jatropha curcas* on the corrosivity of *Jatropha curcas* biodiesel and its blends on carbon steel.

2. Materials and Methods

2.1. Sample Collection, Enumeration, and Isolation of Bacteria. The deposit at the bottom of the container having a two-year-old sample of *Jatropha curcas* biodiesel was collected in a sterile container. The collected sample was serially diluted (10-fold) using sterile distilled water. This is then inoculated on the agar medium by pour plate technique and incubated for 24–48 hours. Petri plates having countable colonies ranging from 30 to 300 are chosen for enumeration and counted (3 strains).

2.2. DNA Extraction, PCR Amplification, and Gene Sequencing. The molecular identification of bacterial isolates was done by 16S rRNA sequencing. Genomic DNA were isolated from the bacterial strains, amplified by polymerase chain reaction (PCR) using universal 16S rRNA primers, cloned, and sequenced with dideoxy nucleotide. The sequences obtained were matched with the previously published sequences available in National Centre for Biological Information (NCBI) using BLAST.

2.3. Corrosion Rate Determination by Mass Loss Method. Commercially available carbon steel sheet was machined into coupons of size $7.5 \times 1.9 \times 0.3$ cm as per ASTM G184 and holes were drilled on the top centre of the coupons. The elemental composition of carbon steel is given in Table 1. *Jatropha curcas* biodiesel (JBD) was purchased from a biodiesel exporter in India and commercial diesel (CD) was purchased from a nearby petrol bunk. The carbon steel coupons were polished with 400, 600, and 800 grit emery paper and then degreased using trichloroethylene.

The following four fuel matrices were used as test media:

(i) CD: 100% commercial diesel,

(ii) B5: 5% JBD and 95% CD (% by volume),

(iii) B20: 20% JBD and 80% CD (% by volume),

(iv) B100: 100% JBD.

The experiment was carried out in the absence (control system) and presence (experimental system) of bacteria. The control system consisted of 800 mL fuel mixture with 2% (v/v) water (500 ppm chloride) to simulate corrosion conditions while the experimental system used was 800 mL fuel mixture with 2% (v/v) water (500 ppm chloride) and 0.5% (v/v) of bacterial inoculum (a load of 1×10^6 CFU/mL).

The mass loss measurements were carried out as per ASTM G1. Previously weighed metal coupons were immersed in the test matrices and agitated using a magnetic stirrer. After 100 h, the coupons were removed and pickled in pickling solutions, washed with water, and dried. Final masses of the coupons in each system were taken and the mean corrosion rates (in triplicate) were calculated and expressed in mils per year (mpy). The corrosion rate was calculated using the following formula:

$$\text{Corrosion Rate (mpy)} = \frac{3.45 \times 10^6 \times \text{mass loss (gram)}}{\text{Density (g/cm}^2) \times \text{Area (cm}^2) \times \text{Time (hour)}}. \tag{1}$$

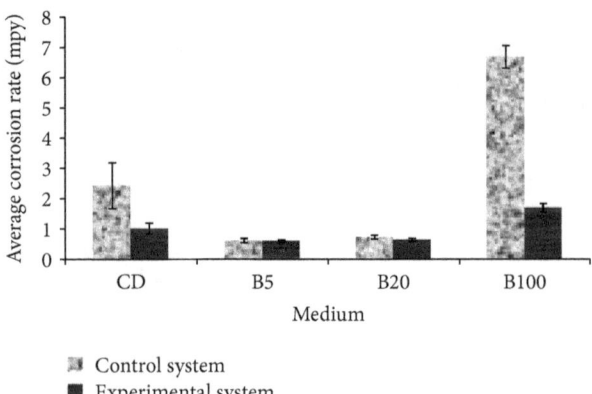

FIGURE 1: Corrosion rates of carbon steel in the control and experimental system.

2.4. Surface Analysis by Scanning Electron Microscopy (SEM). For SEM studies, the surface of the coupons exposed to CD and B100 test matrices for 100 h were chosen. The coupons were exposed to 2.5% glutaraldehyde for 8 h and subsequently washed with a graded series (30%, 50%, and 70% and 100%) of ethanol for dehydration. The samples were then coated with a gold alloy prior to SEM observations. The entire surface area of the coupon was examined to locate sessile bacteria.

3. Results and Discussion

3.1. Isolation of Bacteria. Three bacterial strains were isolated from the sediments formed in a two-year-old jatropha biodiesel left over from a previous research. Preliminary identification of the three bacterial strains obtained indicated that the isolates belonged to the genus *Bacillus* sp.

3.2. 16S rRNA Gene Sequence Analysis. Amplification of targeting bacterial 16S rRNA gene was performed using eubacterial 16S rRNA primers. The 16S rRNA gene was cloned and the isolated plasmids from the clones were subjected to 16S rRNA gene sequencing. The sequences obtained were matched with the previously published sequences available in NCBI (National Centre for Biological Information) using BLAST. Sequence alignment and comparison revealed similarity with *Bacillus pumilus*. The nucleotide sequence data have been deposited in GenBank under the accession numbers *KF410588, KF410589,* and *KF410590* [14]. *Bacillus* species constitute a diverse group of bacteria widely distributed in soil and aquatic environment. *Bacillus pumilus* is a gram-positive, aerobic, rod-shaped, endospore forming bacteria, belonging to the genus *Bacillus*. *Bacillus pumilus* resides in soils and some colonise in the root area of some plants.

3.3. Mass Loss Measurement. The corrosion rates of carbon steel in four test matrices in the control and experimental systems as determined by mass loss method are depicted in Figure 1. It is clear that carbon steel shows the highest

corrosion rates in B100 both in the control and in experimental systems followed by CD, B5, and B20. The corrosion rates of carbon steel in B100 in the control and experimental systems are 6.69 ± 0.3731 mpy and 1.70 ± 0.1386 mpy, respectively. Several studies have shown that corrosion of metals in biodiesel is higher than that in petrodiesel and this may be due to the presence of water content, free fatty acids, and unconverted monoalkyl esters. [5, 15, 16]. It is also noted that the corrosion rate increases with increasing concentration of biodiesel in the blends. The corrosivity of biodiesel blends is found to be lesser compared to that of petrodiesel alone. The same trend was observed by Ambrozin et al. [17]. Also the corrosion rates of carbon steel in all the four fuel matrices used in this study were found to be lesser in the presence of *Bacillus pumilus* compared to the control system and a marked difference is noticed in the corrosion rates of carbon steel in B100 between the control and experimental systems. This may be due to the formation of a protective film/biofilm on the surface of carbon steel where biological activities modify the environmental conditions at the metal/solution interface which may result in the reduction of corrosion rates in the presence of bacteria [18]. Studies have shown that various strains of bacteria like *Staphylococcus Sp.* and *Pseudomonas cichorii* formed corrosion inhibiting biofilm on mild steel [19, 20]. A significant reduction in corrosion rates of brass was achieved by *Bacillus subtilis* bacterial biofilm [21].

Corrosion studies involving *Bacillus pumilus* are limited. Bolton et al. [22] investigated the role of *Bacillus pumilus* isolated from samples taken from corroding galvanised steel pipes conveying water on the corrosion of steel and galvanised steel. The results showed that *Bacillus pumilus* had increased the corrosion of zinc galvanised steel but did not increase the corrosion rate of steel.

3.4. Surface Analysis by Scanning Electron Microscopy. Figure 2 shows the SEM micrographs of carbon steel in CD and B100 in the experimental system after exposure to the bacterial system, without removal of corrosion products on the metal surface. It shows that the coupons are covered with dense and lumpy corrosion products.

4. Conclusion

(i) Carbon steel shows the highest corrosion rates in B100 in both the control and experimental systems.

(ii) The corrosion rates of carbon steel increase with increasing concentration of biodiesel in the blends.

(iii) In the presence of *Bacillus pumilus* carbon steel shows lesser corrosion than that in the control system in all the test matrices.

Competing Interests

The authors declare that there is no conflict of interests regarding the publication of this article.

(a)   (b)

FIGURE 2: SEM images of carbon steel (a) CD and (b) B100 in the experimental system.

Acknowledgments

The authors would like to thank the authorities of Avinashilingam Institute for Home Science and Higher Education for Women, Coimbatore 641 043, India, for providing the necessary facilities for carrying out this work.

References

[1] A. Anisha, H. N. Meenakshi, R. Shyamala, R. Saratha, and S. Papavinasam, "Compatibility of metals in Jatropha oil," in *Proceedings of the NACE Corrosion Conference (CORROSION '11)*, Paper No. 11140, Houston, Tex, USA, 2011.

[2] K. Openshaw, "A review of *Jatropha curcas*: an oil plant of unfulfilled promise," *Biomass and Bioenergy*, vol. 19, no. 1, pp. 1–15, 2000.

[3] W. M. J. Achten, L. Verchot, Y. J. Franken et al., "*Jatropha* biodiesel production and use," *Biomass and Bioenergy*, vol. 32, no. 12, pp. 1063–1084, 2008.

[4] A. S. M. A. Haseeb, M. A. Fazal, M. I. Jahirul, and H. H. Masjuki, "Compatibility of automotive materials in biodiesel: a review," *Fuel*, vol. 90, no. 3, pp. 922–931, 2011.

[5] R. D. Kane and S. Papavinasam, "Corrosion and SCC issues in fuel ethanol and biofuels," in *Proceedings of the NACE Corrosion Conference (CORROSION '09)*, Paper No.09528, Atlanta, Ga, USA, 2009.

[6] S. Maruthamuthu, S. Mohanan, A. Rajasekar et al., "Role of corrosion inhibitor on bacterial corrosion in petroleum product pipelines," *Indian Journal of Chemical Technology*, vol. 12, no. 5, pp. 567–575, 2005.

[7] H. N. Meenakshi, A. Anisha, R. Shyamala, and R. Saratha, "Compatibility of biofuel/diesel blends on storage tank material," *Chemical Science Transactions*, vol. 2, supplement 1, pp. S99–S104, 2013.

[8] D. F. Aktas, J. S. Lee, B. J. Little et al., "Anaerobic metabolism of biodiesel and its impact on metal corrosion," *Energy and Fuels*, vol. 24, no. 5, pp. 2924–2928, 2010.

[9] R. J. Watkinson, *Hydrocarbon Degradation in Microbial Problems and Corrosion in Oil and Oil Products Storage*, The Institute of Petroleum, London, UK, 1984.

[10] W. Wang, P. E. Jenkins, and Z. Ren, "Heterogeneous corrosion behaviour of carbon steel in water contaminated biodiesel," *Corrosion Science*, vol. 53, no. 2, pp. 845–849, 2011.

[11] J. S. Lee, R. I. Ray, and B. J. Little, "An assessment of alternative diesel fuels: microbiological contamination and corrosion under storage conditions," *Biofouling*, vol. 26, no. 6, pp. 623–635, 2010.

[12] B. Klofutar and J. Golob, "Microorganisms in diesel and in biodiesel fuels," *Acta Chimica Slovenica*, vol. 54, no. 4, pp. 744–748, 2007.

[13] G. Sørensen, D. V. Pedersen, A. K. Nørgaard, K. B. Sørensen, and S. D. Nygaard, "Microbial growth studies in biodiesel blends," *Bioresource Technology*, vol. 102, no. 8, pp. 5259–5264, 2011.

[14] S. Malarvizhi, R. Shyamala, and S. Papavinasam, "Assessment of microbiologically influenced corrosion of metals in biodiesel from *Jatropha curcas*," in *Proceedings of the NACE Corrosion Conference (CORROSION '15)*, Paper No. 5772, Dallas, Tex, USA, 2015.

[15] M. A. Fazal, A. S. M. A. Haseeb, and H. H. Masjuki, "Biodiesel feasibility study: an evaluation of material compatibility; performance; emission and engine durability," *Renewable and Sustainable Energy Reviews*, vol. 15, no. 2, pp. 1314–1324, 2011.

[16] M. A. Fazal, A. S. M. A. Haseeb, and H. H. Masjuki, "Degradation of automotive materials in palm biodiesel," *Energy*, vol. 40, no. 1, pp. 76–83, 2012.

[17] A. R. P. Ambrozin, S. E. Kuri, and M. R. Monteiro, "Corrosão metálica associada ao uso de combustíveis minerais e biocombustíveis," *Química Nova*, vol. 32, no. 7, pp. 1910–1916, 2009.

[18] H. A. Videla and L. K. Herrera, "Understanding microbial inhibition of corrosion. A comprehensive overview," *International Biodeterioration and Biodegradation*, vol. 63, no. 7, pp. 896–900, 2009.

[19] S. Ponmariappan, S. Maruthamuthu, and R. Palaniappan, "Inhibition of corrosion of mild steel by *Staphylococcus* sp," *Transactions of the SAEST*, vol. 39, no. 4, pp. 99–108, 2004.

[20] S. Chongdar, G. Gunasekaran, and P. Kumar, "Corrosion inhibition of mild steel by aerobic biofilm," *Electrochimica Acta*, vol. 50, no. 24, pp. 4655–4665, 2005.

[21] D. Örnek, T. K. Wood, C. H. Hsu, and F. Mansfeld, "Corrosion control using regenerative biofilms (CCURB) on brass in different media," *Corrosion Science*, vol. 44, no. 10, pp. 2291–2302, 2002.

[22] N. Bolton, M. Critchley, R. Fabien, N. Cromar, and H. Fallowfield, "Microbially influenced corrosion of galvanized steel pipes in aerobic water systems," *Journal of Applied Microbiology*, vol. 109, no. 1, pp. 239–247, 2010.

5

Effect of Deformation Structure and Annealing Temperature on Corrosion of Ultrafine-Grain Fe-Cr Alloy Prepared by Equal Channel Angular Pressing

Muhammad Rifai ⓘⒹ, **Motohiro Yuasa, and Hiroyuki Miyamoto** ⓘⒹ

Department of Mechanical Engineering, Doshisha University, Kyoto, Japan

Correspondence should be addressed to Muhammad Rifai; rmuhamma@mail.doshisha.ac.jp

Academic Editor: Michael I. Ojovan

The effect of the deformation structure and annealing temperature on the corrosion of ultrafine-grain (UFG) Fe-Cr alloys with 8 to 12% Cr prepared by equal channel angular pressing (ECAP) was investigated with particular emphasis on the stability of the passivation layer. Fe-Cr alloys were processed by ECAP using up to eight passes at 423 K by the Bc route, followed by annealing at temperatures of 473 to 1173 K for 1 h. Passivity appeared in all alloys as a result of ECAP, and the stability of the passivation layer was evaluated by anodic polarization measurements in a 1000 mol·m^{-3} NaCl solution. The stability of the passivation layer increased as the degree of deformation became more extensive with successive ECAP passes, and distinct escalation occurred with the formation of a UFG microstructure. In the early stages of annealing at moderate temperatures, the stability of the passivation layer deteriorated, although no visible grain growth occurred, and this effect increased monotonically with increasing annealing temperature. The high degree of stability of the passivation layer on UFG alloys following ECAP can be attributed to the large number of high-angle nonequilibrium grain boundaries, which may lead to Cr enrichment of the surface region. The deterioration of the passivation layer in the early stages of annealing may be attributed to a change in the grain boundaries to an equilibrium state. The present results show that the superiority of as-ECAPed materials of the Fe-Cr alloy to recovered ones by heat treatment can be achieved with 8–10% Cr as observed in 20% Cr.

1. Introduction

Metal forming processes such as equal channel angular pressing (ECAP) have received considerable attention over the past two decades as methods for producing ultrafine-grain (UFG) bulk materials for structural applications [1, 2]. The UFG microstructure produced by severe plastic deformation has a high density of nonequilibrium grain boundaries, which are characterized by excess grain boundary energy and the presence of long-range elastic stress due to a large number of dislocations [3, 4]. Annealing at moderate temperatures leads to annihilation and rearrangement of these dislocations and causes the grain boundaries to change to an equilibrium state [4]. Such annealing is likely to impact the corrosion behavior of UFG materials because the grain boundary state influences the mechanical and electrochemical properties. Generally, corrosion resistance in Fe-Cr alloys is associated

with the formation of a passivation layer. To produce highly corrosion-resistant stainless steel, the minimum Cr content is 11%, which can be explained in terms of the percolation limit and the selective dissolution of Fe atoms [5, 6] in an aqueous corrosive environment.

In previous studies, it was found that UFG Fe-Cr alloys with 20% Cr processed by ECAP, with a large fraction of nonequilibrium high-angle grain boundaries (HAGBs), exhibited higher corrosion resistance than coarse-grained (CG) material, and this was attributed to enhanced passivation [7–11]. In addition to the chemical composition and the corrosive environment, the deformation structure and the grain boundary state are also expected to play an important role in determining the corrosion resistance. In the present paper, Fe-Cr alloys with Cr contents of 8, 10, and 12% were processed by ECAP, and their corrosion resistance was evaluated by anodic polarization measurements to determine

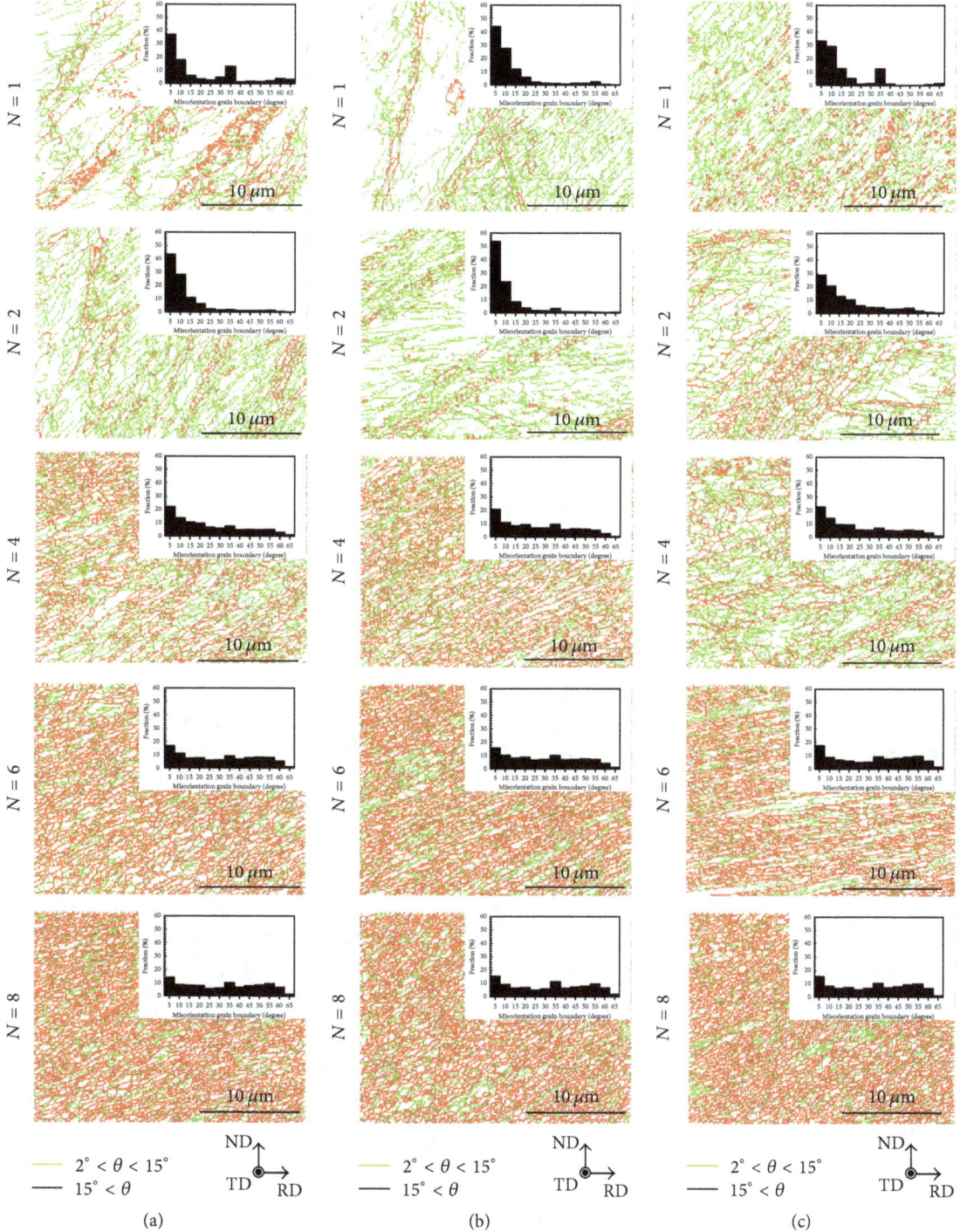

FIGURE 1: Misorientation maps with the inset figure of distributions for Fe-Cr alloy processed by ECAP with (a) 8% Cr, (b) 10% Cr, and (c) 12% Cr.

the stability of the passivation layer. The purposes of the present study were as follows:

(1) To examine the microstructure, hardness, and anodic polarization of UFG Fe-Cr alloys with different Cr content and the effect of annealing

(2) To characterize the influence of post-ECAP annealing on the critical potential

(3) To investigate the effect of the deformation structure and the annealing temperature on the stability of the passivation layer.

2. Experimental Procedure

The material used in the present study was Fe-Cr alloy with low carbon and nitrogen contents and Cr contents of 8, 10, and 12%. Each billet was annealed in an argon atmosphere at 1323 K for 1 h. The initial grain size was about 20–250 μm. Specimens with dimensions of $8 \times 8 \times 120$ mm were machined and subjected to ECAP for up to eight passes at 423 K by the Bc route, while being lubricated with high-temperature fluorine-based grease. After the ECAP process, the billets were annealed in an infrared furnace (ULVAC MILA5000) from 473 to 1173 K for 1 h under a vacuum. An $8 \times 10 \times 2$ mm corrosion test specimen was prepared from a unannealed specimen using a spark-erosion machine [7]. The specimen was coated with epoxy molding to cover the connection between the specimen and cable. The edge area was sealed with premium grade vinyl tape to prevent pitting (crevice) corrosion. The mounted specimen was ground and then polished. Details of the ECAP procedure have been previously published [7]. Field-emission scanning electron microscopy (FE-SEM; JSM 7001F), together with electron backscatter diffraction (EBSD; Oxford Instruments Co.: Model: HKL), was used to observe the grain maps. The EBSD images were processed using INCA™ (Oxford Instruments Co.). The microstructure was examined by field-emission transmission electron microscopy (FE-TEM, JEM 2100F). Samples in the form of thin foils were prepared for TEM analysis and subsequently polished using abrasive paper to a thickness of approximately 100 μm and and then thinned by twin-jet polishing using a Tenupol 5 instrument (Struers Co., Ltd.) with a solution consisting of 40% acetic acid, 30% phosphoric acid, 20% nitric acid, and 10% distilled water. Finally, the specimens were polished using an ion beam in a Gatan 691 precision ion polishing system. An acceleration voltage of 200 kV was used for all TEM observations. Further details regarding the microstructure observation process can be found in our prior publication [7].

Microhardness experiments were performed using a Vickers hardness testing machine (Shimadzu HMV-2) under a load, with a 15 s dwell time, with ten measurements being performed for each sample. X-ray diffraction (XRD) analysis was carried out using a Rigaku SmartLab system with CuKα radiation, an operating voltage of 40 kV, a current of 0.2 A, and a 2θ angle of 30–130°, under continuous scanning mode. Details concerning the setup for ECAP, XRD, and microhardness testing are available elsewhere [7].

Anodic polarization corrosion tests were carried out in a neutral solution of 1000 mol·m^{-3} NaCl at room temperature by potentiodynamic polarization, using a HZ5000 potentiostat at a scan rate of 20 mV·min^{-1}, a corrosion current, and an Ag/AgCl reference electrode. The reference electrode was placed in a saturated 3000 mol·m^{-3} KCl solution (representing a saturated solution). Each sample was immersed in the etchant solution for one hour. The solution was deaerated with argon gas to remove dissolved oxygen. The test process was initialized after the open circuit potential (OCP) of the specimen was stabilized. Details of the corrosion test procedure have also been previously published [7].

Surface analysis was carried out using glow discharge-optical emission spectroscopy (GD-OES; HORIBA GD Profiler 2); the specimens were first ground and dried in air at room temperature. This technique provides accurate

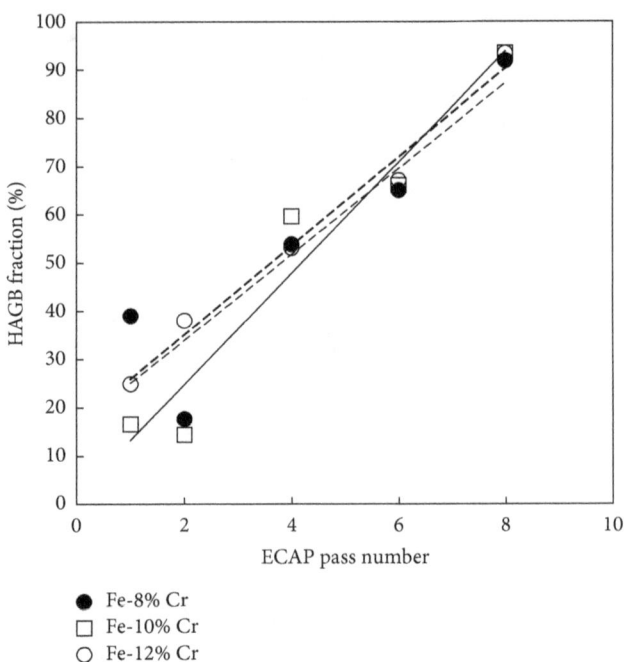

FIGURE 2: Dependence of HAGB fraction for Fe-Cr alloy on a number of ECAP passes.

sputtering sequences, a constant etching rate, excellent sensitivity, and a nanoscale depth resolution. Measurements were performed in synchronous mode using a pulsed RF source at 10 W, with $\varnothing$ 4 mm anodes, under an argon atmosphere at 300 Pa.

3. Results and Discussion

Microstructural observations using EBSD and TEM revealed the deformation structure and the formation of nonequilibrium grain boundaries from one to eight ECAP passes. Figure 1 shows grain boundary misorientation maps and distributions for Fe-Cr alloys following one, two, four, six, and eight ECAP passes in the transverse plane. After one and two passes, low-angle grain boundaries (LAGBs) with misorientations of 2–15° are common, with only a few HAGBs with higher misorientations. In the distributions shown in the insets, most grains have misorientations of 2–10°. As the number of ECAP passes increases, the distribution shifts to higher misorientation angles, and, following eight passes, there is a large fraction of HAGBs. The change of misorientation to higher values may appear due to local crystal orientation in grains during deformation process by ECAP [11]. This effect has been previously reported for Fe-20% Cr alloy following ECAP [11].

Figure 2 plots the HAGB fraction against the number of ECAP passes. It can be seen that as the number of passes increases, the HAGB fraction increases rapidly. Regardless of the Cr content, following eight passes, all specimens had a similar HAGB fraction, indicating a UFG microstructure. This is likely to have an effect on the stability of the passivation layer and the corrosion resistance of the material [11].

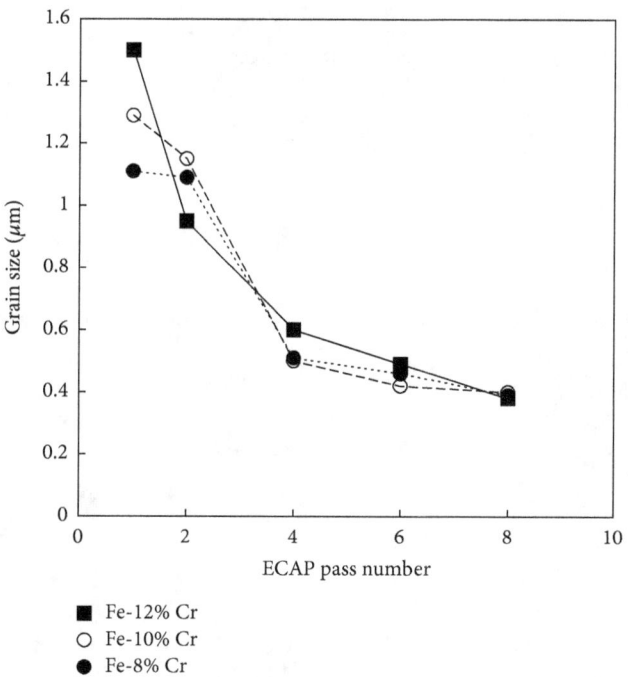

FIGURE 3: Dependence of grain size for Fe-Cr alloy on a number of ECAP passes.

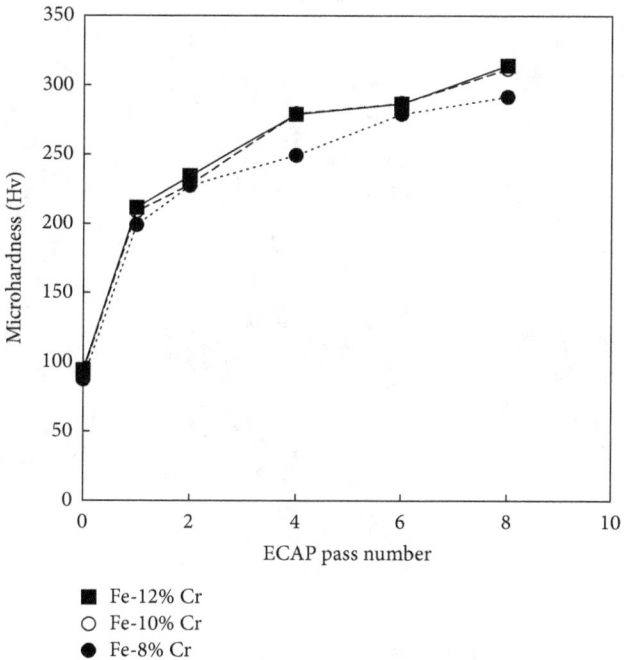

FIGURE 4: Dependence of microhardness for Fe-Cr alloy on a number of ECAP passes.

After a single pass, the microstructure is typical of that produced by shear deformation, with shear bands present inside grains. These bands appear in the EBSD maps in Figure 1 as diagonally inclined lines. It is apparent that the grains are finely subdivided into subgrain after one pass [12]. After four passes, the microstructure is a combination of UFGs and submicron grains. After the first pass, the grains are elongated along the shear deformation bands, but they

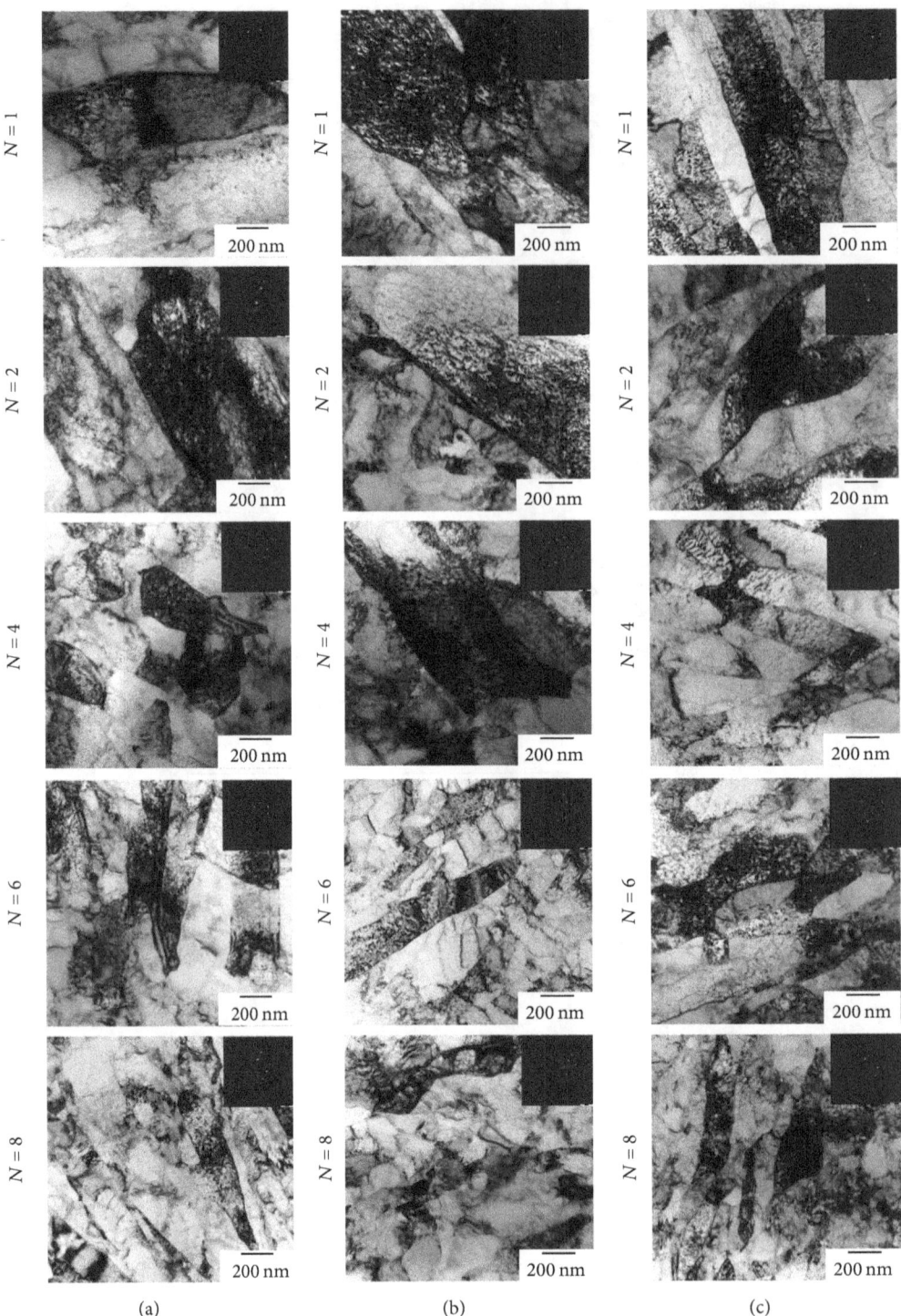

FIGURE 5: TEM images of Fe-Cr processed by ECAP with (a) 8% Cr, (b) 10% Cr, and (c) 12% Cr.

become more refined and equiaxed, in addition to being more uniform in size, after six and eight passes. This is the result of repeated strain along different routes, which activates multiple slip systems and causes shear deformation bands to intersection [7]. From the EBSD maps, the average grain size $d^{-0.5}$ was determined to be 199, 180, and 176 nm for

alloys with 8, 10, and 12% Cr, respectively, following eight ECAP passes. With increasing number of ECAP passes, the grain size decreased for all samples, regardless of Cr content, as seen in Figure 3. Figure 4 shows the dependence of the Vickers microhardness on the number of ECAP passes. The microhardness increases rapidly with increasing number of

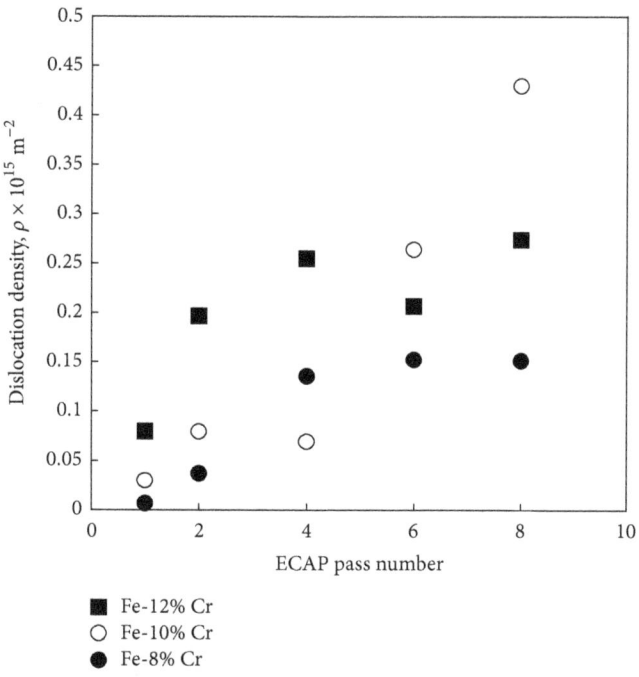

FIGURE 6: Dependence of dislocation density for Fe-Cr alloy on number of ECAP passes.

ECAP passes, due to the higher dislocation density and strengthening due to grain refinement [13, 14].

By increasing the number of ECAP passes or the equivalent strain, the dislocation density increases and a subgrain structure begins to form [15]. This subgrain structure, together with the large number of dislocations and nonequilibrium grain boundaries, leads to work hardening and an increase in the strength of the material due to the difficulty of dislocation movement inside the grains; this has been verified by TEM and XRD measurements [13]. Figure 5 shows TEM images of specimens subjected to one, two, four, six, and eight ECAP passes. After one pass, the grains are elongated and dislocations are present at the grain boundaries. For higher numbers of passes, the nonequilibrium grain boundaries are easy to observe. Grain fragmentation also can be seen in the initial of UFG structure formation. Following a single ECAP pass, the specimen is finely subdivided into grains and dislocations both at grain boundaries and within the grains. The microstructure following the initial deformation process consists of low-angle grain boundaries, as is evidenced by the blurred appearance of the grain boundaries in the TEM images [15]. After two ECAP passes, the microstructure became finer and consisted of elongated grains with more planar boundaries. The dislocation density significantly increased after two passes. After four passes, the microstructure consisted of more equiaxed UFGs with sharper boundaries. The sharpest grain boundaries were observed following eight passes. The insets in Figure 5 show selected area diffraction patterns corresponding to the TEM images. Up to six ECAP passes, the patterns are spot-like, indicating the presence of LAGBs and a high dislocation

density inside grains or at grain boundaries which can be indicated from XRD peak broadening. However, after eight passes, the pattern becomes ring-like, indicating a UFG structure with a large fraction of HAGBs. This is consistent with the TEM observation results, which indicated that a higher fraction of HAGBs was present following eight ECAP passes, whereas a smaller number of passes produced mainly dislocations.

The dislocation density was calculated using a Williamson-Hall plot based on the full width at half maximum of XRD diffraction peaks. Broadening of the peaks occurs due to the presence of dislocations and nonequilibrium grain boundaries containing extrinsic defects and subjected to elastic stress [7]. The calculations were performed using the (110), (200), (211), (220), and (310) diffraction peaks. Figure 6 shows the dependence of the dislocation density on the number of ECAP passes. As a result of grain refinement, the dislocation density increases with increasing number of ECAP passes, and after eight passes it is about three times higher than after one pass. This is in agreement with our previous findings for Fe alloy with 20% Cr [15].

The effect of the deformation structure on the corrosion resistance was investigated in a $1000 \, mol \cdot m^{-3}$ NaCl solution at room temperature by potentiodynamic polarization. Anodic polarization curves for the as-annealed specimen and specimens following one, two four, six, and eight ECAP passes are shown in Figure 7. The results are divided into two groups. In Group A, no passivation layer was present on the surface (intergranular corrosion), whereas in Group B a passivation layer was present (pitting corrosion). Fe-Cr

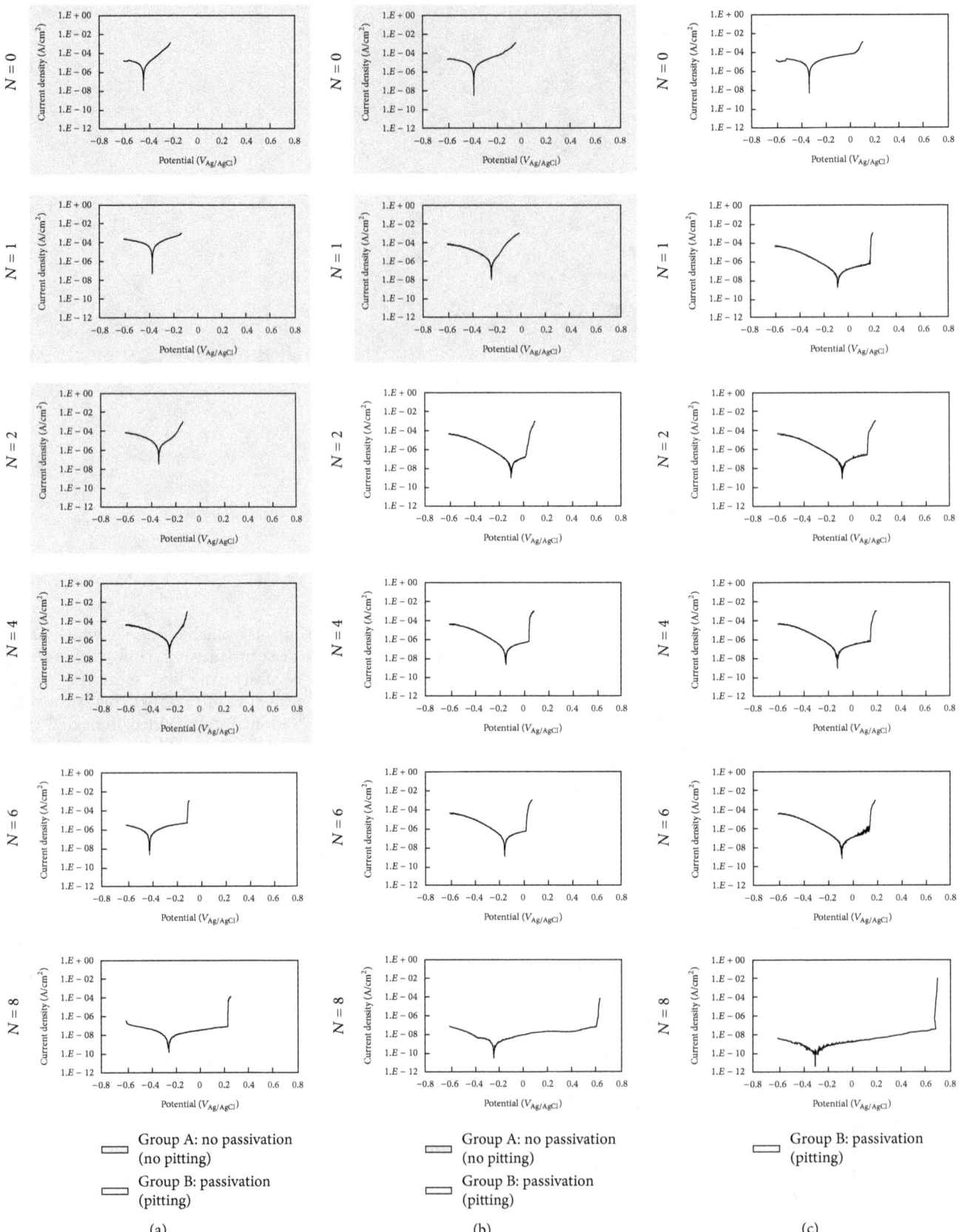

FIGURE 7: Anodic polarization curves of Fe-Cr alloy processed by ECAP with (a) 8% Cr, (b) 10% Cr, and (c) 12% Cr in 1000 mol·m^{-3} NaCl solution.

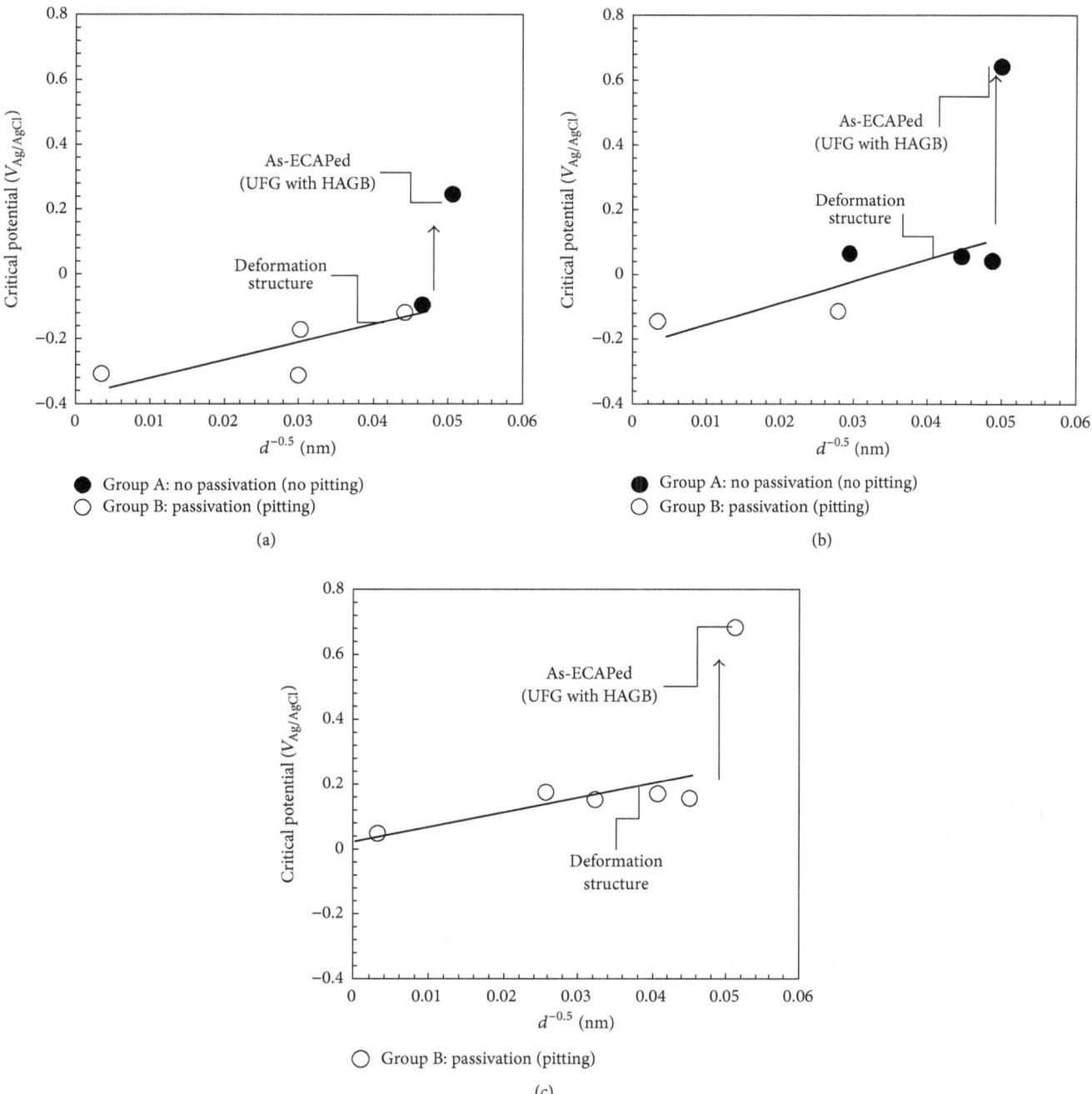

FIGURE 8: Potential at 0.1 mA/cm^2 and grain size of Fe-Cr alloy processed by ECAP with (a) 8% Cr, (b) 10% Cr, and (c) 12% Cr.

alloy with 8 and 10% Cr showed no passivation until four ECAP passes and one pass, respectively. However, the Fe-Cr alloy with 12% Cr showed passivation from the beginning of the grain refinement process. This means that, in the case of 8% Cr, far more grain refinement and a higher dislocation density are required in order to achieve the same level of passivation as that for 10 and 12% Cr. The critical potential, which is defined as the potential required for the anodic current density to reach 0.1 mA·cm^{-2}, increased with the number of passes and became positive, so that the passivation range became longer or nobler.

The critical potential is plotted against the average grain size $d^{-0.5}$ in Figure 8. It is clear that all the samples produced by ECAP have a nobler potential than the as-annealed sample. The polarization results imply that corrosion resistance is enhanced by ECAP regardless of the Cr content. The most

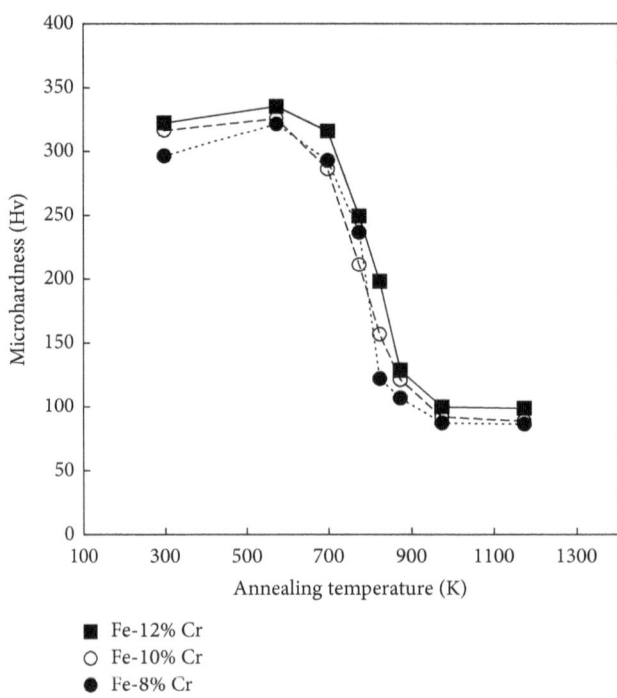

FIGURE 9: Relationship between microhardness and post-ECAP annealing temperature.

extended passivation range, representing higher corrosion resistance, is found following eight ECAP passes. It is interesting to note that even though the Cr content is less than that required for stainless steel, UFG material with 8 and 10% Cr showed improved corrosion resistance due to the stability of the passivation layer in the $1000\,mol\cdot m^{-3}$ NaCl solution. Following eight passes, the specimen exhibited a high fraction of HAGBs instead of mainly dislocations, as can be seen in Figure 5. This means that a high HAGB fraction can increase the critical potential and the stability of the passivation layer, thus reducing the critical Cr content for stainless steel.

To investigate the effect of the state of equilibrium of the grain boundaries on the corrosion resistance, heat treatment was applied to the UFG material. Figure 9 shows the change in the microhardness of UFG samples by eight passes of ECAP, subjected to heat treatments at temperatures of 473–1173 K for 1 h. At 698 K, the initial hardness remains; this corresponds to the recovery stage in which dislocations are annihilated and grain boundaries change to an equilibrium state [16]. By 973 K, the hardness has drastically decreased to around 100 HV, indicating the completion of the recrystallization stage, accompanied by grain growth. Figure 10 shows EBSD misorientation maps for UFG specimens subjected to heat treatments at 573, 698, 773, 873, 973, and 1173 K. The recovery stage can be seen until 698 K, and there is no significant change in grain size. At 698 K, several new grains are seen to have nucleated along the shear bands, and this is related to the recrystallization stage. At 873 K, recrystallized grains grow by annihilating deformed grains. This is the grain growth stage in which the hardness has not changed and

the microstructure showed equiaxed grain appearance. The effect of the annealing temperature on the grain size is shown in Figure 11. In addition, Figure 12 plots anodic polarization curves for USG specimens following heat treatments at different temperatures. These results were also divided into two groups, as described earlier. Regardless of the Cr content, the specimens exhibited stable passivation until the recovery stage, but those with 10 and 12% Cr maintained their passivation to a temperature of 50 K higher than the 8% Cr specimen. The occurrence of pitting is indicated by an abrupt increase in the anodic current at a nobler potential than that for a sample with passivation [17–19]. In Figure 12, it becomes more difficult to identify the pitting potential as the annealing temperature increases. The high anodic current in UFG material with a high density of nonequilibrium grain boundaries may be associated with the stability of the passivation layer. The critical potential at $0.1\,mA/cm^2$ is plotted as a function of grain size in Figure 13. The UFG material shows a more negative OCP than the as-annealed specimen due to higher fraction of nonequilibrium grain boundaries. Here also, the passivation range increases with increasing number of ECAP passes. The surface after anodic polarization was observed by laser microscopy and could be divided into two groups based on the anodic polarization results. The first group (Group A) exhibited no pitting (intergranular corrosion), which was related to the absence of a passivation layer. This group included, for example, the coarse-grained specimens with 8 and 10% Cr seen in Figures 14(a) and 14(b). The second group (Group B) included all UFG specimens and

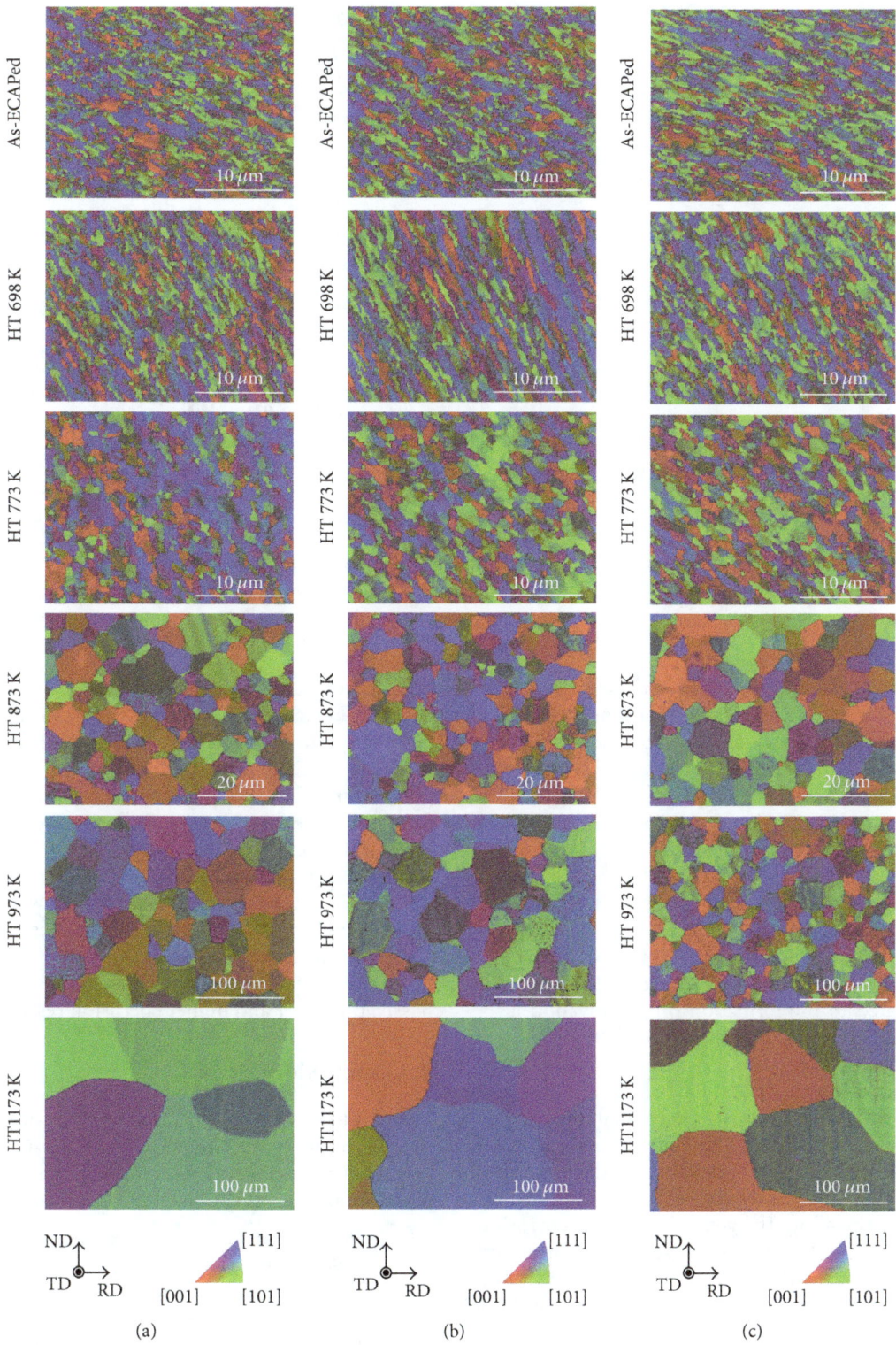

FIGURE 10: Color orientation maps obtained by EBSD after post-ECAP annealing with (a) 8% Cr, (b) 10% Cr, and (c) 12% Cr.

the as-annealed specimen with 12% Cr, as seen in Figures 14(c)–14(f). The anodic polarization tests indicated that the specimen produced using eight ECAP passes with a high density of nonequilibrium grain boundaries had a more stable passivation layer than specimens produced using a smaller number of ECAP passes or the as-annealed specimen.

The present result shows that the superiority of as-ECAPed materials of the Fe-Cr alloy to recovered ones can be

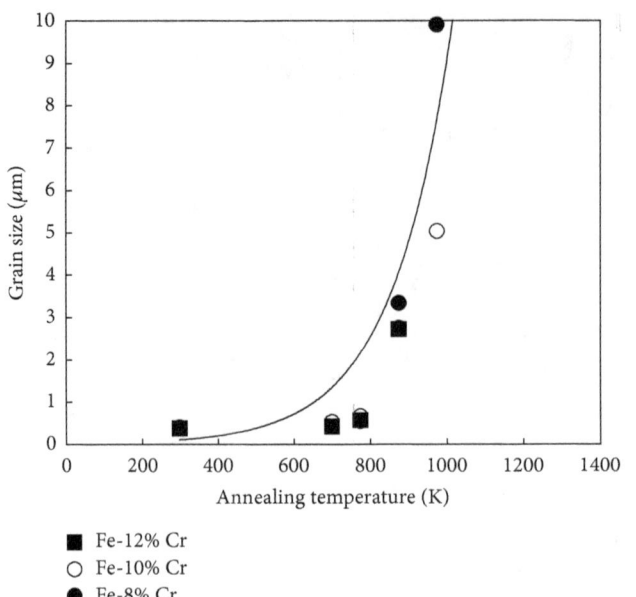

FIGURE 11: Effect of post-ECAP annealing temperature on grain size.

achieved even with 8–10% Cr as observed in 20% Cr [7]. The deformation structure, nonequilibrium grain boundaries, and chemical composition of the Fe-Cr alloy may have a high impact on the corrosion resistance. It is generally considered that the Cr content in particular plays a leading role in corrosion resistance [20–25]. Improved corrosion resistance can be achieved when the Cr content is above the critical value for stainless steel [26]. Asami et al. showed that the Cr content, expressed as Cr/(Fe + Cr), of the passivation layer increased by six times when the Cr content in the alloy was 13–15% [26]. This implies that the corrosion resistance of Fe-Cr alloy depends on the Cr content in the alloy. The corrosion resistance is related to the formation of a passivation layer on the surface of the metal. The addition of Cr to iron leads to an improvement in corrosion resistance by replacing the iron oxide passivation layer with Cr oxide [20]. It is therefore essential to determine the critical Cr content required to improve the corrosion resistance and the pitting resistance [27, 28]. Corrosion of Fe-Cr alloy depends on the properties of the passivation layer, which are mainly determined by the Cr content of the alloy. The general theory of corrosion resistance is based on the selective dissolution of iron and oxidation of Cr manages the formation of the passive film [29, 30]. Previous research suggested that more than 50% Cr is required in the passivation layer in order to make it stable [26]. It was also reported that, by increasing the Cr content in the bulk, the passivation layer showed a significant improvement [31, 32]. In a binary system such as Fe-Cr alloy, corrosion initiation depends on the formation of iron clusters [32]. The size of these clusters influences local dissolution and the occurrence of pitting. However, the iron clusters are also affected by the Cr content in the alloy, which is referred to as the critical value [20–25]. It can be expected that the deformation structure and the presence of nonequilibrium grain

boundaries can influence the critical Cr content in stainless steel. The extended stability of the passivation layer in Fe-Cr alloys with 8 and 10% Cr can be explained by enhanced diffusion of Cr due to the high density of nonequilibrium grain boundaries [24, 33, 34] and also by the passivation layer formed by selective iron dissolution at the surface [5, 6]. The rapid diffusion of Cr was supplied by the Cr stored in the dislocation, grains, and nonequilibrium grain boundaries [35, 36]. Based on the GD-OES Cr profile shown in Figure 15, the passivation layer on UFG material is richer than that in Cr on CG material. This is an indication of enhanced Cr diffusion in the UFG structure, leading to a composition change in the passivation layer and higher corrosion resistance.

4. Conclusion

The effect of Cr content and post-ECAP annealing temperature on the corrosion behavior of UFG Fe-Cr alloy with 8, 10, and 12% Cr was investigated, focusing on the stability of the passivation layer. The following conclusions were obtained:

(1) Passivation characteristics appeared in the anodic polarization results for all UFG alloys after ECAP, and the critical potential for UFG material was higher than that for CG material before ECAP. However, the critical potential decreased was by post-ECAP annealing.

(2) Destabilization of the passivation and a drastic drop in the pitting potential were observed following moderate-temperature annealing, as the grain boundaries changed from a nonequilibrium to an equilibrium state with no significant grain growth.

(3) Nonequilibrium grain boundaries may facilitate the formation of a passivation layer in Fe-Cr alloys with

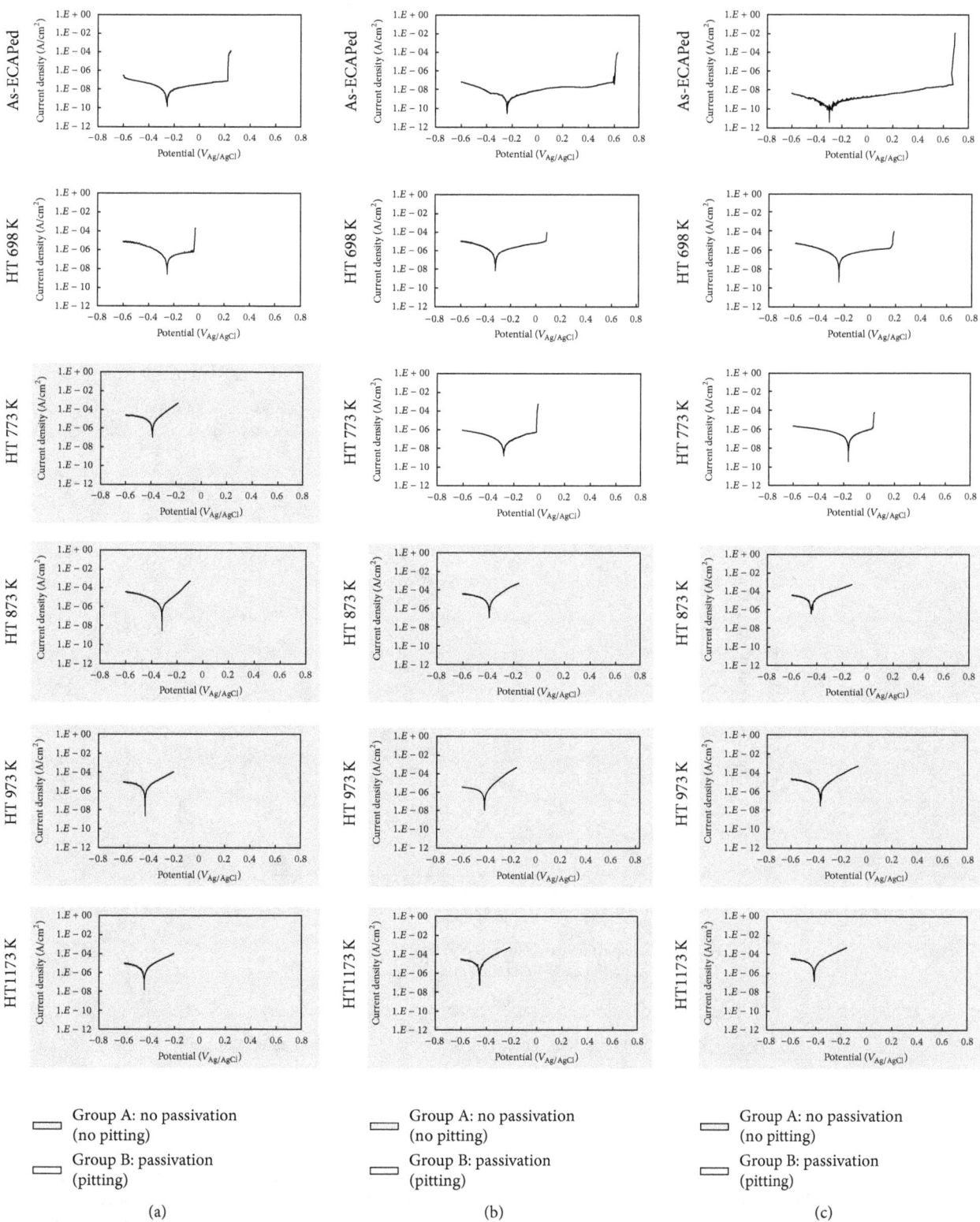

FIGURE 12: Anodic polarization curves for post-ECAP annealed samples in 1000 mol·m^{-3} NaCl solution with (a) 8% Cr, (b) 10% Cr, and (c) 12% Cr.

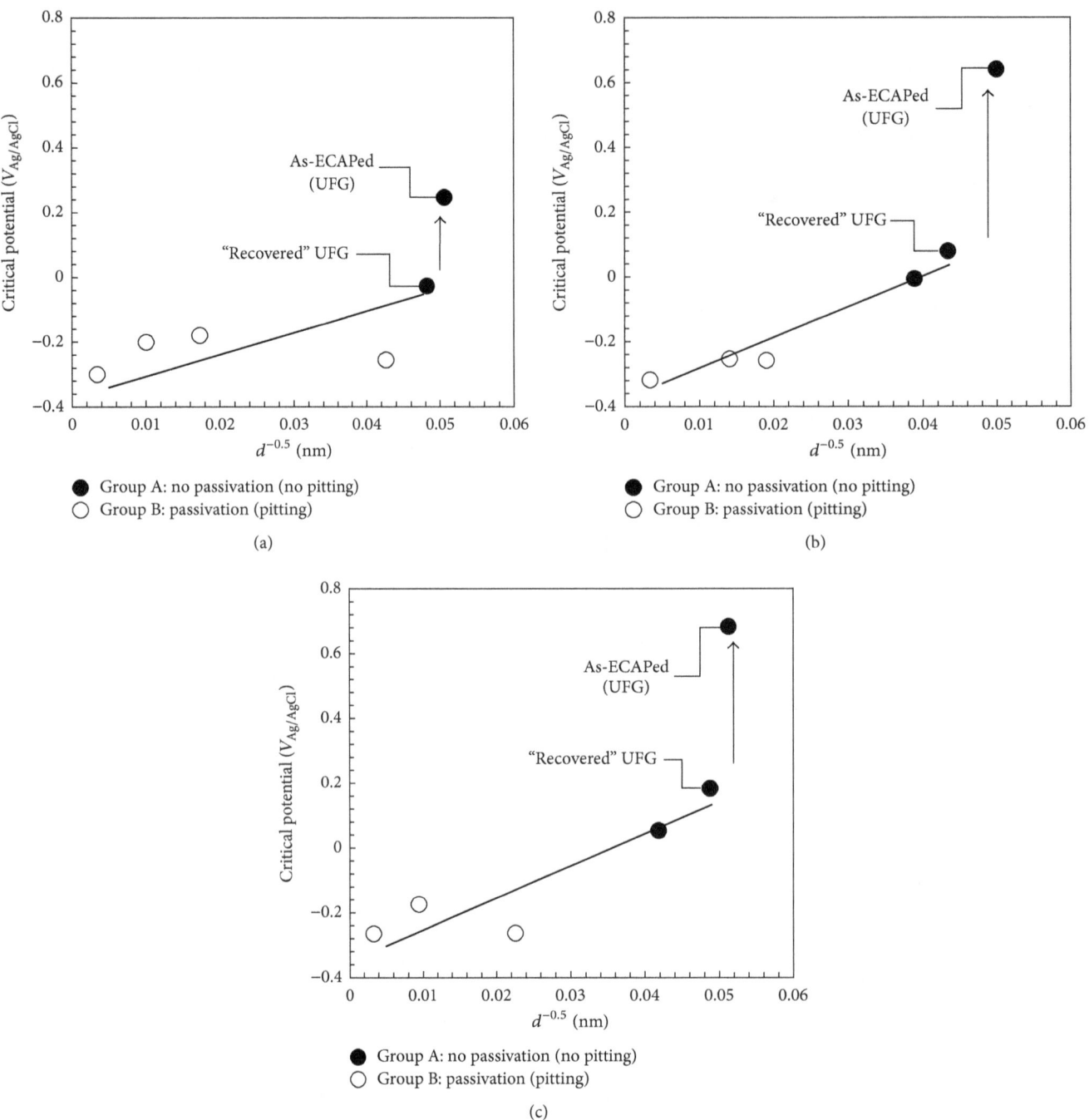

FIGURE 13: Potential at 0.1 mA/cm^2 and grain size of post-ECAP annealed sample.

a Cr content of less than 10%, which is generally considered to be the minimum value.

(4) A UFG specimen produced by eight ECAP passes exhibited a deformation structure with a higher fraction of HAGBs; other specimens mainly contained dislocations. This is consistent with the higher critical potential for the former specimen. This means that a high HAGB fraction can improve the critical potential and the stability of the passivation layer and also reduce the Cr limit required for stainless steel.

Conflicts of Interest

The authors declare that there are no conflicts of interest regarding the publication of this paper.

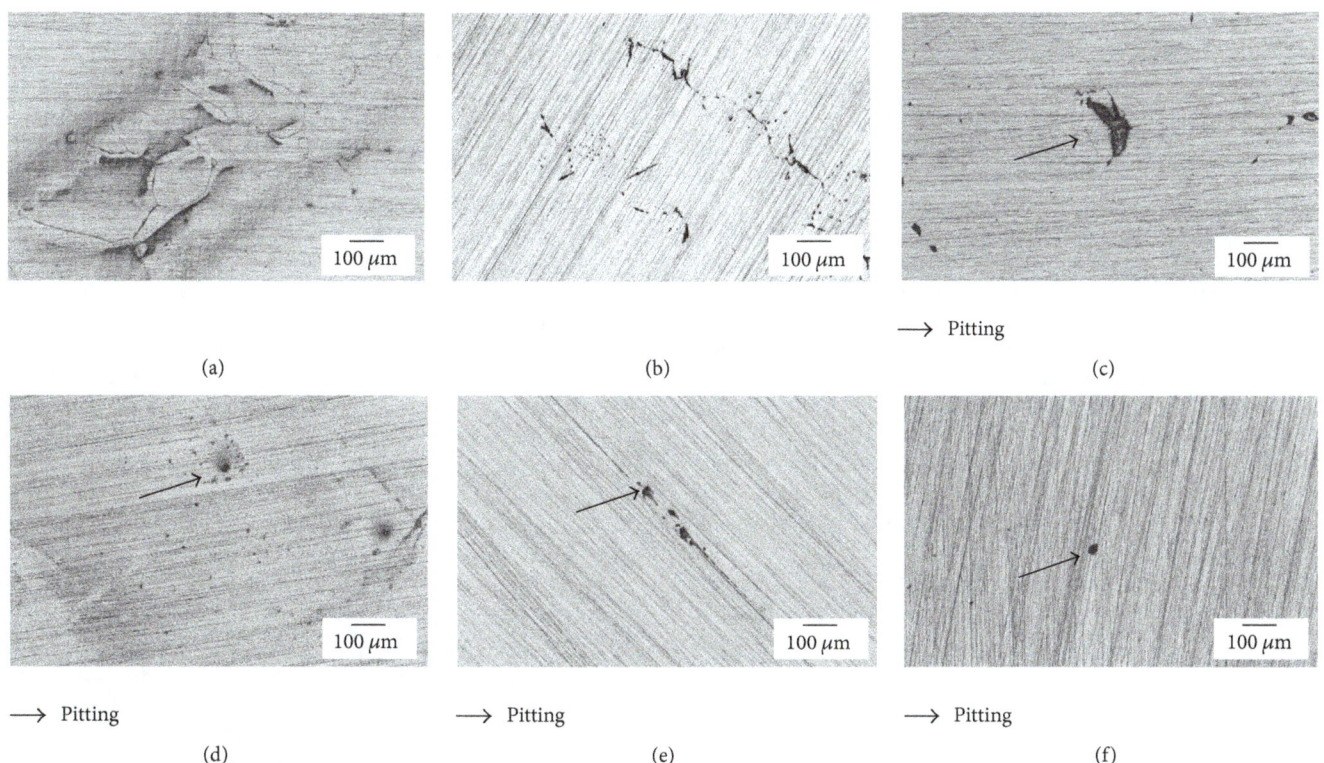

→ Pitting

(a)   (b)   (c)

→ Pitting   → Pitting   → Pitting

(d)   (e)   (f)

FIGURE 14: Surface appearance after anodic polarization test on as-ECAP processed sample with (a) 8% Cr, (b) 10% Cr, and (c) 12% Cr and grain growth stage (CG structure) with (d) 8% Cr, (e) 10% Cr, and (f) 12% Cr by laser microscopy.

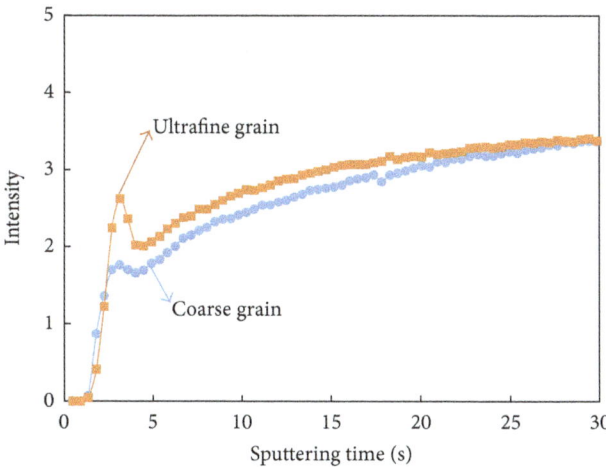

FIGURE 15: Cr profile obtained by GD-OES for Fe-10% Cr.

References

[1] R. Z. Valiev, Y. Estrin, Z. Horita, T. G. Langdon, M. J. Zechetbauer, and Y. T. Zhu, "Producing bulk ultrafine-grained materials by severe plastic deformation," *JOM Journal of the Minerals, Metals and Materials Society*, vol. 58, no. 4, pp. 33–39, 2006.

[2] R. Z. Valiev and T. G. Langdon, "Principles of equal-channel angular pressing as a processing tool for grain refinement," *Progress in Materials Science*, vol. 51, no. 7, pp. 881–981, 2006.

[3] M. W. Grabski, "Mechanical properties of internal interfaces," *Le Journal de Physique Colloques*, vol. 46, no. C4, pp. C4-567–C4-579, 1985.

[4] R. Z. Valiev, V. Y. Gertsman, and O. A. Kaibyshev, "Grain boundary structure and properties under external influences," *Physica Status Solidi (a) – Applications and Materials Science*, vol. 97, no. 1, pp. 11–56, 1986.

[5] K. Sieradzki and R. C. Newman, "A percolation model for passivation in stainless steels," *Journal of The Electrochemical Society*, vol. 133, no. 9, pp. 1979-1980, 1986.

[6] S. Qian, R. C. Newman, R. A. Cottis, and K. Sieradzki, "Validation of a percolation model for passivation of Fe-Cr alloys: Two-dimensional computer simulations," *Journal of The Electrochemical Society*, vol. 137, no. 2, pp. 435–439, 1990.

[7] M. Rifai, H. Miyamoto, and H. Fujiwara, "Effects of strain energy and grain size on corrosion resistance of ultrafine grained Fe-20%Cr steels with extremely low C and N fabricated by ECAP," *International Journal of Corrosion*, vol. 2015, Article ID 386865, 9 pages, 2015.

[8] Z. J. Zheng, Y. Gao, Y. Gui, and M. Zhu, "Corrosion behaviour of nanocrystalline 304 stainless steel prepared by equal channel angular pressing," *Corrosion Science*, vol. 54, no. 1, pp. 60–67, 2012.

[9] Y. Gui, Z. J. Zheng, and Y. Gao, "The bi-layer structure and the higher compactness of a passive film on nanocrystalline 304 stainless steel," *Thin Solid Films*, vol. 599, pp. 64–71, 2016.

[10] M. Pisarek, P. Kedzierzawski, T. Płociński, M. Janik-Czachor, and K. J. Kurzydłowski, "Characterization of the effects of hydrostatic extrusion on grain size, surface composition and

the corrosion resistance of austenitic stainless steels," *Materials Characterization*, vol. 59, no. 9, pp. 1292–1300, 2008.

[11] M. Rifai, H. Miyamoto, and H. Fujiwara, "The Effect of ECAP Deformation Route on Microstructure, Mechanical and Electrochemical Properties of Low CN Fe-20% Cr Alloy," *Materials Sciences and Applications*, vol. 5, no. 08, p. 568, 2014.

[12] S. S. Kumar, M. Vasanth, V. Singh, P. Ghosal, and T. Raghu, "An investigation of microstructural evolution in 304L austenitic stainless steel warm deformed by cyclic channel die compression," *Journal of Alloys and Compounds*, vol. 699, pp. 1036–1048, 2017.

[13] S. V. Muley, A. N. Vidvans, G. P. Chaudhari, and S. Udainiya, "An assessment of ultra fine grained 316L stainless steel for implant applications," *Acta Biomaterialia*, vol. 30, pp. 408–419, 2016.

[14] J. R. Weertman, "Hall-Petch strengthening in nanocrystalline metals," *Materials Science and Engineering: A Structural Materials: Properties, Microstructure and Processing*, vol. 166, no. 1-2, pp. 161–167, 1993.

[15] M. Rifai, H. Miyamoto, and H. Fujiwara, "Effect of ECAP deformation route on the degree of anisotropy of microstructure of extremely low CN Fe-20mass% Cr alloy," *Metals*, vol. 4, no. 1, pp. 55–63, 2014.

[16] A. Belyakov, T. Sakai, H. Miura, R. Kaibyshev, and K. Tsuzaki, "Continuous recrystallization in austenitic stainless steel after large strain deformation," *Acta Materialia*, vol. 50, no. 6, pp. 1547–1557, 2002.

[17] H. P. Leckie and H. H. Uhlig, "Environmental factors affecting the critical potential for pitting in 18–8 stainless steel," *Journal of The Electrochemical Society*, vol. 113, no. 12, pp. 1262–1267, 1966.

[18] J. Horvath and H. H. Uhlig, "Critical potentials for pitting corrosion of Ni, Cr-Ni, Cr-Fe, and related stainless steels," *Journal of the Electrochemical Society*, vol. 115, no. 8, pp. 791–795, 1968.

[19] G. S. Frankel, "Pitting corrosion of metals: a review of the critical factors," *Journal of the Electrochemical Society*, vol. 145, no. 6, pp. 2186–2198, 1998.

[20] R. K. Gupta and N. Birbilis, "The influence of nanocrystalline structure and processing route on corrosion of stainless steel: A review," *Corrosion Science*, vol. 92, pp. 1–15, 2015.

[21] M. Naka, K. Hashimoto, and T. Masumoto, "Effect of metalloidal elements on corrosion resistance of amorphous iron-chromium alloys," *Journal of Non-Crystalline Solids*, vol. 28, no. 3, pp. 403–413, 1978.

[22] R. K. Gupta, R. K. S. Raman, C. C. Koch, and B. S. Murty, "Effect of nanocrystalline structure on the corrosion of a Fe20Cr alloy," *International Journal of Electrochemical Science*, vol. 8, no. 5, pp. 6791–6806, 2013.

[23] R. S. Raman, R. K. Gupta, and C. C. Koch, "Resistance of nanocrystalline vis-à-vis microcrystalline Fe-Cr alloys to environmental degradation and challenges to their synthesis," *Philosophical Magazine*, vol. 90, no. 23, pp. 3233–3260, 2010.

[24] R. K. Gupta, R. K. Singh Raman, and C. C. Koch, "Electrochemical characteristics of nano and microcrystalline Fe–Cr alloys," *Journal of Materials Science*, vol. 47, no. 16, pp. 6118–6124, 2012.

[25] R. K. Gupta, R. K. Singh Raman, and C. C. Koch, "Fabrication and oxidation resistance of nanocrystalline Fe10Cr alloy," *Journal of Materials Science*, vol. 45, no. 17, pp. 4884–4888, 2010.

[26] K. Asami, K. Hashimoto, and S. Shimodaira, "An XPS study of the passivity of a series of iron-chromium alloys in sulphuric acid," *Corrosion Science*, vol. 18, no. 2, pp. 151–160, 1978.

[27] J. S. Noh, N. J. Laycock, W. Gao, and D. B. Wells, "Effects of nitric acid passivation on the pitting resistance of 316 stainless steel," *Corrosion Science*, vol. 42, no. 12, pp. 2069–2084, 2000.

[28] C. O. Olsson and D. Landolt, "Passive films on stainless steels—chemistry, structure and growth," *Electrochimica Acta*, vol. 48, no. 9, pp. 1093–1104, 2003.

[29] A. R. Brooks, C. R. Clayton, K. Doss, and Y. C. Lu, "On the role of Cr in the passivity of stainless steel," *Journal of The Electrochemical Society*, vol. 133, no. 12, pp. 2459–2464, 1986.

[30] I. O. Wallinder, J. Lu, S. Bertling, and C. Leygraf, "Release rates of chromium and nickel from 304 and 316 stainless steel during urban atmospheric exposure-A combined field and laboratory study," *Corrosion Science*, vol. 44, no. 10, pp. 2303–2319, 2002.

[31] T. P. Hoar, "The production and breakdown of the passivity of metals," *Corrosion Science*, vol. 7, no. 6, pp. 341–355, 1967.

[32] D. E. Williams, R. C. Newman, Q. Song, and R. G. Kelly, "Passivity breakdown and pitting corrosion of binary alloys," *Nature*, vol. 350, no. 6315, pp. 216–219, 1991.

[33] R. A. Andrievski, "Review stability of nanostructured materials," *Journal of Materials Science*, vol. 38, no. 7, pp. 1367–1375, 2003.

[34] C. C. Koch, "Structural nanocrystalline materials: an overview," *Journal of Materials Science*, vol. 42, no. 5, pp. 1403–1414, 2007.

[35] Y. R. Kolobov, G. P. Grabovetskaya, M. B. Ivanov, A. P. Zhilyaev, and R. Z. Valiev, "Grain boundary diffusion characteristics of nanostructured nickel," *Scripta Materialia*, vol. 44, no. 6, pp. 873–878, 2001.

[36] R. Würschum, S. Herth, and U. Brossmann, "Diffusion in nanocrystalline metals and alloys—a status report," *Advanced Engineering Materials*, vol. 5, no. 5, pp. 365–372, 2003.

6

The Compressive Strength and Resistivity toward Corrosion Attacks by Chloride Ion of Concrete Containing Type I Cement and Calcium Stearate

Agus Maryoto [ID],[1] **Buntara Sthenly Gan,**[2]
Nor Intang Setyo Hermanto,[1] **and Rachmad Setijadi**[3]

[1]*Department of Civil Engineering, Engineering Faculty, Jenderal Soedirman University, Jl. Mayjen Sungkono KM 5, Blater, Purbalingga, Jawa Tengah, Indonesia*
[2]*Department of Architecture, College of Engineering, Nihon University, 1 Nakagawara, Koriyama, Fukushima, Japan*
[3]*Department of Geology Engineering, Engineering Faculty, Jenderal Soedirman University, Jl. Mayjen Sungkono KM 5, Blater, Purbalingga, Jawa Tengah, Indonesia*

Correspondence should be addressed to Agus Maryoto; agus_maryoto1971@yahoo.co.id

Academic Editor: Susai Rajendran

This study aims to determine the effect of calcium stearate on concrete. Three kinds of concrete quality are studied, namely, 20, 30, and 40 MPa. Tests performed in the laboratory comprise a compressive strength test and an infiltration test of chloride ion content. The specimens used were cylinders with a diameter of 150 mm and height of 300 mm. The chloride ion infiltration test was carried out on a cube with sides of 150 mm. The infiltration of ions into the concrete was examined at depths of 1, 2, 4, 6, and 8 cm. Four dosages of calcium stearate were added to the concrete, namely, 0, 0.25, 1.27, and 2.53% for 20 MPa concrete; 0, 0.21, 1.07, and 2.48% for 30 MPa concrete; and 0, 0.19, 0.90, and 1.87% for 40 MPa concrete. The results of compressive strength tests indicate that the amount of calcium stearate that could be safely applied to the concrete was 0.25% of the weight of cement. On the other hand, the infiltration of chloride ions at a depth of 6 cm from the unprotected concrete surface decreased by 87, 69, and 113% for the 20, 30, and 40 MPa concrete, respectively, compared to concrete without calcium stearate. The test shows that the use of calcium stearate in concrete significantly increases its resistivity against corrosion attacks because, in the absence of chloride ions, the process of corrosion does not take place in the concrete.

1. Introduction

Generally, steel reinforcement within concrete will not corrode, since concrete mostly has a high pH of around 12.5. A high pH or alkaline property occurs in concrete when cement is mixed with water. Because of this alkaline property, the surface of the steel in the concrete forms a passive layer that protects the steel from corrosion attacks. Steel begins to corrode when the passive layer is damaged. The concrete pH decreases when the chloride ion attacks. The decrease in pH in the concrete is caused by carbonation [1], infiltration of chloride [2, 3], magnesium salts and sulfate, and acid attack by bacteria. Finally, the infiltration of chloride ion in the concrete also decreases the compressive strength [4, 5].

The sulfate and chloride contained in seawater [6] are very corrosive substances. When they interact directly with iron, rust is formed. In the meantime, many infrastructures are built in corrosive environments. Girder bridges in river estuaries, reinforced concrete pier in ports, basement-retaining walls, and hazardous waste collection pools are constructed from reinforced concrete structures that are attacked by corrosion. Although the outside appearance of concrete is a solid material with no cavities, in fact, concrete is a material that has millions of capillaries. These capillaries are formed from the surface of the concrete to a certain depth. These capillaries connect with each other so that they reach the surface of the reinforcement in the concrete. The diameter of these concrete capillaries is of microsize. The capillary

TABLE 1: Mix design of concrete.

Material	Unit Weight (kg/m^3)		
	20 MPa	30 MPa	40 MPa
Slump (cm)	10 ± 2	10 ± 2	10 ± 2
Cement	395	468	535
Crushed stone	1222	1085	870
Sand	605	700	750
Free water	205	205	210
Water–cement ratio (%)	52	44	0.39
Gmax (mm)	20	20	20
s/a (%)	32	39	45
Calcium stearate	0, 1, 5, 10	0, 1, 5, 10	0, 1, 5, 10

diameters were divided into three groups by Thomas and Jennings 2014 [7]. The first group is capillary pores with a size of 10 nm to 10 μm, the second is a porous gel with sizes of 0.5 to 10 nm, and the last is interlayer spaces with capillary sizes smaller than 0.5 nm.

Capillary sizes are very small; however, corrosive compounds, that is, sulfate and chloride [8], can still seep into the concrete because the size of this corrosive compound is less than the capillary diameter. The process of entry of chloride ions into the concrete can occur due to the pressure difference. The liquid pressure outside the concrete is higher, causing the sulfate-containing fluid and chloride to seep into the concrete. In addition, the difference in viscosity of corrosive compounds outside and inside the concrete can also activate the infiltration of the compound into the concrete. Similarly, differences in humidity and temperature in the environment and within the concrete structure are also the cause of penetration of corrosive compounds into reinforced concrete structures. The transport of sea vapor to the land by wind, where it collides with the reinforced concrete structure, is the first stage of the process of chloride ion infiltration [9]. Environmental conditions such as differences in pressure, humidity, and temperature and the physical condition of reinforced concrete that contains many capillaries greatly affect the speed of corrosion of reinforced concrete [10].

The rapidity of the corrosion process in reinforced concrete can be activated by repeated loads, prestressing force [11, 12], shock loads, cyclic drying-wetting loads [13, 14], and shrinkage cracking. The speed of the process can be measured from the velocity with which chloride ions permeate the concrete. Several previous studies [15, 16] have researched and determined a constant to be used to predict the lifespan of concrete structures [17] when corroded. This research is very useful for calculating the capacity of the structures. They [15–17] can determine when the reinforced concrete structure can still be used without repair, when it must be repaired, and when it should be immediately demolished. Unfortunately, the resistance of concrete to corrosion has not been improved yet.

Increased corrosion resistance can be done by protecting the reinforcement with fine materials such as fly ash [18, 19] or an impermeable surface coating [20], by cathodic protection with an electric current on the reinforcement, or by green inhibitors [21]. It can also be done by coating the surface [22] of the concrete with a membrane or paint so that the concrete becomes impermeable. The disadvantages of this type of protection are that the life of the membrane or coating is only a few years. Another type of protection is achieved by increasing the contact angle between concrete and water by mixing butyl stearate into the concrete. The use of butyl stearate in concrete has been shown to significantly reduce the corrosion rate in concrete reinforcement by the accelerated corrosion method [23–27]. However, it has not been clearly proven whether the corrosion is caused by the infiltration of chloride ions into the concrete. This study aims to determine the effect of calcium stearate on the quality of 20, 30, and 40 MPa concrete with regard to the infiltration of chloride ions. Since concrete compressive strength is a very important mechanical property in the design of reinforced concrete structures, the effect of calcium stearate on concrete compressive strength is also observed and discussed in depth.

2. Experimental Program

2.1. Preliminary Test. The fine aggregate employed in the study is sand from Merapi mountain and the coarse aggregate used is a granite stone produced by a crushed stone company from Banyumas region, Indonesia. Prior to conducting the main test, the physical properties of crushed stone and sand were tested. The physical properties of crushed stone and sand that were analyzed included the specific gravity, volume weight, clay content, and fineness modulus. In addition, crushed stone was also tested for wear resistance using Los Angeles equipment. The sand properties were a specific gravity of 2.75, volume weight of 1.39 ton/m^3, clay content of 1.13%, and fineness modulus of 2.2. Furthermore, the specific gravity, volume weight, clay content, and fineness modulus of crushed stone were 2.61, 1.49 tonne/m^3, 0.75, and 5.69, consecutively. The wear resistance of the crushed stone was 19%. The wear resistance met the SNI requirement for crushed stone, which should be less than 30%. Based on these physical properties, a concrete mixture design with quality values of 20, 30, and 40 MPa was arranged. The concrete workability was determined to be 10 +/− 2 cm. The complete concrete mix design is shown in Table 1. In order to ensure that the design of the concrete mixture

TABLE 2: The quantity of specimens.

No	Code	Quality of concrete (MPa)	Quantity of specimens	
			Compressive strength	Infiltration of chloride ions
1	St-0.00-20	20	3	3
2	St-0.25-20	20	3	3
3	St-1.27-20	20	3	3
4	St-2.53-20	20	3	3
5	St-0.00-30	30	3	3
6	St-0.21-30	30	3	3
7	St-1.07-30	30	3	3
8	St-2.48-30	30	3	3
9	St-0.00-40	40	3	3
10	St-0.19-40	40	3	3
11	St-0.90-40	40	3	3
12	St-1.87-40	40	3	3

met the desired compressive strength of the concrete, two compressive strength test objects were taken and tested at 28 days for each concrete quality. The type of cement utilized was a Portland pozzolan cement. The water was from a deep well near to the Laboratory of Civil Engineering, Jenderal Soedirman University, Purbalingga, Indonesia.

In Table 1, Gmax is the maximum size of aggregate, w/c is the water to cement ratio by weight, and s/a is the sand to the total aggregate ratio by volume.

2.2. Specimens.

Two types of specimens were used in this study for the compressive strength test and chloride ion infiltration test. The compressive strength test specimens were cylindrical in shape with a diameter of 150 mm and height of 300 mm. The specimens for testing the infiltration of chloride ions were in the form of a cube with sides of 150 mm. Table 2 shows the total number and type of specimens used in the study. Three types of concrete mix designs were used: 20, 30, and 40 MPa. Each quality of concrete had four variations of mix design with different calcium stearate contents. The calcium stearate contents in concrete with a quality of 20 MPa were 0, 0.25, 1.27, and 2.53% of the weight of cement. The concrete with a quality of 30 MPa had calcium stearate contents of 0, 0.21, 1.07, and 2.48% of the weight of cement. For the concrete with a quality of 40 MPa, the calcium stearate contents were 0, 0.19, 0.9, and 1.87% of the weight of cement. In total, 36 test specimens were used for the compressive strength test and 36 were used for the chloride ion infiltration test. In Table 2, the first two letters, "St", indicate the calcium stearate content. The three numbers after them indicate the content of calcium stearate in the concrete, where 0.25 means that the calcium stearate content in each $1\,m^3$ of concrete was 0.25% by weight of cement. The last two numbers of the code represent the quality of concrete, namely, "20" for 20 MPa concrete, "30" for 30 MPa concrete, and "40" for 40 MPa concrete. Each code refers to three specimens for the compressive strength test and three for the chloride ion infiltration test.

2.3. Procedure for Making Specimens.

The equipment used for making the concrete specimens included cylinder molds with a diameter of 150 mm and height of 300 mm, cubic molds with sides of 150 mm, a tamping rod with a diameter of 16 mm and length of 60 mm, rubber hammers, slump gauges, and concrete mixers. Preparation and curing of specimens were carried out in accordance with SNI 2493: 2011 [28]. The coarse and fine aggregates were conditioned at room temperature between 20 and 30°C. Before agitation was initiated, the coarse aggregate and part of the free water were introduced into the concrete mixer. The next step was to turn on the concrete mixer. Fine aggregate, cement, calcium stearate, and water were put into a rotating concrete mixer. The concrete mixer was allowed to rotate safely for 3 minutes and was then paused for 3 minutes. The gate of the concrete mixer was covered to avoid evaporation during the break period. The concrete mixer was then operated again for 2 minutes to make the concrete mixture more homogeneous. Water was added so that the target fresh concrete slump was achieved.

The slump testing conducted after the concrete mixing process was based on SNI 03-1972-1990 [29]. The equipment used for slump testing was a conical cut mold with a bottom diameter of 203 mm, a top diameter of 102 mm, and a height of 305 mm, a tamping rod of steel with a diameter of 16 mm and a length of 600 mm, a steel plate with a flat and waterproof surface, a concave spoon, and a measuring gauge. The slump testing was done by filling the mold in three layers. Each layer contained fresh concrete filling as much as one-third of the volume of the mold and compacted with a tamping rod 25 times. After this had been completed, the surface of the test bar was flattened with a stick, and then the mold was lifted slowly and perpendicularly upward. The slump testing process from filling fresh concrete into the mold until the mold was removed should be finished within 2.5 minutes. The value of the slump can be obtained by measuring the difference between the average height of the fresh concrete and the height of the mold.

Concrete test specimens can be made immediately after the slump testing is complete. Fresh concrete is put into a

FIGURE 1: The process of compression testing.

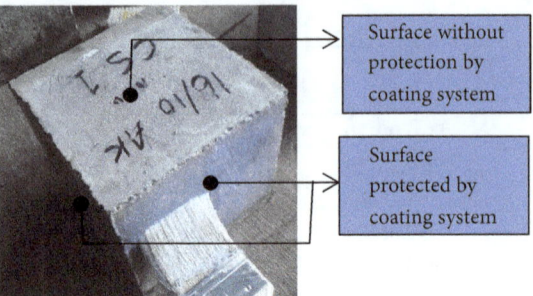

FIGURE 2: The process of surface treatment using coating system.

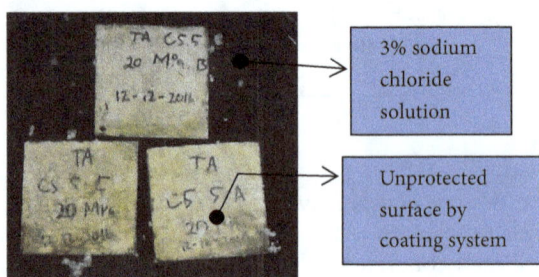

FIGURE 3: The process of soaking the specimen using 3% sodium chloride solution.

cylinder and cube mold in three layers. Each layer contains a volume of fresh concrete filling as much as one-third of the volume of the mold. Each layer is compacted with the tamping rod 25 times. The next part of the process is flattening of the surface of the specimen and closing it with plastic sheets to avoid rapid evaporation. The specimen mold was opened after the concrete was 24 ± 8 hours old. The hardened concrete specimen was then treated by soaking it in water at a temperature of 23 ± 1.7°C according to SNI 2493: 2011 [28]. The specimens were removed from the water pond after the concrete aged 28 days.

2.4. *Compressive Strength.* Dirt and water were cleaned from the specimens using a damp cloth. The weight and diameter of the cylindrical specimens were measured. A capping of sulfur mortar was installed with a melting pot on each end surface of the cylinder. The cylindrical specimen was put centrally on the compression machine. The compression machine was run with constant load additions ranging from 2 to 4 kg/cm² per second. Loading was continued until the cylindrical specimen collapsed. The maximum load at the collapse of the specimen was noted. The compressive strength of the concrete could be calculated by using (1). The process of the compression test can be observed in Figure 1.

$$\text{Compressive Strength} = \frac{P}{A} \text{ MPa} \tag{1}$$

In the equation, P is the maximum load in Newtons and A is the average area of the cylinder.

2.5. *Infiltration of Chloride Ions.* After 28 days of curing treatment in water with a temperature of around 23°C, the water was removed from all of the surfaces of the cubed specimen of 150 mm × 150 mm × 150 mm to dry them. A coating system was applied on five of the cube's surfaces

so that water could not infiltrate from the surface of the concrete. Only one surface of the cube was not protected by the coating system. The surface protection with the coating system can be seen in Figure 2. The coating system material was a synthetic polymer with a specific gravity of about 1.05. The coating system used in this study had a tensile strength of 1.2–1.5 N/mm².

In order to find out the infiltration of chloride ions in the concrete, the cubic specimens were immersed in water containing 3% sodium chloride. An aqueous solution of 3% NaCl was used to represent sea water. The immersion time was three months. Surfaces that were not protected by the coating system were placed in the upper position so that the water and chloride ions could effectively infiltrate into the concrete. The cubes which are soaking in 3% NaCl solution can be observed in Figure 3.

The next process is to drill a specimen of the concrete cube to the depths of 1, 2, 4, 6, and 8 cm from the surface of the concrete exposed to the solution of chloride. Figure 4 shows the drilling depth of concrete from the surface that was not protected by the coating system. The drilling results in the form of concrete powder can be seen in Figure 5. The amount of concrete powder taken from the drilling process was around 15 g. The concrete powder taken from each drilling depth was wrapped in plastic. Every cubic specimen was drilled in five depths: 1, 2, 4, 6, and 8 cm. Five concrete powders from each cubic specimen were analyzed.

The concrete powder was then analyzed to determine the amount of chloride that had seeped into the concrete. The wet analysis method was chosen and used in this study. The complete chloride content analysis step is as follows.

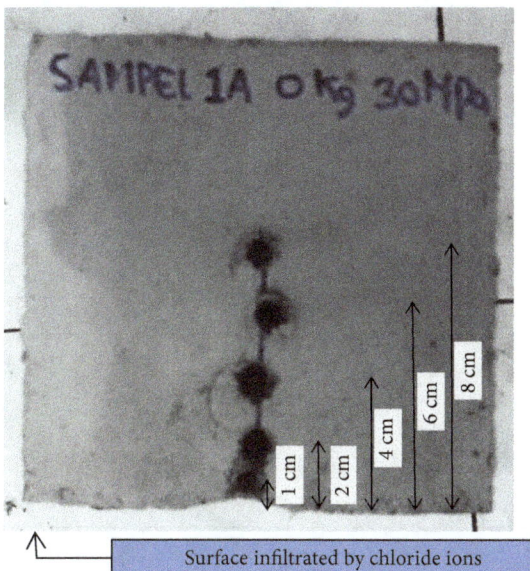

FIGURE 4: Drilling depth from the unprotected surface by the coating system.

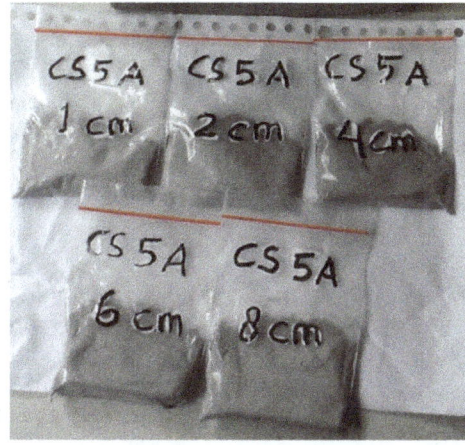

FIGURE 5: The concrete powder obtained by drilling the concrete specimen with a quality of 30 Mpa.

(1) Weigh the 5 g samples and put them into a 300 cc measuring cup.

(2) Add 50 cc of water and heat it in a water bath for 1 hour while stirring occasionally.

(3) Filter with filter paper number 40 and insert the filtrate into a 300 cc measuring cylinder.

(4) Wash the sediment with hot water as many as eight to ten times until the filtrate volume is 200 cc.

(5) Let the solution cool down and then add five drops of brome phenol blue indicator, so the color becomes blue.

(6) Add HNO3 (1 + 65) until the blue color turns yellowish green or the pH becomes 3.2.

(7) Titrate with Hg (NO3)2 0.014 N until the color becomes purple.

(8) Record the required volume of the pen.

(9) Conduct the blank test as follows.

(10) Prepare 200 cc of distilled water in a 300 cc measuring cup.

(11) Repeat steps (5), (6), and (7).

(12) Calculate the chloride content in the concrete powder using the formula below.

$$Cl^- = \frac{f \times V}{5000} \times 100\%, \qquad (2)$$

where f is the equivalence of Hg(NO3)2 to Cl^- (weight in milligrams of Hg(NO3)2 divided by volume in cubic centimeters of Cl^-) and V is the volume of the penetrant.

2.6. Manufacture of Reagents

(1) Take 4.84 g of $Hg(NO_3)_2$ and 0.5 cc of concentrated HNO3 dissolved in 1 L of distilled water.

(2) Heat NaCl p.a. to a temperature of 500–600°C.

(3) Cool it down in the H_2SO_4 desiccator.

(4) Take as much as 1,648 g and dissolve it in distilled water.

(5) Put it in a 1 L flask to an official seal mark precisely.

(6) This solution contains 1 mg of Cl^- per cubic centimeter of the solution.

(7) For standardization, put the 20 cc Hg $(NO_3)_2$ standard into a 300 cc measuring cylinder and then dilute it to 100 cc with distilled water. Add five drops of brome phenol blue indicator and then HNO3 (1 + 65) until the color exactly turns yellow, with a pH equal to 3.1. Titrate with Hg (NO3) 2 until the color becomes purple. Record the volume (V) of neutralizer and determine its equivalence factor (f). $f = V/20$ cc HgNO3/mg NaCl.

(8) Create a brome phenol blue indicator. Weigh 0.5 g of diphenyl carbazole, 0.05 g of brome phenol blue, and 0.12 g of xylene cyanol and then dissolve it into 100 cc of 99.5% alcohol.

(9) Make HNO3 (1 + 65) by mixing 1 ml of HNO3 (p) in a glass beaker containing 65 mL of water and then homogenize it.

3. Results and Discussion

3.1. Preliminary Test of Compressive Strength. A preliminary test of compressive strength was conducted in order to ensure that the proposed mix design met the expected concrete quality. The results can be observed in Table 3. It shows that the compressive strength results reached the expected concrete quality. The average compressive strengths of concrete were 23.6, 31.6, and 49.0 MPa.

TABLE 3: Average compressive strength during preliminary test.

No	Quality of Concrete	Compressive Strength		
		Specimen 1 (MPa)	Specimen 2 (MPa)	Average (MPa)
1	20 MPa	26.3	21.00	23.6
2	30 MPa	30.7	32.4	31.6
3	40 MPa	50.6	47.4	49.0

TABLE 4: Percentage reduction of compressive strength.

	20 MPa concrete			30 MPa concrete			40 MPa concrete		
Calcium stearate content (% by cement weight)	0.25	1.27	2.53	0.21	1.07	2.48	0.19	0.90	1.87
Reduction of compressive strength (%)	7.0	17.2	36.4	-10.1	26.1	15.7	3.7	2.6	11.1

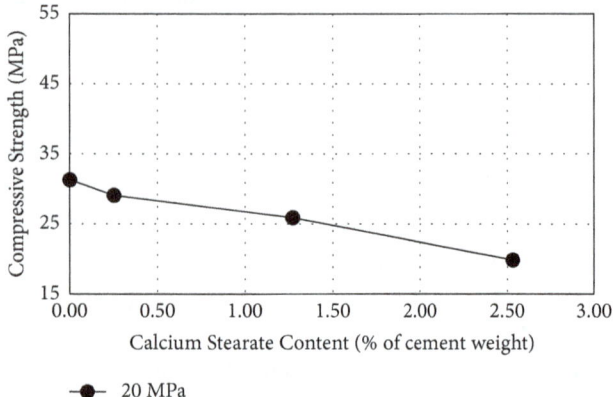

FIGURE 6: The compressive strengths of concrete grade 20 MPa.

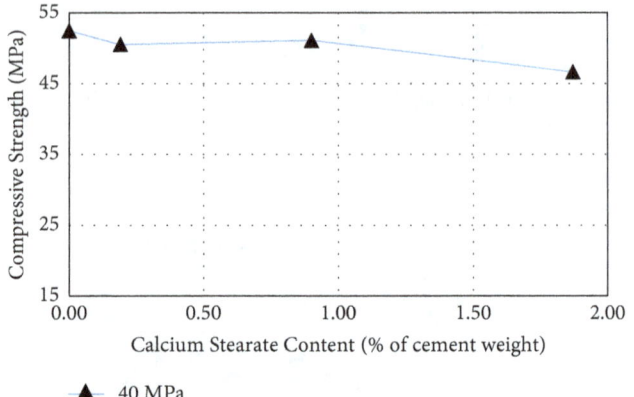

FIGURE 8: Compressive strength of concrete grade 40 MPa.

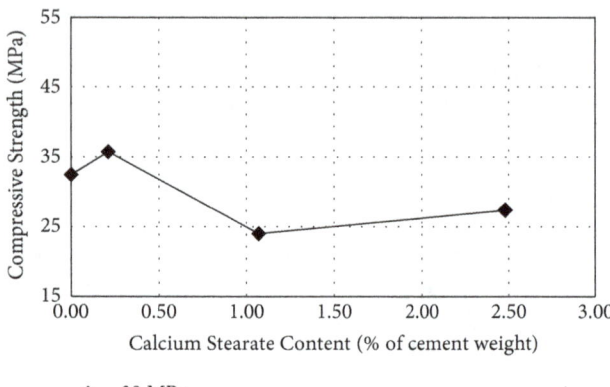

FIGURE 7: The compressive strengths of concrete grade 30 MPa.

3.2. Compressive Strength.

The results of the compressive strength tests are shown in Figures 6, 7, and 8. Figure 6 presents the average compressive strength of three specimens for the 20 MPa concrete quality. Based on Figure 6, it can be confirmed that the addition of calcium stearate to the 20 MPa concrete quality decreased the compressive strength. The smallest decrease occurred with the addition of 0.25% calcium stearate by weight of cement and was equal to 7.0%. Furthermore, the influence on the compressive strength decreased by 17.2 and 35.4% when 1.27 and 2.53% calcium stearate by weight of cement were added to the concrete. The percentage reduction in compressive strength of the concrete increased when the calcium stearate content added to the concrete increased.

Slightly different from the concrete with a quality of 20 MPa, the compressive strength of the concrete with a quality of 30 MPa increased slightly when 0.21% calcium stearate by weight of cement was added. However, when the calcium stearate content added to the concrete increased, the compressive strength dropped dramatically. For the concrete with a quality of 40 MPa, the trend of compressive strength decreased due to the effect of adding calcium stearate, as the concrete quality of 20 MPa. The percentage reduction in compressive strength of concrete due to the addition of calcium stearate can be seen in Table 4.

Figure 9 shows the sharpness of the reduction of compressive strength of concrete due to the addition of calcium stearate. After the trend line of the data has been solved, equations for the trend line are then obtained for the 20, 30, and 40 MPa concrete. The equation for the 20 MPa concrete quality is $y = -4.2701x + 30.86$, and those for the 30 and 40 MPa concrete are $y = -2.9801x + 32.66$ and $y = -2.6565x + 52.127$, respectively. Based on Figure 9, it can be

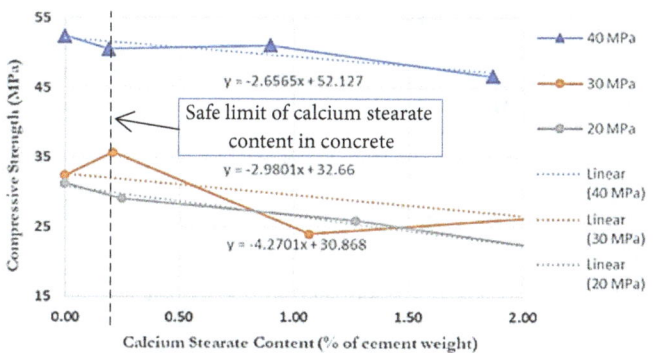

FIGURE 9: The tendency of compressive strength reduction in the concrete.

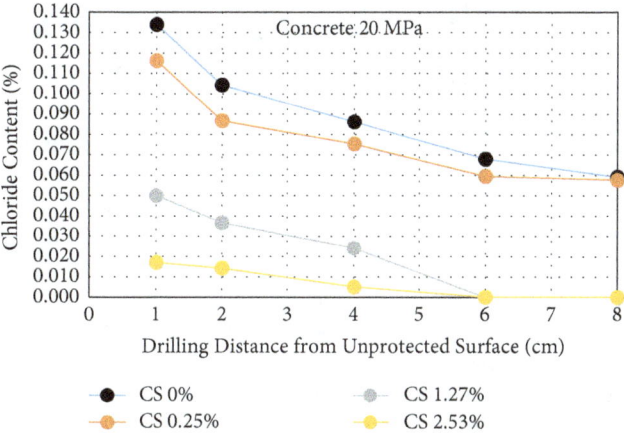

FIGURE 10: Chloride content in 20 MPa concrete.

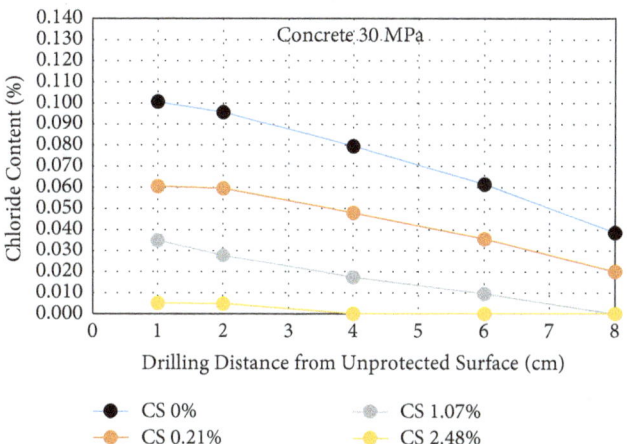

FIGURE 11: Chloride content in 30 MPa concrete.

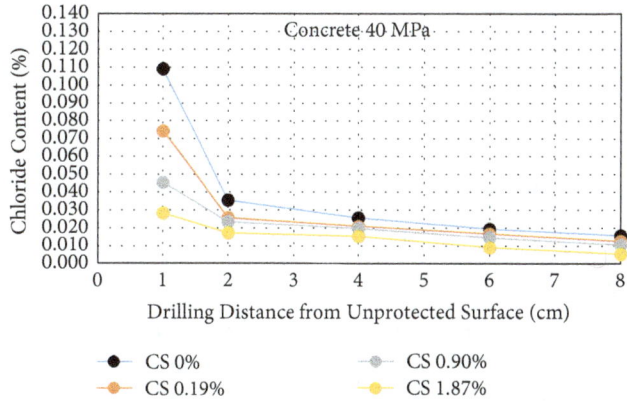

FIGURE 12: Chloride content in 40 MPa concrete.

seen that the use of calcium stearate in the higher quality concrete did not decrease the compressive strength as sharply as it did in the low-quality concrete. The lowest slope among the lines is for the 40 MPa concrete, followed by the 30 MPa concrete and then the 20 MPa concrete.

The decrease in the compressive strength of concrete is most likely caused by the reaction of calcium stearate, water, and cement during the reaction of cement hydration. These three materials in the concrete form a wax-like compound. The wax-like compound is not as strong as the calcium silicate hydrate (CSH) bond. CSH is a compound that is formed when cement and water react. It is understood that when the concrete contains a small amount of cement, the effect of compressive strength on the concrete declines very sharply due to the use of calcium stearate. However, this is not the case when the amount of cement in the concrete is very high, and the use of calcium stearate in concrete does not have a significant influence on the compressive strength.

Another interesting point from Table 4 is that the use of less than 0.25% calcium by weight of cement leads to a decrease of compressive strength of concrete that is still below 10% if compared to concrete without calcium stearate. On page 41 in Section 5.6.3.3.b of SNI 2847: 2013 [30], it is mentioned that the compressive strength of concrete can still be considered satisfactory if the reduction of the compressive strength of concrete does not exceed 10% of the planned concrete compressive strength. Hence, it can be concluded that calcium stearate can be safely utilized in concrete up to a maximum calcium stearate content of 0.25% of the weight of cement.

3.3. Infiltration of Chloride Ions.

The amount of infiltration of chloride ions into the 20, 30, and 40 MPa concrete strengths can be seen in Figures 10, 11, and 12. These figures show the relationship of the drilling depth from the unprotected concrete surface with the chloride ion infiltration on the horizontal axis and the infiltration of chloride ion content on the vertical axis. The horizontal axis shows the depth of chloride ion infiltration entering the concrete. In Figure 10, it can be seen that the deeper the drilling from the surface of the concrete exposed to chloride attack, the greater the decrease in the number of chloride ions. The decrease occurs

very drastically when the drilling depth reaches 8 cm. The difference in the content of chloride ions that seeped to depths of 1 and 8 cm from the surface became smaller than 56%. This behavior occurs not only in concrete with 0% calcium stearate content but also in concrete with calcium stearate contents of 0.25, 1.27, and 2.53%.

It can also be seen that, in the concrete with calcium stearate contents (the line marked CS) of 0, 0.25, 1.27, and 2.53, increasing the addition of calcium stearate in the concrete significantly reduces the ingress of chloride ions into the concrete. The same behavior of a decrease in chloride ion infiltration in 20 MPa concrete quality was also observed in the 30 and 40 MPa concrete samples. These behaviors can be observed in Figures 11 and 12. Generally, this indicates that the use of calcium stearate in all concrete grades decreases the infiltration of chloride ions. This is because the cement, calcium stearate, and water react to form a wax-like material. This wax-like material has a hydrophobic property. Because of this property, there is an increase of the contact angle between the concrete surface and water. The increase of the contact angle makes it more difficult for the water to ingress into the concrete. A further effect is that the infiltration of chloride ions into the concrete also decreases notably. Another condition that causes a decrease in the infiltration of chloride ions is the very fine physical properties of calcium stearate. The calcium stearate granules are finer than the cement grains. Hence, calcium stearate granules can fill the smaller-diameter pores and capillaries in concrete. As a consequence, the concrete becomes more solid and more difficult to penetrate by water.

On comparing Figures 10, 11, and 12, it can be seen that the higher the concrete quality is, the more the rate of chloride ion infiltration decreases. This is very easy to understand because concrete with a higher quality usually has a low water-cement ratio. With a low–water-cement ratio, the amount of water that evaporates during the hydration process is very small. The effect of the evaporation of this very small amount of water is that the capillaries formed in the concrete become small. This means that the concrete becomes more solid.

4. Conclusions

Based on the results discussed above, some conclusions are drawn.

(1) The use of calcium stearate in concrete can decrease the compressive strength of concrete. The amount that can be added safely is no more than 0.25% of the weight of the cement.

(2) The higher the amount of calcium stearate added to the concrete, the lower the chloride ion infiltration. This is due to the hydrophobic nature of the material formed from the reaction between cement, calcium stearate, and water.

(3) When the concrete powder is taken from depths further from the surface of the concrete exposed to NaCl solution, the chloride ion content decreases.

Conflicts of Interest

The authors declare that they have no conflicts of interest. Funding of this manuscript did not lead to any conflicts of interest with the funder.

Acknowledgments

This work was supported by LPPM, Jenderal Soedirman University, Purwokerto, Central Java, Indonesia, and funding through the International Research Collaboration Scheme, BLU Unsoed 2017. The authors would also like to thank Yusak, Anugerah, and Azhar for their assistance in conducting this study.

References

[1] T. Saeki, "Effect of carbonation on chloride penetration in concrete," in *Proceedings of the Third International RILEM Workshop on Testing and Modelling Chloride Ingress into Concrete*, pp. 1–14, Madrid, Spain, September 2002.

[2] C. Sosdean, L. Marsavina, and G. De Schutter, "Experimental and numerical investigations of the influence of real cracks on chloride ingress in concrete," *De Gruyter*, vol. 16, no. 2, pp. 151–161, 2016.

[3] W. L. Jin, Y. D. Yan, and H. L. Wang, "Chloride diffusion in the cracked concrete," *Fracture Mechanics of Concrete and Concrete Structures Assessment, Durability, Monitoring and Retrofitting of Concrete Structures*, pp. 880–887, 2010.

[4] J. Bai, S. Wild, and B. B. Sabir, "Chloride ingress and strength loss in concrete with different PC-PFA-MK binder compositions exposed to synthetic seawater," *Cement and Concrete Research*, vol. 33, no. 3, pp. 353–362, 2003.

[5] W. Sutrisno, I. K. Hartana, P. Suprobo, E. Wahyuni, and D. Iranata, "Cracking process of reinforced concrete induced by non-uniform reinforcement corrosion," *Jurnal Teknologi*, vol. 79, no. 3, pp. 1–6, 2017.

[6] J. Kim, W. J. McCarter, B. Suryanto, S. Nanukuttan, P. A. M. Basheer, and T. M. Chrisp, "Chloride ingress into marine exposed concrete: A comparison of empirical- and physically-based models," *Cement and Concrete Composites*, vol. 72, pp. 133–145, 2016.

[7] J. Thomas and H. Jennings, *The Science of Concrete*, Northwestern University, Evanston, Ill, USA, 2006.

[8] N. M. Noor, H. Hamada, Y. Sagawa, and D. Yamamoto, "Effect of crumb rubber on concrete strength and chloride ion penetration resistance," *Jurnal Teknologi*, vol. 77, no. 32, pp. 171–178, 2015.

[9] A. A. Shakir and A. A. Mohammed, "Durability property of clay ash, quarry dust and billet scale bricks," *Journal of Engineering Science and Technology*, vol. 10, no. 5, pp. 591–605, 2015.

[10] A. A. Castañeda, F. Corvo, J. J. Howland, and T. Pérez, "Atmospheric corrosion of reinforced concrete steel in tropical coastal regions," *Engineering Journal*, vol. 17, no. 2, 2013.

[11] A. Maryoto and T. Shimomura, "Effect of prestressed force and size of reinforcement on corrosion crack width in concrete member," *Journal of Engineering Science and Technology*, vol. 12, no. 10, pp. 2664–2676, 2017.

[12] A. Maryoto, N. I. S. Hermanto, Y. Haryanto, S. Waluyo, and N. A. Anisa, "Influence of prestressed force in the waste tire

reinforced concrete," *Procedia Engineering*, vol. 125, pp. 638–643, 2015.

[13] H. Ye, C. Fu, N. Jin, and X. Jin, "Influence of flexural loading on chloride ingress in concrete subjected to cyclic drying-wetting condition," *Computers and Concrete*, vol. 15, no. 2, pp. 183–198, 2015.

[14] S. B. Allampallewar and A. Srividya, "Corrosion performance of reinforced concrete member along Indian coasts: Effect of temperature & relative humidity," *International Journal of Performability Engineering*, vol. 4, no. 3, pp. 285–292, 2008.

[15] N. Bester, "Mechanisms and modelling of chloride ingress in concrete," *Durability and Condition Assessment of Concrete Structures*, pp. 1–6, 2014.

[16] C. Andrade, M. Castellote, and R. D'andrea, "Chloride aging factor of concrete measured by means of resistivity," in *Proceedings of the International Conference Durability of Building Materials and Components*, pp. 1–8, Porto, Portugal, 2011.

[17] C. Evans and M. G. Richardson, "Service life of chloride-contaminated concrete structures," in *Technical Session on Material, Concrete Research in Ireland Colloquium*, pp. 131–138, 2005.

[18] A. Sofi and B. R. Phanikumar, "Durability properties of fibre-reinforced pond ash-modified concrete," *Journal of Engineering Science and Technology*, vol. 11, no. 10, pp. 1385–1402, 2016.

[19] P. M. Mrudul, T. Upender, M. Balachandran, and K. M. Mini, "Study on silica infused recycle aggregate concrete using design of experiments," *Journal of Engineering Science and Technology*, vol. 12, no. 4, pp. 958–971, 2017.

[20] T. Bibi, J. Mirza, S. Khan, H. Hamid, Z. Fida, and M. M. Tahir, "Resistance of concrete protective coatings in different chemical environments," *Jurnal Teknologi*, vol. 74, no. 4, pp. 183–189, 2015.

[21] A. S. Abdulrahman and M. Ismail, "Assessment of green inhibitor on the crystal structures of carbonated concrete," *Jurnal Teknologi*, vol. 69, no. 3, pp. 1–9, 2014.

[22] T. S. W. Plesser, "Effect of surface treatment on chloride ingress and carbonation in concrete structure," COIN Project 3, 2008.

[23] A. Maryoto, "Improving Microstructures of Concrete Using Ca(C18H35O2)2," *Procedia Engineering*, vol. 125, pp. 631–637, 2015.

[24] A. Maryoto, B. S. Gan, and H. Aylie, "Reduction of chloride ion ingress into reinforced concrete using a hydrophobic additive material," *Jurnal Teknologi*, vol. 79, no. 2, pp. 65–72, 2017.

[25] A. Maryoto, "Resistance of Concrete with Calcium Stearate due to Chloride Attack Tested by Accelerated Corrosion," *Procedia Engineering*, vol. 171, pp. 511–516, 2017.

[26] A. Maryoto, B. S. Gan, N. I. S. Hermanto, and R. Setijadi, "On the water absorption and corrosion rate of concrete using calcium stearate," *Journal of Engineering and Applied Sciences*, vol. 12, no. 20, pp. 5233–5238, 2017.

[27] A. Maryoto, B. S. Gan, N. I. S. Hermanto, and R. Setijadi, "Protection of Corrosion Attack in Reinforced Concrete Due to ChlorideIon using Calcium Stearate," *Journal of Engineering and Applied Sciences*, vol. 12, no. 6, pp. 7965–7970, 2017.

[28] Standar Nasional Indonesia, *Tata Cara Pembuatan Dan Perawatan Benda Uji Beton Di Laboratorium*, Badan Standar Nasional, Jakarta, Indonesia, 2011.

[29] Standar Nasional Indonesia, *Metode Pengujian Slump Beton*, Badan Standar Nasional, Jakarta, Indonesia, 1990.

[30] Standar Nasional Indonesia, *Persyaratan Beton Struktural Untuk Bangunan Gedung*, Badan Standar Nasiona, Jakarta, Indonesia, 2013.

7

Numerical Simulation of Galvanic Corrosion between Carbon Steel and Low Alloy Steel in a Bolted Joint

Rachid Radouani, Younes Echcharqy, and Mohamed Essahli

Laboratory of Applied Chemistry and Environment, Faculty of Science and Technology, University of Hassan 1, Settat, Morocco

Correspondence should be addressed to Rachid Radouani; r.radouani@gmail.com

Academic Editor: Jerzy A. Szpunar

The galvanic corrosion of a bolt joint combining carbon steel end plate and low alloy steel bolt was investigated electrochemically in a 1 M HCl solution. The corrosion parameters of the joint components were used for numerical simulation using Comsol Multiphysics software to analyze the galvanic corrosion behavior at the contact zone between the head bolt and the end plate. In this research work we evaluate the variation of the corrosion rate in the steel end plate considered as the anode, in order to determine the lifetime of the bolted assembly used in steel structures. Three materials (20MnCr5, 42CrMo4, and 32CrMoV13) and three bolts (M12, M16, and M20) were tested in two thicknesses of electrolyte 1 M HCl ($e = 1$ mm, $e = 20$ mm). It is found that the corrosion rate of the anode part (end plate) is higher for 32CrMoV13 materials and it increases if both diameter of the bolt and thickness of the electrolyte increase (Cr(M20) > Cr(M16) > Cr(M12) and Cr($e = 20$ mm) > Cr($e = 1$ mm)). This corrosion rate is higher in the contact area between the bolt head and the end plate, and it decreases if we move away from this contact area.

1. Introduction

Galvanic corrosion can simply be defined as the corrosion that occurs as a result of one metal being in contact with another in a conducting, corrosive environment. The corrosion is stimulated by the potential difference that exists between the two metals: the more noble material acting as a cathode where some oxidizing species is reduced and the more active metal, which corrodes, acting as the anode. The anodic reaction is, by definition, some form of metal dissolution; the cathodic reaction is, in the majority of practical situations, either oxygen reduction or hydrogen evolution, or a combination of both. Many factors affecting galvanic corrosion are already discussed to determining whether or not galvanic corrosion will occur in a particular instance and if so at what rate; it is important when considering the theory of galvanic corrosion to be aware of these factors including electrode potential, reaction kinetics, alloy composition, protection film characteristics, bulk solution environment, total geometry, and type of joint [1].

There is a high incidence of past scientists taking an interest in corrosion to understand what causes it and what limits or accelerates the process. Numerous studies have been conducted; some take a more global outlook [2], whereas some take a more focused approach [3]. The study conducted in [2] looked at many different galvanic couples commonly used in seawater applications. The study focused on developing reasonable models for systems experiencing varying periods of exposure to the corrosive environment.

Simulation of galvanic corrosion between magnesium and aluminum has been performed by Lacroix et al. [4], Deshpande [5–7], Jia et al. [8], and Trinh et al. [9] who have studied the corrosion of magnesium alloys in contact to mild steel under static conditions. The publications of Murer et al. [10–12] and Shi and Kelly [13] in this context also gave an extended insight into the topic, especially in the very important choice of boundary conditions. New studies of Sun et al. [14], who applied the mathematical approach of Yan et al. [15] to the modeling of deposit formation under seawater conditions, clearly introduce a possible way of a useful model built up for the mentioned purpose. The following studies and results are based on the progress achieved by them. Basic galvanic current density computations were modified by layer growth aspects leading to time dependent variations in the electrochemical response of the electrodes.

In the work of Höchez [18], simulation of galvanic corrosion between Mg and Al on a geometry like a self-pierce punch rivet joint has been performed. Starting from the initial model setup and related to the scientific challenge, the mathematical model requires fundamental assumptions. As always in modeling action they define the quality and accuracy of simulation results. The analysis via computer simulations showed the ability to study parameters influencing corrosion performance like the affected surface fraction by a reaction product.

In the works of Johnson and Abbott [19] and Xu et al. [20], the impact of the nature of the materials on the corrosion rate of mild steel in a galvanic coupling was studied. These articles show that the corrosion rate of mild steel changes if the material is changed in the galvanic corrosion. The addition of Cr in steel alloys improves the corrosion resistance of the steel. It is found that as the Cr content increases, the corrosion rate of the mild steel decreases.

Finite element analysis (FEA) of potential and current distributions in galvanic systems has long been studied in the literature [21–27]. Such studies are often carried out to investigate fundamental effects of electrolyte geometry [26, 28, 29], electrode kinetics [25], and unique part geometries [21]. Early modeling attempts have provided an analytical solution [26, 30] for the galvanic current and potential distribution around a galvanic junction involving two dissimilar metals kept side by side and covered by a corroding solution. These theoretical predictions were also later validated with experimental measurement of the potential and current density across the junction. Recently, semianalytical and numerical methods [29, 31] were developed to solve for current density distribution on anode and cathode in a galvanic couple and similar electrochemical phenomena such as electrodeposition. Galvanic corrosion is found to be better modeled with a continuum approach than a lattice model since the curvature of the interface is an important parameter in determining the electrochemical dissolution rate.

In our work the finite element method was used to perform a parametric study of the galvanic corrosion of a column-beam bolted assembly used in steel structures. This study takes into consideration the material of the cathode "bolt," the thickness of the electrolyte "e," the dimensions of the bolt head, and the distance from the contact area between the head bolt and the end plate in order to determine the parameters which can cause the maximum corrosion rate of the anode "end plate."

2. Materials and Methods

2.1. Bolted Joint and Materials.
Three types of materials used for the bolt were tested to predict the corrosion rate of the end plate. Thus three types of galvanic connections were studied in a bolted joint. The material composition of the elements of the assembly is stated in Table 1. The bolt was made by three types of low alloy steel which have the higher corrosion potential, and the metal of the end plate is the carbon steel (S235JR), Table 2. Galvanic corrosion therefore took place in the contact zone between end plate and head bolt in each bolted joint. In other words, the end plate acted as anode, and the bolt becomes cathode.

TABLE 1: Bolt and end plate materials.

Bolted joint	Bolt	End plate
1	20MnCr5	S235JR
2	42CrMo4	S235JR
3	32CrMoV13	S235JR

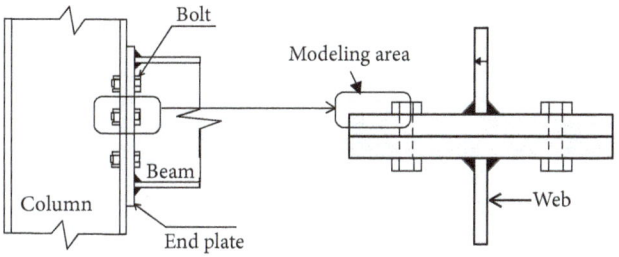

FIGURE 1: Column-beam bolted joint with end plate.

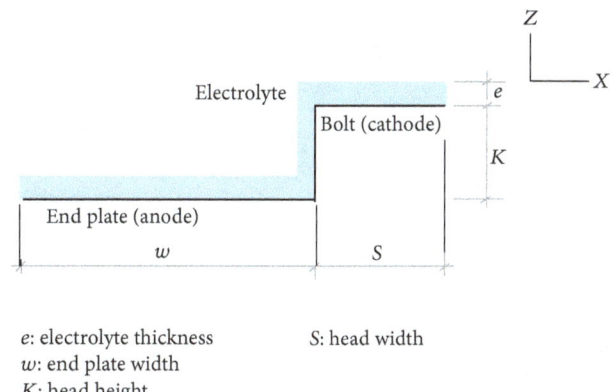

e: electrolyte thickness S: head width
w: end plate width
K: head height

FIGURE 2: Geometric characteristics of the modeling area.

2.2. Geometry and Modeling.
The bolted assembly is column-beam type and it is used in steel structures, Figure 1. The developed model is Cartesian 2D type and it is used for a representation of the plane modeling of the elements entering into the galvanic corrosion, Figure 2. The modeled part is the area of contact between the bolt head, the end plate, and the electrolyte. The width of the anode (end plate) is $w = 26.5$ mm and the dimensions of the bolt head depend on the size of the bolt (Table 3).

2.3. Study Parameters.
The analysis of galvanic corrosion is parametric, taking into account the effect of the bolt material, the bolt size, the electrolyte thickness, and the immersion time on the corrosion rate of the end plate, Table 4 and Figure 3.

Three types of low alloys were used for the bolt (20MnCr5, 42CrMo4, and 32CrMoV13) and will represent the cathode. For the bolt size three diameters were taken in this study (M12, M16, and M20); the dimensions that will be taken into consideration during the galvanic analysis are the dimensions of the bolt head as it is the part that will be in contact

TABLE 2: Chemical composition of materials.

	C	Cr	Mo	V	Mn	Si	P	S	Cu
S235JR	0.17	-	-	-	1.4	-	0.035	0.035	0.4
42CrMo4	0.38–0.45	0.9–1.2	0.15–0.3	-	0.6–0.9	0.4	0.025	0.035	-
32CrMoV13	0.32	3	1	0.2	0.35	0.25	<0.05	<0.05	-
20MnCr5	0.17–0.22	1–1.3	-	-	1.1–1.4	0.4	0.025	0.035	-

TABLE 3: Bolt size.

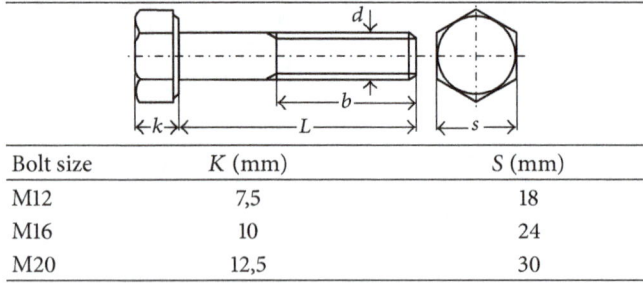

Bolt size	K (mm)	S (mm)
M12	7,5	18
M16	10	24
M20	12,5	30

TABLE 4: Study parameters for the galvanic connection.

Bolt material	Bolt size	Electrolyte thickness	Immersion time
20MnCr5	M12		1 month
42CrMo4	M16	1 mm	6 months
32CrMoV13	M20	20 mm	12 months

with the end plate. Galvanic corrosion will be studied in two thicknesses of the 1 M HCl electrolyte ($e = 1$ mm, $e = 20$ mm) and for the following period range: 1 month, 6 months, and 12 months.

3. Method and Boundaries Conditions

3.1. Analysis of the Electrode Potential Distribution. The present analysis is based on a model galvanic corrosion couple consisting of two elements shown in Figure 2. the couple consists of a cathodic element (bolt) of a width "S" in the $+X$ direction and a height "K" in $+Z$ direction; the width of the anode (plate) is "w" in the $-X$ direction. The couple is covered by an electrolyte of depth "e" in the $+Z$ direction. The electrolyte is bounded by perfect insulators at $z = e$, $x = -w$, and $x = +S$

The various current versus electrode potential relationships for the anodic and cathodic reactions are assumed to be subject to activation control with logarithmic (Tafel) polarization behavior. Thus the net cathodic current density per unit length of corrosion couple $i_c(x)$ on the bolt at an electrode potential E_x is

$$i_c(x)$$
$$= i_{0(\text{bolt})} \left\{ \exp\left(\frac{E_{\text{bolt}} - E_x}{\beta_{\text{bolt}}} \right) - \exp\left(\frac{E_{\text{bolt}} - E_x}{-\alpha_{\text{bolt}}} \right) \right\}, \quad (1)$$

where E_{bolt} is the free corrosion potential of element A, $i_{0(\text{bolt})}$ is the free corrosion current density, and α_{bolt} and β_{bolt} are the Tafel parameters for the anodic and cathodic reactions,

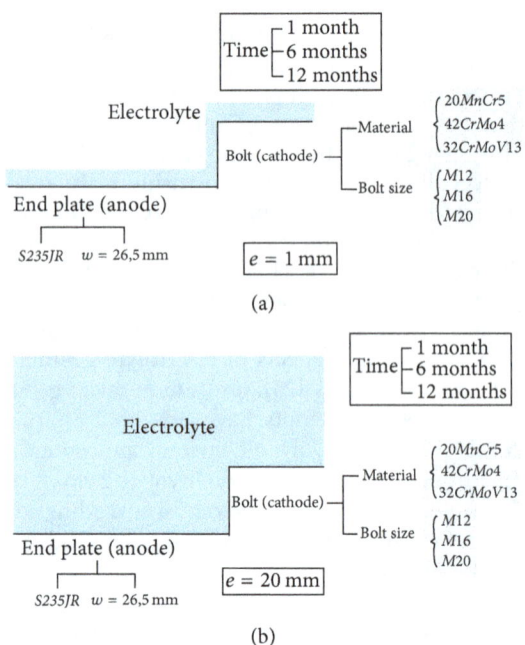

FIGURE 3: Study parameters for the galvanic connection. (a) $e = 1$ mm, (b) $e = 20$ mm.

respectively. Similarly for the end plate, the net anodic current density per unit length of corrosion couple $i_a(x)$ on end plate at an electrode potential E_x is

$$i_a(x) = i_{0(\text{end plate})} \left\{ \exp\left(\frac{E_{\text{end plate}} - E_x}{-\alpha_{\text{end plate}}} \right) \right.$$
$$\left. - \exp\left(\frac{E_{\text{end plate}} - E_x}{\beta_{\text{end plate}}} \right) \right\}, \quad (2)$$

where $E_{\text{end plate}}$ is the free corrosion potential of end plate, $i_{0(\text{end plate})}$ is the free corrosion current density, and $\alpha_{\text{end plate}}$ and $\beta_{\text{end plate}}$ are the Tafel parameters for the anodic and cathodic reactions, respectively.

The equation governing the potential distribution in the electrolyte becomes Laplace's equation (for constant conductivity):

$$\nabla^2 \emptyset = 0. \quad (3)$$

The governing equation for the potential must be solved using software based on finite element method (FEM); this equation must be subjected to appropriate boundary conditions. At any electrode-electrolyte interface, both anodic and cathodic processes take place simultaneously and the current density through the interface is the result of the

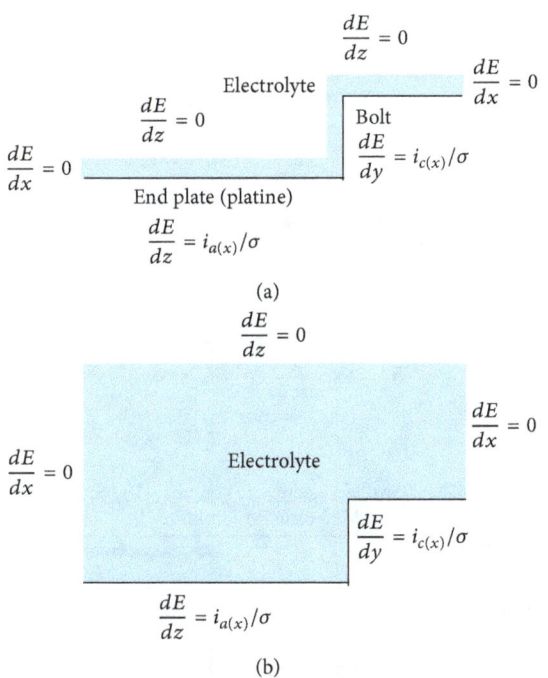

FIGURE 4: Boundary conditions. (a) $e = 1$ mm, (b) $e = 20$ mm.

electronic exchanges in both processes. In a galvanic couple, the cathodic process dominates in the more noble metal (the bolt) while the anodic process is dominant in the less noble metal (the end plate).

3.2. Boundary Conditions. The solution of (3) is based on the boundary conditions shown in Figure 4.

For $z = 0$ the normal derivative of electrode potential E is given by

$$\frac{dE}{dz} = \frac{i_{c(x)}}{\sigma} \quad \text{For} \; -w < x < 0,$$

$$\frac{dE}{dz} = \frac{i_{a(x)}}{\sigma} \quad \text{For} \; 0 < x < S, \tag{4}$$

where σ is the conductivity of the corrosion electrolyte.

Since there is no current flow normal to the insulating boundaries,

$$\frac{dE}{dx} = 0 \quad \text{at} \; x = -w, \; x = +S,$$

$$\frac{dE}{dz} = 0 \quad \text{at} \; z = e. \tag{5}$$

The numerical solution of the Laplace equation requires the definition of the electrochemical parameters of the anode, the cathode, and the electrolyte in a 1 M HCl electrolyte whose characteristics are cited in the Table 5 [16]. These parameters are determined from experimental tests already carried out in the literature [17] and they are cited in Table 6.

4. Results and Discussion

4.1. Electrolyte Potential. Once the electrochemical parameters were defined, the geometry was applied, the boundary

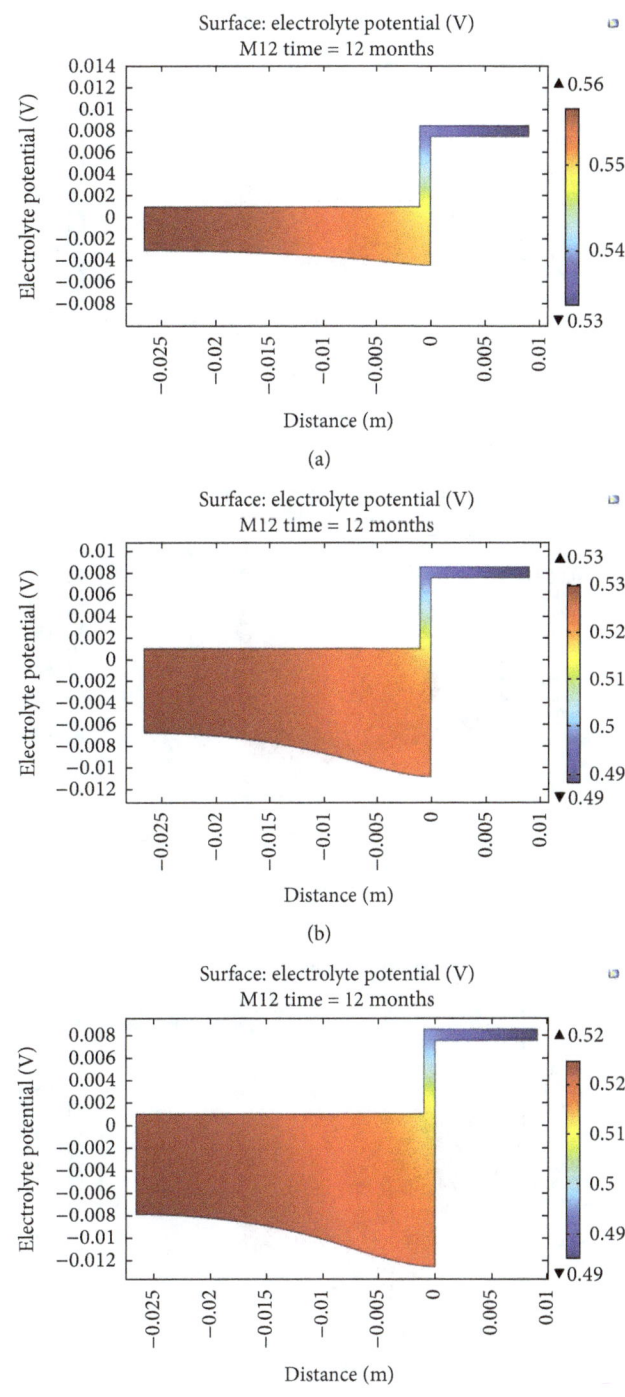

FIGURE 5: Electrolyte potential for M12 bolt and electrolyte thickness, $e = 1$ mm at 12 months with deformed geometry. (a) 20MnCr5, (b) 42CrMo4, (c) 32CrMoV13.

conditions and governing equation were applied, and the appropriate mesh was found for the geometries, the model for galvanic corrosion was solved. The resolution of Laplace's equation enables us to present electrolyte potential on the surfaces of electrodes bolt/end plate. Figures 5, 6, and 7 show us the surface plot of electrolyte potential with deformed

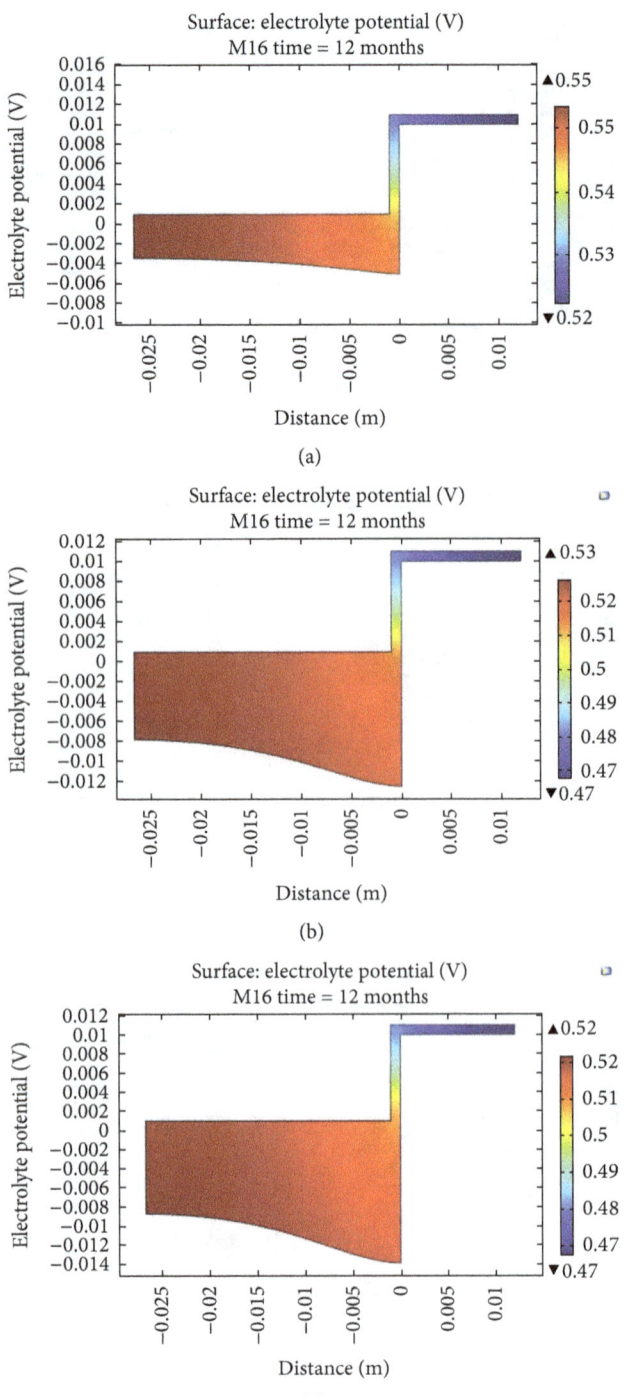

FIGURE 6: Electrolyte potential for M16 bolt and electrolyte thickness, $e = 1$ mm at 12 months with deformed geometry. (a) 20MnCr5, (b) 42CrMo4, (c) 32CrMoV13.

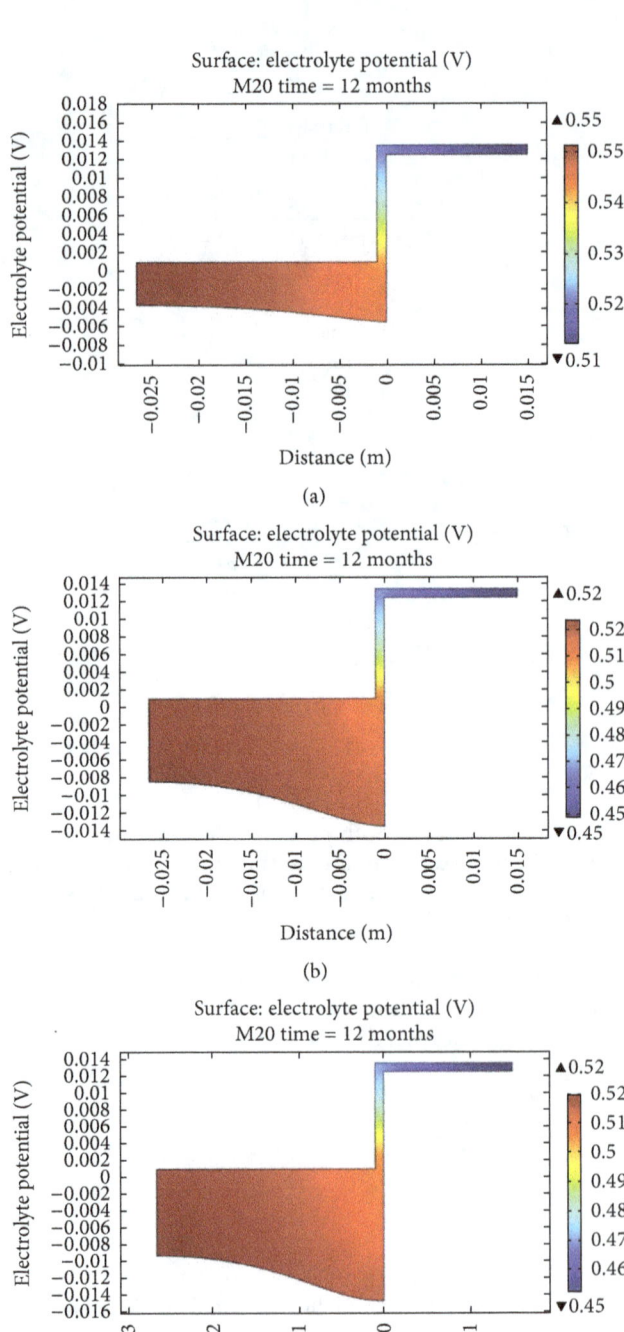

FIGURE 7: Electrolyte potential for M20 bolt and electrolyte thickness, $e = 1$ mm at 12 months with deformed geometry. (a) 20MnCr5, (b) 42CrMo4, (c) 32CrMoV13.

TABLE 5: Electrolyte characteristics [16].

Conductivity (S/m)	33.2 (S/m)
Low thickness (mm)	1 (mm)
High thickness (mm)	20 (mm)

TABLE 6: Electrochemical parameters in 1 M HCl [17].

	β_a (mV/dec)	β_c (mV/dec)	i_0 (A/m^2)	E_{eq} (mV)
S235JR	65	122	3.55	−553
42CrMo4	209.9	249.9	7,1	−495
32CrMoV13	263.7	173.1	10	−500
20MnCr5	106.2	124.3	1,6	−488

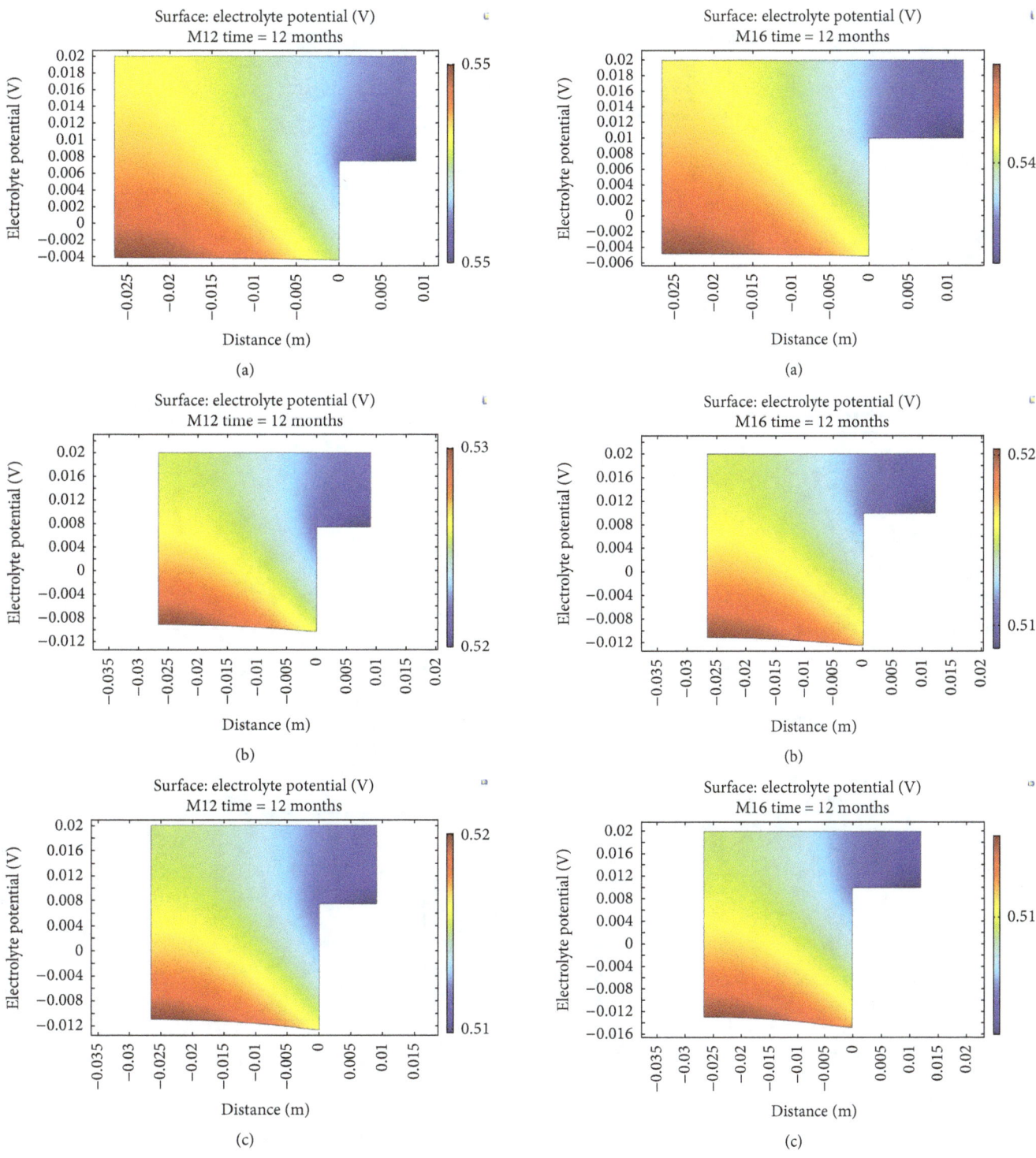

FIGURE 8: Electrolyte potential for M12 bolt and electrolyte thickness, $e = 20$ mm at 12 months with deformed geometry. (a) 20MnCr5, (b) 42CrMo4, (c) 32CrMoV13.

FIGURE 9: Electrolyte potential for M16 bolt and electrolyte thickness, $e = 20$ mm at 12 months with deformed geometry. (a) 20MnCr5, (b) 42CrMo4, (c) 32CrMoV13.

geometry at different bolt sizes and different materials of bolt. It can be seen that the potential of the electrolyte increases if the bolt size increases: $E(M20) > E(M16) > E(M12)$. These potential values also depend on the bolt material: $E(32CrMoV13) > E(42CrMo4) > E(20MnCr5)$.

The potential of the electrolyte also depends on the thickness of the electrolyte; it increases if the thickness of the electrolyte increases, and this applies to the two parameters: bolt size and bolt material, $E(e = 20$ mm$) > E(e = 1$ mm$)$, Figures 8, 9, and 10.

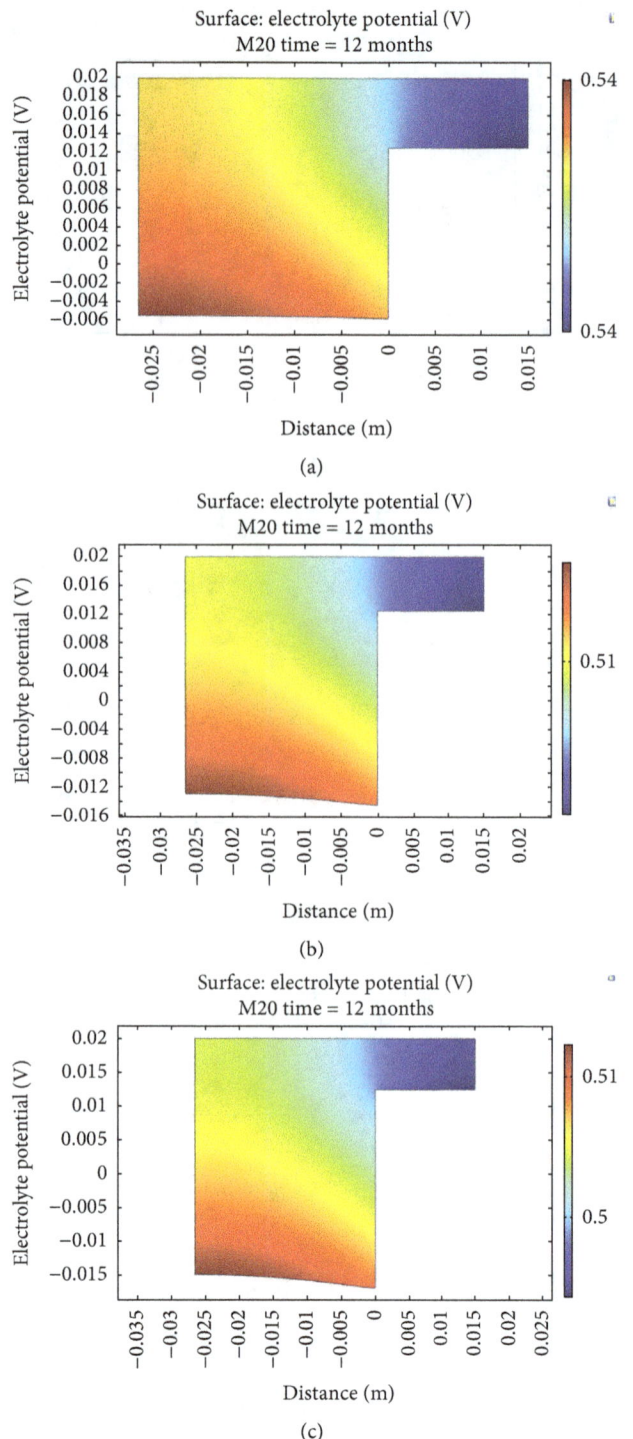

FIGURE 10: Electrolyte potential for M20 bolt and electrolyte thickness, $e = 20$ mm at 12 months with deformed geometry. (a) 20MnCr5, (b) 42CrMo4, (c) 32CrMoV13.

4.2. Corrosion Rate C_R.

The corrosion rate was calculated using the following relation [32]:

$$V_c = \frac{I_{\text{corr}} \times M}{z \times F \times \rho},$$ (6)

where V_c is the corrosion rate (cm/yr), I_{corr} is the corrosion current density (A/cm^2), M is the molar mass

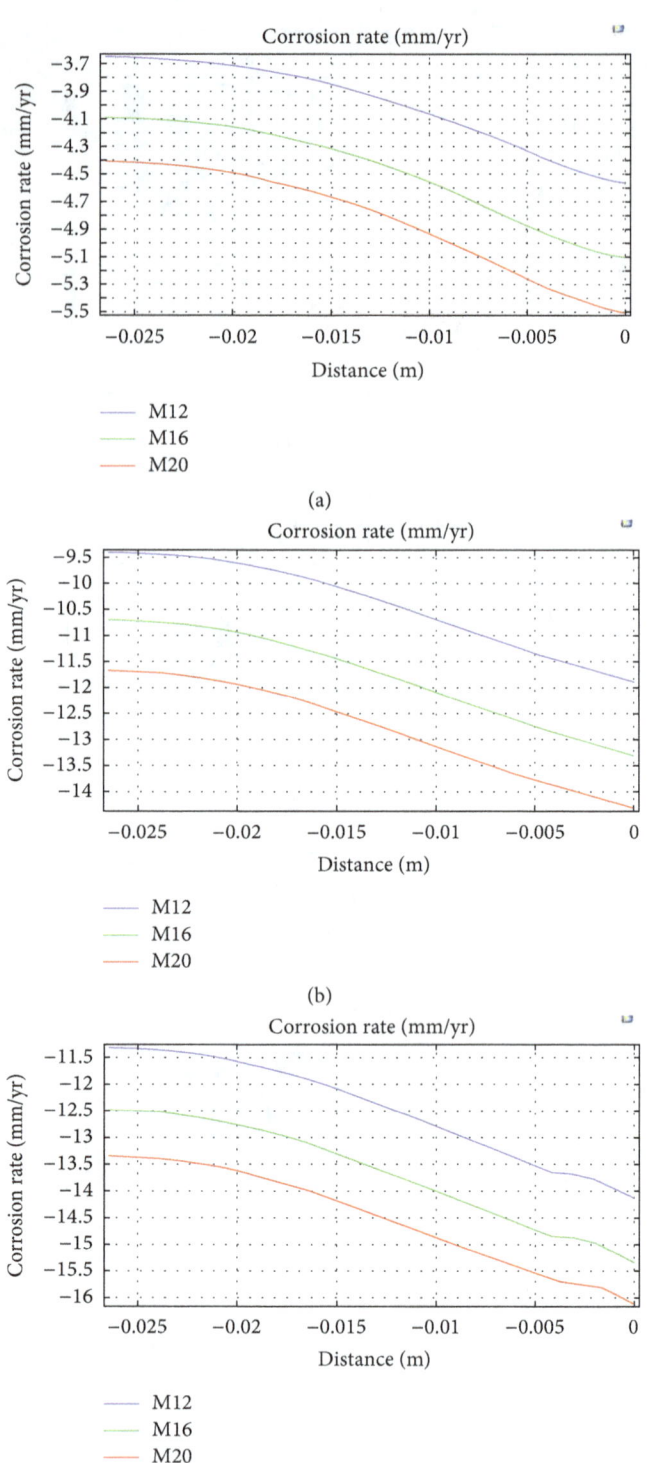

FIGURE 11: Corrosion rate for electrolyte thickness, $e = 1$ mm. (a) 20MnCr5, (b) 42CrMo4, (c) 32CrMoV13.

$M(\text{Fe}) = 55.85$ g/mol, z is the valence of iron $z = 2$, F is the faraday constant $F = 96500$ A·s/mol, and ρ is the density of steel $\rho = 7.87$ g/cm^3.

Table 7 and Figures 11 and 12 show that the corrosion rate depends on the bolt size, the bolt material, and the electrolyte thickness. It can be seen that the corrosion rate

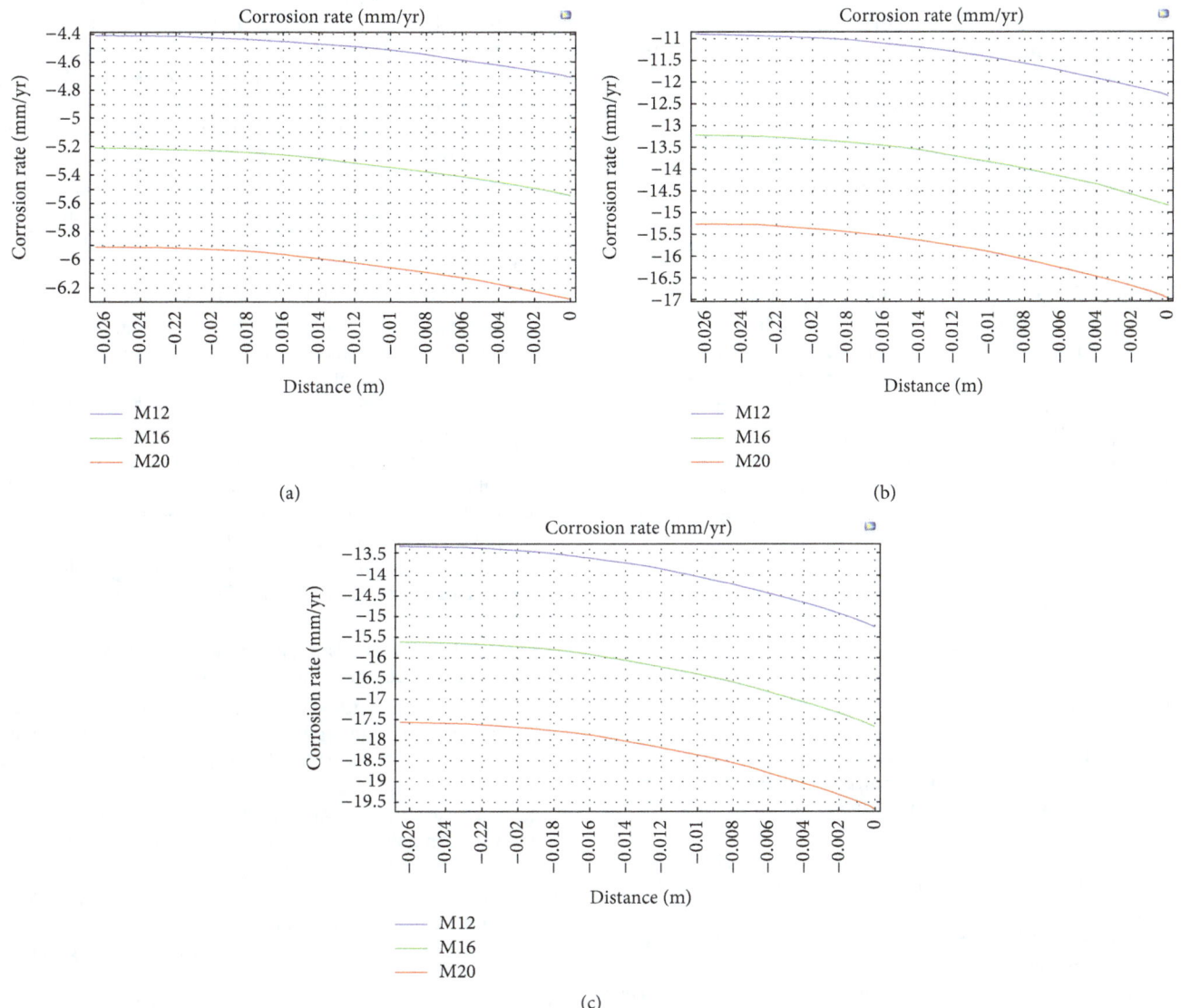

FIGURE 12: Corrosion rate for electrolyte thickness, $e = 20$ mm. (a) 20MnCr5, (b) 42CrMo4, (c) 32CrMoV13.

TABLE 7: Corrosion rate values as a function of bolt size, material, and electrolyte thickness.

| Bolt size | Bolt material | Corrosion rate C_R of end plate (mm/yr) | | | |
| | | Electrolyte thickness $e = 1$ mm | | Electrolyte thickness $e = 20$ mm | |
		$x = 0$	$x = -26.5$ mm	$x = 0$	$x = -26.5$ mm
M12	20MnCr5	4.55	3.65	4.70	4.40
	42CrMo4	11.80	9.40	12.30	10.90
	32CrMoV13	14.10	11.30	15.25	13.30
M16	20MnCr5	5.10	4.10	5.55	5.20
	42CrMo4	13.25	10.70	14.80	13.20
	32CrMoV13	15.30	12.50	17.70	15.60
M20	20MnCr5	5.50	4.40	6.30	5.90
	42CrMo4	14.40	11.70	17.00	15.25
	32CrMoV13	16.10	13.40	19.60	17.55

of the end plate increases if the bolt size increases: $C_R(M20)$ > $C_R(M16)$ > $C_R(M12)$. These corrosion rate values depend also on the bolt material: $C_R(32CrMoV13)$ > $C_R(42CrMo4)$ > $C_R(20MnCr5)$.

The corrosion rate of the end plate depends also on the thickness of the electrolyte; it increases if the thickness of the electrolyte increases, and this applies to the two parameters: bolt size and bolt material, $C_R(e = 20\,mm)$ > $C_R(e = 1\,mm)$.

5. Conclusion

The choice of bolts in galvanic bolted joints of steel structures must take into account the effect of the bolt size and bolt material. It is shown in this study that as the size of the bolt increases, the corrosion rate increases. Also if the steel alloys of the bolt change, the corrosion rate changes. The increasing of the corrosion rate in the process of galvanic corrosion of the end plate leads to a reduction in lifetime of the bolted assembly. Hence a great interest must be shown by engineers/designers in the choice of the bolts when designing the steel structures to ensure long lifetime.

Conflicts of Interest

The authors declare no conflicts of interest.

References

[1] J. W. Oldfield, "Electrochemical theory of galvanic corrosion," in *Galvanic Corrosion*, H. P. Hack, Ed., pp. 5–22, American Society for Testing and Materials, Philadelphia, Pa, USA, 1988, ASTM STP 978.

[2] H. P. Hack and J. R. Scully, "Galvanic corrosion prediction using long- and short-term polarization curves," *Corrosion*, vol. 42, no. 2, pp. 79–90, 1986.

[3] J. F. Yan, S. N. Pakalapati, T. V. Nguyen, and R. E. White, "Mathematical modeling of cathodic protection using the boundary element method with a nonlinear polarization curve," *Journal of the Electrochemical Society*, vol. 139, no. 7, pp. 1932–1936, 1992.

[4] L. Lacroix, C. Blanc, N. Pébère, B. Tribollet, and V. Vivier, "Localized approach to galvanic coupling in an aluminum-magnesium system," *Journal of The Electrochemical Society*, vol. 156, no. 8, pp. C259–C265, 2009.

[5] K. B. Deshpande, "Validated numerical modelling of galvanic corrosion for couples: Magnesium alloy (AE44)-mild steel and AE44-aluminium alloy (AA6063) in brine solution," *Corrosion Science*, vol. 52, no. 10, pp. 3514–3522, 2010.

[6] K. B. Deshpande, "Experimental investigation of galvanic corrosion: Comparison between SVET and immersion techniques," *Corrosion Science*, vol. 52, no. 9, pp. 2819–2826, 2010.

[7] K. B. Deshpande, "Numerical modeling of micro-galvanic corrosion," *Electrochimica Acta*, vol. 56, no. 4, pp. 1737–1745, 2011.

[8] J. X. Jia, G. Song, and A. Atrens, "Experimental measurement and computer simulation of galvanic corrosion of magnesium coupled to steel," *Advanced Engineering Materials*, vol. 9, no. 1-2, pp. 65–74, 2007.

[9] D. Trinh, P. Dauphin Ducharme, U. Mengesha Tefashe, J. R. Kish, and J. Mauzeroll, "Influence of edge effects on local corrosion rate of magnesium alloy/mild steel galvanic couple," *Analytical Chemistry*, vol. 84, no. 22, pp. 9899–9906, 2012.

[10] N. Murer, N. Missert, and R. Buchheit, Towards the Modeling of Microgalvanic Corrosion in Aluminum Alloys: the Choice of Boundary Conditions, COMSOL Users Conference, 2008.

[11] N. Murer, R. Oltra, B. Vuillemin, and O. Néel, "Numerical modelling of the galvanic coupling in aluminium alloys: A discussion on the application of local probe techniques," *Corrosion Science*, vol. 52, no. 1, pp. 130–139, 2010.

[12] N. Murer, N. A. Missert, and R. G. Buchheit, "Finite element modeling of the galvanic corrosion of aluminum at engineered copper particles," *Journal of The Electrochemical Society*, vol. 159, no. 6, pp. C265–C276, 2012.

[13] Y. Shi and R. G. Kelly, ECS Transactions, 41, 155, 2012.

[14] W. Sun, G. Liu, L. Wang, T. Wu, and Y. Liu, "An arbitrary Lagrangian-Eulerian model for studying the influences of corrosion product deposition on bimetallic corrosion," *Journal of Solid State Electrochemistry*, vol. 17, no. 3, pp. 829–840, 2013.

[15] J. F. Yan, T. V. Nguyen, R. E. White, and R. B. Griffin, "Mathematical modeling of the formation of calcareous deposits on cathodically protected steel in seawater," *Journal of The Electrochemical Society*, vol. 140, 733 pages, 1993.

[16] V. G. Artemov, A. A. Volkov, N. N. Sysoev, and A. A. Volkov, "Conductivity of aqueous HCl, NaOH and NaCl solutions: Is water just a substrate?" *EPL (Europhysics Letters)*, vol. 109, no. 2, 2015.

[17] Okba Belahssen. Thesis: Etude comparative du comportement chimique et tribologique des aciers nitrurés.

[18] D. Höchez, "Simulation of corrosion product deposit layer growth on bare magnesium galvanically coupled to aluminum," *Journal of The Electrochemical Society*, vol. 162, no. 1, pp. C1–C11, 2015.

[19] K. E. Johnson and J. S. Abbott, "Bimetallic corrosion effects on mild steel in natural environments," *British Corrosion Journal*, vol. 9, no. 3, pp. 171–176, 1974.

[20] Q. Xu, K. Gao, W. Lv, and X. Pang, "Effects of alloyed Cr and Cu on the corrosion behavior of low-alloy steel in a simulated groundwater solution," *Corrosion Science*, vol. 102, pp. 114–124, 2016.

[21] J. Scully and H. P. Hack, ASTM STP 978 Galvanic Corrosion, 136, 1988.

[22] C. Wagner, "Theoretical analysis of the current density distribution in electrolytic cells," *Journal of The Electrochemical Society*, vol. 98, no. 3, pp. 116–128, 1951.

[23] F. J. Presuel-Moreno, H. Wang, M. A. Jakab, R. G. Kelly, and J. R. Scully, "Computational modeling of active corrosion inhibitor release from an Al-Co-Ce metallic coating," *Journal of The Electrochemical Society*, vol. 153, no. 11, Article ID 002611JES, pp. B486–B498, 2006.

[24] J. S. Lee, M. L. Reed, and R. G. Kelly, "Combining rigorously controlled crevice geometry and computational modeling for study of crevice corrosion scaling factors," *Journal of The Electrochemical Society*, vol. 151, no. 7, pp. B423–B433, 2004.

[25] E. Kennard and J. T. Waber, "Mathematical Study of Galvanic Corrosion: Equal Coplanar Anode and Cathode with Unequal Polarization Parameters," *Journal of The Electrochemical Society*, vol. 117, no. 7, pp. 880–885, 1970.

[26] J. T. Waber and B. Fagan, "Mathematical Studies on Galvanic Corrosion IV. Influence of Electrolyte Thickness on the Potential and Current Distributions over Coplanar Electrodes Using Polarization Parameters," *Journal of The Electrochemical Society*, vol. 103, no. 1, pp. 64–72, 1956.

[27] J. W. Fu, "Technical note: a finite element analysis of corrosion cells," *Corrosion*, vol. 38, no. 5, pp. 295-296, 1982.

[28] J. T. Waber, "Mathematical Studies of Galvanic Corrosion VI. Limiting Case of Very Thin Films," *Journal of The Electrochemical Society*, vol. 103, no. 10, pp. 567–570, 1956.

[29] P. Doig and P. E. J. Flewitt, "A finite difference numerical analysis of galvanic corrosion for semi-infinite linear coplanar electrodes," *Journal of The Electrochemical Society*, vol. 126, no. 12, pp. 2057–2063, 1979.

[30] E. McCafferty, "Distribution of potential and current in circular corrosion cells having unequal polarization parameters," *Journal of The Electrochemical Society*, vol. 124, no. 12, pp. 1869–1878, 1977.

[31] A. C. West, S. Mayer, and J. Reid, "A superfilling model that predicts bump formation," *Electrochemical and Solid-State Letters*, vol. 4, no. 7, pp. C50–C53, 2001.

[32] N. Perez, *Electrochemistry and Corrosion Science*, Kluwer Academic Publishers, New York, NY, USA, 2004.

8

Corrosion Behavior of Brazed Zinc-Coated Structured Sheet Metal

A. Nikitin, L. Schleuss, R. Ossenbrink, and V. Michailov

Department of Joining and Welding Technology, Brandenburg University of Technology Cottbus-Senftenberg, Konrad-Wachsmann-Allee 17, 03046 Cottbus, Germany

Correspondence should be addressed to R. Ossenbrink; ralf.ossenbrink@b-tu.de

Academic Editor: Yu Zuo

Arc brazing has, in comparison to arc welding, the advantage of less heat input while joining galvanized sheet metals. The evaporation of zinc is reduced in the areas adjacent to the joint and improved corrosion protection is achieved. In the automotive industry, lightweight design is a key technology against the background of the weight and environment protection. Structured sheet metals have higher stiffness compared to typical automobile sheet metals and therefore they can play an important role in lightweight structures. In the present paper, three arc brazing variants of galvanized structured sheet metals were validated in terms of the corrosion behavior. The standard gas metal arc brazing, the pulsed arc brazing, and the cold metal transfer (CMT®) in combination with a pulsed cycle were investigated. In experimental climate change tests, the influence of the brazing processes on the corrosion behavior of galvanized structured sheet metals was investigated. After that, the corrosion behavior of brazed structured and flat sheet metals was compared. Because of the selected lap joint, the valuation of damage between sheet metals was conducted. The pulsed CMT brazing has been derived from the results as the best brazing method for the joining process of galvanized structured sheet metals.

1. Introduction

Galvanized sheet metals have good corrosion resistance to atmospheric corrosion and therefore they are widely used in the automotive industry [1]. However, during the arc welding of such sheet metals, the protective zinc layer is badly damaged, so that even the zinc anode protection does not prevent the atmospheric steel corrosion. The zinc coating evaporates as a result of the large heat input of arc welding processes. Furthermore, zinc vapor and zinc oxides lead to pores in the weld metal and an unstable arc behavior [2]. These problems are reduced by bonding or brazing. During bonding, there is no or less heat input and thus the protective zinc layer remains intact. However, often the strength of the adhesive bond is insufficient [3]. The arc brazing advantage is the higher joint strength, which can reach the strength of the base material, and significantly reduced heat input compared to the arc welding. The disadvantages of welding and bonding are simultaneously solved [4]. The technology of cold metal transfer (CMT) is a further development of arc welding and brazing processes, which even more reduces the heat input into the joined components [5, 6].

The lightweight design plays a key role in the automotive industry to achieve climate targets and to reduce energy consumption. Structuring of sheet metals increases their stiffness and reduces their weight. After the structuring process by hydroforming, the corrosion resistance of zinc-coated sheet metals slightly increased, which was studied in [7]. Galvanized structured sheet metals could be applied in various fields such as automotive and construction industry [8] which also reflects the application fields of the arc brazing. Structured sheet metals were already qualified for further processing, especially in the field of metal forming [9, 10], cutting [11, 12], and welding [13–17]. Their mechanical properties were studied in [18, 19]. The corrosion behavior of galvanized structured sheet metals was studied in nonwelded joints in [7, 20]. Consequently, there is a need to validate the brazed joints corrosion behavior of galvanized structured sheet metals.

TABLE 1: Geometrical characteristics of the structured sheet metals.

Geometrical characteristics [mm]			
SW	SL	SH	BW
33	36	3.1	2

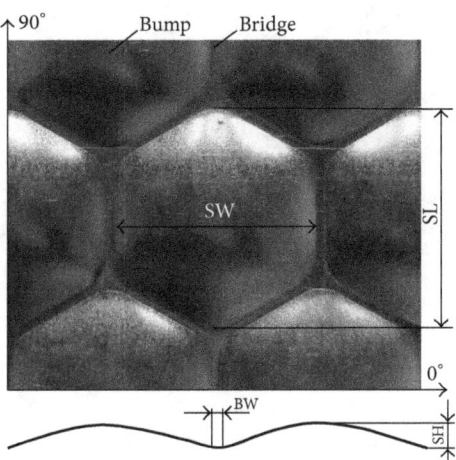

FIGURE 1: Zink-coated structured sheet metal with structure directions 0 and 90°.

2. Experimental Details

2.1. Structured Sheet Metals. Structured sheet metals have a hexagonal structure. The production method is hydroforming [10]. The sheet metals consist of stiffening elements with hexagonal bumps, whose dimensions are shown in Figure 1 and described in Table 1 with specific values.

Due to the asymmetry relating to the sheet plain, there are two structural locations: the structure location is "positive" if the curvature extends upwards and "negative" if the curvature extends inwards. In [20], a small difference of the corrosion behavior between the structure locations is described. All specimens were brazed in the positive structure location and in the structure direction 0°. Therefore, a top and a bottom side of brazed samples, respectively, coincide with the positive and negative structure location of the structured sheet metal.

In the present paper, the atmospheric corrosion of the protective coat damaging by galvanized sheets brazing was studied. The estimated locations for corrosion of the brazed joint are shown in Figure 2. Areas 1 and 2 are located beside the brazed joint on the top side of the joint. Area 3 is on the seam bottom side. Because of the probable zinc layer damage, these three areas are most susceptible against corrosion due to the brazing process heat input. At area 4, the risk of corrosion is high, not only due to protective layer destruction, but also because of the gap existence by overlapping where corrosive media can be accumulated and the corrosion processes accelerated.

For corrosion tests, thin sheet metals were used. The steel type is DX56D + Z. This is hot-dip galvanized low carbon steel with 10 μm coating thickness. The sheet thickness was 0.5 mm.

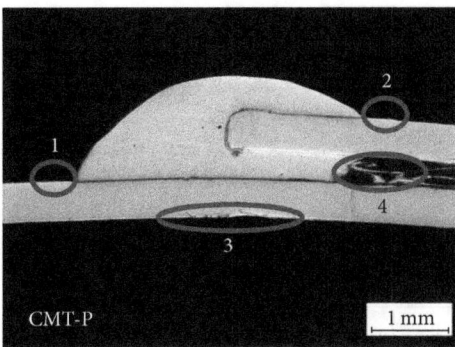

FIGURE 2: The probable locations of the corrosion occurrence.

2.2. Brazing Process. Mechanized brazing was carried out by robot system Kuka 30 HA (high accuracy). First, brazing tests were conducted on the flat sheet metals of similar thickness and similar material to determine the initial parameters. It turned out that the flat sheet metals could be qualitatively brazed without special requirements by linear torch controlling normal to the sheet plain. When transferring these parameters on structured sheet metals, the three-dimensional structure caused a continuous variation of the brazing conditions, especially on the contact tube spacing and the incidence angle between the brazing torch and the plate. The brazing with the same parameters showed bad quality of the brazed joints. The contact tube spacing (the distance between torch and sheet) has a significant influence on the brazed joint quality. These effects are described for the GMA welding in [13]. In [11], it was reported that an altitude correction in dependence on the arc voltage at the plasma cutting of the structured sheet metals has a positive effect on the cutting quality. However, since a height correction at the pulse arcs cannot be used as a control variable, the structure height values were measured along the brazing zone and transferred to the robot program. Consequently, the robot path has been realized with a constant distance between the torch and the sheet metal.

The gas metal arc (GMA) brazing of the structured sheet metals was investigated by three different processes variants [21]. The parameters for the standard gas metal arc brazing (hereinafter GMAB), the pulsed arc brazing (hereinafter GMAB-P), and the cold metal transfer (CMT) in combination with a pulsed cycle (hereinafter CMT-P) were determined. The standard GMAB involves the conventional arc brazing with drop transfer at the short circuit. The main disadvantages of this method due to the short circuit are increasing energy consumption and spattering. The pulsed arc brazing leads to a drop transfer which is almost free from the short circuit and free of spattering. The third examined process variant CMT brazing is based on a reversing wire supply, which moves the brazing wire with a frequency up to 90 Hz back and forward. At first, the arc ignites at the wire tip. Then, the forward movement of the wire begins at the same time with drop formation at the wire tip. By the contact due to the forward movement of the wire, the drop at the wire tip is transferred to the sheet and the arc is extinguished,

TABLE 2: Process parameters for GMA brazing of the structured and flat sheet metal.

(a) Brazing parameters for the structured sheet metals

Brazing method	GMAB	GMAB-P	CMT-P
Current [A]	74	39	37
Voltage [V]	13.5	16.5	13.6
Wire speed [m/min]	3.5	2.2	2.3

(b) Brazing parameters for the flat sheet metals

Brazing method	GMAB	GMAB-P	CMT-P
Current [A]	82	54	40
Voltage [V]	13.5	17.9	13.6
Wire speed [m/min]	4.1	2.8	2.5

TABLE 3: Physical properties of the filler material [21].

Filler material	Melting point [°C]	Tensile strength [MPa]	Elongation [%]	Hardness [HB]
CuSi3	910–1026	330–370	40	80–90

so that the metal transfer almost without current occurs. Subsequently, the wire is pulled back after the extinction of the arc. The described variant of CMT brazing was improved by the pulse technique, so that during the drop formation more filler material is melted by the increased current pulse.

In Table 2, the brazing parameters for flat and structured sheet metals are presented. For gas metal arc brazing, a filler material CuSi3 with diameter 1.0 mm was used. Table 3 shows the relevant physical properties of the filler material. Due to the low melting temperature, this filler wire is used very often for the GMA brazing of electrolytic or hot-dip galvanized sheet metals.

2.3. Corrosion Testing. Corrosion tests were carried out by applying a salt spray chamber with 1000-liter volume; the model CT 1000-S MF was made by VLM GmbH. In this work, the climate change test according to VDA 621-415 specification [22] was selected as accelerated corrosion test. The test cycle consists of 1 day of salt spray exposure according to [23], 4 days of condensation cyclic climate according to [24], and 2 days of ambient climate. The duration of one cycle was one week. The number of cycles depends on the corrosion resistance of the specimen coatings and was 5 weeks. The results of this test are used for comparison of the corrosion resistance of various protective coatings. The corrosive medium was a 5 wt.% aqueous sodium chloride solution with pH of 6.7. The solution temperature was 25°C. This is necessary to ensure the independence of the results from the chamber and the comparability with the results of other studies.

Because of the lap joint, the arrangement of the specimen in the specimen holder must be considered. In Figure 3, the arrangements are shown. For the all-round analysis, it is necessary to carry out the corrosion tests for all 6 arrangement variants. To reduce the experimental scope, the corrosion tests were conducted for one arrangement of the

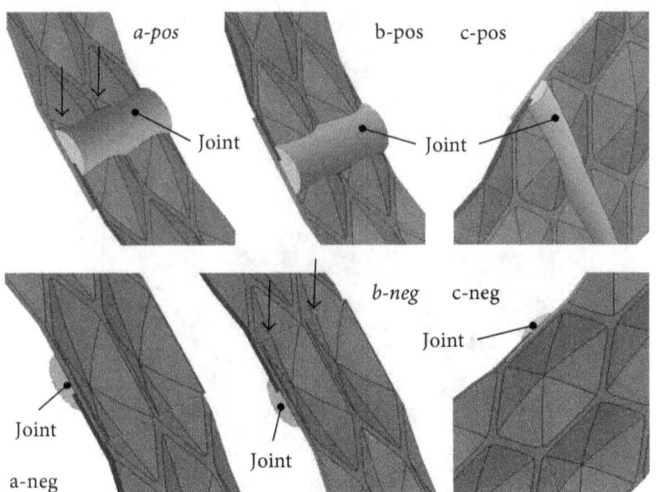

FIGURE 3: Variants of the arrangement of the specimen.

samples in positive and negative structure locations with the brazing seam on the top and bottom side, respectively. The reason of selecting this sample arrangement consists in the assertion that every bump plays a barrier role which accumulates the moisture and therefore provides additional conditions for corrosion [20]. According to the standard [23], the specimens must be placed at $(20 \pm 5)°$ angle to the vertical position. The moisture and corrosion products flow down at such angle along the sample surface. In the arrangement of the a-pos brazed joint is a bigger barrier for that than the b-pos. By the use of the arrangement of c-pos the brazed joint does not build a barrier for the corrosive medium. This means that the arrangement of a-pos leads to the strongest corrosion conditions. For the bottom side of the specimen, the situation is reversed. There is no brazed joint and only the edge of the second sheet in the arrangement b-neg is considered as a barrier to the salt solution. In addition, this position gives higher corrosion load for gaps between sheet metals. In an ongoing research, only variants of the specimen arrangement with bigger relief were selected for experiments: variant a-pos for the top seam side and variant b-neg for the bottom seam side.

For a conclusive analysis, it is necessary to investigate at least 3 samples for every brazing variant. Besides the structured samples, the smooth samples as a reference were brazed and investigated by the climate change test. As Table 4 shows, a total of 36 samples were tested.

The geometric specimen parameters were 150×70 mm. The examination of corroded specimens was carried out weekly. Pictures of the samples were taken. The corrosion products were not removed during the test; otherwise, the natural corrosion process would be interrupted and therefore only a visual inspection was used as an evaluation method. To increase the capacity of the evaluation, a graphical analysis of the corroded areas by using the program Leica Application Suite (LAS) Phase Expert was conducted. The program identifies the corrosion areas as percentage of the evaluation area (Figures 4 and 5) by regions of homogeneous color.

TABLE 4: Number of specimens.

Brazing methods	Type of the sheet metal			
	Flat		Structured	
	Sheet thickness			
	0.5 mm		Bump: 0.48 mm	
			Bridge: 0.49 mm	
	Positive	Negative	Positive	Negative
GMAB	3	3	3	3
GMAB-P	3	3	3	3
CMT-P	3	3	3	3
Sum	18		18	
	36			

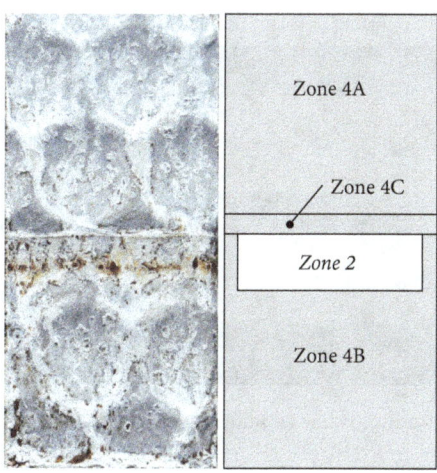

FIGURE 5: Picture of the structured specimen (bottom seam side) after the first week of the climate change test and scheme of the evaluation areas.

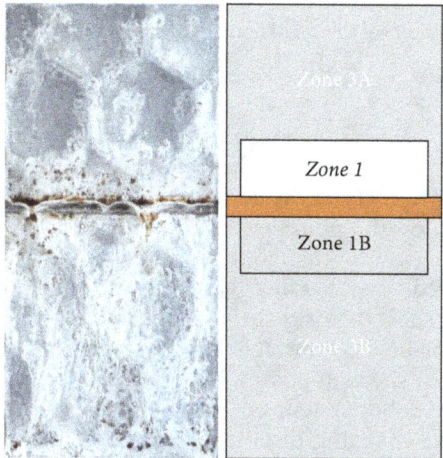

FIGURE 4: Picture of the structured specimen (top seam side) after the first week of the climate change test and scheme of the evaluation areas.

Primarily, the thresholds for homogeneous colors of "red" rust could be clearly defined.

2.4. Evaluation Method. For examination, the areas near the brazed joint were used (zone 1), because the CuSi3 seam is not corroded. These zones included the full width of a sample excluding the edges. The height of the zone was set at the level that was determined for all samples as detected maximum damage range, so that all corroded areas were in this zone. Zone 1B under the joint was not used because it showed a consistent picture for all structured samples. The surface without zinc in this zone is corroded slower than in zone 1. This may be explained by the fact of decreasing of moisture level due to the brazed seam, which acts as a barrier. Frequently, the accumulated water in zone 1 flows across the seam and then some spots of "red" corrosion appear in zone 1B (Figure 4). Therefore, this area cannot be used for comparison. Zone 3 has no heat influence due to the brazing process and that is why there are no "red" rust areas after the first test week.

The negative structure location has no barrier seam of moisture and corrosion products. However, there is another barrier that is smaller and farther from the place where a heat source was. This is the edge of the bottom sheet in the lap joint where the moisture is accumulated (zone 4C). However, there is no zinc coating damage due to the heat input; therefore, this zone is not important for the brazed joint corrosion investigation. Zone 2 is located directly below the brazed seam. Its borders were found by determining the maximum damaged surface.

All samples were cut before by laser beam. However, this leads to a low heat input into the edges of the samples. Therefore, side areas in 5 mm from the edge were not used in the examination. All sheets had comparable surfaces. Examination area was 750 mm^2 for every specimen.

During the corrosion test, two types of corrosion products arise on the surface of the galvanized sheet metals. The first form is "white" rust, generally consisting of zinc oxide, zinc hydroxide, and zinc carbonate products. These products are produced by oxidation of the zinc coating in the chlorine-containing conditions. This process leads to the destruction of the protective zinc layer and steel begins to corrode. The result is the oxidation of the iron and the formation of brownish iron(II) and iron(III) oxides, or so-called "red" rust. The "white" rust has no influence on the mechanical properties of components and just changes the surface appearance. The program software for color detection was used for the examination of corrosion-damaged surface. The surface analysis was done by the evaluation of the "red" rust, because the area of "white" rust provides no clear statements on the corrosion of the brazed joint. The "white" rust is also an indication for the corrosion-damaged surface. Some "red" rust may develop under the "white" rust and cannot be detected by the program software.

Previous investigations [7] on zinc-coated structured sheet metals of the same type and material showed that "red" rust in magnitude of 22–26% develops under the "white" rust for short exposure times (240 h) and rapidly decreases to magnitude of 3–9% for longer exposure times. There are no significant differences between flat and structured sheet

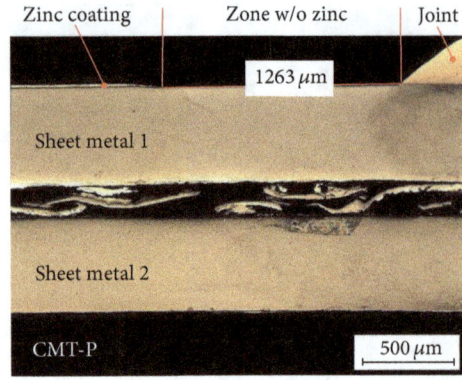

Figure 6: Minimum zone without zinc coating near the brazed joint.

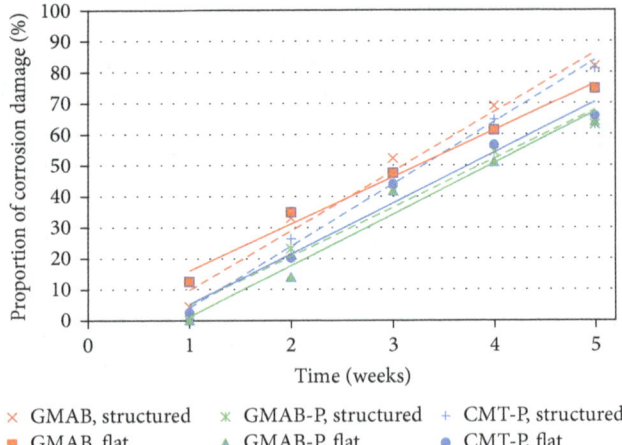

× GMAB, structured   × GMAB-P, structured   + CMT-P, structured
■ GMAB, flat   ▲ GMAB-P, flat   ● CMT-P, flat

Figure 8: Results of the corrosion test for zone 2 (bottom seam side).

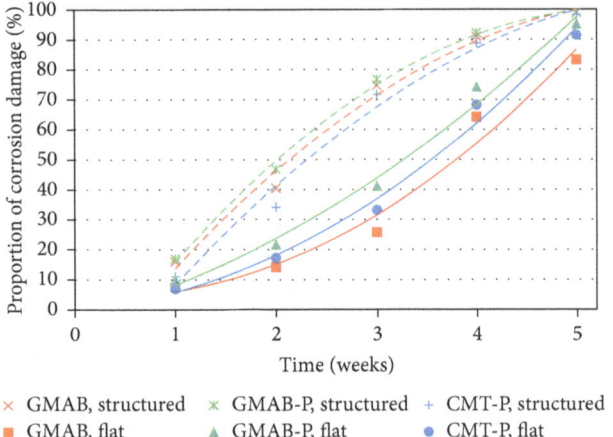

× GMAB, structured   × GMAB-P, structured   + CMT-P, structured
■ GMAB, flat   ▲ GMAB-P, flat   ● CMT-P, flat

Figure 7: Results of corrosion test for zone 1 (top seam side).

metals. Transferring these results to this investigation, the relative positions and distances for the corrosion development of different brazing procedures remain unchanged and can be compared to each other.

3. Results and Discussion

After the brazing process, all samples had areas without zinc coating near the brazed seam. The metallographic studies confirm this fact (Figure 6). The width of this area has a direct influence on the corrosion resistance of the compound. However, due to the uneven evaporation of the zinc layer, the width of the area could not be determined exactly and the "red" rust area after the corrosion test was evaluated.

The test results were shown in Figure 7. As the curves for each brazing method are practically similar (the average deviation is less than 5%), only the main values were used in the further comparisons for the examination. The results are converted into trend lines by approximation.

The diagram shows the increasing corrosion damage proportion in zone 1 during a period of 5 weeks. This zone 1 is located near the seam. First, the big difference between structured and flat samples is observed. In zone 1, there are more "red" corrosion products on the structured specimens

than on the flat ones. As a result, this leads to the deterioration of the compound corrosion behavior. According to [20], the corrosion products on the nonjoined flat sheet metals appear faster than on the nonbrazed structured sheet metals. This behavior seems to change, if the structured sheet metals were brazed. The brazing has a great influence on the structured sheet corrosion behavior near the compound. This can be explained by unstable brazing even when the three-dimensional sheets topology is considered by the program of the used robot system for brazing. These paths take into account the changing distance from the torch to the sheet plain. Furthermore, bumps have an influence on the inert gas flow and generate the variation between rising and falling brazing positions [13]. This affects the quality of the brazed joint.

After the first week, all samples already had the "red" corrosion products near the seam. This means that the cathodic protection of the zinc coating is no longer effective. After the fifth test week, all samples in zone 1 were practically completely covered with "red" rust.

As Figure 7 shows, there is no big difference between the corrosion behaviors of the brazed joints produced by various brazing methods. This is valid for the flat sheet metals (maximum difference is 15.5% by 4.7% deviation) as well as for the structured sheet metals (maximum difference is 12.7% by 6.2% deviation). However, the pulse brazed samples show the worst results, regardless of the sheet structure. The CMT pulse brazing shows the best results for structured sheet metals. For the flat reference samples, the standard GMA brazing is the most suitable method.

The results of the corrosion test for zone 2 are shown in Figure 8. The results of structured and flat samples showed practically no difference. Figure 8 clearly shows a major trend for the bottom seam side of the compound. The standard GMA brazing with the largest current values leads to greater heat input. Therefore, the samples prepared by the standard GMA brazing show the worst corrosion resistance on the bottom seam side. This is valid for all samples irrespective of their structure. On average, the corrosion resistance of the standard-brazed compounds is up to 12%

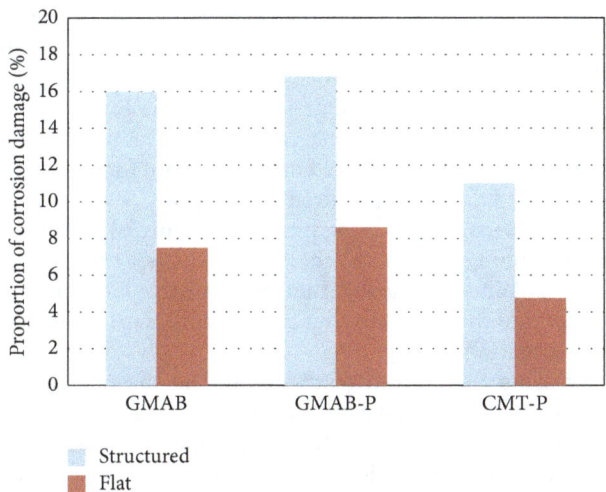

FIGURE 9: Results of the corrosion test after the first week for zone 1 (top seam side).

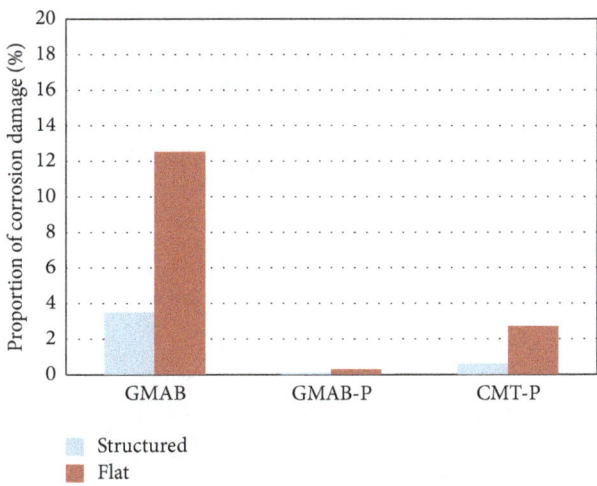

FIGURE 10: Results of the corrosion test after the first week for zone 2 (bottom seam side).

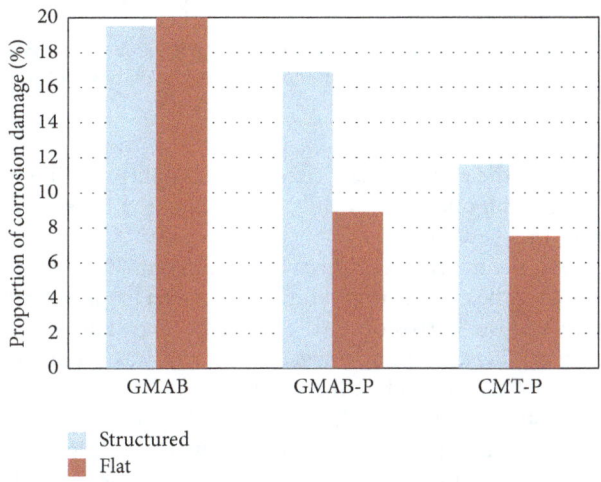

FIGURE 11: Totalized results of the corrosion test after the first week for zones 1 and 2.

lower for flat samples and up to 11.6% lower for structured samples compared to the best results for flat and structured samples brazed by pulsed GMAB. The pulsed GMA brazing modification through CMT technology demonstrates worse corrosion behavior of all samples.

Zone 2 is corroded less than zone 1, because on the bottom compound side the zinc layer was affected only by a low heat input. It was found that after the first test week a big amount of "red" rust on the bottom compound side has developed only on the standard-brazed samples.

The requirements for brazed compounds in industry consist in obtaining compound corrosion resistance comparable to the corrosion resistance of the base material. This could not be accurately detected by the corrosion test according to VDA 621-415 because the conditions of this test are very harsh. After the second week of the climate change test, the "red" rust appears on the sheet surface without thermal influence. Near the brazed seam, the spots of the "red" rust are already observed after the first test week. This means that the results after the first week are the most interesting from the brazing methods comparison point of view.

Figure 9 shows results of corrosion test for positive structure location arranged samples after the first week. It is clear that there is almost no difference in the results of the samples "GMAB" and "GMAB-P" regardless of the sheet topology. It is important to underline that the results of CMT-P samples are much better for both structured and flat samples. It should be mentioned that after the first week of the corrosion test the proportion of the corrosion damage for structured samples in zone 1 is almost two times higher than for flat samples.

The results of the corrosion test for zone 2 on the bottom seam side, which are shown in Figure 10, allow the comparison of the samples "GMAB-P" and "CMT-P." Both methods show similar low corrosion damage values of the brazed structured samples. For the flat specimens, the corrosion damage is a little higher. Zone 2, in contrast to zone

1, demonstrates a radically different correlation between flat and structured specimens.

To make a better evaluation of these conflicting results, the data were summarized for both zones (Figure 11). This diagram shows that the standard GMA brazing leads to the worst corrosion resistance of the brazed compound compared to the pulsed GMA and pulsed CMT brazing.

There is almost no difference between the GMAB-P and CMT-P brazing of flat sheet metals from the compound corrosion behavior point of view. For structured sheet metals, the CMT-P is the best brazing method according to the climate change test.

The lap joint has a special disadvantage. At the lap joint, a gap between the joined sheets forms, so that corrosive fluids can flow into the gap. This can cause corrosion in that place. The arrangement of the specimen with negative structure let the moisture leak, so that the corrosion medium could flow into the gap. That is why visual inspection of corrosion in the overlap interior has been done for every brazing variant after

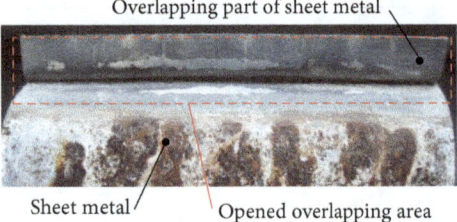

FIGURE 12: Picture of the gap surface in the lap joint of the GMAB brazed flat sheet metal after 5 weeks of the corrosion test.

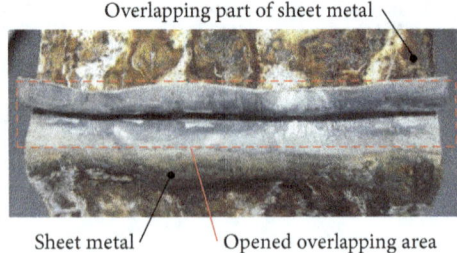

FIGURE 13: Picture of the gap surface in the lap joint of the GMAB-P brazed structured sheet metal after 5 weeks of the corrosion test.

5 weeks of the climate change test. For this purpose, the gap area has been opened manually by bending the overlapping part of the sheet metal.

First, the flat samples were reviewed. Figure 12 shows an example of the zinc coating condition between flat sheets after 5 weeks of the corrosion test. The "red" corrosion was not presented there. The undamaged zinc layer was retained in place. All flat samples had the same look.

For structured samples, no "red" rust was also found in the overlaps irrespective of the brazing methods (Figure 13). The main conclusion is the fact that atmospheric corrosion has no critical influence on the lap joint of galvanized sheet metals regardless of the plate topology and the brazing method.

4. Conclusions

The corrosion brazed joint behavior of galvanized structured sheet metals is affected by the brazing parameters. However, the greatest difference of the corrosion behavior was seen in the results for structured and flat sheet metals on the top side of the joint: the structured sheet metals are corroded much stronger than joints of the flat sheet metals. On the bottom side of the brazed compounds, the situation is quite the opposite. In general, the corrosion damage on the negative side of the brazed joint was significantly lower than on the top side (with the exception of flat samples "GMAB").

The process of conventional GMA brazing is characterized by a bigger heat input in comparison to the pulsed GMA brazing and the pulsed CMT process. Therefore, the specimens of the standard brazing process have reduced corrosion resistance.

The brazed top compound side shows the worst corrosion behavior by using pulsed GMA brazing method. The opposite

was true for the bottom side: the corrosion behavior was the best. Therefore, this brazing method is better in comparison with the standard brazing process.

The pulsed CMT brazing showed on the both top and bottom sides almost the best results. This method can be recommended as the preferred brazing variant for the joining of galvanized flat and structured sheet metals.

Regardless of the brazing process, the corrosion between the overlapping galvanized sheet metals is not critical. This can be explained by the fact that the corrosive medium does not flow into the gap. And that is why there is no big load on the zinc coating.

By using the bottom seam side of the sheet metal as a functional surface, the best way of arc brazing is pulsed GMA brazing. In all other cases, the pulsed CMT brazing is the best brazing process to join the galvanized structured sheet metals.

Competing Interests

The authors declare that they have no competing interests.

Acknowledgments

This work was supported by the Brandenburg Ministry of Science, Research and Culture (MWFK) as part of the International Graduate School at Brandenburg University of Technology (BTU) Cottbus-Senftenberg.

References

[1] P. Maass and P. Peissker, *Handbuch Feuerverzinken*, Wiley, Hoboken, NJ, USA, 2012.

[2] T. Shinoda, Y. Takeuchi, and T. Shimizu, "Effect of surface active element on porosity formation by GMA welding of zn plated steel for automobile industry," in *Exploiting Advances in Arc Welding Technology*, pp. 63–74, Woodhead Publishing, 1999.

[3] K. Graf, L. Schleuß, R. Ossenbrink, and V. Michailov, "Auswirkungen von strukturierten Fügepartnern auf geklebte Verbindungen," in *DVS-Berichte, Band 275*, pp. 472–477, DVS Media GmbH, Düsseldorf, Germany, 2011.

[4] R. Killing, "MIG brazing on thin sheet metals," in *Metallbau Heft 9/1999*, pp. 52–54, Callwey, Munich, Germany, 1999.

[5] P. Kah, R. Suoranta, and J. Martikainen, "Advanced gas metal arc welding processes," *International Journal of Advanced Manufacturing Technology*, vol. 67, no. 1-4, pp. 655–674, 2013.

[6] G. Trommer, "Cold Metal Transfer and pulsed arc welding processes separately used or in combination, in order to obtain an wide power range," *Sudura*, vol. 20, no. 1, pp. 25–29, 2010.

[7] E. Kornienko, R. Ossenbrink, and V. Michailov, "Corrosion resistance of zinc-coated structured sheet metals," *Corrosion Science*, vol. 69, pp. 270–280, 2013.

[8] A. Sterzing, *Bewertung von Leichtbaupotenzial und Einsatzfähigkeit wölbstrukturierter Feinbleche [Ph.D. thesis]*, Chemnitz University of Technology, 2005.

[9] M. Hoppe, *Umformverhalten strukturierter Feinbleche [Ph.D. thesis]*, Brandenburg University of Technology (BTU) Cottbus, Senftenberg, Germany, 2002.

[10] V. Malikov, R. Ossenbrink, B. Viehweger, and V. Michailov, "Analytical and numerical calculation of the force and power

requirements for air bending of structured sheet metals," *Key Engineering Materials*, vol. 473, pp. 602–609, 2011.

[11] L. Schleuß, Th. Richter, R. Ossenbrink, and V. Michailov, "Plasma cutting of structured sheet metals in comparison with laser beam cutting," *Journal of Materials Science and Engineering B*, vol. 5, no. 3-4, pp. 135–144, 2015.

[12] I. Sasse, L. Schleuß, R. Ossenbrink, and V. Michailov, "Cutting of structured sheet metals with high power fiber laser," in *Proceedings of the International Conference on Innovative Technologies (IN-TECH '11)*, pp. 485–488, 2011.

[13] L. Schleuß, K. Springer, J. Zoeke, R. Ossenbrink, and V. Michailov, "Qualifizierung von wärmearmen MSG-Schweißverfahren für Leichtbaukonstruktionen aus strukturierten Halbzeugen," in *JOIN-EX 2012: Internationaler Fachkongress der Schweiss-und Verbindungstechnik*, 2012.

[14] I. Sasse, L. Schleuß, R. Ossenbrink, and V. Michailov, "Joining of structured sheet metals—remote laser beam welding in comparison with resistance spot welding," *Welding and Cutting*, vol. 13, no. 1, pp. 43–47, 2014.

[15] L. Schleuß, R. Ossenbrink, and V. Michailov, "Widerstandspunktschweißen strukturierter bleche—schweißbereiche, prüfung, anwendung," in *12. Kolloquium Widerstandsschweißen und Alternative Verfahren, Halle (Saale)*, pp. 54–59, 2010.

[16] I. Sasse, L. Schleuß, R. Ossenbrink, and V. Michailov, "Fügen strukturierter Bleche—Remote-Laserstrahlschweißen im Vergleich zum Widerstandspunktschweißen," in *DVS-Berichte*, Band 296, pp. 98–103, DVS Media, Düsseldorf, Germany, 2013.

[17] L. Schleuß, A. Brobeck, R. Ossenbrink, R. Polzin, and V. Michailov, *Rollennahtschweißen Strukturierter Feinbleche*, Abschlussbericht. Leibniz Information Centre for Science and Technology (TIB), Hanover, Germany, 2015.

[18] S. Fritzsche, R. Ossenbrink, and V. Michailov, "Experimental characterisation of structured sheet metal," *Key Engineering Materials*, vol. 473, pp. 404–411, 2011.

[19] F. Kazak, L. Schleuß, R. Ossenbrink, V. Michailov, and S. Weiß, "Particularities of testing structured sheet metals in 3-point bending tests," *Materials Testing*, vol. 58, no. 6, pp. 495–500, 2016.

[20] E. Kornienko, *Korrosionsverhalten von strukturierten Blechen [Ph.D. thesis]*, Brandenburg University of Technology (BTU) Cottbus-Senftenberg, Senftenberg, Germany, 2014.

[21] A. Kliche, *MIG-Löten von strukturierten Blechen [Ph.D. thesis]*, Brandenburg University of Technology (BTU) Cottbus-Senftenberg, Senftenberg, Germany, 2013.

[22] VDA-Prüfblatt 621-415: Anstrichtechnische Prüfungen, Prüfung des Korrosionsschutzes von Kraftfahrzeug-lackierungen bei zyklisch wechselnder Beanspruchung, 1982.

[23] ISO, "Corrosion tests in artificial atmospheres—salt spray tests," EN ISO 9227, 2006.

[24] "Paints and varnishes—determination of resistance to humidity—part 2: procedure for exposing test specimens in condensation-water atmospheres, 2005," EN ISO 6270-2, 2005.

9

Corrosion and Leaching Behaviours of Sn-0.7Cu-0.05Ni Lead-Free Solder in 3.5 wt.% NaCl Solution

Jan-Ervin C. Guerrero,[1] Drexel H. Camacho ⓘ,[1] Omid Mokhtari,[2] and Hiroshi Nishikawa[2]

[1]*Chemistry Department, De La Salle University, 2401 Taft Avenue, 0922 Manila, Philippines*
[2]*Joining and Welding Research Institute, Osaka University, 11-1 Mihogaoka, Ibaraki, Osaka 567-0047, Japan*

Correspondence should be addressed to Drexel H. Camacho; drexel.camacho@dlsu.edu.ph

Academic Editor: Arvind Singh

The corrosion and leaching behaviour of a new ternary Sn-0.7Cu-0.05Ni alloy in 3.5 wt.% NaCl solution is reported herein. Potentiodynamic polarization measurements show that Sn-0.7Cu-0.05Ni has the highest corrosion rate. Results of the 30-day Sn leaching measurement show that Sn-Cu-Ni joint has slight decrease attributed to the formation of thin passivation film after 15 days. The leaching amounts of Sn are observed to be higher in solder joint than in solder alloy due to the galvanic corrosion happening on the surface. EDS and XRD results of the corroded surface confirm that the corroded product is made up of oxides of tin.

1. Introduction

Solder, a fusible metal alloy used to join metal pieces, is a staple material in the electronics industry. It functions as an adhesive or joining material to provide electrical continuity between the active silicon die, the substrate, and the printed wiring boards [1]. Solder alloys made up of tin and lead (Sn-Pb) predominate the manufacturing and electronics industries for years. However, due to the bad effects of lead and its compounds to human health and the environment [2], its use has been widely limited if not banned.

Global lead-free regulations started with the US banning lead (Pb) in gasoline additives, plumbing, and construction [3]. Japan pioneered the use of lead-free solders in the electronics industry and the European Union (EU) passed the restriction of the use of certain hazardous substances (ROHS) in electrical and electronic equipment banning the use of lead, mercury, and cadmium, among others [4]. Thus, there was a critical necessity to look for alternatives to Pb solders for the electronics industry. For solders to be a good environmentally benign substitute to Pb, it should be cost-effective and should mimic the properties of Pb, that is, low melting temperature (around 183°C), good mechanical, thermal, and electrical properties. Almost every lead-free solder

in the market is Sn-based solders such as Sn-Ag-Cu and Sn-Cu [1]. At present, there is no common standard for lead-free electronic products since each country or region has its own recommended lead-free solders. In the US, the National Electronic Manufacturing Initiative, Inc., (NEMI) recommends Sn-3.9Ag-0.6Cu as the lead-free solder. Japan, through the Japan Electronic and Information Technology Association (JEITA), recommends the use of Sn-3.0Ag-0.5Cu solder alloy, while the European Consortium recommends Sn-3.8Ag-0.5Cu as the lead-free solder. Overall, the most common type of lead-free solder (LFS) used worldwide is the alloy made up of Sn-Ag-Cu. Alternative Pb-free solder systems such as Sn-Ag, Sn-Bi, Sn-Zn, and Sn-Cu have been developed [1, 5, 6] but uncertainties in their integrity and reliability limited their use in consumer products [7]. New types of lead-free solders are being used by some companies in Japan; Panasonic and Hitachi use Sn-Ag-Bi, Sharp uses Sn-Bi, and Sony uses Sn-Ag-Bi-Cu and Sn-Ag-Bi-Ge solder alloys in their products [3, 4]. New lead-free solder alloys are being studied [8–10] and reliable solders for general and specific applications are still highly sought.

The ternary alloy Sn-0.7Cu-0.05Ni has been shown to be a potential Pb-free solder [11] where the wettability of the alloy was shown to be comparable with some other lead-free

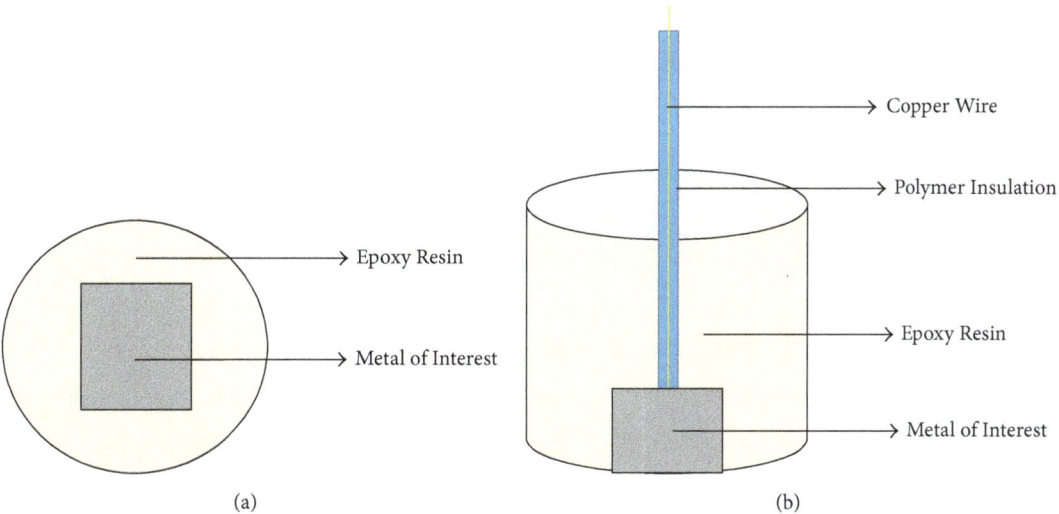

FIGURE 1: Schematic diagram of the working electrode. (a) Bottom view; (b) side view.

solders [12]. The low experimental wetting angle is a consequence of the addition of Ni in Sn-0.7% wt. Cu. Increasing the amount of Ni also decreases the void percentage and Cu_3Sn growth in Sn-Cu solder alloy [13]. The tensile properties and wettability in solder joints with Cu also improved as a result of adding 0.05 wt.% Ni [14]. The addition of 0.05 wt.% Ni to Sn-0.7Cu solder was shown to effectively reduce the formation of intermetallic compound (IMC) layer at the interface during the reflow process and for inhibiting the growth of IMC during the aging process [15].

To study further the potential of Sn-0.7Cu 0.05Ni solder alloy, there is a need to assess its corrosion behaviour as it is an important factor to consider in formulating new solder materials [16]. The presence of moisture and corrosive salts/ions triggers the corrosion activity of these solder metals affecting the form, fit, and function of the electronic device [17, 18]. Corrosion behaviours of lead-free solders are mainly studied using sodium chloride (NaCl) electrolyte to simulate the seawater condition using different types of lead-free solders [19–24]. Li et al. [20] reported that lead-free solders such as Sn-3.5Ag, Sn-0.7Cu, and Sn-3.8Ag-0.7Cu exhibit better corrosion resistance than Sn-Pb solder in 3.5 wt.% NaCl solution, where Sn-3.5Ag solder was found to be the most resistant among them. Lead-free solder exhibits better corrosion resistant because it exhibits lower passivation current density, lower corrosion current density after the breakdown of the passivation film, and a more stable passivation film on the surface compared to Sn-Pb solder. By investigating the corrosion behaviours of new lead-free solders, the fatigue life of the material can be predicted. A cursory survey of the literature reveals zero investigation on the corrosion behaviour of the new ternary Sn-0.7Cu-0.05Ni alloy. The goal of the present study is to investigate the corrosion resistance of Sn-0.7Cu-0.05Ni as compared to commercially used solder alloys and metal and study the leaching behaviour of tin in the alloys and in their corresponding joints in 3.5 wt% NaCl solution.

2. Materials and Methods

2.1. Materials

2.1.1. Preparation of Working Electrode. The compositions of the solder alloys used in the study are Sn-0.7 wt.% Cu-0.05 wt.% Ni, Sn-3.0 wt.% Ag-0.5 wt.% Cu and Sn-0.8 wt.% Cu, 60 wt.% Sn-40 wt.% Pb, and pure Sn metal. The alloys were cut using low-speed cut-off saw into square blocks (approximate dimension of $0.5 \times 0.5 \times 0.5$ cm). Each metal piece was attached to an insulated Cu wire by hand soldering using the same metal alloy to provide electrical connections and then cold-mounted using epoxy resin as shown in Figure 1. The specimens were ground using SiC paper/sand paper up to 1200 grit, rinsed with distilled water, and cleaned in an ultrasonic cleaner for 2 mins. The surface area exposed to the test solution was 0.5 ± 0.2 cm^2.

2.1.2. Preparation of Solder Alloy and Solder Joint for Leaching Measurement. Solder joints were prepared by joining a previously acid-cleaned (HNO_3/MeOH) copper plate ($5 \times 10 \times 2$ mm) with similarly sized lead-free solder plate using a solder paste. The coupled metals were heated on a hot plate at 230°C until the solders stuck on the Cu plate. For leaching experiments, the samples were prepared as shown in Figure 2.

2.1.3. Preparation of Solution. The corrosion test was carried out at room temperature in air-saturated aqueous solution of 3.5 wt.% NaCl prepared by dissolving 17.5 grams of analytical grade NaCl dissolved in deionized water to make a 500 mL solution.

2.2. Methods

2.2.1. Potentiodynamic Polarization Method–Tafel Plot. Electrochemical measurements were carried out in a single compartment cell using a standard three-electrode setup: Ag/AgCl (3 M KCl) as a reference electrode, a platinum sheet

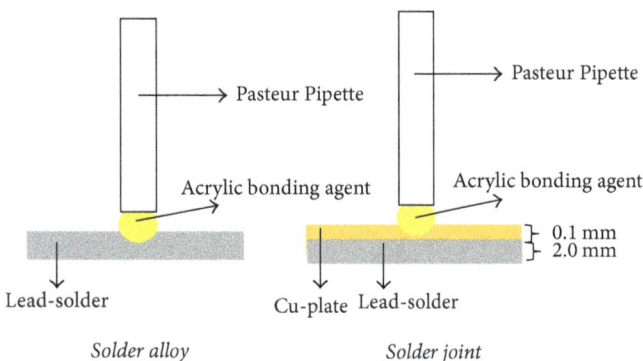

FIGURE 2: Preparation of samples for leaching measurements.

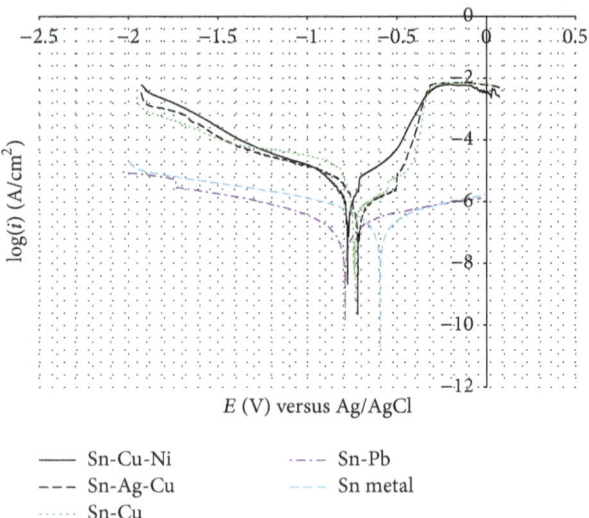

FIGURE 3: Potentiodynamic polarization curves of Sn and solder alloys in 3.5 wt.% NaCl solution under N_2 atmosphere. Scan range: from −1500 to 500 mV; scan rate: 1.00 mV/s; step potential: 0.45 mV.

(0.64 cm^2) as a counter electrode, and the lead-free solders as the working electrode. All the measurements were done under N_2 conditions. Metrohm Autolab PGSTAT 128 N was used as an electrochemical interface to control and record the potential. The samples were immersed in 250 mL corrosive medium inside the cell at room temperature for 180 s to attain a steady-state potential or open circuit potential (OCP). Potentiodynamic polarization curves were determined at −1500 to 500 mV range distinctive for the lead-free solder used on the study at an ASTM scan rate of 1.00 mV/s; step potential is at 0.45 mV and is presented in the form of typical polarization curves log I versus voltage. From the Tafel plot, the corrosion current (I_{CORR}), corrosion potential (E_{CORR}), polarization resistance (R_p), and corrosion rate (CR) of the metal alloys were calculated using the corrosion rate, Tafel slope method (Autolab application note COR02).

2.2.2. Leaching Measurements. For the leaching/dissolution measurement, the method by Cheng et al. [25] was adopted with slight modification. Small test cells containing 15 mL of test solutions were placed in a temperature-controlled oil bath at 45 ± 3°C. Every 3 days, oxygen was injected into the test solutions by gas flowmeter with a flow rate of 46 mL/min for 5 mins to keep the solution under low saturation of oxygen. Two mL of fresh test solutions was added to the test cell every 7 days or when the solution is reduced to keep the solution at 15 mL. The testing periods were done for 7, 15, and 30 days. After immersion for the subscribed period, 15 mL was taken and diluted into a 25 mL volumetric flask with deionized water. If there exists some precipitation in the test solution, sodium phosphate solution was used to dissolve the precipitates. The concentration of Sn in the diluted solution was analyzed using Atomic Absorption Spectrophotometer (Shimadzu AA-6300 Atomic Absorption Spectrophotometer) and the leaching amount per surface area of each element in the solders and their joints were calculated.

2.2.3. Surface Morphology and Elemental Analysis. The corroded surfaces of the samples were investigated using field-emission scanning electron microscopies (FESEM) (JEOL JSM-5310) and SEM (JEOL JSM-6500F) both equipped

with an energy dispersive X-ray spectroscopy (EDS). X-ray diffraction (XRD) measurements were performed with a diffractometer (Rigaku Ultima IV) using Cu-Kα radiation (λ = 1.5405 Å) at an accelerating voltage of 40 kV. The diffracted beam was scanned in steps of 0.02° across a 2θ range of 20–90°.

3. Results and Discussion

3.1. Potentiodynamic Polarization Method–Tafel Plot. The Sn-0.7CuNi0.05 (referred to herein as Sn-Cu-Ni) in 3.5 wt.% NaCl solution showed a Tafel plot positioned at (E_{corr}) −0.78 mV versus Ag/AgCl (Figure 3). As compared to Sn-3.0Ag-05Cu (referred to herein as Sn-Ag-Cu) and Sn-0.8Cu (referred to herein as Sn-Cu) alloys, there is an observed shift of corrosion potential E_{corr} of lead-free solders to a less negative value following the sequence Sn-Cu-Ni < Sn-Cu < Sn-Ag-Cu. During the anodic polarization process, a stable passivation film is formed on the surface of the metal alloys (*vide infra*). This passivation film determines the corrosion behaviour of the solder in the given media. The shift to a less negative potential signifies a formation of a more stable passivation film [26], which protects the solders and increases their corrosion resistance. The polarization resistance (R_p) data (Table 1) showed that Sn-Pb solder has the highest R_p (805.250 KΩ) and lowest corrosion rate (0.1949 × 10^{-2} mm/yr) compared to the lead-free solders. Between the three lead-free solder alloys, Sn-Cu-Ni showed the highest corrosion rate and Sn-Ag-Cu showed the lowest corrosion rate in 3.5 wt.% NaCl solution.

3.1.1. Microstructure Characterization of the Lead-Free Solders. The microstructure of solders before potentiodynamic polarization (Figure 4(a)) is a smooth surface showing the polishing lines and EDS analysis confirms the elemental

TABLE 1: Summary of corrosion parameters of metal alloys in 3.5 wt.% NaCl solution.

Sample	E_{corr} (V) versus Ag/AgCl	I_{corr} (μA/cm^2)	R_p (kΩ)	CR (mm/yr) $\times 10^{-2}$
Sn-Cu-Ni	−0.78	8.98	14.94	4.87
Sn-Ag-Cu	−0.73	3.06	43.86	1.71
Sn-Cu	−0.75	3.91	28.26	2.07
Sn-Pb	−0.79	0.39	805.25	0.19
Sn	−0.59	0.55	406.02	0.23

E_{corr}: corrosion potential, I_{corr}: corrosion current density, R_p: polarization resistance, and CR: corrosion rate.

TABLE 2: Surface element concentration of different solders before potentiodynamic polarization in 3.5 wt.% NaCl solution.

	Surface element composition (atom%)				
	Sn	Ag	Cu	Ni	Pb
Sn-Cu-Ni	98.59	-	1.27	0.14	-
Sn-Ag-Cu	96.74	2.74	0.52	-	-
Sn-Cu	98.97	-	1.38	-	-
Sn-Pb	59.03	-	-	-	40.97
Sn	100.00	-	-	-	-

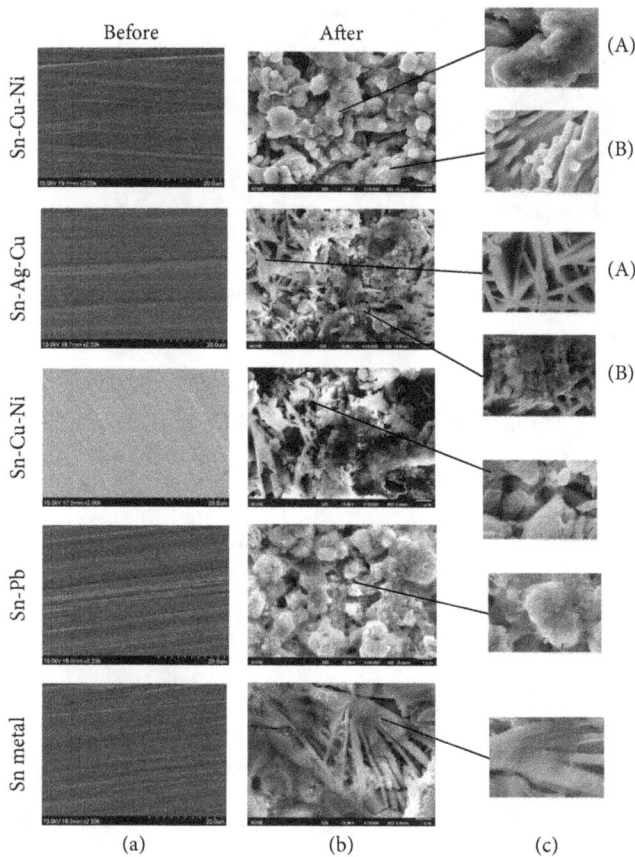

FIGURE 4: The microstructure of the metal surfaces on different solders before (a), after (b), and distinct morphologies on specified points (c) after potentiodynamic polarization tests in 3.5 wt.% NaCl solution.

composition of each solder (Table 2). After potentiodynamic polarization in 3.5 wt.% NaCl solution (Figure 4(b)), the microstructure of corroded Sn-Cu-Ni alloy had a porous flake-like surface while the Sn-Ag-Cu and Sn-Cu solders showed fibrous network oriented randomly on the surface in agreement with the observations of Li et al., [20]. Notable is the flake-like microstructures in Sn-Cu-Ni, which is similar in some respect to the Sn-Pb alloy. The EDS analysis on the surface of the corroded samples on each specified point was determined (Figure 4(c) and Table 3) and revealed the presence of Sn, O, and Cl indicating that the corrosion products are composed of the oxides and chlorides of tin. In conventional electrochemical reaction, Sn acts as an anode and reacts with Cl$^-$ from the medium to form SnCl$_2$, which results in pitting and severed dissolution of Sn [27].

The cross-section of the metals after potentiodynamic polarization (Figure 5) showed visible corrosion layers distinct from the bulk metal. EDS analysis of the cross-section area (Table 4) for both lead and lead-free solders showed the top layer having similar corrosion products composed of Sn-rich and O-rich areas with <1% Cl. The Sn-Pb solder showed an outer layer rich in O while the inner layer was rich in Sn. The formation of oxides on the surface of the Sn-Pb solder indicates the presence of a stable passivation film that protects the metal from corroding. This explains why Sn-Pb solder has a better corrosion resistance than lead-free solders. In the case of the lead-free solders, since they have a different composition than Sn-Pb, the thin layer of oxides that formed on the surface of the alloy is not enough to protect the metal from the electrochemical process.

TABLE 3: Surface element concentration of different solders *after* potentiodynamic polarization in 3.5 wt.% NaCl solution on specific points as indicated in Figure 4(c).

	Surface element concentration (atom%)						
	Sn	Ag	Cu	Pb	Ni	Cl	O
Sn-Cu-Ni (A)	20.40	-	1.90	-	-	3.50	74.20
Sn-Cu-Ni (B)	56.70	-	0.80	-	-	0.20	42.30
Sn-Ag-Cu (A)	23.60	-	-	-	-	8.70	67.70
Sn-Ag-Cu (B)	32.30	-	0.20	-	-	12.10	55.40
Sn-Cu	20.50	-	0.10	-	-	10.70	68.70
Sn-Pb	28.40	-	-	-	-	0.60	71.00
Sn	29.90	-	-	-	-	15.00	55.10

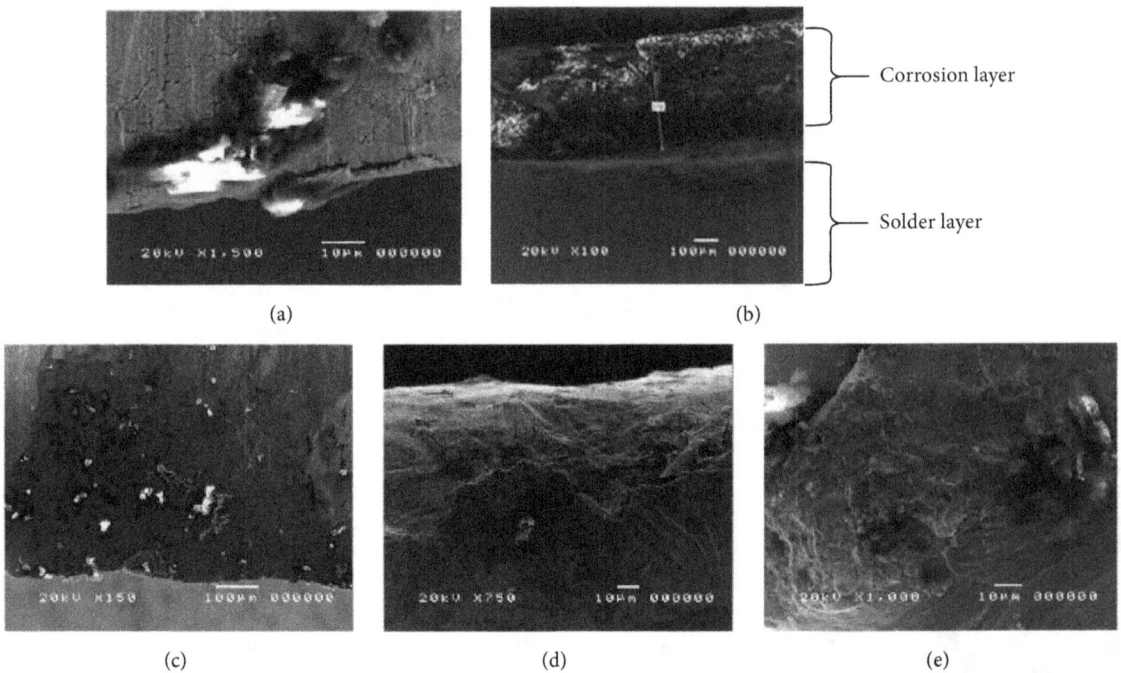

FIGURE 5: Cross-section image of the solders after potentiodynamic polarization test in 3.5 wt.% NaCl solution. (a) Sn; (b) Sn-Pb; (c) Sn-Ag-Cu; (d) Sn-Cu-Ni; (e) Sn-Cu.

3.1.2. Phase Composition Analysis of Corrosion Product on the Surface of the Solder Sample.

The XRD diffractograms of corrosion products (Figure 6) showed corresponding XRD peaks attributed to tin oxides (SnO and SnO_2) and some chlorides. When compared with the XRD results of Yan and Xian [28] peaks for tin hydroxide ($Sn(OH_2)$) were also observed particularly for Sn-Pb and Sn metal [25]. These oxides effectively protect the metal. The oxides, however, were not visibly detected in Pb-free solders indicating nil or minimal formation of the passivation layer.

During potentiodynamic polarization measurements, the reduction of oxygen in the neutral aqueous solution (1) initially occurs [20]

$$4e^- + O_2 + 2H_2O \longrightarrow 4OH^-. \tag{1}$$

Once the current reaches 10 mA/cm^2, bubbles were observed from the solution caused by the evolution of hydrogen from the cathode.

$$2H_2O + 2e^- \longrightarrow H_2 + 2OH^-. \tag{2}$$

The following plausible anodic reactions would occur [17, 19]:

$$Sn \longrightarrow Sn^{2+} + 2e^- \tag{3}$$

$$Sn^{2+} \longrightarrow Sn^{4+} + 2e^- \tag{4}$$

$$Sn + 2OH^- - 2e^- \longrightarrow Sn(OH)_2 \tag{5}$$

$$Sn + 2OH^- - 2e^- \longrightarrow SnO + H_2O \tag{6}$$

$$2Sn + O_2 + 6H_2O \longrightarrow 2Sn(OH)_2 \tag{7}$$

$$Sn(OH)_2 + 2OH^- - 2e^- \longrightarrow Sn(OH)_4 \tag{8}$$

$$2Sn^{2+} + 6H_2O + O_2 \longrightarrow 2Sn(OH)_4 + 4H^+. \tag{9}$$

Formation of tin oxides is thermodynamically favourable thus tin hydroxides can dehydrate easily to form SnO and SnO_2 [25, 29]. Although not detected in XRD due to its trace concentration relative to Sn, the potential corrosion products

TABLE 4: Cross-section element concentration of different solders after Potentiodynamic polarization tests in 3.5 wt.% NaCl solution.

| | Cross-section element concentration (atom%) | | | | | | |
	Sn	Ag	Cu	Ni	Pb	Cl	O
Sn-Cu-Ni							
Top layer	63.99	-	0.60	0.22	-	0.92	33.91
Inner layer	98.59	-	1.27	0.14	-	-	-
Sn-Ag-Cu							
Top layer	40.95	1.12	0.17	-	-	0.15	57.27
Inner layer	97.36	2.01	0.63	-	-	-	-
Sn-Cu							
Top layer	64.58	-	0.72	-	-	0.60	33.35
Inner layer	99.30	-	0.70	-	-	-	-
Sn-Pb							
Top layer	6.85	-	-	-	3.91	0.91	88.03
Inner layer	60.16	-	-	-	39.84	-	-
Sn							
Top layer	32.62	-	-	-	-	0.37	66.41
Inner layer	99.85	-	-	-	-	0.15	-

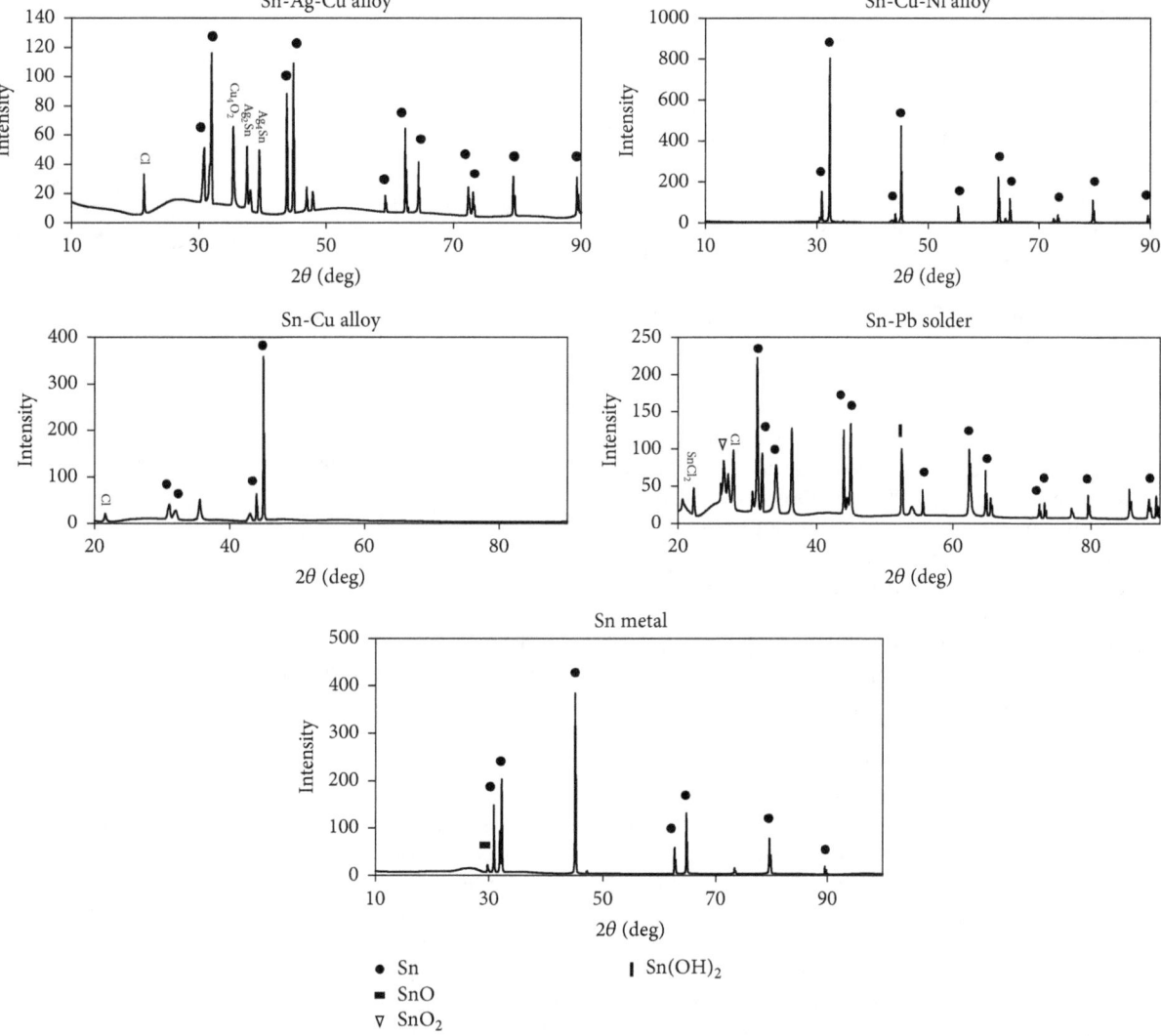

FIGURE 6: XRD patterns of surface product on different solder material after potentiodynamic polarization test.

of copper and nickel are metal chlorides. Corrosion of copper is dependent on the presence of chloride ions either through direct formation of cupric chloride

$$Cu \longleftrightarrow Cu^+ + e^- \qquad (10)$$

$$Cu^+ + 2Cl^- \longleftrightarrow CuCl_2^- \qquad (11)$$

or through the electro-dissolution of copper [30]

$$Cu + Cl^- \longleftrightarrow CuCl + e^- \qquad (12)$$

$$CuCl + Cl^- \longleftrightarrow CuCl_2^-. \qquad (13)$$

The anodic polarization of copper alloys in NaCl solution lowers the fracture stresses as measured in slow strain experiments making copper alloys under cyclic stresses to have lower service lives in chloride solutions [31]. Thus, chlorides penetrate into the material crevices allowing for corrosion of susceptible atoms such as copper to occur. For nickel corrosion, the initial step is nickel hydration to facilitate dissolution and this can occur when water transport into the material is enhanced. The transport of water is dependent on chloride environment. This was observed when we compared the potentiodynamic polarization of Sn-Cu-Ni alloy in neutral chloride-containing medium ($E_{corr} = -0.776$ V versus Ag/AgCl, $I_{corr} = 8.982$ μA/cm^2) and in acidic chloride-free electrolyte (0.1 M HNO$_3$; $E_{corr} = -0.356$ V versus Ag/AgCl, $I_{corr} = 4.34$ μA/cm^2). A sudden shift of corrosion potential to a less negative value is observed attributed to the formation of the passivation film. The absence of chloride in HNO$_3$ medium promotes barrier creation preventing hydration and corrosion to occur. This suggests that Sn-Cu-Ni alloy has high corrosion resistance in an acidic chloride-free environment. Chlorides even in small amounts can break the protective films initiating the occurrence of corrosion.

3.2. Soldering Properties and Leaching Measurements. During the preparation of the solder joints with copper substrates, a better surface finish of the joints was observed when Sn-Cu-Ni solder was used compared to Sn-Cu alloys. The design of adding Ni to common Sn-Cu alloy improved the soldering property [32]. These observations were consistent with the unique morphology of the intermetallic compound (IMC) layer reported by Harcuba et al. [33]. The IMC was described as noncompact islands of solder entrapped in the IMC phase and that the IMC layer contains more Ni than the solder. Nickel addition promoted the significant acceleration of the growth kinetics of the IMC layer.

The solder joints and the raw Pb-free solder alloys were subjected to corrosive media immersion to determine the leaching behaviour (Figure 7). The amount of Sn leached correlates with the corrosion rate of the alloys [25, 29, 34]. Since the results of potentiodynamic polarization test showed that all of the three Pb-free solders had higher corrosion rate than the Sn-Pb solder and Sn metal, leaching amounts of Sn were done only for the Pb-free solders. The leaching amounts of Sn after immersion in 3.5 wt.% NaCl solution after 30 days are observed to be greater for solder joints compared to its corresponding alloy. Sn-Ag-Cu alloy had the highest leaching

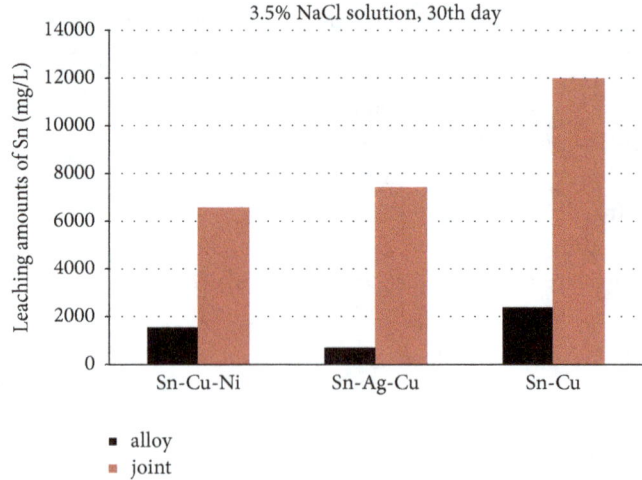

FIGURE 7: Atomic Absorption Spectroscopy (AAS) results on Tin (Sn) leaching from lead-free solder alloy and joint after 30 days.

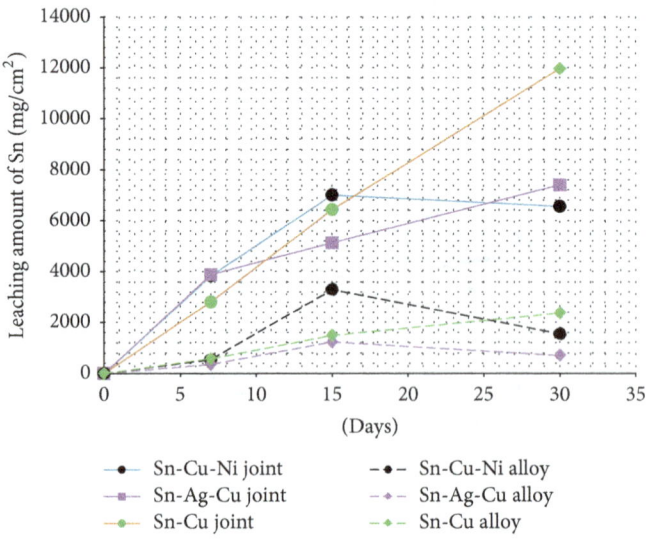

FIGURE 8: Leaching kinetics of Sn from Sn-Ag-Cu, Sn-Cu-Ni, and Sn-Cu alloy and their joints in 3.5 wt.% NaCl solution.

of Sn for the solder alloys followed by Sn-Cu alloy while Sn-Cu-Ni had the lowest leaching amount after 30 days. For solder joints, leached Sn follows the order Sn-Cu > Sn-Ag-Cu > Sn-Cu-Ni.

Leaching kinetics of Sn from the metals (Figure 8) show that solder joints gave a higher amount of leached Sn compared to their alloy counterparts. The use of Cu-substrate in the solder joint setup affects the leaching behaviour of Sn. This is because dissimilar metals having different oxidation potentials in contact with each other experience galvanic corrosion [25, 29], which does not occur in solder alloys alone. According to Lao et al., [29] the current density of the solder joint is almost twofold higher than its corresponding solder alloy. Thus, the galvanic cell (anode = Sn; cathode = Cu) accelerates the leaching amount of Sn from the joint.

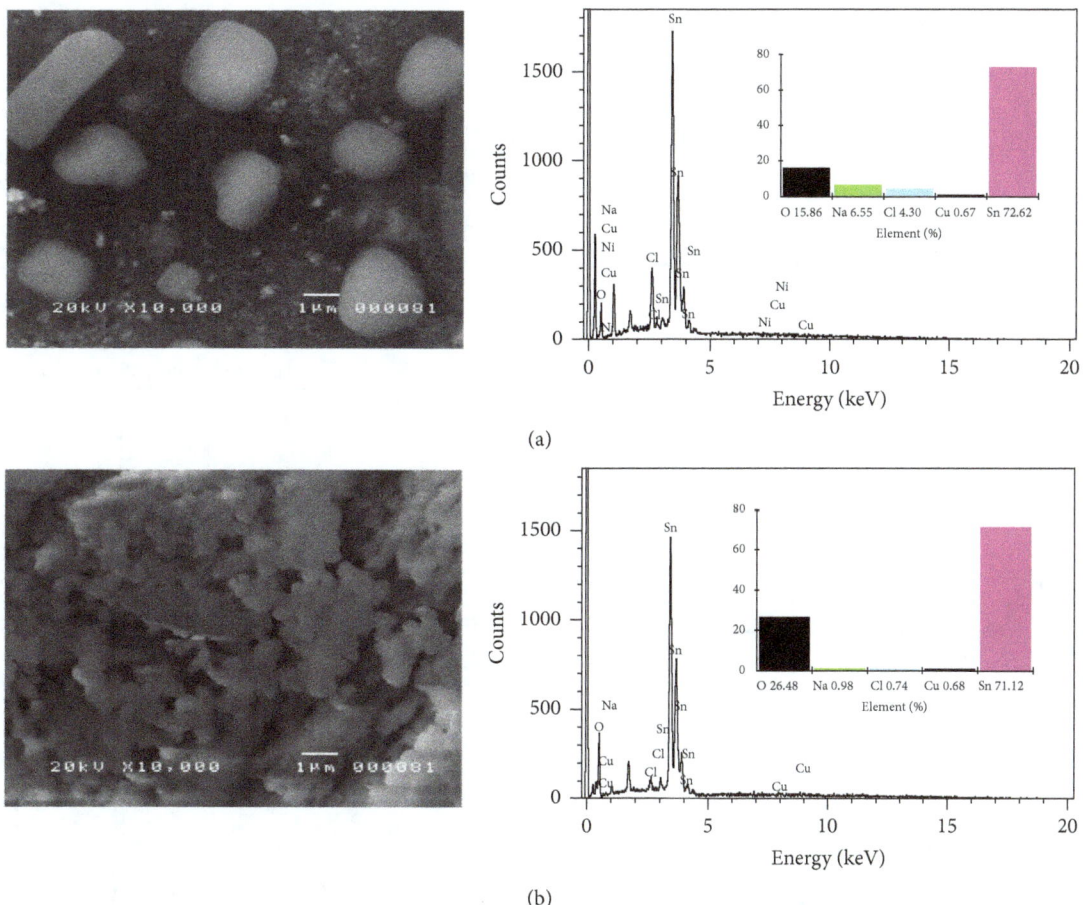

FIGURE 9: Surface morphology analysis of Sn-Cu-Ni (a) solder alloy and (b) solder joint and their corresponding EDS spectra after 30-day leaching measurements in 3.5 wt.% NaCl solution.

The microstructure analysis of Sn-Ni-Cu solder after 30-day immersion in NaCl solution (Figure 9) showed a more corroded surface for the joint and microstructure is composed of platelet-like materials on the surface. EDS analysis on the surface showed high amounts of Sn and O suggesting that the corrosion products are composed primarily of oxides of Sn and small amounts of chlorides. This is further confirmed by the detection of XRD patterns specific to SnO_2 (Figure 10). Similarly, corresponding peaks for $Sn(OH)_2$ were detected from Yan and Xian [28].

3.2.1. Microstructure of Corrosion Products on Solder Surface. The microstructures of the corrosion surface were monitored as the exposure to the corrosive media increases. Generally, as the alloy (Figure 11) was exposed to the salt solution, the surface changed dramatically from smooth to porous with the formation of plate-like structures attributed to the weakly bonded corrosion products that form on the surface. These compounds, which have been identified in EDS and XRD as oxides of tin are typical corrosion products. The formation of these compounds is a result of a charge transfer reaction between Sn metal, salt ion, atmospheric O_2, and H_2O during anodic polarization following the mechanism proposed by Mohran et al. [35]. These corrosion products

are weakly held by the bulk metal and can easily chip-off into the solution in agreement with the leaching experiments. The microstructural changes that can be observed for Sn-Cu-Ni are the noticeable degree of high corrosion product (platelet-like structure) after 7-day exposure compared to Sn-Ag-Cu and Sn-Cu. The corrosion products are formed along the cracks, which allow the water to seep-in triggering the corrosion reaction. As the exposure time increases, the entire surface was covered with corrosion products.

For the solder joints, the microstructure changes were likewise monitored (Figure 12). A more dramatic morphology can be observed characterized by branched crystallites, sponge-like structure, networked branches, and platelet-like materials. The formation of these microstructures occurs within 7-day exposure and increases gradually as time increases. This is the consequence of galvanic reaction that occurs since the solder joints are in contact with the Cu metal as substrate. The galvanic reaction triggers the very fast corrosion process. The EDS analysis of these microstructures reveals O and Sn-rich surface indicating oxides of Sn formation, which was further confirmed by XRD as SnO_2 (Figure 10).

Generally, the surfaces of the alloys show some corrosion product as evidenced by the plate-like structures. The surfaces

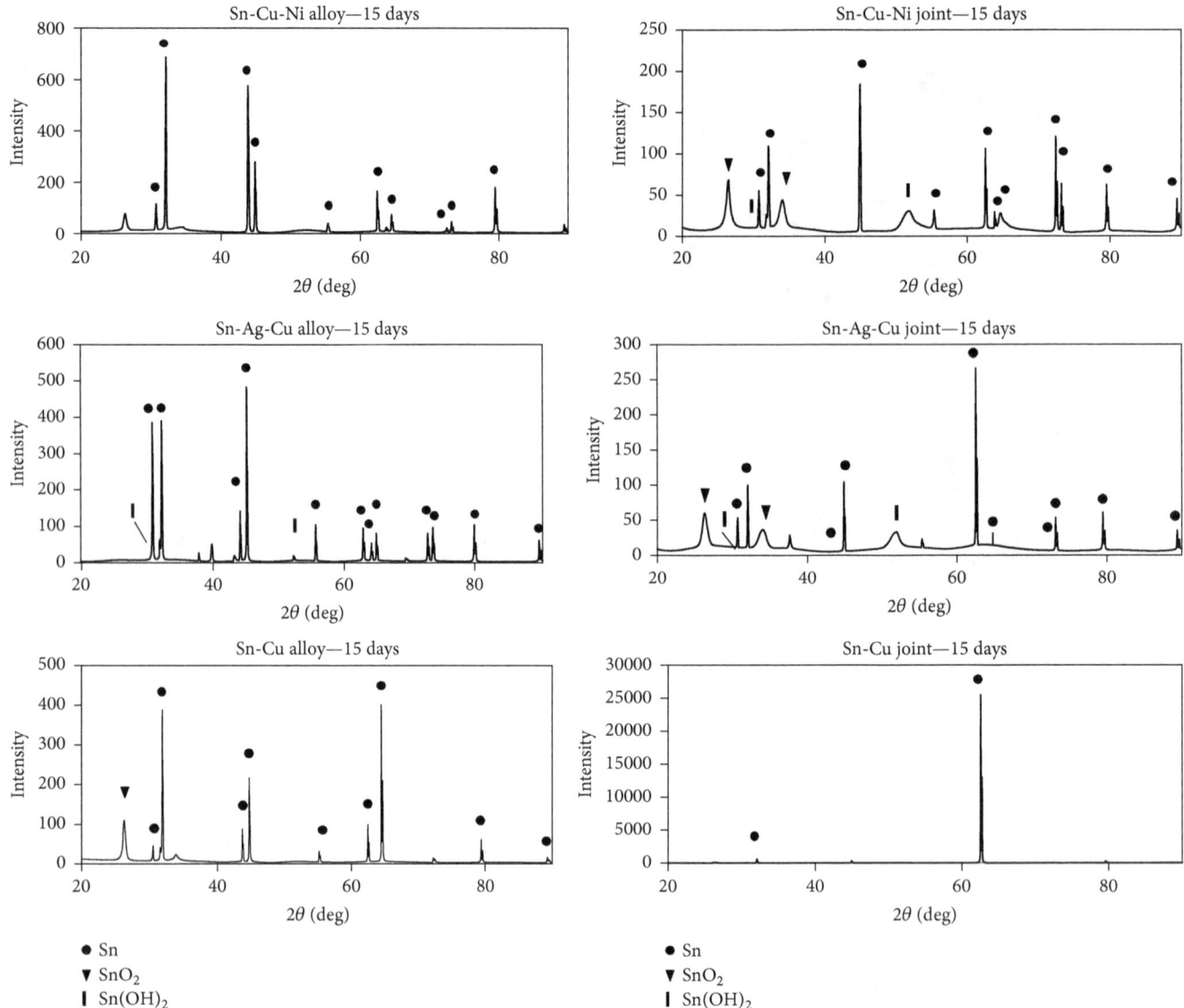

FIGURE 10: XRD patterns of surface product of Sn-Ag-Cu, Sn-Cu-Ni, and Sn-Cu solder alloy and joint after 15-day immersion in 3.5 wt.% NaCl solution.

are covered by a small degree of corrosion layer suggesting minimal corrosion on the alloy surface. However, the surface of the joint showed a high degree of corrosion with the entire surface covered with corrosion products forming nanocrystals and sponges. Similar morphology was reported by Lao et al., [29] in the leaching measurement of Sn-0.75Cu solder alloy and joint in simulated soil solutions. The corrosion layer formed on the surface of the alloy forms a protecting film that has a shielding effect on the Sn ion transport between the solder and the solution. However, in the case of the solder joint, the loosely bound corrosion layer can break-off easily. The corrosion products that made up the surface would no longer protect the substrate resulting in plenty of leaching amounts of Sn coming from the solder joint.

The result of potentiodynamic polarization and leaching tests in 3.5 wt.% NaCl solution is conflicting. Results for the Tafel extrapolation show that Sn-Cu-Ni has the highest

corrosion rate, while the 30-day immersion of the solder alloy and joint in NaCl solution showed the lowest amount of leached Sn for all the solders understudied. This was attributed to the rapid analysis in an electrochemical process, where corrosion is forcedly induced on the surface of the metal by applying voltage. A rapid measurement can influence the formation of the passivation film on the surface of the metal. On the other hand, the long exposure time with the corrosive media in leaching experiments allows an actual electrochemical process to occur and measured progressively. In these conditions, the formation of corrosion product is gradual and there is longer timeframe for the stabilization and formation of the passivation film.

These results are significant since, in electronic processes, the solder alloy is not used alone. Rather it is used in contact with other metals mostly Cu in most electronic substrates. These lead-free solders therefore, once in a corrosive media

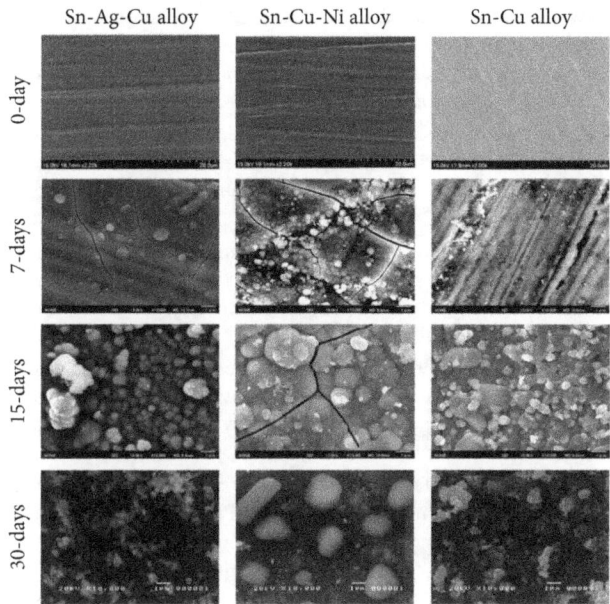

FIGURE 11: SEM image of lead-free solder alloy after 7, 15, and 30 days leaching measurement in 3.5 wt.% NaCl solution.

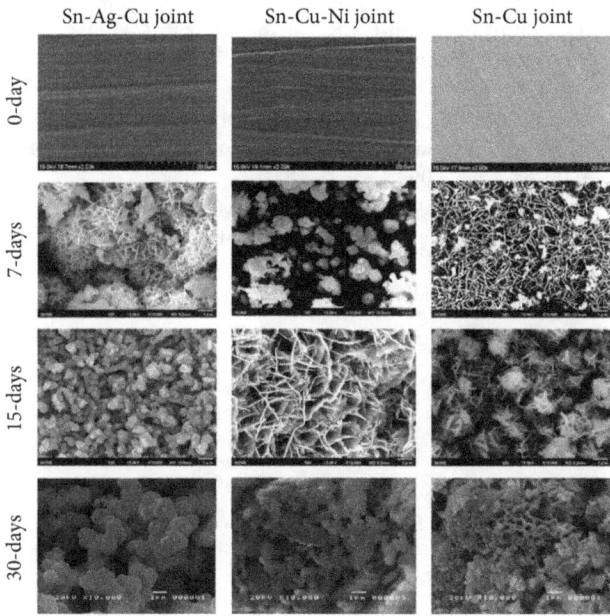

FIGURE 12: SEM image of lead-free solder joint after 7, 15, and 30 days leaching measurement in 3.5 wt.% NaCl solution.

such as aqueous NaCl atmosphere, are prone to corrosion even within 7 days of exposure. The corrosion reaction is triggered by the galvanic reaction that occurs when this happens inside an electronic gadget. The interconnections and electrical connectivity are lost affecting the form, features, and function of the device.

4. Conclusions

The corrosion behaviour of Sn-Cu-Ni lead-free solder was described for the first time. In 3.5 wt.% NaCl environment,

potentiodynamic polarization test revealed that the new Sn-Cu-Ni solder has higher corrosion rate compared to Sn-Cu and Sn-Ag-Cu. However, longer exposure of the Sn-Cu-Ni alloy and joint to the corrosive medium for up to 30 days showed the lowest leaching rate of Sn compared to Sn-Cu and Sn-Ag-Cu solders. Generally, alloys have lower leaching rate compared to the corresponding joint due to the galvanic reaction occurring in the joint setup. For chloride-containing solution, the corrosion and leaching behaviour of Sn-Ag-Cu is better compared to Sn-Ni-Cu and Sn-Cu. Further study is underway in acidic and basic media and will be reported elsewhere.

Conflicts of Interest

There are no conflicts of interest related to this paper.

Acknowledgments

This work was supported by the Accelerated Science and Technology Human Resource Development Program (ASTHRDP) of the Department of Science and Technology, Science Education Institute (DOST-SEI). Jan-Ervin C. Guerrero is grateful for the Japan Sakura Science Program.

References

[1] M. Abtew and G. Selvaduray, "Lead-free solders in microelectronics," *Materials Science and Engineering: A Structural Materials: Properties, Microstructure and Processing*, vol. 27, no. 5, pp. 95–141, 2000.

[2] S. Tong, Y. von Schirnding, and T. Prapamontol, "Environmental lead exposure: a public health problem of global dimensions," *Bulletin of the World Health Organization*, 2000, http://www.who.int/bulletin/archives/78(9)1068.pdf.

[3] J. H. Lau and K. Liu, "Global Trends in Lead-free soldering," *International Journal of Advanced Packaging Technology*, vol. 13, no. 2, pp. 25–28, 2004.

[4] S. Garesan and M. Pecht, *Lead Free Electronics*, IEEE Press/John Wiley & Sons. Inc, Hoboken, NJ, USA, 2006.

[5] K. Suganuma, "Advances in lead-free electronics soldering," *Current Opinion in Solid State & Materials Science*, vol. 5, no. 1, pp. 55–64, 2001.

[6] N. Chawla, "Thermomechanical behaviour of environmentally benign Pb-free solders," *International Materials Reviews*, vol. 54, no. 6, pp. 368–384, 2009.

[7] K. Zeng and K. N. Tu, "Six cases of reliability study of Pb-free solder joints in electronic packaging technology," *Materials Science and Engineering: Reports*, vol. 38, no. 2, pp. 55–106, 2002.

[8] D. K. Mu, S. D. McDonald, J. Read, H. Huang, and K. Nogita, "Critical properties of Cu6Sn5 in electronic devices: recent progress and a review," *Current Opinion in Solid State & Materials Science*, vol. 20, no. 2, pp. 55–76, 2016.

[9] H. R. Kotadia, P. D. Howes, and S. H. Mannan, "A review: on the development of low melting temperature Pb-free solders," *Microelectronics Reliability*, vol. 54, no. 6-7, pp. 1253–1273, 2014.

[10] G. Zeng, S. McDonald, and K. Nogita, "Development of high-temperature solders: review," *Microelectronics Reliability*, vol. 52, no. 7, pp. 1306–1322, 2012.

[11] H. Nishikawa, J. Y. Piao, and T. Takemoto, "Effect of Ni addition on interfacial reaction between Sn-Cu solder and Cu base metal," in *Joining of Advanced and Specialty Materials VII, Proceedings from Materials Solutions 2004 on Joining of Advanced and Specialty Materials*, T. J. Lienert, Ed., vol. 35, pp. 208–211, Columbus, OH, USA, October 2006.

[12] B. L. Silva, N. Cheung, A. Garcia, and J. E. Spinelli, "Evaluation of solder/substrate thermal conductance and wetting angle of Sn–0.7 wt%Cu–(0–0.1 wt%Ni) solder alloys," *Materials Letters*, vol. 142, pp. 163–167, 2015.

[13] J.-W. Yoon, B.-I. Noh, B.-K. Kim, C.-C. Shur, and S.-B. Jung, "Wettability and interfacial reactions of Sn-Ag-Cu/Cu and Sn-Ag-Ni/Cu solder joints," *Journal of Alloys and Compounds*, vol. 486, no. 1-2, pp. 142–147, 2009.

[14] L. Yang, J. Ge, Y. Zhang, J. Dai, H. Liu, and J. Xiang, "Investigation on the Microstructure, Interfacial IMC Layer, and Mechanical Properties of Cu/Sn-0.7Cu-xNi/Cu Solder Joints," *Journal of Electronic Materials*, vol. 45, no. 7, pp. 3766–3775, 2016.

[15] H. Nishikawa, J. Y. Piao, and T. Takemoto, "Interfacial reaction between Sn-0.7Cu (-Ni) solder and Cu substrate," *Journal of Electronic Materials*, vol. 35, no. 5, pp. 1127–1132, 2006.

[16] S. Farina and C. Morando, "Comparative corrosion behaviour of different Sn-based solder alloys," *Journal of Materials Science: Materials in Electronics*, vol. 26, no. 1, pp. 464–471, 2014.

[17] K. Pietrzak, M. Grobelny, K. Makowska et al., "Structural aspects of the behavior of lead-free solder in the corrosive solution," *Journal of Materials Engineering and Performance*, vol. 21, no. 5, pp. 648–654, 2012.

[18] M. Reid and L. F. Garfias-Mesias, "Corrosion of electronics: lead-free initiatives," in *Uhlig's Corrosion Handbook*, R. W. Revie, Ed., pp. 565–570, John Wiley & Sons, Inc, Hoboken, NJ, USA, 3rd edition, 2011.

[19] K.-L. Lin and T.-P. Liu, "The electrochemical corrosion behaviour of Pb-free Al-Zn-Sn solders in NaCl solution," *Materials Chemistry and Physics*, vol. 56, no. 2, pp. 171–176, 1998.

[20] D. Li, P. P. Conway, and C. Liu, "Corrosion characterization of tin-lead and lead free solders in 3.5 wt.% NaCl solution," *Corrosion Science*, vol. 50, no. 4, pp. 995–1004, 2008.

[21] W. R. Osório, E. S. Freitas, J. E. Spinelli, and A. Garcia, "Electrochemical behavior of a lead-free Sn-Cu solder alloy in NaCl solution," *Corrosion Science*, vol. 80, pp. 71–81, 2014.

[22] M. Liu, W. Yang, Y. Ma, C. Tang, H. Tang, and Y. Zhan, "The electrochemical corrosion behavior of Pb-free Sn-8.5Zn-XCr solders in 3.5 wt.% NaCl solution," *Materials Chemistry and Physics*, vol. 168, pp. 27–34, 2015.

[23] A. Kamarul Asri and E. Hamzah, "Corrosion behaviour of lead-free and Sn-Pb solders in 3.5wt% NaCl," *Advanced Materials Research*, vol. 686, pp. 250–260, 2013.

[24] U. S. Mohanty and K.-L. Lin, "Electrochemical corrosion behaviour of Pb-free Sn-8.5Zn-0.05Al-XGa and Sn-3Ag-0.5Cu alloys in chloride containing aqueous solution," *Corrosion Science*, vol. 50, no. 9, pp. 2437–2443, 2008.

[25] C. Q. Cheng, F. Yang, J. Zhao, L. H. Wang, and X. G. Li, "Leaching of heavy metal elements in solder alloys," *Corrosion Science*, vol. 53, no. 5, pp. 1738–1747, 2011.

[26] A. Wierzbicka-Miernik, J. Guspiel, and L. Zabdyr, "Corrosion behavior of lead-free SAC-type solder alloys in liquid media," *Archives of Civil and Mechanical Engineering*, vol. 15, no. 1, pp. 206–213, 2015.

[27] A. Sharma, S. Das, and K. Das, "Electrochemical corrosion behavior of CeO$_2$ nanoparticle reinforced Sn-Ag based lead free nanocomposite solders in 3.5wt.% NaCl bath," *Surface and Coatings Technology*, vol. 261, pp. 235–243, 2015.

[28] Z. Yan and A.-P. Xian, "Corrosion of Ga-doped Sn-0.7Cu solder in simulated marine atmosphere," *Metallurgical and Materials Transactions A: Physical Metallurgy and Materials Science*, vol. 44, no. 3, pp. 1462–1474, 2013.

[29] X.-D. Lao, C.-Q. Cheng, X.-H. Min et al., "Corrosion and leaching behaviors of Sn-based alloy in simulated soil solutions," *Transactions of Nonferrous Metals Society of China*, vol. 26, no. 2, pp. 581–588, 2016.

[30] F. Arjmand and A. Adriaens, "Influence of pH and chloride concentration on the corrosion behavior of unalloyed copper in NaCl solution: a comparative study between the micro and macro scales," *Materials*, vol. 5, no. 12, pp. 2439–2464, 2012.

[31] B. D. Craig and D. S. Anderson, Eds., *Handbook of Corrosion Data*, ASM International, Materials Park, OH, USA, 2nd edition, 1995.

[32] T. Ventura, C. M. Gourlay, K. Nogita, T. Nishimura, M. Rappaz, and A. K. Dahle, "The influence of 0-0.1 wt.% Ni on the microstructure and fluidity length of Sn-0.7Cu-xNi," *Journal of Electronic Materials*, vol. 37, no. 1, pp. 32–39, 2008.

[33] P. Harcuba, M. Janeček, and M. Slámová, "The effect of Cu and Ni on the structure and properties of the IMC formed by the reaction of liquid Sn-Cu based solders with Cu substrate," in *Proceedings of Contributed Papers (WDS '08)*, part III, pp. 220–224, Prague, Czech, June 2008.

[34] M. Mori, K. Miura, T. Sasaki, and T. Ohtsuka, "Corrosion of tin alloys in sulfuric and nitric acids," *Corrosion Science*, vol. 44, no. 4, pp. 887–898, 2002.

[35] H. S. Mohran, A.-R. El-Sayed, and H. M. Abd El-Lateef, "Anodic behavior of tin, indium, and tin-indium alloys in oxalic acid solution," *Journal of Solid State Electrochemistry*, vol. 13, no. 8, pp. 1279–1290, 2009.

10

Corrosion Behavior of Carbon Steels in CCTS Environment

M. Cabrini, S. Lorenzi, and T. Pastore

Department of Engineering and Applied Sciences, University of Bergamo, Viale Marconi 5, Dalmine, 24044 Bergamo, Italy

Correspondence should be addressed to S. Lorenzi; sergio.lorenzi@unibg.it

Academic Editor: Ksenija Babic

The paper reports the results of an experimental work on the effect of steel microstructures on morphology and protectiveness of the corrosion scale formed in water saturated by supercritical CO_2. Two HSLA steels were tested. The microstructures were modified by means of different heat treatments. Weight loss was measured after exposure at CO_2 partial pressure of 80 bar and 60°C temperature. The morphology of the scale was analyzed by means of scanning electron microscope (SEM) energy-dispersive X-ray spectroscopy (EDX). Cathodic potentiodynamic tests were carried out on precorroded specimens for evaluating the effect of preformed scales on cathodic polarization curves in CO_2 saturated sulphuric acid solution at pH 3, which is the value estimated for water saturated by supercritical CO_2. The results are discussed in order to evaluate the effect of iron carbide network on scale growth and corrosion rate. Weight loss tests evidenced average corrosion rate values in the range 1–2.5 mm/y after 150-hour exposure. The presence of thick siderite scale significantly reduces the corrosion rate of carbon steel. A slight decrease of the corrosion rate was observed as the scale thickness increases and moving from martensite to microstructures containing carbides.

1. Introduction

Environmental pollution, produced by industrial and civil activities, represents one of the most serious problems for climatic changes and human health. Combustion processes of fossil fuels contribute to the atmospheric level of carbon dioxide that is the most important one of the greenhouse gases, largely responsible for the "enhanced greenhouse effect."

CCTS (Carbon Capture, Transport, and Storage) technologies for capturing waste CO_2 from combustion gases of fossil fuels used for the production of energy, its compression and liquefaction, transport in pipelines, and storage in the deep underground sites become increasingly important. Despite the significant ecological benefits, there are still unsolved issues, mainly related to economics and risks of accidental release into the atmosphere of large amounts of CO_2. Plant reliability requires accurate material selection and a deep knowledge of material corrosion in the presence of very high pressure of wet CO_2. The transport in supercritical conditions will be necessarily carried out by means of existing carbon steel pipelines to reduce and maximize the amount of the mass flux [1].

Although dry carbon dioxide is not aggressive for carbon steel, the presence of water and other pollutants can stimulate corrosion [2]. The CO_2 corrosion, usually named sweet corrosion, occurs by hydration of CO_2 to carbonic acid in the aqueous phase. It has widely been studied in the Oil and Gas industry in which considerable amount of data were collected from plants, at CO_2 partial pressures up to 10 bar.

The role of steel microstructure on the sweet corrosion can mostly be ascribed to the formation and quality of protective corrosion scales. However, such effect has not widely been studied in CCTS systems. Supercritical CO_2 partial pressures cause very high corrosion rates in early exposure but supersaturation conditions for precipitating the protective scale are more easily reached, owing to the high concentration of carbonate species. Compared to the large number of literature works on sweet corrosion in Oil and Gas, only few works are devoted to the supercritical conditions, above 73.9 bar and 31°C [3–8]. Steel microstructure plays an important role on corrosion scale adhesion. In ferritic-pearlitic steels, the selective attack of ferrite grain due to the galvanic coupling between ferrite and Fe_3C and the consequent internal acidification [9] leaves steel behind a network of iron carbides that can act as reinforcement into the film and enhance the adhesion on substrate. Thus, the presence and morphology of carbide phase (i.e., size and distribution) in the scale may be crucial [9].

The aim of the work is the study of the corrosion scale morphology on steel with different microstructures, covering coarse ferrite-pearlite, fine ferrite-pearlite, martensite, and tempered martensite, by means of scanning electron microscope (SEM) energy-dispersive X-ray spectroscopy (EDX). Cathodic potentiodynamic tests were carried out on precorroded specimens for evaluating the effect of preformed scales on cathodic polarization curves in CO_2 saturated sulphuric acid solution at pH equal to that estimated in the literature for water saturated with supercritical CO_2.

2. Experimental

2.1. Material and Specimens. The steel chemical compositions are reported in Table 1 and microstructures are shown in Figure 1. Steel A is a hot rolled API 5L grade X65 with a banded microstructure containing ferrite and pearlite whereas Steel B is not standard grade microalloyed HSLA steel. The microstructure of Steel B was modified by means of five different heat treatments: annealing for 2 hours at 950°C (A), normalization (N), water quenching (WQ), water quenching and tempering for 1.5 hours at 350°C (WQ-T1), and water quenching and tempering for 1.5 hours at 600°C (WQ-T2).

After heat treatment, disks (∅ 20 mm × 2 mm) for weight loss and electrochemical tests were machined from heat-treated bars. Steel A was tested only in as-received conditions (sample A). Steel B samples are indicated with the letter B followed by the heat treatment. The microstructures and Vickers hardness are summarized in Table 2 and Figure 2.

2.2. Tests in Autoclave. The tests were performed at 60°C (±1°C), 80 bar CO_2 partial pressure for 150 hours.

Before exposure, the specimens were grinded with silicon carbide emery paper up to 1000 grit and then cleaned with acetone in ultrasonic bath. A PTFE shaft was used to grant electrical insulation between the specimens and the holder. The specimen holder was placed in a 6 L autoclave filled with about 4 L distilled water. All the specimens with different heat treatments (12 specimens) were fully dipped into water. Several pressurization and depressurization cycles with nitrogen were carried out after specimens immersion to achieve oxygen contents well below 0.2 ppm. Low pressure (about 2 bar) CO_2 pressurization and depressurization cycles were executed for stripping nitrogen from the solution. The autoclave was then heated up to 60°C and, finally, it was pressurized with CO_2. CO_2 was directly pumped inside the autoclave by means of a gas-booster equipped with a preheater regulated at 40°C in order to directly achieve supercritical condition.

2.3. Corrosion Tests. The specimens for both weight loss tests and potentiodynamic tests were 15 mm diameter and 5 mm height cylinders. X-ray analysis specimens were of 50 mm diameter and 5 mm height. At least four specimens for each condition were immersed. Half of the specimens were used for weight loss tests and the others were used for the cathodic potentiodynamic tests. After the X-ray analysis, SEM observation has been performed to evaluate the corrosion

TABLE 1: Chemical composition of steels.

Steel	C	Mn	Si	P	S	Ni	Cr	Mo
A	0.27	0.70	0.32	0.02	0.04	0.10	0.10	0.09
B	0.06	1.94	0.30	0.01	—	0.29	0.04	0.25

scale morphology. SEM observations have been also carried out on metallographic sections grinded with emery paper and polished up to 1 μm with diamond paste. Nital 2% metallographic etching was considered to evidence microstructure.

Corrosion rate was evaluated by weight loss measurements after pickling in 6 N hydrochloric acid inhibited with 3 g/L hexamethylenetetramine at 60°C for 1 minute. The weight loss data were corrected according to ASTM G1 standard to take into account about the bare metal removal due to pickling.

Cathodic Potentiodynamic Tests. Potentiodynamic tests were performed on disk specimens after exposure in supercritical CO_2 saturated water. The specimens were rapidly dried and stored in order to maintain the scale. The electrochemical tests were performed in one-liter ASTM G5 standard cell by using a sample holder with exposed area of 1 cm^2, a standard calomel reference electrode (SCE), and two graphite counter electrodes. Before tests, the open circuit potential was monitored for 30 minutes. The tests were performed at room temperature in H_2SO_4 diluted solution (10^{-3} Mole/L) saturated with flowing CO_2 at 1 bar (pH 3). The free corrosion potential was measured for 300 seconds after the immersion of specimens. The tests were carried out at 10 mV/minute scan rate from free corrosion potential (E_{cor}) to 0.50 V of cathodic polarization. The curves were modified to take into account about the ohmic drop in the electrolyte and into the scale by means of electrochemical impedance spectroscopy measurements at high frequency (10^3–10^4 Hz).

3. Results

3.1. Corrosion Rate. Table 3 reports the corrosion rates after autoclave tests. The mean value was calculated on two specimens for Steel B. The standard deviation was about 0.02 mm/y for Steel A specimens. The corrosion rate is about 2 mm/y and slightly increases from ferritic-pearlitic to martensitic microstructures.

3.2. Morphologies. At the end of the exposure, all the specimens are totally covered by a thick scale of siderite (Figure 2) confirmed by XRD spectra (Figure 3). The high thickness masked the X-ray peaks of ferrite, pearlite, and martensite of steel substrate. Two or more layers of iron carbonate crystals of different dimension compose the scale.

Figure 4 shows the aspect of the scale formed on Steel A as a function of the ferrite-pearlite bands orientation. Corrosion attack penetrates inside the steel following preferential path along pearlite islands if the bands are oriented perpendicularly to the exposed area. The film mainly grows towards the steel and large cavities can be noticed between the ferrite bands. With the pearlite bands parallel to the

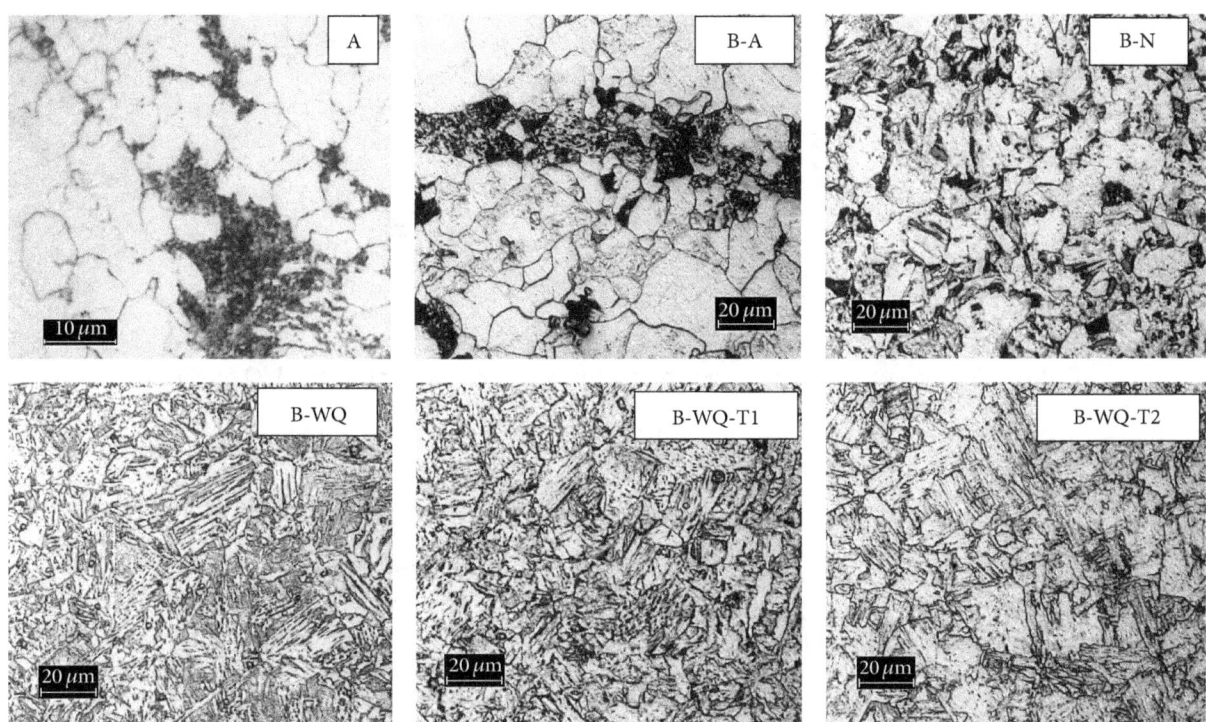

FIGURE 1: Microstructures of steels (Nital 2 etching).

TABLE 2: Heat treatments and microstructure of the tested steels.

Steel	Heat treatment	Microstructure	HV
A	Hot rolling	Ferrite-pearlite	170
B-A	Annealing	Coarse ferrite-pearlite	135
B-N	Normalizing	Ferrite-pearlite	196
B-WQ	Water quenching (WQ)	Martensite	270
B-WQ-T1	WQ and tempering at 350°C	Low-temperature-tempered martensite	245
B-WQ-T2	WQ and tempering at 600°C	Fully tempered martensite	230

TABLE 3: Average corrosion rates (CR) from weight loss measurement and scale thickness.

Sample	A	B-A	B-N	B-WQ	B-WQ-T1	B-WQ-T2
CR (mm/a)	2.35	1.89	1.86	2.18	2.11	1.87
Scale thickness (μm)	80–100	60–100	40–70	30–100	40–50	40–70

exposed surface, the morphology shows large discontinuities, which assume an elongated shape and tend to detach the scale. The corrosion rate reaches the highest level (Table 3). Similar behavior was evidenced on B-A specimens (Figure 5). However, it should be underlined that the dimension of the bands is lower than Steel A due to the very low carbon content. Moreover, the scale was less porous on the B-N specimens, which have more fine perlite microstructure and much less evident bands (Figure 6).

Pearlite is not present in quenched steels (B-WQ) and anisotropy of microstructure with respect to the rolling direction was eliminated by the heat treatment. The scale is less porous (Figure 7) but the corrosion attack propagates along the martensite laths and undissolved martensite is incorporated in the scale. This effect is more pronounced for the full martensitic specimens (B-WQ). In this last case, the scale is thinner than the ferritic-pearlitic samples, and slightly higher corrosion rate was observed.

3.3. Cathodic Potentiodynamic Tests.
The potentiodynamic tests were carried out in dilute sulfuric acid saturated with CO_2 at pH 3 as suggested by Sim et al. [10] to evidence the shielding effect of the scale on the cathodic process.

Figure 8 shows the potentiodynamic curves of the steels. All bare specimens, regardless of steel or heat treatment, showed coincident curves with hydrogen limiting current density around $3 \cdot 10^{-4}$ A/cm^2. The scales produce two main variations. It decreases hydrogen diffusion limiting current density more than two order magnitudes and rises corrosion potential. The systematic increasing of free corrosion potentials denotes an effect of the scale that is more accentuated on anodic curve than cathodic curves.

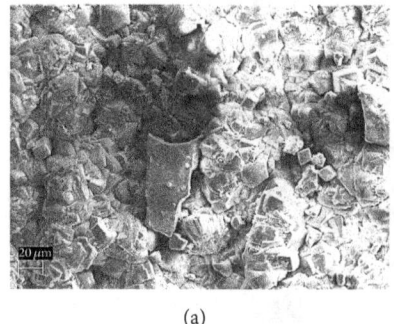 

(a)   (b)

FIGURE 2: SEM image of scale after exposure in autoclave: (a) B-N specimen, (b) B-WQ specimen.

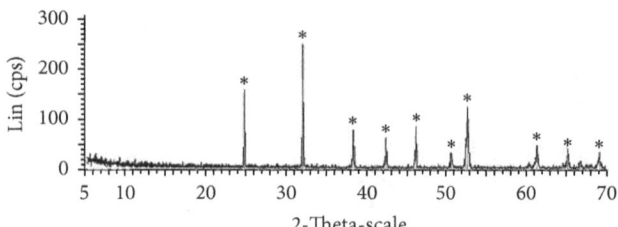

FIGURE 3: XRD spectra of scale on specimens B-A and B-WQ.

4. Discussion

4.1. Corrosion Mechanism. The corrosion of carbon steel in the presence of CO_2 involves the anodic oxidation of iron to ferrous ions (1) and the cathodic process of hydrogen evolution (2):

$$Fe \longrightarrow Fe^{2+} + 2e^- \qquad (1)$$

$$2H^+ + 2e^- \longrightarrow H_2 \qquad (2)$$

The corrosivity of CO_2 derives from the chemical reaction of hydration of CO_2 in the electrolyte to carbonic acid (3) and its dissociation ((4) and (5)) that produces hydrogen ions and from the direct reduction of carbonic acid (6) [11]:

$$CO_2 + H_2O \rightleftarrows H_2CO_3 \qquad (3)$$

$$H_2CO_3 \rightleftarrows HCO_3^- + H^+ \qquad (4)$$

$$HCO_3^- \rightleftarrows CO_3^{2-} + H^+ \qquad (5)$$

$$2H_2CO_3 + 2e^- \longrightarrow H_2 + 2HCO_3^- \qquad (6)$$

The reduction of carbonic acid becomes important at pH > 5. The local alkalization due to H^+ consumption increases the concentration of dissociated species of CO_2.

The corrosion rate after 24-hour exposure is in the range from 10 to 28 mm/years, as reported by Nešić and Lee, Cui et al. [12], and Cabrini et al. [13, 14]. The corrosion rates observed in this work are very lower than these values, owing to the long exposure time that promotes the formation of a protective scale of siderite according to reaction (7) [15]:

$$Fe^{2+} + CO_3^{2-} \rightleftarrows FeCO_3 \qquad (7)$$

The reduction of an order of magnitude of the corrosion rate of the specimens was also observed in previous works for exposure time higher than 150 hours, CO_2 partial pressure ranging from 40 to 135 bar, and 60°C temperature [13, 14].

4.2. Scale Precipitation. The morphologies of the scale after the exposure in autoclave show a composite nature with inner layer of big crystals of siderite and external layer of very small crystals (Figure 2) due to variation of the precipitation mechanism with buildup of corrosion product in the solution.

The precipitation of the scale occurs when the concentrations of ferrous and carbonate ions exceed the solubility product, which is function of temperature and ionic strength [16]. Exposure time promotes the scale formation because the concentration of Fe^{2+} ions inside the solution quickly increases due to very high initial corrosion rates. In addition, a high volume of gas in equilibrium with the solution grants the supply of the carbonate species depleted by the cathodic reaction.

The conditions at metal/solution interface are quite different from the bulk solution due to active steel corrosion. The limit of solubility is easily reached very close to the metal surface and scale begins to precipitate. Several authors emphasize the importance of cathodic species diffusion from the bulk solution to the metal surface and vice versa, the reaction products from the interface to the bulk solution, to achieve the conditions that favor iron carbonate precipitation [11, 15, 16].

As the steel corrodes, the concentration of Fe^{2+} in solution increases and a decrease of the diffusion rate of these ions from the reaction interface to the bulk solution occurs. The diffusion rate of CO_3^{2-}/HCO_3^- ions produced by the cathodic reaction also decreases to preserve electroneutrality at the interface. Therefore, the precipitation of the siderite can take place at Fe^{2+} ion concentration lower than supersaturation of the bulk solution, but the corrosion products layer is stable only in the case of continuous Fe^{2+} ions supply by corrosion.

The siderite crystals nucleation rate is high and does not affect the film formation: during their growth on the steel surface to the bulk solution, they tend to form a continuous but very porous film. The balance between the consumption of

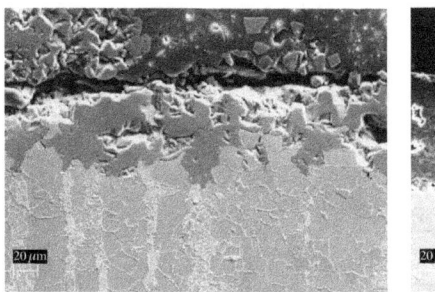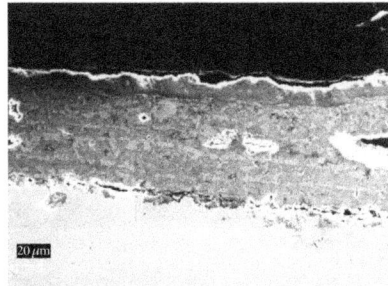

FIGURE 4: Image of the scale on surfaces of specimen A with different ferrite-pearlite bands orientation.

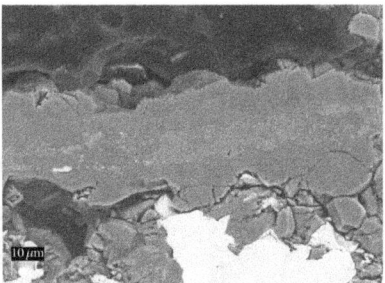

FIGURE 5: Carbide bands in the scale of B-A specimens.

ferrous ions and their production will maintain the supersaturation conditions at metal/scale interface that are necessary to stimulate the siderite crystal growth. The H^+ consumption due to the cathodic reaction promotes the dissociation of carbonic acid, moving the equilibrium of CO_2 hydration and dissociation to the right. The corrosion rate decreases as the scale grows and so the ions concentration falls below the supersaturation conditions, promoting the protective film dissolution and increasing the Fe^{2+} concentration in the bulk solution.

Afterwards, once oversaturation conditions are achieved at the scale-solution interface, the scale can also grow for direct reprecipitation from bulk solution.

When supersaturation conditions are reached in the bulk solution, small crystals of siderite precipitate on the outer surface of the scale and seal the porosity.

Choi and Nešić [7] report a model for estimating mutual solubilities of CO_2 and water in the two coexisting phases and calculating the concentration of corrosive species in the free water at various pressure and temperatures. Based on this model, the content of carbonic acid, bicarbonate, and carbonate ions at 80 bar CO_2 partial pressure and $60°C$ can be assumed equal to 2–2.5, 0.6–0.7, and $7–7.5 \cdot 10^{-8}$ mmole/L, respectively, with pH in the range 3.1–3.2. Solubility product of iron carbonate is equal to $8.24 \cdot 10^{-13}$ at $60°C$ according to Braun [17]. The limit for iron carbonate precipitation can be calculated by considering solubility product of iron carbonate and on the concentrations evaluated by the model of Choi and Nešić, through the following relation:

$$K_{sFeCO_3} = \frac{[Fe^{2+}] \cdot [CO_3{}^{2-}]}{[FeCO_3]} \qquad (8)$$

For the testing conditions assumed in this work, the precipitation begins once Fe^{2+} ions concentration exceeds 11 mmole/L.

Considering an autoclave volume of 4 liters and the total exposed area of the specimens equal to about 10 cm^2, this concentration is reached in about 136 hours at 20 mm/years constant corrosion rate, which is compatible with the time required for the formation of the scale.

Actually, tests carried out at short time exposure showed the presence of noncontinuous carbonate scale probably due to the fact that corrosion rate just after immersion on bare steel is higher than the mean value measured at 24-hour exposure [13, 14].

4.3. Effect of the Scale on Polarization Curve of Steel in Acid Solution. Increasing of the corrosion potential due to the presence of the scale was observed in the potentiodynamic tests (Figure 8). The siderite scale is not conductive and acts as a barrier that covers the metal surface, while cementite can act as cathode. Nešić et al. reported that the main effect of protective iron carbonate films in CO_2 corrosion is to cover the metal surface and make it unavailable for corrosion rather than act as an effective diffusion barrier [16].

A pure shielding effect should not affect the corrosion potential because both anodic and cathodic areas decrease. Furthermore, a reduction of diffusion transport of hydrogen ion, which reduces the hydrogen limiting current, tends to decrease the free corrosion potential and not to increase it.

The systematic increasing of free corrosion potentials during potentiodynamic test in acid solution indicates that there is an effect of the scale on anodic curves. Han et al. [18] studied the electrochemical behavior of steel in the presence of CO_2, $HCO_3{}^-$, or carbonate ions, founding a sort of pseudopassivation strictly dependent upon the pH:

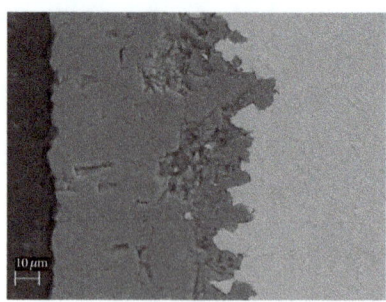

FIGURE 6: Scale morphology on B-N specimens.

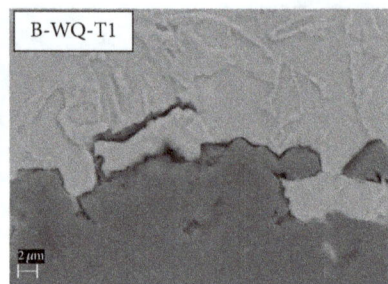
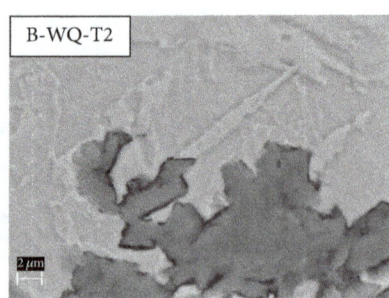

FIGURE 7: Scale morphology on quenched and tempered steels.

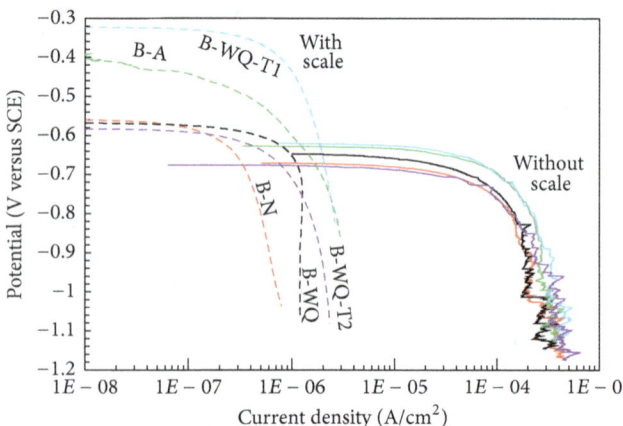

FIGURE 8: Effect of scale formed in autoclave test in supercritical CO_2 on cathodic potentiodynamic curve in sulphuric acid solution at pH 3.

the higher the pH, the higher the protectiveness of $FeCO_3$ layer and pseudopassivation. Significant increase in open circuit potential and a decrease of the corrosion rate were observed at pH 8, owing to the formation of a mixed film of $FeCO_3$ and magnetite (Fe_3O_4). Li et al. reported that the pseudopassivation could not be achieved at pH 5.6 [19]. At pH 3, the scale is mainly constituted by siderite, which does not promote a passive or pseudopassive state.

The increase of the corrosion potential could be ascribed to iron ion concentrations on metal surface in the scale higher than bare metal, producing high overvoltage of anodic process. On surface without any scale, the fast cathodic process (Figure 8) induces hydrogen evolution that stirs the

diffusion layer and reduces the accumulation of iron ions at the metal solution surface.

However, it was noted that the martensitic samples showed the lowest increases of the corrosion potential in the presence of the scale and the sample with microstructures containing carbide was characterized by the highest increases of corrosion potential. Cathodic reaction could also take place on carbides inside the scale, especially on continuous carbides networks. Such a conductivity counteracts the barrier effect on cathodic process.

4.4. Effect of Steel Microstructure. The role of cementite in the scale formation is complex. Crolet et al. [9] hypothesized a corrosion mechanism based on the local galvanic couple between ferrite and cementite and the action of pearlite favors on scale stability. Nešić et al. evidence that the film protectiveness depends on the porosity more than the scale thickness [16].

Figure 9 shows the variation of corrosion rate as a function of average thickness of scale and steel microstructure. The corrosion rate decreases by 20% moving from specimens without any carbides to microstructure containing carbides and from scale thickness of about 40 micrometers to 80 micrometers. However, it must be outlined that the results were obtained on steel with very low carbon content.

5. Conclusion

The paper reports the results of electrochemical and weight loss test on steels with different microstructures exposed to CCTS environment. The analysis of the morphology of corrosion scales is also presented. Weight loss tests evidenced average corrosion rate values in the range 1–2.5 mm/y after

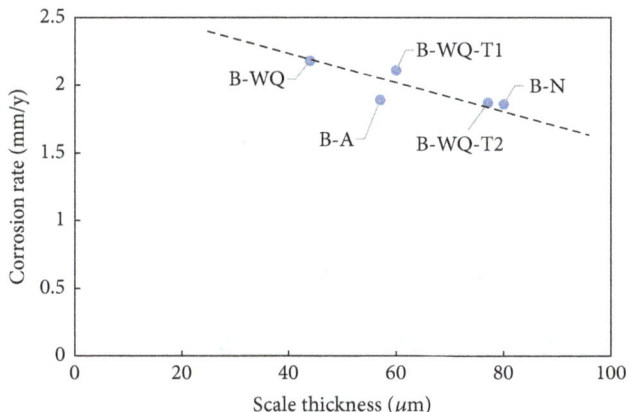

FIGURE 9: Correlation between corrosion rate and average scale thickness.

150-hour exposure. The values agree with the literature data and are significantly lower than expected in the absence of protective scale. All the specimens evidenced the presence of thick scale of corrosion products that significantly reduces the corrosion rate of carbon steel. Great siderite crystals, at the steel-scale interface, and fine crystals, at the scale/solution interface, composed the scale. Fine crystals partially seal the scale porosities, reducing corrosion rate.

Potentiodynamic cathodic curves in sulphuric acid on precorroded specimens covered by the scaled form during test in autoclave evidenced a reduction of the hydrogen limiting current and an increase of corrosion potential in the presence of the scale for all the microstructures. A slight decrease of the corrosion rate was observed as the scale thickness increases and moving from martensite to microstructures containing carbides.

Conflict of Interests

The authors declare that there is no conflict of interests regarding the publication of this paper.

Acknowledgment

This research was financed by Italy® Project of University of Bergamo.

References

[1] F. Edelvik, B. Graver, L. Torbergsen, and O. Saugerud, "Development of a guideline for safe, reliable and cost efficient transmission of CO_2 in pipelines," *Energy Procedia*, vol. 1, no. 1, pp. 1579–1580, 2009.

[2] B. Sass, H. Farzan, R. Prabhakar et al., "Considerations for treating impurities in oxy-combustion flue gas prior to sequestration," *Energy Procedia*, vol. 1, no. 1, pp. 535–542, 2009.

[3] M. Seiersten, "Corrosion of pipeline steels in supercritical CO_2/water mixtures," in *Proceedings of the European Corrosion Congress (EUROCORR '00)*, London, UK, September 2000.

[4] M. Seiersten, "Materials selection for separation, transportation and disposal of CO_2," in *Proceedings of the CORROSION*, Paper no. 01042, NACE International, Houston, Tex, USA, March 2001.

[5] S. Hesjevik, S. Holsen, and M. Seiersten, "Corrosion at high CO_2 pressure," Paper 03345, CORROSION, NACE, Houston, Tex, USA, 2003.

[6] Y.-S. Choi and S. Nešić, "Corrosion behavior of carbon steel in supercritical CO_2-water environments," in *Proceedings of the CORROSION*, Paper no. 09256, NACE International, Atlanta, Ga, USA, March 2009.

[7] Y.-S. Choi and S. Nešić, "Determining the corrosive potential of CO_2 transport pipeline in high pCO_2-water environments," *International Journal of Greenhouse Gas Control*, vol. 5, no. 4, pp. 788–797, 2011.

[8] F. Ayello, K. Evans, R. Thodla, and N. Sridhar, "Effect of impurities on corrosion of steel in supercritical CO_2," in *Proceedings of the CORROSION*, Paper no. 10193, NACE International, San Antonio, Tex, USA, March 2010.

[9] J. L. Crolet, N. Thevenot, and S. Nesic, "Role of conductive corrosion products in the protectiveness of corrosion layers," *Corrosion*, vol. 54, no. 3, pp. 194–203, 1998.

[10] S. Sim, P. Corrigan, I. Cole, and N. Birbilis, "Use of aqueous solutions to simulate supercritical CO_2 corrosion," *Corrosion*, vol. 68, no. 4, pp. 045004-1–045004-11, 2012.

[11] M. Nordsveen, S. Nešić, R. Nyborg, and A. Stangeland, "A mechanistic model for carbon dioxide corrosion of mild steel in the presence of protective iron carbonate films—part 1: theory and verification," *Corrosion*, vol. 59, no. 5, pp. 443–456, 2003.

[12] Z. D. Cui, S. L. Wu, C. F. Li, S. L. Zhu, and X. J. Yang, "Corrosion behavior of oil tube steels under conditions of multiphase flow saturated with super-critical carbon dioxide," *Materials Letters*, vol. 58, no. 6, pp. 1035–1040, 2004.

[13] M. Cabrini, S. Lorenzi, T. Pastore, and M. Redaelli, "Corrosion rate of carbon steel in condensed water and humid gas at high CO_2 pressure," in *Proceedings of the European Corrosion Congress (EUROCORR '13)*, Estoril, Portugal, September 2013.

[14] M. Cabrini, S. Lorenzi, T. Pastore, and M. Radaelli, "Corrosion rate of high CO_2 pressure pipeline steel for carbon capture transport and storage," *La Metallurgia Italiana*, vol. 106, no. 6, pp. 21–27, 2014.

[15] S. Nešić and K.-L. J. Lee, "A mechanistic model for carbon dioxide corrosion of mild steel in the presence of protective iron carbonate films—part 3: film growth model," *Corrosion*, vol. 59, no. 7, pp. 616–628, 2003.

[16] S. Nešić, M. Nordsveen, R. Nyborg, and A. Stangeland, "A mechanistic model for carbon dioxide corrosion of mild steel in the presence of protective iron carbonate films—part 2: a numerical experiment," *Corrosion*, vol. 59, no. 6, pp. 489–497, 2003.

[17] R. D. Braun, "Solubility of iron(II) carbonate at temperatures between 30 and 80°," *Talanta*, vol. 38, no. 2, pp. 205–211, 1991.

[18] J. Han, D. Young, H. Colijn, A. Tripathi, and S. Nešić, "Chemistry and structure of the passive film on mild steel in CO_2 corrosion environments," *Industrial and Engineering Chemistry Research*, vol. 48, no. 13, pp. 6296–6302, 2009.

[19] W. Li, B. Brown, D. Young, and S. Nesic, "Investigation of pseudo-passivation on mild steel in CO_2 corrosion," *Corrosion Science*, vol. 70, no. 3, pp. 294–302, 2014.

11

The Effect of Graphene on the Protective Properties of Water-Based Epoxy Coatings on Al2024-T3

T. Monetta, A. Acquesta, A. Carangelo, and F. Bellucci

Department of Chemical Engineering, Materials and Industrial Production, University of Napoli Federico II, Piazzale Tecchio 80, 80125 Napoli, Italy

Correspondence should be addressed to T. Monetta; monetta@unina.it

Academic Editor: Flavio Deflorian

0.5 and 1% wt. of graphene nanoflakes were added to an anticorrosive additives-free water-based epoxy resin applied to Al2024-T3 samples. Calorimetric (DSC) and adhesion (cross-cut test) tests indicated that the presence of graphene did not affect the polymerization process of the resin or its adhesion to the substrate while it had some effect on its wettability. Electrochemical Impedance Spectroscopy (EIS) results obtained suggested that the addition of a small amount of graphene greatly enhanced the protective properties of the epoxy coating, retarding electrolytes absorption and reducing the total amount of adsorbed water. The latter occurrence suggests that the graphene effect on coating performances is related to both extended diffusion pathway length and graphene/matrix interaction due to the unique properties of graphene.

1. Introduction

The study of graphene effect when used as a nanofiller in polymer matrices to form advanced multifunctional materials is one of the most promising research fields in various areas of application [1–4]. Loading of graphene to polymeric materials has shown a significant increase in the electrical and thermal properties and the mechanical properties of nanocomposites [5–11]. Few papers, however, addressed the development of graphene nanocomposites with the aim of increasing the protective properties of organic coatings against corrosion phenomena [12–17]. Reports on epoxy/graphene nanocomposite coatings are less common [14, 18–20] and very few researchers described the use of a waterborne resin [15, 21–24]; however, the environmentally friendly solutions that take into account the new regulations about the emission of volatile organic compounds [25–28] are of considerable interest. As it is well known, several types of green surface treatments, or coating systems, have been investigated to decrease the aluminium corrosion rate when exposed to an aggressive environment [28–36]. In this study, the effect of graphene on protective properties of epoxy coating has been evaluated by incorporating reduced concentrations of nanoparticles (i.e., 0.5% and 1% by weight) into

waterborne epoxy resin and applied to 2024-T3 aluminium alloys sample. The protective properties of the modified epoxy coatings were investigated by using Electrochemical Impedance Spectroscopy (EIS) and differential scanning calorimetric (DSC) techniques. Furthermore, the adhesion between the coating and the metallic substrate was evaluated by means of cross-cut test and the effect of the filler on the wettability of the coating was also addressed in this investigation.

2. Materials and Methods

Graphene particles, used in this work (C500, Cometox, Italy), are made of few layers with a width of less than $2\,\mu m$, an average thickness of about $2\,nm$, and a surface area of $500\,m^2/g$, as reported by the supplier. The epoxy system (Wapex 660, Sikkens, Italy) is a commercial waterborne resin, without corrosion inhibitors. It contains, in addition to other additives (such as emulsifying and dispersing agents and surfactants) normally used in the production cycles of commercial resins, 63% wt. of solid, mainly TiO_2. The volume ratio of epoxy resin to hardener used was $4:1$ as indicated by the manufacturer. The safety data sheets for

TABLE 1: Components of water-based epoxy resin.

	Common names	CAS number	% weight	Acronym
Component A	Polyamine epoxy-resin adduct Tetraethylenepentamine	112-57-2	≥10, <20	TEPA
	Bisphenol A-co-epichlorohydrin	25068-38-6	≥0.25, <1	DGEBA
	Bisphenol-F epichlorohydrin	28064-14-4	≥0.25	DGEBF
Component B	Bisphenol A-co-epichlorohydrin	25068-38-6	≥50, <75	DGEBA
	Bisphenol-F epichlorohydrin	28064-14-4	≥20, <25	DGEBF
	2,3-Epoxypropyl neodecanoate	26761-45-5	≥2.5, <25	

TABLE 2: Nomenclature used to identify the different specimens.

Sample	Nomenclature
Unfilled coating	EP
Coating containing 0.5% wt. of graphene	EG05
Coating containing 1% wt. of graphene	EG1

"component A" and "component B" of the resin report the presence of chemical compounds listed in Table 1.

The Al2024-T3 substrates, having dimensions of 20 cm × 10 cm × 0.5 cm, were degreased by acetone and dried with air pressure before painting. In order to evaluate the influence of the dispersed graphene on the polymer matrix, specific amounts of graphene nanofiller were chosen, in particular 0.5% and 1% wt. The nomenclature adopted to individuate the samples is reported in Table 2.

The dispersion of the graphene nanoflakes in the epoxy resin has been obtained, using an ultrasonic bath with a frequency of 50 Hz for 20 min, blending component A of the resin with graphene, by means of a hermetic box, keeping the container refrigerated. The hardener was added to the graphene/epoxy blend and mixed for another 20 min by using a mechanical stirrer. Finally, the hybrid coating was applied to the aluminium substrate with a spiral bar applicator. The samples were cured at 150°C for 10 min. The dry thickness of the cured coatings was 27 ± 1.3 μm, measured with an Elcometer Dualscope Mpor-Fp (IMCD Italia Spa, Italy).

In order to evaluate the effect of the nanofiller on the epoxy matrix properties, a thermal analysis was carried out taking 10 mg of samples. Three scans (heating, cooling, and heating again) were carried out from 30 to 250°C with a heating rate of 10°C/min, for each specimen. Tests were performed in a dry nitrogen atmosphere by using a Mettler-Toledo DSC12E (Mettler-Toledo Spa, Italy) apparatus.

The protective properties of the coatings were investigated by EIS following ISO 16773-2016. A conventional electrochemical cell was used including a saturated calomel reference electrode (SCE), platinum as counter electrode, and the coated aluminium sample as working electrode. A frequency response analyzer (FRA), in conjunction with a potentiostat/galvanostat, 1255 and 1286 Solartron (Photo Analytical S.r.l., Italy), respectively, was employed. The electrolyte used was an air saturated 3.5 wt.% NaCl aqueous solution and the area exposed was of about 5 cm². Measurements were carried out at open circuit potential (OCP) over a frequency range from 10^5 to 0.02 Hz with an amplitude sinusoidal voltage of 10 mV up to an immersion time of 21 days.

The effect of the graphene on substrate/coating adhesion was analyzed by a Cross Hatch Cutter (Sheen Instruments, Italy), following ASTM D3359-09, whereas wettability was investigated by means of water contact angle (WCA) test using an OCA 15 EC (DataPhysics Instruments GmbH, Filderstadt, Germany). In particular, water droplets, with a volume equal to 3,5 μL, were dispensed on the surface of specimens. The WCA was evaluated on an average of 50 measurements taken on different points of the surface. Finally, all measurements reported in this paper were carried out at room temperature and repeated at least three times to ensure reproducibility and accuracy.

3. Results and Discussion

To investigate the effect of graphene on the transition glass temperature, T_g, and hence on the physical properties of epoxy matrix, DSC measurements were carried out on the basic resin, EP, and on the epoxy-graphene coatings, EG05 and EG1. Findings obtained were reported in Figure 1. As can be seen from this figure, both the residual cure and T_g (at about 100°C) were not affected by the graphene content. These results suggest that the presence of a small amount of graphene does not affect either the glass transition temperature or the curing of the epoxy matrix. Therefore, the effect of the low amount of graphene on the protective behavior of the coating, as will be discussed in the subsequent paragraph, must not be attributed to a physical modification of the matrix of the epoxy or to a different curing process.

Good adhesion between the coating and the substrate is a "desideratum" property. In fact, it is well known that water molecules at the metal/coating interface may decrease the coating adhesion accelerating the corrosion of the metallic substrate [37]. Thus, poor adhesion permits aggressive ions to accumulate at the coating/metal interface and to trigger a degradation process. Cross-cut adhesion tests (Figure 2) highlighted that the loading of graphene nanoflakes had no effect on the coating/substrate adhesion. In fact, no peelings were observed in any of the coatings, receiving 4B rating (less than 5% peeling) according to the ASTM standard.

In order to study the influence of graphene on the wettability of the coating, water contact angle measurements were carried out (Figure 3).

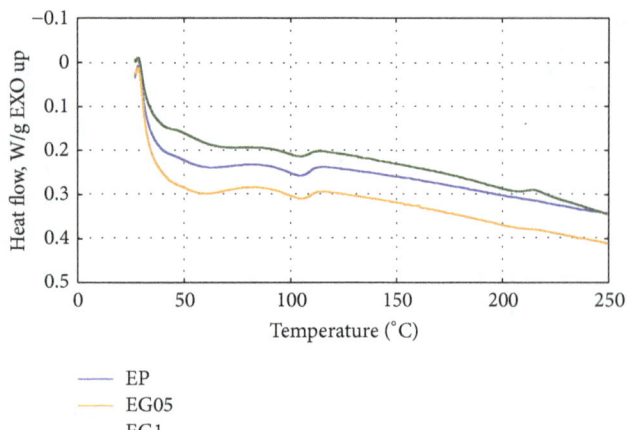

— EP
— EG05
— EG1

FIGURE 1: DSC thermograms of the unfilled coating (EP) and the epoxy-graphene coating containing 0.5% wt. (EG05) or 1% wt. (EG1) of the nanofiller.

TABLE 3: Water contact angle average values of all tested specimens.

Sample	EP	EG05	EG1
Water contact angle (degree)	$65,1 \pm 0,5$	$71,3 \pm 0,8$	$81,2 \pm 0,9$

The measured water contact angle (see Table 3) of the unfilled epoxy coating is about 65,1°; when the filler is dispersed in the epoxy resin, the contact angle increased to 71.3° (EG05 sample) and 81,2° (EG1 sample). This rise is linked to the amount of nanoflakes loaded (i.e., by increasing the filler content, an increase of the contact angle is observed). Thus, graphene nanoflakes confer some hydrophobic character to the epoxy resin contributing to the reduction of the amount of water uptake.

EIS measurements were performed in order to investigate the effect of graphene addition on the coatings' performance, when exposed to an aerated 3.5% wt. NaCl aqueous solution up to 21 days. Findings obtained are presented as Bode and phase angle plots in Figures 4, 5, and 6.

The impedance modulus trend of the EP sample, shown in Figure 4(a), indicates poor corrosion protection offered by the no-loaded epoxy coating. In fact, after only one day of exposure to the test solution, the impedance modulus at low frequency (attributed to the coating resistance) is slightly less than 10^7 $\Omega \cdot cm^2$. The latter value is generally regarded as the lower limit beyond which the protective properties of coatings are considered poor. This result shows the limited protective properties of the coating due to the absence of anticorrosive pigments as well as to the water and the electrolytes that penetrate rapidly into the coating. After 4 days of immersion, the impedance modulus at low frequency has shifted to the value of about 4×10^6 Ω cm^2, while it was possible to detect a new phenomenon that is developing at the substrate/metallic interface, namely, the initiation of a corrosion process. This effect becomes even more evident with exposure time and is attributed to the absorption of water and electrolytes by the coating with the development of local anodic area at the metallic substrates [38]. For prolonged immersion time (14–21 days), a new

phenomenon is displayed, characterized by an increase of the impedance at low frequency (see data at 21 days) that is currently attributed to the formation of corrosion products at the interface filling the pores of the coating [36].

The latter impedance data were confirmed by the phase angle plot that better emphasizes these effects [38]. Thus, looking at data shown in Figure 4(b), it is possible to observe that the phase angle plot, after one day of exposure to the test solution, exhibits a value of about 80 degrees at high frequency, while a minimum value (3 degrees) is observed at a frequency of about 0.25 Hz. Moreover, the increase of the phase angle observed in the lowest frequency range (10^{-2}–10^0 Hz) suggests the initiation of the corrosion process at the alloy/coating interface. Therefore, it can be assumed that after 24 h of exposure the solution has penetrated the coating, reaching the metallic substrate where corrosion phenomena started to develop [38]. The subsequent shift of the phase angle at high frequency, as a function of immersion time, is also consistent with further degradation of the coating, while the wide variation of the phase angle observed in the medium-high frequency range suggests a pronounced effect of corrosion products on the coating structure.

Data obtained by the EG05 specimen, shown in Figure 5, exhibit a quite different behavior compared to the unfilled sample. The impedance modulus decreases slowly with time (Figure 5(a)), suggesting a better stability of the loaded epoxy coating. Namely, the impedance modulus displays a little bit higher value after one day of immersion compared to that exhibited by the unloaded epoxy, suggesting better protective properties of the coating. For further immersion in the test solution up to 21 days, a slight decrease of the impedance modulus is continuously observed. It is worth mentioning that no increase in the impedance modulus is detected in the time interval of 14–21 days, as it was reported for the base coating, EP. Therefore, the addition of 0.5% wt. of graphene improves the EIS response of the epoxy loaded coating.

The phase angle data, shown in Figure 5(b), further support the beneficial effect of graphene on the stability of the epoxy filled coating. Thus, looking at the data shown in this figure, it is possible to observe that the phase angle plot, after one day of exposure, exhibits a value of about 80 degrees at high frequency, while the minimum value is observed at a frequency of about 0.1 Hz. Moreover, the increase of the phase angle observed in the lowest frequency range (10^{-2}–10^{-1} Hz) is much less pronounced when compared with the data obtained for base coating, suggesting that the initiation of the corrosion process at the alloy/coating interface involves restricted metallic area [38]. Furthermore, it is also worth mentioning the shift of the phase angle versus the low frequency range in the time interval of 7–14 days. These findings can be attributed to the corrosion products at the interface filling the coating porosity, leading to a more effective and compact coating [37, 38]. Finally, in the time interval of 14–21 days, the degradation of the coating clearly appears as suggested by the shift at high frequency of the phase angle plot. Data reported in Figure 5(b) also support the idea that the filled epoxy appears quite stable as a function of the immersion time in the test solution even if

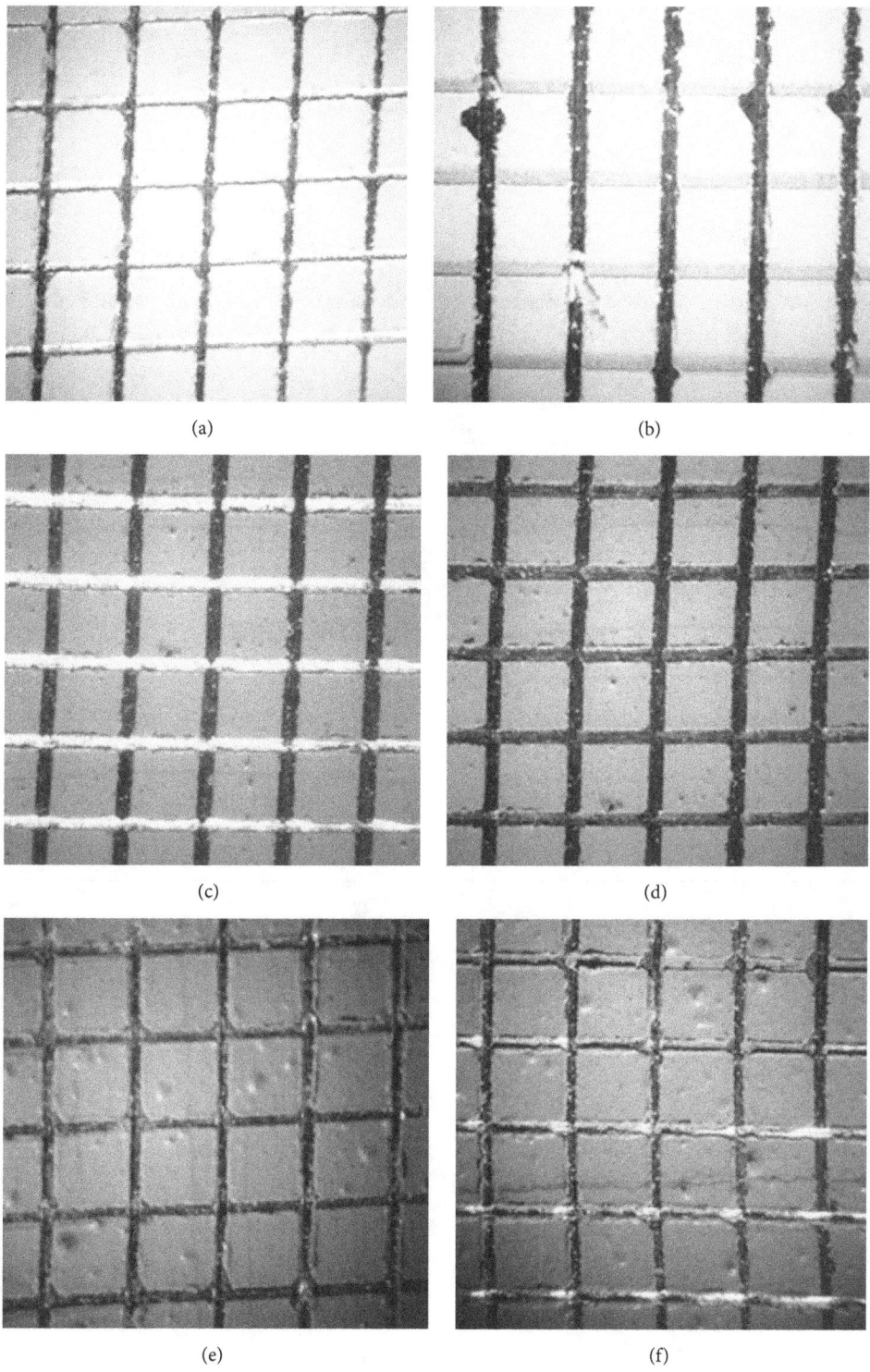

(a)

(b)

(c)

(d)

(e)

(f)

FIGURE 2: Cross-cut test on the EP sample (a) before and (b) after the test, on the EG05 sample (c) before and (d) after the test, and on the EG1 sample (e) before and (f) after the test.

the coating was fully penetrated by the electrolyte at one day of immersion.

The impedance modulus of the EG1 sample (Figure 6(a)) showed the highest impedance value (about $6 \times 10^7 \, \Omega \, cm^2$), as a function of the immersion time, further suggesting the beneficial effect of the addition of 1% graphene to the epoxy matrix as far as the protective properties of this coating are concerned.

An overview of the phase angle plot (Figure 6(b)) reveals further beneficial effects due to the addition of graphene to

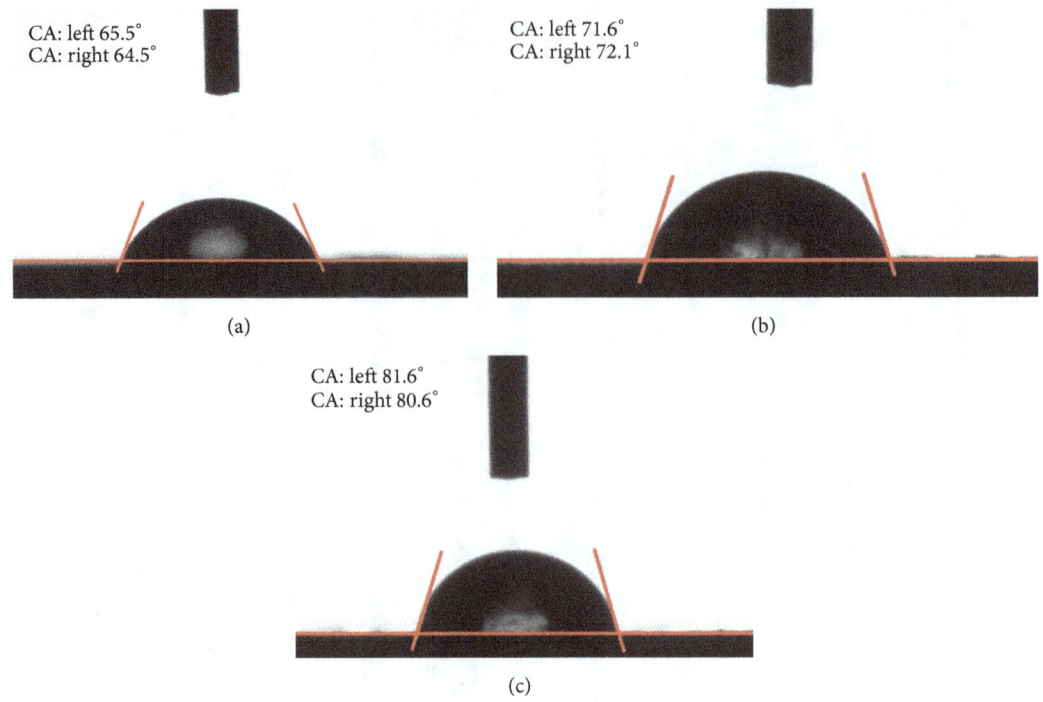

FIGURE 3: Water contact angle pictures of (a) EP, (b) EG05, and (c) EG1 coatings.

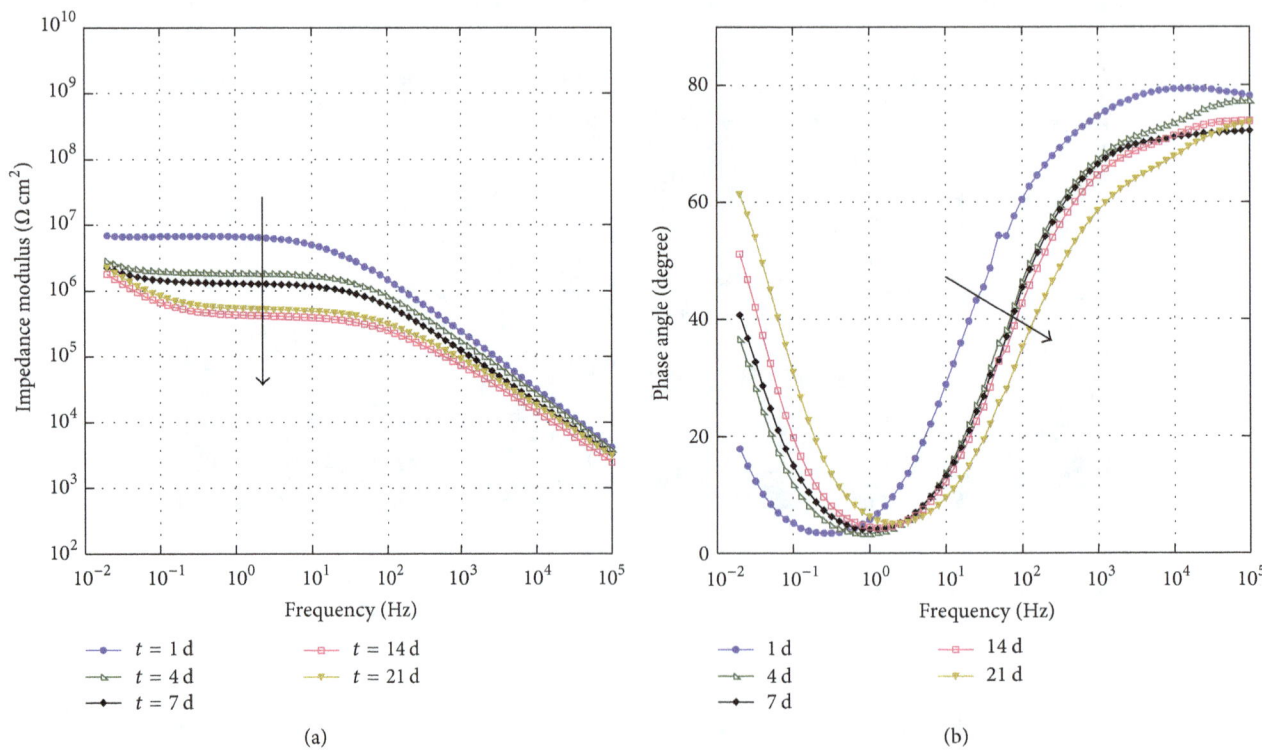

FIGURE 4: (a) Impedance modulus and (b) phase angle plots of the unfilled coating (EP) in 3.5% wt. NaCl aqueous solution.

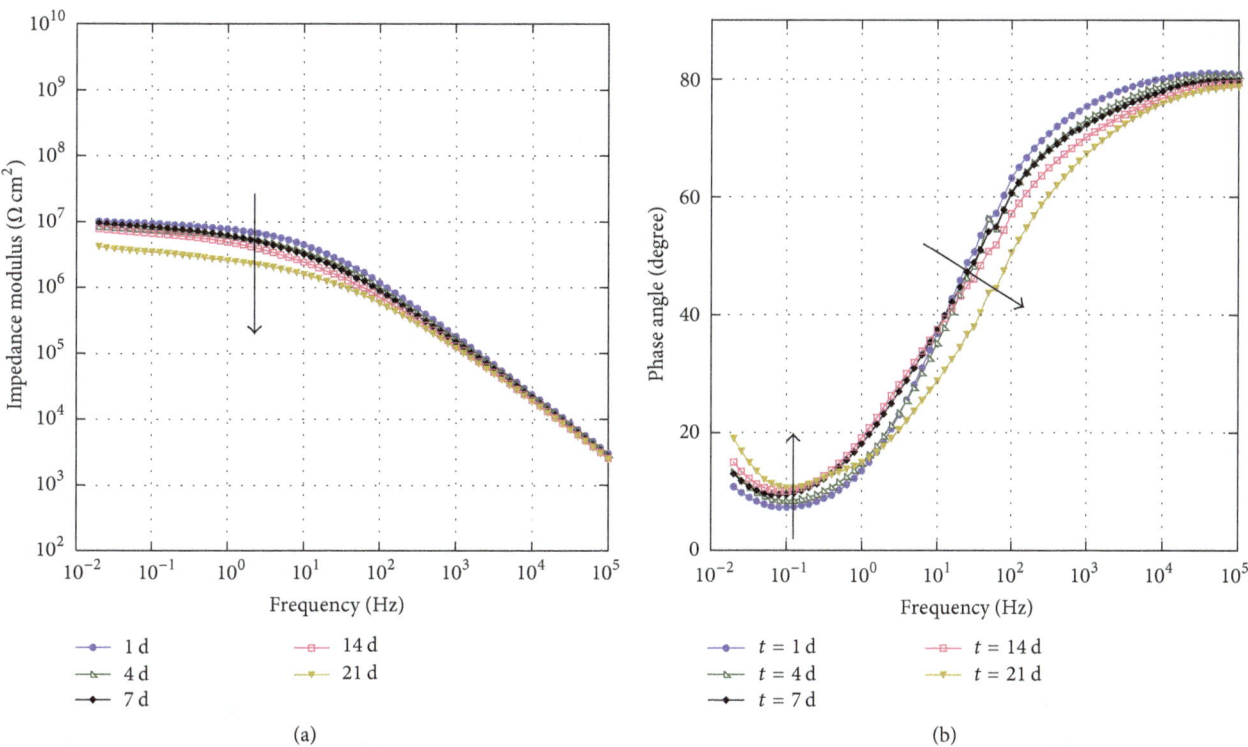

FIGURE 5: (a) Impedance modulus and (b) phase angle plots of epoxy-graphene coating loaded with 0.5% wt. (EG05) of graphene nanofiller in a 3.5% wt. NaCl aqueous solution.

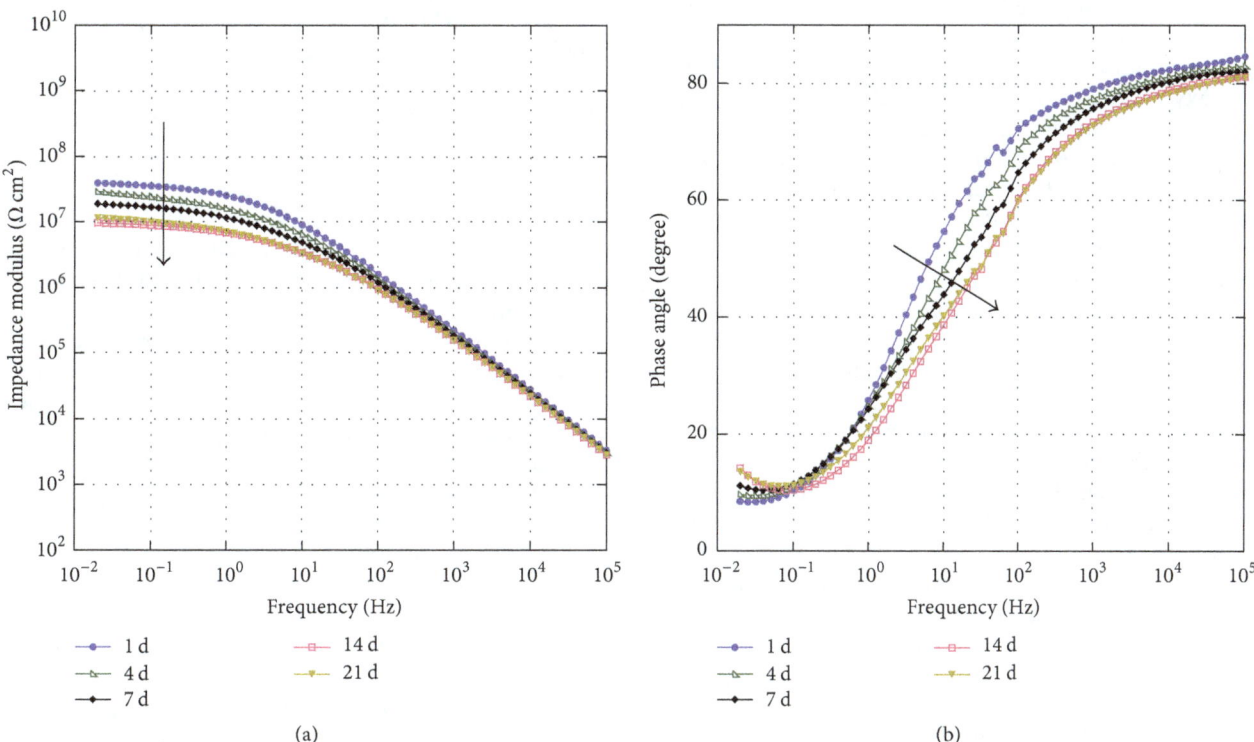

FIGURE 6: (a) Impedance modulus and (b) phase angle plots of epoxy-graphene coating loaded with 1% wt. (EG1) of graphene nanofiller in a 3.5% wt. NaCl aqueous solution.

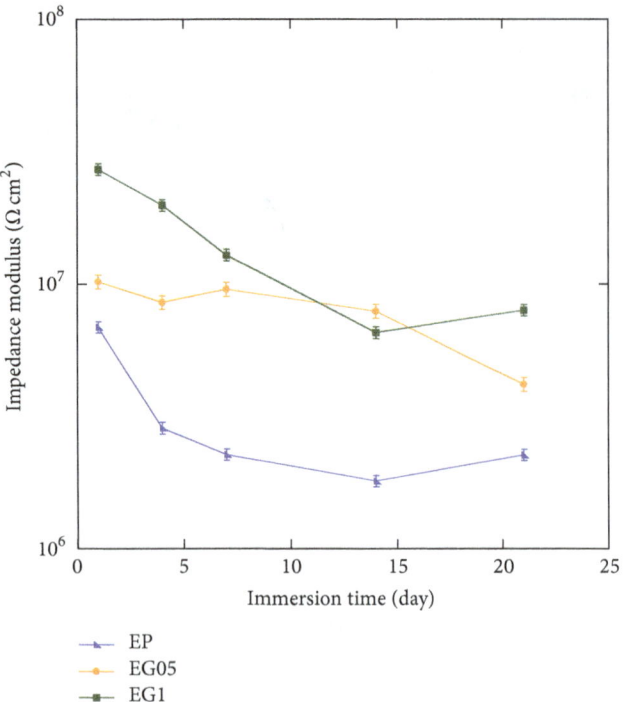

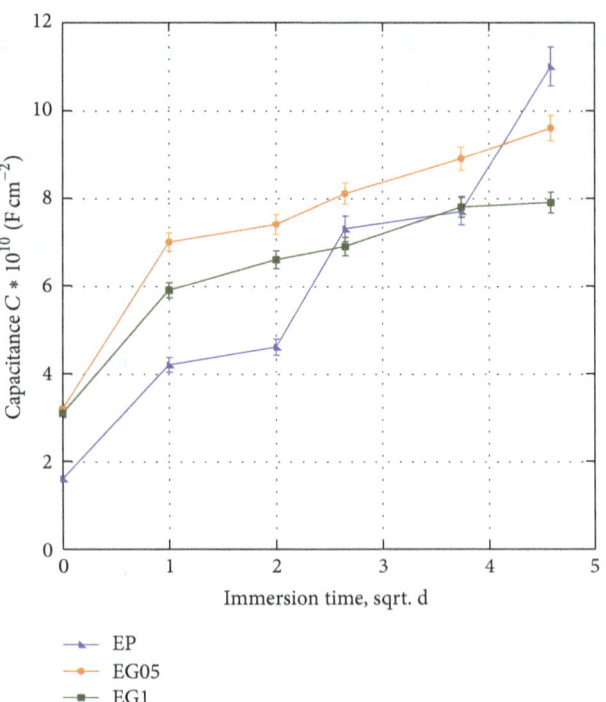

FIGURE 7: Impedance modulus values of unfilled coating (EP) and coating filled with 0.5% wt. and 1% wt. of graphene nanofiller (EG05 and EG1, resp.) at 0.02 Hz.

FIGURE 8: Coating capacitance for unfilled coating (EP) and coating filled with 0.5% wt. (EG05) and 1% wt. (EG1) of graphene resins.

the base epoxy as reported in the following. Data at very low frequency upon 4 days of immersion do not clearly show a minimum as observed in the base and 0.5% wt. filled epoxy coating. Therefore, no full penetration of the electrolyte occurs in the above time interval, suggesting a compact behavior of the filled epoxy at least for the first 4 days. A slight tendency to increase in the phase angle appears in the time interval of 4–7 days of exposure, while a clear minimum is observed after 14 days, suggesting the initiation of the corrosion process at the metallic interface. In addition, all phase angle data as a function of the immersion time are shifted to the low frequency, suggesting better protective properties as far as the addition of graphene nanofillers is concerned.

Meaningful impedance data related to the effect of graphene on the protective properties of the epoxy coating evaluated at 0.02 Hz are reported in Figure 7. As can be seen from this figure, the filled nanocoatings exhibited impedance modulus greater than the unfilled base epoxy for all exposure times showing the noticeable effect due to the use of graphene. In addition, EG1 showed an impedance modulus of one order of magnitude greater than that exhibited by the base epoxy for all exposure times.

Further insight into the beneficial effect of graphene on the protective properties of the base epoxy can be obtained by the knowledge of water transport since water permeation is closely related to the metallic substrate corrosion rate. The permeation of water in a coating is usually determined by the

increase of the capacitance of the paint film as a function of the exposure time [39, 40].

Capacitance values can be obtained by the impedance data by using the following equation [41]:

$$C = \frac{1}{(2\pi f_i Z_i)} = \frac{1}{(2\pi f Z'')},\qquad(1)$$

where f_i (equal to 6.5 kHz) is a frequency value where the slope of the impedance modulus curve is equal to -1 and Z_i is the imaginary part (or reactive component) of the impedance. Findings obtained are reported in Figure 8 as a function of the square root of t.

As can be seen from this figure, all investigated coatings show an increase of the capacitance with exposure time that is currently attributed to the uptake of water. However, Figure 8 shows also a different shape of transient highlighting a different mechanism of water adsorption and transport into the nanocoatings. Namely, the base coating capacitance exhibits a step-like increase in the time interval of 1–16 days, whereas a smooth-like behavior is shown only by the 0.5% wt. filled epoxy in the time interval of 1–4 days. Quite different is the behavior of the 1% wt. nanofilled epoxy; in fact, no step-like increase was observed in this case and an almost steady state is shown at a later stage of immersion, suggesting a quasi-Fickian behavior. By using the theory of diffusion modified for an applied film to a metallic substrate [40], the water diffusion coefficient through the coating is proportional to the initial slope of the capacitance curve when it is plotted versus the square root of the time, t. Inspection of data reported in Figure 8 reveals that the slope of EG1 is lower than those exhibited by the base and the EG05. This finding suggests that the addition of 1% wt. of graphene to the base epoxy modifies both the mechanism of water uptake and water diffusivity, thus reducing the accumulation of water at the metallic interface. Furthermore, from the values of capacitance extrapolated at time zero, the relative permittivity of the coatings can be obtained. These values are 4.93, 10.21, and 9.70 F/cm^2 for the base and for the 0.5 and 1% filled coatings, respectively. These values are in agreement with literature data for an epoxy resin [42] and with the effect of "carbon" filler on the relative permittivity of polymeric materials that is attributed to the polarizing effect of graphene [42]. Data reported above are in agreement with literature data suggesting that the increased performances of organic coatings are attributed to the barrier effect played by graphene that induces a water diffusivity decrease through the coating [43, 44]. In other words, the effect of graphene seems to be due, simply, to an increase of pathway length of water through the coating, acting as other nanoparticles used as a filler. Data reported in Figure 8 show that the change in capacitance of the epoxy coating assumes a step-like behavior, displaying a sharp increase at the beginning of the test, followed by a quasi-steady-state plateau and a subsequent step rise. However, nanocomposite coatings, after the first 4 days of immersion, show a slight increase, day by day, to the end of immersion time, reaching a steady value, indicating a different water absorption process. Data reported in this paper suggest that the observed behavior of the nanofilled epoxy can be attributed to a different mechanism of water adsorption and transport through the coating rather than to a simple barrier effect. This assumption has not been highlighted until now and needs to be verified by collecting more data.

4. Conclusions

In this work, the properties of graphene loaded water-based epoxy coating, when applied to Al2024-T3 sample, were evaluated. Data obtained in this investigation showed that the structural properties and the adhesion to the metallic substrate were not affected by the presence of graphene, while nanoflakes addition led to a slight increase of the hydrophobic character of the coating. Furthermore, EIS data collected in this paper showed that loaded coatings exhibit improved performances toward corrosion protection of the metallic substrate. This behavior was attributed both to a barrier and to a different mechanism of water transport through the filled epoxy coating.

Conflicts of Interest

The authors declare that there are no conflicts of interest regarding the publication of this paper.

References

[1] S. Kumar, S. Raj, S. Jain, and K. Chatterjee, "Multifunctional biodegradable polymer nanocomposite incorporating graphene-silver hybrid for biomedical applications," *Materials & Design*, vol. 108, pp. 319–332, 2016.

[2] R. B. Ladani, S. Wu, A. J. Kinloch et al., "Multifunctional properties of epoxy nanocomposites reinforced by aligned nanoscale carbon," *Materials & Design*, vol. 94, pp. 554–564, 2016.

[3] C. Wang, Y. Lan, W. Yu, X. Li, Y. Qian, and H. Liu, "Preparation of amino-functionalized graphene oxide/polyimide composite films with improved mechanical, thermal and hydrophobic properties," *Applied Surface Science*, vol. 362, pp. 11–19, 2016.

[4] S. Zhao, H. Chang, S. Chen, J. Cui, and Y. Yan, "High-performance and multifunctional epoxy composites filled with epoxide-functionalized graphene," *European Polymer Journal*, vol. 84, pp. 300–312, 2016.

[5] H. Kim, Y. Miura, and C. W. Macosko, "Graphene/polyurethane nanocomposites for improved gas barrier and electrical conductivity," *Chemistry of Materials*, vol. 22, no. 11, pp. 3441–3450, 2010.

[6] S. H. Song, K. H. Park, B. H. Kim et al., "Enhanced thermal conductivity of epoxy-graphene composites by using non-oxidized graphene flakes with non-covalent functionalization," *Advanced Materials*, vol. 25, no. 5, pp. 732–737, 2013.

[7] B. Tang, G. Hu, H. Gao, and L. Hai, "Application of graphene as filler to improve thermal transport property of epoxy resin for thermal interface materials," *International Journal of Heat and Mass Transfer*, vol. 85, pp. 420–429, 2015.

[8] Y. Wang, J. Yu, W. Dai et al., "Enhanced thermal and electrical properties of epoxy composites reinforced with graphene nanoplatelets," *Polymer Composites*, vol. 36, no. 3, pp. 556–565, 2015.

[9] A. S. Wajid, H. S. T. Ahmed, S. Das, F. Irin, A. F. Jankowski, and M. J. Green, "High-performance pristine graphene/epoxy composites with enhanced mechanical and electrical properties," *Macromolecular Materials and Engineering*, vol. 298, no. 3, pp. 339–347, 2013.

[10] F. Wang, L. T. Drzal, Y. Qin, and Z. Huang, "Mechanical properties and thermal conductivity of graphene nanoplatelet/epoxy composites," *Journal of Materials Science*, vol. 50, no. 3, pp. 1082–1093, 2015.

[11] B. Zhang, R. Asmatulu, S. A. Soltani, L. N. Le, and S. S. A. Kumar, "Mechanical and thermal properties of hierarchical composites enhanced by pristine graphene and graphene oxide nanoinclusions," *Journal of Applied Polymer Science*, vol. 131, no. 19, Article ID 40826, 2014.

[12] K.-C. Chang, M.-H. Hsu, H.-I. Lu et al., "Room-temperature cured hydrophobic epoxy/graphene composites as corrosion inhibitor for cold-rolled steel," *Carbon*, vol. 66, pp. 144–153, 2014.

[13] K.-C. Chang, W.-F. Ji, M.-C. Lai et al., "Synergistic effects of hydrophobicity and gas barrier properties on the anticorrosion property of PMMA nanocomposite coatings embedded with graphene nanosheets," *Polymer Chemistry*, vol. 5, no. 3, pp. 1049–1056, 2014.

[14] D. Liu, W. Zhao, S. Liu, Q. Cen, and Q. Xue, "Comparative tribological and corrosion resistance properties of epoxy composite coatings reinforced with functionalized fullerene C60 and graphene," *Surface and Coatings Technology*, vol. 286, pp. 354–364, 2016.

[15] S. Liu, L. Gu, H. Zhao, J. Chen, and H. Yu, "Corrosion resistance of graphene-reinforced waterborne epoxy coatings," *Journal of Materials Science & Technology*, vol. 32, no. 5, pp. 425–431, 2016.

[16] M. Rajabi, G. R. Rashed, and D. Zaarei, "Assessment of graphene oxide/epoxy nanocomposite as corrosion resistance coating on

carbon steel," *Corrosion Engineering Science and Technology*, vol. 50, no. 7, pp. 509–516, 2015.

[17] B. Ramezanzadeh, S. Niroumandrad, A. Ahmadi, M. Mahdavian, and M. H. M. Moghadam, "Enhancement of barrier and corrosion protection performance of an epoxy coating through wet transfer of amino functionalized graphene oxide," *Corrosion Science*, vol. 103, pp. 283–304, 2016.

[18] B. Ramezanzadeh, A. Ahmadi, and M. Mahdavian, "Enhancement of the corrosion protection performance and cathodic delamination resistance of epoxy coating through treatment of steel substrate by a novel nanometric sol-gel based silane composite film filled with functionalized graphene oxide nanosheets," *Corrosion Science*, vol. 109, pp. 182–205, 2016.

[19] Z. Yu, H. Di, Y. Ma et al., "Preparation of graphene oxide modified by titanium dioxide to enhance the anti-corrosion performance of epoxy coatings," *Surface and Coatings Technology*, vol. 276, pp. 471–478, 2015.

[20] Z. Yu, L. Lv, Y. Ma, H. Di, and Y. He, "Covalent modification of graphene oxide by metronidazole for reinforced anti-corrosion properties of epoxy coatings," *RSC Advances*, vol. 6, no. 22, pp. 18217–18226, 2016.

[21] L. Gu, S. Liu, H. Zhao, and H. Yu, "Facile preparation of water-dispersible graphene sheets stabilized by carboxylated oligoanilines and their anticorrosion coatings," *ACS Applied Materials & Interfaces*, vol. 7, no. 32, pp. 17641–17648, 2015.

[22] W. Xiao, Y. Liu, and S. Guo, "Composites of graphene oxide and epoxy resin assuming a uniform 3D graphene oxide network structure," *RSC Advances*, vol. 6, no. 90, pp. 86904–86908, 2016.

[23] N. Yousefi, X. Lin, Q. Zheng et al., "Simultaneous in situ reduction, self-alignment and covalent bonding in graphene oxide/epoxy composites," *Carbon*, vol. 59, pp. 406–417, 2013.

[24] T. Monetta, A. Acquesta, and F. Bellucci, "Graphene/epoxy coating as multifunctional material for aircraft structures," *Aerospace*, vol. 2, no. 3, pp. 423–434, 2015.

[25] "Council Directive 1999/13/EC of 11 March 1999 on the limitation of emissions of volatile organic compounds due to the use of organic solvents in certain activities and installations," *Official Journal of the European Union*, no. L 085, pp. 0001–0022, 1999.

[26] F. Andreatta, M. Bortolotto, A. Lanzutti, L. Paussa, D. Bravin, and L. Fedrizzi, "Environmentally friendly conversion coating for aluminium alloy AA6014," in *Proceedings of 18th International Corrosion Congress*, pp. 355–366, November 2011.

[27] F. Deflorian, S. Rossi, and M. Fedel, "Aluminium components for marine applications protected against corrosion by organic coating cycles with low environmental impact," *Corrosion Engineering Science and Technology*, vol. 46, no. 3, pp. 237–244, 2011.

[28] C. Sinagra, F. Bravaccino, C. Bitondo et al., "Green Technology for Surface Treatments of Aluminium Foil for Flexible Packaging," *Key Engineering Materials*, vol. 710, pp. 186–191, 2016.

[29] C. Bitondo, A. Bossio, T. Monetta, M. Curioni, and F. Bellucci, "The effect of annealing on the corrosion behaviour of 444 stainless steel for drinking water applications," *Corrosion Science*, vol. 87, pp. 6–10, 2014.

[30] A. Carangelo, M. Curioni, A. Acquesta, T. Monetta, and F. Bellucci, "Cerium-based sealing of anodic films on AA2024T3: effect of pore morphology on anticorrosion performance," *Journal of the Electrochemical Society*, vol. 163, no. 14, pp. C907–C916, 2016.

[31] T. Monetta, A. Acquesta, V. Maresca et al., "Characterization of aluminum alloys environmentally friendly surface treatments for aircraft and aerospace industry," *Surface and Interface Analysis*, vol. 45, no. 10, pp. 1522–1529, 2013.

[32] A. Scala, A. Squillace, T. Monetta, D. B. Mittonb, D. Larsonb, and F. Belluccia, "Corrosion fatigue on 2024T3 and 6056T4 aluminum alloys," *Surface and Interface Analysis*, vol. 42, no. 4, pp. 194–198, 2010.

[33] J. Carneiro, J. Tedim, S. C. M. Fernandes et al., "Chitosan-based self-healing protective coatings doped with cerium nitrate for corrosion protection of aluminum alloy 2024," *Progress in Organic Coatings*, vol. 75, no. 1-2, pp. 8–13, 2012.

[34] H. A. Fetouh, T. M. Abdel-Fattah, and M. S. El-Tantawy, "Novel plant extracts as green corrosion inhibitors for 7075-t6 aluminium alloy in an aqueous medium," *International Journal of Electrochemical Science*, vol. 9, no. 3, pp. 1565–1582, 2014.

[35] M. Gobara, H. Kamel, R. Akid, and A. Baraka, "Corrosion behaviour of AA2024 coated with an acid-soluble collagen/hybrid silica sol-gel matrix," *Progress in Organic Coatings*, vol. 89, pp. 57–66, 2015.

[36] T. Monetta, D. B. Mitton, and F. Bellucci, "Protective properties of organic coatings on plasma-treated cold rolled aluminum," *Electrochemical and Solid-State Letters*, vol. 7, no. 11, pp. B39–B41, 2004.

[37] L. De Rosa, T. Monetta, and F. Bellucci, "Moisture uptake in organic coatings monitored with EIS," *Materials Science Forum*, vol. 289–292, no. 1, pp. 315–326, 1998.

[38] F. Mansfeld, "Use of electrochemical impedance spectroscopy for the study of corrosion protection by polymer coatings," *Journal of Applied Electrochemistry*, vol. 25, no. 3, pp. 187–202, 1995.

[39] F. Bellucci and L. Nicodemo, "Water transport in organic coatings," *Corrosion*, vol. 49, no. 3, pp. 235–247, 1993.

[40] L. Nicodemo, F. Bellucci, A. Marcone, and T. Monetta, "Water and oxygen transport as performance parameters of paint films," *Journal of Membrane Science*, vol. 52, no. 3, pp. 393–403, 1990.

[41] G. W. Walter, "A review of impedance plot methods used for corrosion performance analysis of painted metals," *Corrosion Science*, vol. 26, no. 9, pp. 681–703, 1986.

[42] C. C. Ku and R. Liepins, *Electrical Properties of Polymers: Chemical Principles*, Hanser Publishers, 1987.

[43] Y. Su, V. G. Kravets, S. L. Wong, J. Waters, A. K. Geim, and R. R. Nair, "Impermeable barrier films and protective coatings based on reduced graphene oxide," *Nature Communications*, vol. 5, article 4843, 2014.

[44] B. M. Yoo, H. J. Shin, H. W. Yoon, and H. B. Park, "Graphene and graphene oxide and their uses in barrier polymers," *Journal of Applied Polymer Science*, vol. 131, no. 1, Article ID 39628, 2014.

12

Chemical Composition of Corrosion Products of Rebar Caused by Carbonation and Chloride

Jundi Geng⊙, **Junzhe Liu**⊙, **Jiali Yan, Mingfang Ba, Zhimin He, and Yushun Li**

Faculty of Architectural, Civil Engineering and Environment, Ningbo University, Ningbo 315211, China

Correspondence should be addressed to Junzhe Liu; junzheliu@163.com

Academic Editor: Michael J. Schütze

The microstructures of steel bars were studied by X-ray photoelectron spectroscopy (XPS), and the mechanism of corrosion of steel bars under the corrosion factors was elucidated. The results show that the passivation film and corrosive surface of the steel surface in the solution of the chloride-containing salt were coarser and the surface state was denser. The main corrosion products are FeOOH and FeO. The surface of the steel immersed in the simulated carbonized solution had loose pores. The main components are FeOOH, Fe_3O_4, and Fe_2O_3. The surface of the steel bar has a large amount of yellowish brown corrosion products in the simulated carbonization and chloride salt. The surface of the corrosion products was stripped and the main components are FeOOH, Fe_3O_4, and $FeCl_3$, where the content of FeOOH is as high as 60%. The peak value of iron is gradually increased from the simulated chloride salt solution to the carbonized solution to the combined effect of carbonation and chloride salt; the iron oxide content is increased and corrosion of steel is obviously serious.

1. Introduction

Chloride salt and carbonization are the main causes of steel corrosion, as evidenced by the failure of a large number of reinforced concrete structures as well as a large number of concrete works under double factor of carbon dioxide and chloride ion [1–4]. In general, the steel surface of the passive film in the steel is stable due to the presence of overbased concrete pore solution. When the outside carbon dioxide and chloride ions infiltrated into the concrete, the pH of the pore solution decreased and chloride content increased, leading to the destruction of the passive film corrosion [5–9]. Therefore, the main barrier to prevent corrosion of steel is the passive film.

There are many factors affecting the destruction of passivation film, including the surface conditions of steel bars, alloy composition, iron phase composition, and other material factors as well as the permeability of concrete, the concentration of chloride ion, the PH of solution, temperature and humidity, and other environmental factors [10–12]. The properties of the steel passivation film, that is, the thickness, composition, and stability, are influenced by the polarization potential, the polarization time, and the ion concentration in the medium, and the microstructure characteristics of passivation films are related to passivation potential and passivation time. The reason for the corrosion of steel is finally due to the change of the composition and structure of passivation film [13–16]. It can be seen that there is an urgent need to clarify the failure process of passivation film under carbonation and chloride corrosion and clarify the damage mechanism of passivation film of steel reinforcement under the action of corrosion factor so as to improve the obtuse environment on the surface of rebar in concrete [17].

Ghods used X-ray photoelectron spectroscopy (XPS) to study the passive oxide film of carbon steel in saturated calcium hydroxide solution and the effect of chloride on the film properties showing that chloride exposure decreased the thickness of the oxide films and changed their stoichiometry such that near the film/substrate interface Fe^{3+}/Fe^{2+} ratio increased.

In this paper, the key factors influencing the passive film structure of the surface of rebar in concrete are taken as the starting points, and the composition and microstructure characteristics of the corrosion on the surface of the rebar

TABLE 1: The ratio of simulated concrete pore solution.

Species	Uncarbonized simulated pore solution			Carbonized simulated pore solution	
Reagents	Ca(OH)$_2$	NaOH	KOH	Na$_2$CO$_3$	NaHCO$_3$
Mol/L	0.001	0.2	0.6	0.0015	0.03

are studied under the action of carbonation and chloride corrosion so as to provide the theoretical basis to improve the corrosion resistance of steel through optimizing the composition of the steel surface oxide layer.

2. Experiment

2.1. Materials and Specimen Preparation. The 10 mm diameter steel bars are cut into 2 mm steel thin section, which are polished with 400-mesh, 500-mesh, and 800-mesh sandpaper, wiped clean with 95% alcohol, and placed in a desiccator for rusting X-ray photoelectron spectroscopy (XPS) analysis. Three simulated concrete pore solution environments were used. Table 1 shows the composition of the precarbonized concrete pore solution and the simulated carbonated concrete pore solution. Test was divided into three groups' preparation: (1) 3% sodium chloride is mixed with the concrete hole solution to simulate the chloride salt corrosion environment; (2) pore solution is carbonized to simulate a carbonized corrosive environment; (3) carbonation hole solution was performed by adding 3% sodium chloride to simulate the carbonation and chloride salt composite corrosion environment [18].

Pour 300 ml of the three prepared simulated concrete hole solutions into three covered glass flasks, placing two steel thin sections in each of the glass flasks and tightening the cap. After six months, the steel thin section was removed from the solution, rinsed with deionized water and acetone, and then blown dry with argon and stored in an Ar-filled container. The samples were determined by X-ray diffraction analysis and X-ray photoelectron spectroscopy within two hours.

2.2. Instruments

X-Ray Photoelectron Spectroscopy (XPS). XPS used Mg target, X-ray emission current of 20 mA, high-voltage X-ray source 10 kV, doubler voltage 2.8 kV, full spectrum energy 100 eV, and semispectral energy 50 eV. Scan was performed 20 times; each step time was 10 ms. Ar$^+$ was sputtered 5 s before each test starting so as to remove the effects of surface contaminants. The sputtering speed of Ar$^+$ is about 3 nm/min. The data was subjected to fractional fitting analysis using CasaXPS 2.3.16 after XPS testing at 0 nm and 5 nm. The peak curves for all elements are calibrated with C1s, with a binding energy of 284.6 ev.

X-Ray Diffractometer (XRD), D8 Advance Davinci from Bruker, Germany, was used. Using Cu and Kα1 ray, the tube voltage is 40 Kv; the tube current is 40 mA. Continuous scanning was performed with a scan range of 20° to 70°, a scan rate of 8°/min, and a step size of 0.02°.

3. Analysis and Discussion

3.1. XPS Analysis of Steel Corrosion Products. Figure 1 shows the XPS full scan spectra of steel bar corroded after 6 months of immersion in various simulated corrosion hole solutions. The main components of steel bar corroded in the hole solution are iron, oxygen, carbon, and chlorine. Among them, oxygen, carbon, and iron peak are strong, indicating that the chemical composition of corrosion is mainly iron oxide.

There is no carbon peak in the simulated chloride salt solution, while the peak of carbon in Figure 1(a) is very strong. On the one hand, carbon comes from the steel itself and on the other hand it comes from the dissolution of carbon dioxide during the test. The Fe 2p binding energy of steel corroded in the simulated chloride salt solution was 709.96 ev, the Fe 2p binding energy of the steel corroded in the simulated carbonation solution was 710.76 ev, and the binding energy of Fe 2p in the simulated carbonation and chloride salt solution was 711.30 ev. This shows that in these three solutions the components of steel corrosion are iron oxide, but the specific composition is different. From simulated chloride salt solution to simulated carbonation solution and then to the simulated carbonization and chloride salt composite solution, the peak value of Fe in steel corrosion products gradually increases, indicating that the content of iron oxide gradually increases and the corrosion of steel reinforcement becomes more serious. The Fe 2p orbital has a bimodal structure due to the splitting of the spin into two energy levels (i.e., Fe 2p 1/2 and Fe 2p 3/2). The corrosive iron compounds can be divided into four classes of Fe-1, Fe-2, Fe-3, and Fe-4, corresponding to elemental Fe, Fe$_3$O$_4$/FeO (Fe^{2+}), Fe$_2$O$_3$/FeOOH (Fe^{3+}), and FeCl$_3$ (Fe^{3+}). The XPS peak fitting of the Fe element of the steel bar corroded after immersing the steel bar in different simulated hole solution for 6 months is shown in Figure 2.

The peak curves of Fe 2p in the simulated corrosion of chloride salt solution fit four peak curves, and the corresponding compounds are FeO and FeOOH, respectively. By fitting, it can be concluded that the binding energy of the peak position and curve peak area of Fe element in corrosion solution of steel bar in chloride salt solution are shown in Table 2, the relative content of FeO is 32.3%, and the relative content of FeOOH is 67.7%.

The peak curves of Fe 2p in the simulated carbonation solution fit three peak curves, and the corresponding compounds are Fe$_3$O$_4$ and FeOOH, respectively. By fitting, it can be concluded that the binding energy of the peak position and curve peak area of Fe element in corrosion solution of steel bar in chloride salt solution are shown in Table 3, the relative content of Fe$_3$O$_4$ is 39.9%, and the relative content of FeOOH is 60.1%.

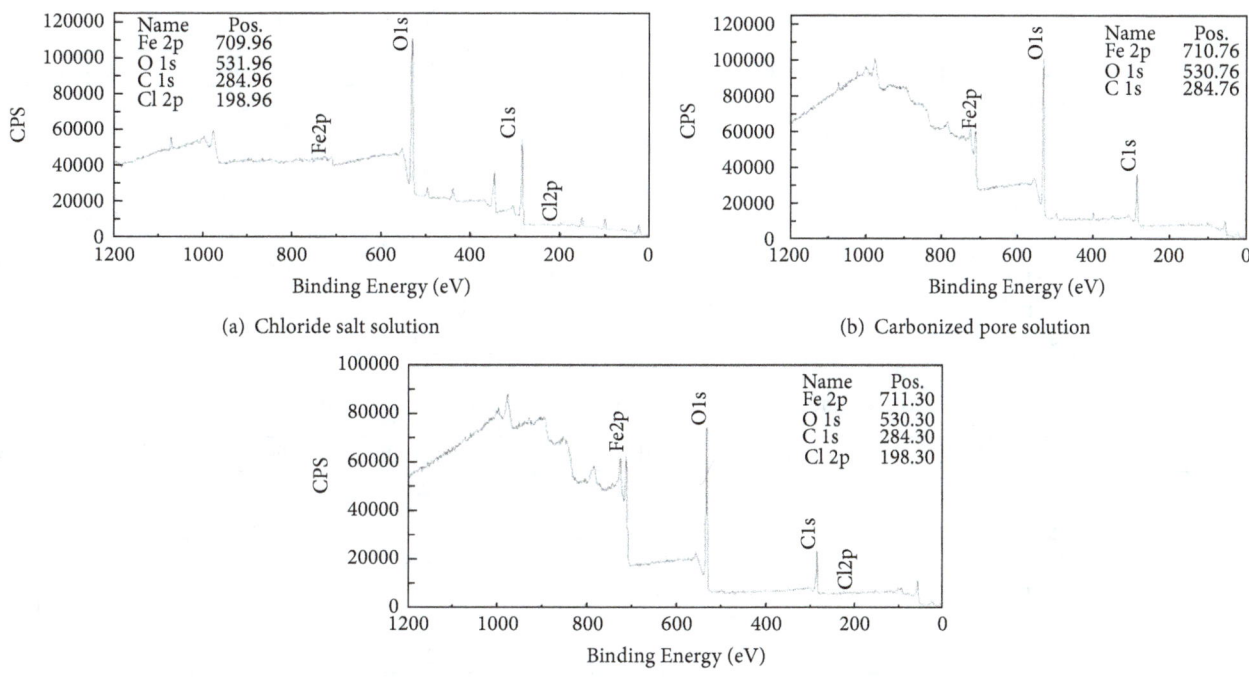

(a) Chloride salt solution

(b) Carbonized pore solution

(c) Carbonization and chloride salt composite pore solution

FIGURE 1: XPS full scanning diagrams of reinforcement corrosion in simulated solutions.

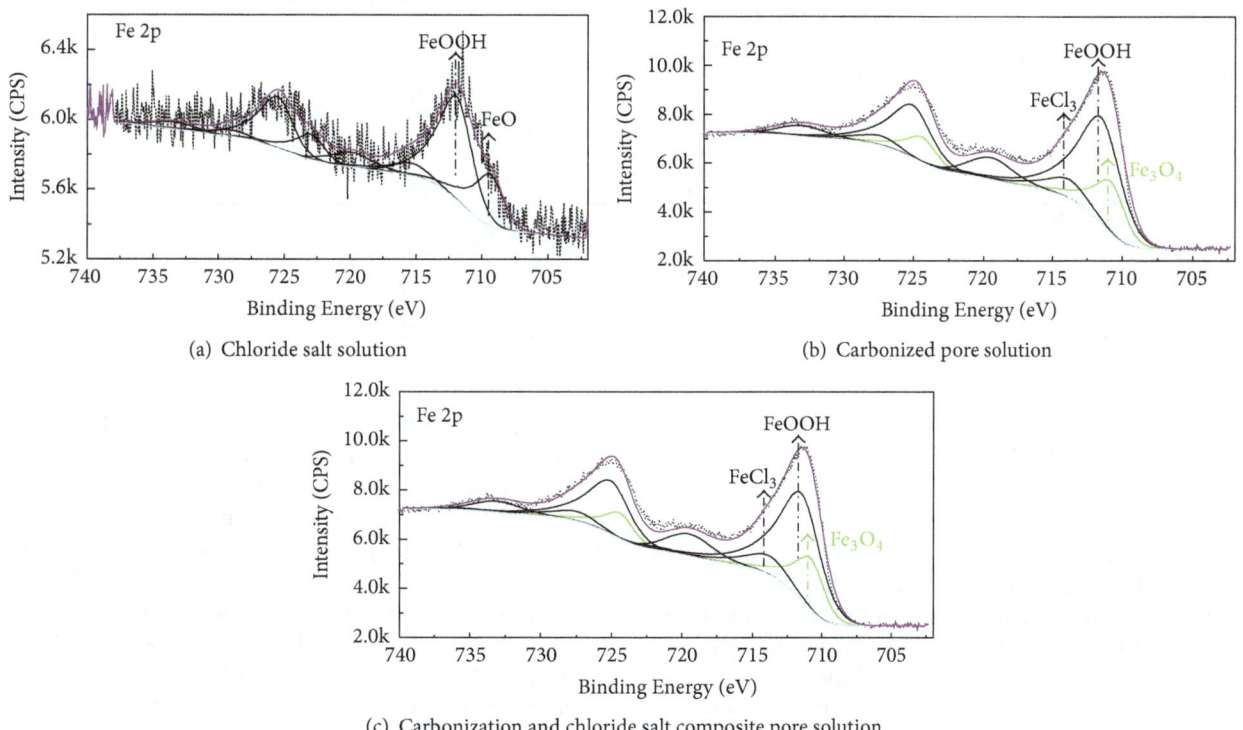

(a) Chloride salt solution

(b) Carbonized pore solution

(c) Carbonization and chloride salt composite pore solution

FIGURE 2: XPS scanning diagrams of Fe element in simulated solutions.

The peak curves of Fe 2p in carbonization and chloride salt composite pore solution fit five peak curves, and the corresponding compounds are Fe_3O_4, FeOOH, and $FeCl_3$, respectively. By fitting, it can be concluded that the binding energy of the peak position and curve peak area of Fe element in corrosion solution of steel bar in chloride salt solution are

TABLE 2: XPS scanning data of Fe element in the simulate chloride salt solution.

Ingredient	Energy level of Fe 2p	Binding energy of peak position	Half width	Peak area	Relative content
FeO	Fe 2p3/2	709.4 ev	2.71	2037.3	32.3%
	Fe 2p1/2	723 ev	2.71	1024.3	
	Fe 2p3/2	715.4 ev	3.9	893.8	
	Fe 2p1/2	729 ev	3.9	449.2	
FeOOH	Fe 2p3/2	711.9 ev	2.4	6215	67.7%
	Fe 2p1/2	725.5 ev	2.4	3125	
	Fe 2p3/2	719.9 ev	3.14	196.9	
	Fe 2p1/2	733.5 ev	3.14	99	

TABLE 3: XPS scanning data of Fe element in the simulate carbonized solution.

Ingredient	Energy level of Fe 2p	Binding energy of peak position	Half width	Peak area	Relative content
Fe_3O_4	Fe 2p3/2	710.8 ev	2.3	825.7	39.9%
	Fe 2p1/2	724.4 ev	2.3	920.6	
FeOOH	Fe 2p3/2	711.5 ev	2.4	1120.1	60.1%
	Fe 2p1/2	725.1 ev	2.4	206.6	
	Fe 2p3/2	719.5 ev	3.14	1336.9	
	Fe 2p1/2	733.1 ev	3.14	206.6	

TABLE 4: XPS scanning data of Fe in the combined effect of carbonation and chloride salt.

Ingredient	Energy level of Fe 2p	Binding energy of peak position	Half width	Peak area	Relative content
Fe_3O_4	Fe 2p3/2	710.8 ev	2.3	906.7	19.3%
	Fe 2p1/2	724.4 ev	2.3	305.1	
FeOOH	Fe 2p3/2	711.5 ev	2.4	2230.6	69.1%
	Fe 2p1/2	725.1 ev	2.4	312.3	
	Fe 2p3/2	719.5 ev	3.14	2323.1	
	Fe 2p1/2	733.1 ev	3.14	343.1	
$FeCl_3$	Fe 2p3/2	713.9 ev	2.7	409.9	11.6%
	Fe 2p1/2	727.5 ev	2.7	102.4	
	Fe 2p3/2	721.9 ev	3.2	468.4	
	Fe 2p1/2	735.5 ev	3.2	130.3	

shown in Table 4, the relative content of Fe_3O_4 is 19.3%, the relative content of FeOOH is 69.1%, and the relative content of $FeCl_3$ is 11.6%.

3.2. XRD Analysis of Corroded Steel. Reinforced corroded surface with 5000 times magnification of the scanning electron microscope (SEM) and corresponding corrosive X-ray diffraction spectrum (XRD) after immersing for 6 months in the chloride salt simulated pore solution is shown in Figure 3(a). Most of the surface of the reinforced concrete in the simulated chloride salt solution is not corroded, and the surface is relatively flat and locally convex, indicating that part of the surface is covered with corrosives and the corrosion layer is a thin layer. The steel surface enlarged 5K times and holes were found in some places, indicating the passivation film of the reinforcing steel in the simulated concrete pore solution is slowly penetrated by chloride ions and begins to corrode. Analysis by JADE6.5 software shows that there are two clear main peaks at 44.6° and 64.9° in 2θ of steel of

chloride salt solution. The results show that there are many noncorroded areas in the simulated salt solution of chloride salt. Compared with other standard cards, it is found that the corrosion products in this solution are mainly FeOOH and FeO, which is consistent with the results of the previous XPS analysis.

Reinforced corroded surface with 5000 times magnification of the scanning electron microscope (SEM) and corresponding corrosive X-ray diffraction spectrum (XRD) after immersing for 6 months in the simulation of carbide pore solution is shown in Figure 3(b). Steel surface generated a yellow-black rust, but the surface is relatively smooth, with no rust protrusions. The steel surface enlarged 5K times and it was found that reinforced steel surface has been corroded and loose porous, with stick shape, and there are three clear main peaks at 36.8°, 44.7°, and 65.1° in 2θ of steel. Compared with the XRD standard card, the main diffraction peak of Fe_3O_4 was found, and the diffraction peak of the simple Fe was not found. Compared with Figure 3(a),

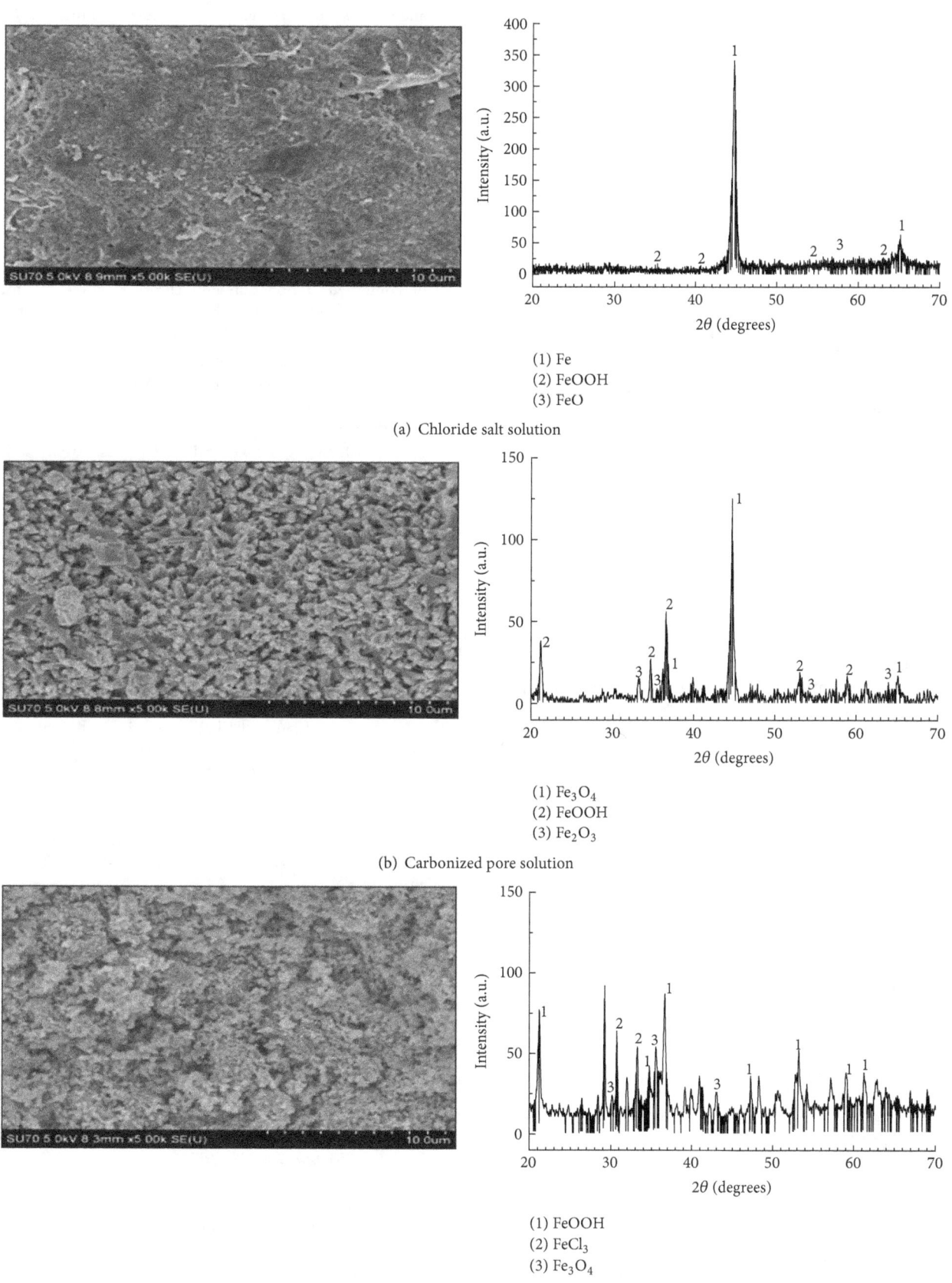

(a) Chloride salt solution

(b) Carbonized pore solution

(c) Carbonization and chloride salt composite pore solution

FIGURE 3: Morphology and composition of corroded surface of steel bar.

there were other minor peaks. This shows that the surface of the rebar immersed in the simulation of carbide pore solution has been completely corroded with the formation of FeOOH and Fe_2O_3, which is a good supplement to XPS analysis.

Reinforced corroded surface with 5000 times magnification of the scanning electron microscope (SEM) and corresponding corrosive X-ray diffraction spectrum (XRD) after immersing for 6 months in carbonation and chloride salt composite pore solution is shown in Figure 3(c). The corroded steel bars are brown in color and are in the shape of a layer with uneven thickness distribution and spalling phenomenon. The peak value of the corrosion products is conspicuous with respect to Figures 3(a) and 3(b). The main diffraction peak compared with the standard card shows FeOOH, subpeak was found in $FeCl_3$ and Fe_3O_4 in the corrosion, and the intensity has been enhanced. This shows that the corrosion of reinforcing steel is more serious under the condition of chloride salt and carbonized composite, which is consistent with the result of XPS analysis.

4. Conclusion

(1) Steel passivation film and rust coexist in chloride salt simulated concrete pore solution, the surface state is more dense, and the main components of corrosion are FeOOH and FeO. Being reinforced with yellow and black rust corrosion and loose porous in the simulation of carbide pore solution, the main ingredients are FeOOH, Fe_3O_4, and Fe_2O_3.

(2) There is a lot of brown rust on the surface of the corrosive and the surface of corrosive was exfoliated in carbonation and chloride salt composite pore solution. The main components are FeOOH, Fe_3O_4, and $FeCl_3$, where FeOOH content is more than 60%.

(3) The ratio of Fe^{3+}/Fe^{2+} increases as chloride ions increase or carbonization increases. And the ratio is maximum under the combined action of chloride ion and carbonization.

Conflicts of Interest

The authors declare that they have no conflicts of interest.

Acknowledgments

This work was sponsored by the National Natural Science Foundation of China (51478227, 51778302) and by K. C. Wong Magna Fund in Ningbo University.

References

[1] A. Poursaee, "Temperature dependence of the formation of the passivation layer on carbon steel in high alkaline environment of concrete pore solution," *Electrochemistry Communications*, vol. 73, pp. 24–28, 2016.

[2] A. Scott and M. G. Alexander, "Effect of supplementary cementitious materials (binder type) on the pore solution chemistry and the corrosion of steel in alkaline environments," *Cement and Concrete Research*, vol. 89, pp. 45–55, 2016.

[3] R. Liu, L. Jiang, J. Xu, C. Xiong, and Z. Song, "Influence of carbonation on chloride-induced reinforcement corrosion in simulated concrete pore solutions," *Construction and Building Materials*, vol. 56, pp. 16–20, 2014.

[4] T. V. Shibaeva, V. K. Laurinavichyute, G. A. Tsirlina, A. M. Arsenkin, and K. V. Grigorovich, "The effect of microstructure and non-metallic inclusions on corrosion behavior of low carbon steel in chloride containing solutions," *Corrosion Science*, vol. 80, pp. 299–308, 2014.

[5] Y. Wang, S. Nanukuttan, Y. Bai, and P. A. M. Basheer, "Influence of combined carbonation and chloride ingress regimes on rate of ingress and redistribution of chlorides in concretes," *Construction and Building Materials*, vol. 140, pp. 173–183, 2017.

[6] T. Bellezze, G. Roventi, E. Barbaresi, N. Ruffini, and R. Fratesi, "Effect of concrete carbonation process on the passivating products of galvanized steel reinforcements," *Materials and Corrosion*, vol. 62, no. 2, pp. 155–160, 2011.

[7] H. Luo, H. Su, C. Dong, and X. Li, "Passivation and electrochemical behavior of 316L stainless steel in chlorinated simulated concrete pore solution," *Applied Surface Science*, vol. 400, pp. 38–48, 2017.

[8] J. Williamson and O. B. Isgor, "The effect of simulated concrete pore solution composition and chlorides on the electronic properties of passive films on carbon steel rebar," *Corrosion Science*, vol. 106, pp. 82–95, 2016.

[9] H. B. Gunay, P. Ghods, O. B. Isgor, G. J. C. Carpenter, and X. Wu, "Characterization of atomic structure of oxide films on carbon steel in simulated concrete pore solutions using EELS," *Applied Surface Science*, vol. 274, pp. 195–202, 2013.

[10] A. S. Yaro, K. R. Abdul-Khalik, and A. A. Khadom, "Effect of CO_2 corrosion behavior of mild steel in oilfield produced water," *Journal of Loss Prevention in the Process Industries*, vol. 38, pp. 24–38, 2015.

[11] E. Sosa, R. Cabrera-Sierra, M. T. Oropeza et al., "Electrochemically grown passive films on carbon steel (SAE 1018) in alkaline sour medium," *Electrochimica Acta*, vol. 48, no. 12, pp. 1665–1674, 2003.

[12] Y. Tang, S. Wang, Y. Xu, and J. Ni, "Influence of calcium nitrite on the passive films of rebar in simulated concrete pore solution," *Anti-Corrosion Methods and Materials*, vol. 64, no. 3, pp. 265–272, 2017.

[13] A. Alhozaimy, R. R. Hussain, A. Al-Negheimish, R. Al-Zaid, and D. D. N. Singh, "Effect of simulated concrete pore solution chemistry, chloride ions, and temperature on passive layer formed on steel reinforcement," *ACI Materials Journal*, vol. 111, no. 4, pp. 411–421, 2014.

[14] Y. Zhou and Y. Zuo, "Surface films on plasma nitrided stainless steel subjected to passivation treatments," *Applied Surface Science*, vol. 353, pp. 924–932, 2015.

[15] J. Z. Liu, W. Sun, J. B. Chen et al., "Carbonation rate and microstructural characteristics of hardened cement paste with high alkali content," *Journal of the Chinese Ceramic Society*, vol. 42, no. 7, pp. 1005–1010, 2014.

[16] P. Ghods, O. B. Isgor, J. R. Brown, F. Bensebaa, and D. Kingston, "XPS depth profiling study on the passive oxide film of carbon steel in saturated calcium hydroxide solution and the effect of chloride on the film properties," *Applied Surface Science*, vol. 257, no. 10, pp. 4669–4677, 2011.

[17] R. Blair, B. Pesic, J. Kline, I. Ehrsam, and K. Raja, "Threshold chloride concentrations and passivity breakdown of rebar steel in real concrete solution at different pH conditions with the addition of glycerol," *Acta Metallurgica Sinica (English Letters)*, vol. 30, no. 4, pp. 376–389, 2017.

[18] J. Liu, M. Ba, Y. Du, Z. He, and J. Chen, "Corrigendum to "Effects of chloride ions on carbonation rate of hardened cement paste by X-ray CT techniques" [Constr. Build. Mater. 122 (2016) 619–627]," *Construction and Building Materials*, vol. 131, p. 800, 2017.

13

Effect of Rare-Earth Elements on the Corrosion Resistance of Flux-Cored Arc-Welded Metal with 10CrNi3MoV Steel

Kai Wang[ID]**,**[1,2] **Qinghua Lu,**[1] **Zexin Jiang,**[3] **Yaoyong Yi,**[2] **Jianglong Yi**[ID]**,**[2] **Ben Niu,**[2] **Jinjun Ma,**[3] **and Huiping Hu**[1]

[1]*Foshan University, Foshan, China*
[2]*Guangdong Welding Institute (China-Ukraine E. O. Paton Institute of Welding), Guangzhou, China*
[3]*Guangzhou Shipyard International Co., Ltd, Guangzhou, China*

Correspondence should be addressed to Kai Wang; hfutwk927@gmail.com

Academic Editor: Flavio Deflorian

We modified the content of rare-earth elements (REE) in the flux-cored wire used to produce welds of high-strength low-alloy (HSLA) steel. The effect of REE addition on the microstructure as well as on the mechanical and electrochemical properties of the welded metal (WM) was investigated. REE-modified welded metals show very different responses during electrochemical impedance spectroscopy and the potentiodynamic polarization tests. The results indicate that the addition of REE of 0.3 wt.% facilitates a more uniform microstructure and improves both mechanical properties and corrosion resistance in welded metals.

1. Introduction

Welding is a common joining method, which is commercially used worldwide. The welding process for the manufacturing of a steel ship, for example, takes up 30%~40% of the overall production time. Flux-cored arc welding (FCAW) is a semiautomatic or automatic arc-welding process that involves the fusion of a flux-cored wire metal with a base metal. FCAW in particular is one of the most widely used welding techniques in the modern shipbuilding industry. The properties of FCAW joints are largely determined by the used welding consumables and the base metals used during the welding process.

10CrNi3MoV steel is a typical high-strength low-alloy steel (HSLA) that combines strength, toughness, and weld-ability thanks to Q-tempered processing. This type of steel is generally used as hull material in high-performance marine vehicles [1]. Due to the high-performance requirements of the welded joints in these vehicles, both low-temperature impact toughness and corrosion resistance of the welded metal determine the overall performance of 10CrNi3MoV steel.

One of the most effective approaches to improve tough-ness and corrosion resistance of a welded metal is to alter the welded metal composition by introducing alloying elements via the flux-cored wire to act as FCAW filler metals [2]. Generally, rare-earth elements (REE) are considered the most suitable microalloying elements for any mature alloy system. A minor addition of REE into the steel can significantly purify the liquid steel and modify any inclusions, which improves the steel properties [3–5]. The correct REE content can improve both impact toughness and temper brittleness of the welded components. This is due to its effect on grain refinement, grain boundary cleaning, and the suppression of grain boundary embrittlement [6]. Theoretical calculations show [7–9] that REE can increase the grain boundary cohesion in steel, which leads to toughening of the material via grain boundary segregation. Furthermore, the addition of REE can also enhance the high-temperature ductility of steel [10, 11]. Tomita [12] found that the addition of REE to vacuum-melted AF1410 steel can stabilize complex-inclusions and improve toughness. Gao et al. substantially improved the toughness of H13 steel by adding 0.015 wt.% REE, which favors finer and more dispersive inclusions [13]. The addition of REE to welded metals to SAW DH32 steel facilitates both inclusions and a small lattice disregistry, which improves elongation, tensile strength, and impact toughness of welded metals [14, 15]. It has also been reported that REE can improve the corrosion resistance of low-carbon steels [16, 17].

TABLE 1: Chemical composition of REE-Si-Fe (wt. %).

Ce	La	Nd	Pr	Sm	Ca	Si	Fe
12.56	4.43	1.41	0.55	4.53	3.1	40.9	Re

TABLE 2: Selected content of REE-Si-Fe for the weld samples in flux-cored wire (flux wt.%).

Weld Sample	REE-Si-Fe addition
REE0	-
REE1	0.3
REE2	0.5
REE3	0.7
REE4	1.0

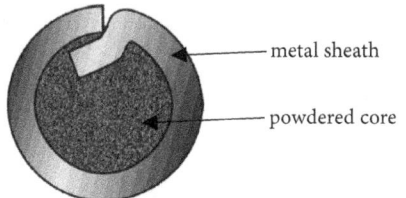

FIGURE 1: Schematic of a flux-cored wire.

However, the effects of REE on the features and properties of welded metal for 10CrNi3MoV steel, in particular, have received little attention, and no general agreement on the correct amount of REE in welded metal has been reported.

It is likely that REE elements can enhance the properties of welded metal through microstructural control during FCAW. In this work, the effect of REE addition to the flux-cored wire on both the microstructure and the properties of welded metal was investigated. In particular, we focused on the correlation between REE content and the resulting properties of the joints.

2. Experimental

The 10CrNi3MoV steel plates were arc-welded using flux-cored wires with different REE concentrations. REE (consisting of Ce-rich rare-earth ferrosilicon, REE-Si-Fe). The chemical composition of REE-Si-Fe is shown in Table 1. The flux-cored wires consist of a metal sheath and a powdered core; see Figure 1. They were prepared after the shaping of a cold-rolled strip and the filling of the hollow core with a powdered mixture and a XZ-YCX8 flux-cored wire production machine. In this study, the basic powdered core belongs to a rutile-fluorite alloying system with a filling ratio of 20%. The diameter of the finished wire was 1.2 mm. The chosen REE contents for the powdered cores are shown in Table 2. Five group samples with different REE concentrations in the welded metals were fabricated in the laboratory using identical welding conditions. The chemical composition and mechanical properties of the 10CrMo3NiV steel base metal are shown in Tables 3 and 4. The conditions and parameters for the FCAW process are listed in Table 5. The drawing of the FCAW grooves and welded metal samples for mechanical testing were prepared according to the Chinese standard GB/T17493-2008; see Figure 2.

The chemical composition of the welded metals (except for carbon, C) was determined using a JY ULTIMA inductive coupled plasma emission spectrometer. The C content was determined using a LECO CS600 carbon-sulphur spectrometer. The phases within welded metals were identified with a D/max-IIIA X-ray diffractometer (XRD). Each welded metal sample was observed and analyzed with a DMM-440D optical microscope (OM) and a JEOLJXA-8100 scanning electron microscope (SEM) and its accessory EDX (OXFORD-7412) after grinding, polishing and etching with a 4% natal solution. The microhardness of the welded metal was measured using a HVS-1000 micro Vickers hardness tester with a load of 0.3 kgf, using four points on the same circumference of the 1/2 radius for each sample. Samples for electrochemical impedance spectroscopy (EIS) and potentiodynamic polarization (PP) testing were taken from the center of the welded metals along the longitudinal direction. This was done via linear cutting into squares of 10 mm×10 mm×0.5 mm. Before the electrochemical tests, the samples were polished with 2000-grid sandpaper, degreased with acetone, washed with distilled water, and blow-dried. The working areas of the tested samples were 1cm^2, and the remaining areas of the samples were sealed with wax. The EIS and PP tests were carried out using an Ametek P4000 electrochemical workstation with a three-electrode system. The working electrode was the tested sample (1 cm^2 working area), the auxiliary electrode was a Pt plate, and the reference electrode was saturated calomel electrode (SCE). The electrolyte was a 3.5 wt.% NaCl water solution. The working electrode was soaked in the electrolyte in open-circuit-potential (OCP) mode for 30 min before testing. The EIS mode was set as a frequency of 10^{-2} Hz ~ 10^5 Hz versus OCP with an AC drive signal amplitude of ±5 mV. The following potentiodynamic polarization mode was chosen for a scanning potential range of -1.0 V ~ +1.5 V versus OCP with a scanning speed of 1 mV/s. The polarized samples were observed with a SEM.

3. Results and Discussion

3.1. Chemical Composition Analysis. The chemical compositions of the welded metals (only key elements) are listed in Table 6. The REE addition causes significant changes of the C and Ni content but only with slight variations in the Si, Mn, and Mo content of the welded metals. Furthermore, there are rare changes in the Al and Ti content of the welded metals.

3.2. Mechanical Properties of the Welded Metals. Table 7 and Figure 3 show the mechanical properties of the FCAW-welded metals with different REE added. In Figure 3(a), the values for tensile strength, yield strength, and elongation ratio increase suddenly after the REE addition of 0.3%. The above test results, however, indicate a gradual decrease when the REE content exceeds 0.3%, which is lower than for the REE-free welded metal (REE0) before the REE addition of 1.0%. Figure 3(b) shows the low-temperature impact energy for the welded metals. The impact energy for all welded metals with added REE is higher than for REE-free welded metals,

TABLE 3: Chemical composition of 10CrNi3MoV steel (wt. %).

C	Si	Mn	Ni	Mo	Cr	V	P	S
0.11	0.31	0.39	2.72	0.23	1.05	0.08	0.010	0.005

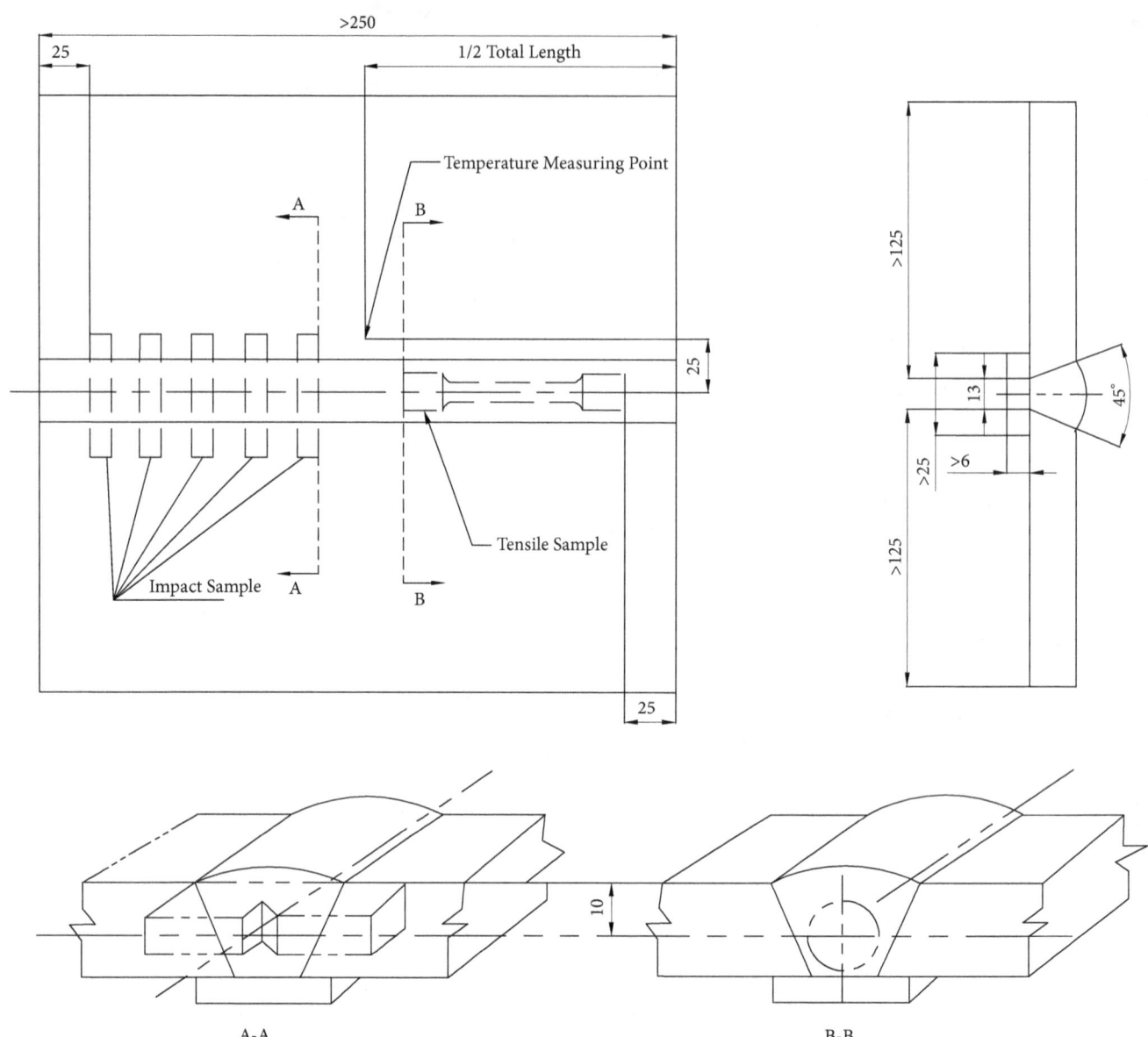

FIGURE 2: Schematic illustrating the grooves in the samples (mm).

TABLE 4: Mechanical properties of 10CrNi3MoV steel.

Yield Strength/MPa	Tensile Strength/MPa	Elongation/%	A_k -20°C/J
590~745	670~850	≥16	≥80

and it increases significantly as the REE content increases to 0.3%. However, the impact energy of the welded metals causes a slight decrease followed by a further REE content increase to 1.0%. The microhardness of the relevant samples is very low. The slight fluctuation may be due to the presence

of C in the welded metals [18]. This is because a higher C content in the REE1 and REE2 welded metals was detected; see Table 6. In other words, adding REE improves indeed the mechanical properties of the welded metals significantly. The correct addition of the right amount of REE can improve the strength of the welded metals, while REE adding over 0.7% reduces the strength.

3.3. Microstructural Analysis. Figure 4 shows the optical morphologies of the FCAW-welded metals with different REE additions. It is hard to identify any prior austenite

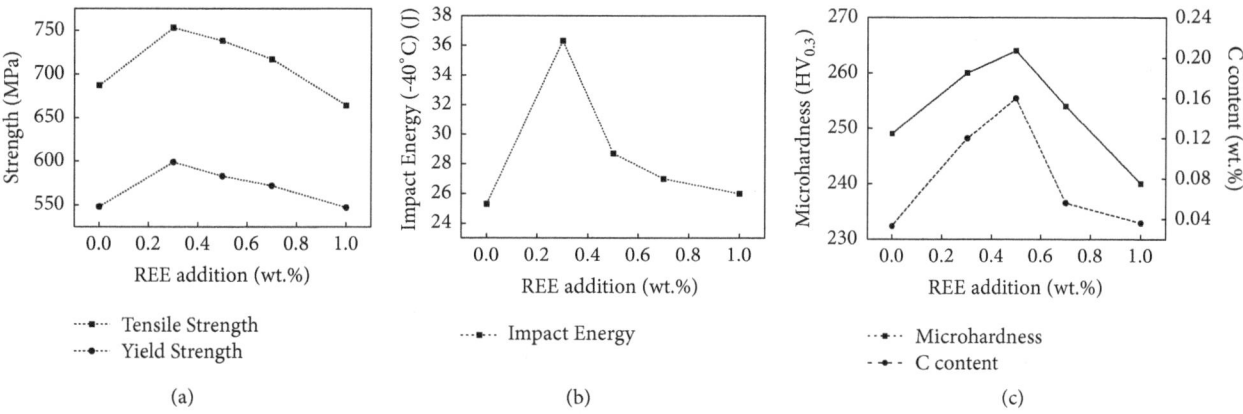

FIGURE 3: Mechanical properties of FCAW-welded metals for different REE contents: (a) tensile properties; (b) low-temperature impact toughness; and (c) microhardness.

TABLE 5: Welding parameters for FCAW.

Plate form and dimensions (mm)	Flat, 300 ×150 ×14
Welding Machine	EWM
Power Mode	DCEP
Electrode	As-prepared wires
Protective gas	M21 (80% Ar + 20% CO_2)
Gas flow rate (L/min)	17 ~ 20
Preheating temperature (°C)	150
Current (A)	240
Voltage (V)	26
Interpass temperature (°C)	150 ~ 180
Welding speed (mm/s)	6.3 ~ 7.9
Cooling after welding	Air

grain boundaries (PAGB), which means a complete phase transformation (to ferrite) occurred during cooling. The microstructure of REE-free welded metal consists of proeutectoid ferrites (PF), acicular ferrites (AF), and a small amount of bainites (B); see Figure 4(a). No bainite was formed in the welded metals with REE added; see Figures 4(b)–4(e). This indicates that a microstructure of different forms of ferrites, i.e., AF, lath ferrites (LF), and granular ferrite (GF), was formed. After REE addition of 0.3% and 0.5%, as shown in Figures 3(b) and 3(c), the microstructure of the welded metals becomes more refined. In addition, the PF size is limited due to inhibited nucleation of PF. Furthermore, there is increased nucleation and AF growth due to the added REE. Some small LF formed due to the continuous growth of AF. The total amount of grain boundaries increases clearly because of the formation of AF. There is a high density of dislocations in the interior of the acicular ferrites, with few low-angle boundaries. These require higher energy for the microcracks to cross the AF, which improves crack growth-resistance [19]. Because there are many large-angle grain boundaries among adjacent grains, which can increase the resistance of dislocation motion and the plastic deformation, the formation of AF increases both strength and low-temperature impact toughness [20, 21]; see Figure 3. When

the REE content REE exceeds 0.5%, the inhibiting property of PF nucleation is restored. This is a result of the pollution of the grain boundaries by REE, which appears as disproportionate growth of PF [22]; see Figures 4(d) and 4(e). The REE addition of 0.3% and 0.5% can refine the grains by promoting the formation of AF effectively. This improves the mechanical properties of welded metals. When the REE content exceeds 0.5%, AF is limited with its growth promotion of LF and PF. This leads to a reduction in strength and low-temperature affects the toughness of the welded metals.

If second-phase particles are of suitable size with a uniform distribution, they could act as nuclei during the solidification process. Studies [23, 24] show that particles finer than 0.6 μm facilitate the nucleation of ferrite, while even more effective nucleation occurs in the 0.2~0.6 μm range. Figure 5 shows the SEM images of the welded metals with different amounts of REE added. The number of second-phase particles in the REE-free welded metal is small, and their size is relatively large (up to 3.3 μm in diameter); see the dark spots in Figure 5(a). It is clear that the number of second-phase particles increases significantly, while the average size decreases gradually REE for higher REE content; see Figures 5(b)–5(e). This is substantially different from the REE-free welded metal, based on the measurement of more than 100 particles. Following a further increase of the REE content (up to 1.00%), the number of second-phase particles continues to increase with a discrete distribution and reduced average size. Some second-phase particles (on a microscale) are beginning to appear in the field of view.

Both the number of second-phase particles and the fraction of different grain sizes were counted using the SEM images. The statistical results are shown in Figure 6. For a higher REE content, the size range for the second-phase particles becomes gradually smaller. Some oversized particles (up to 3.3 μm in the REE-free welded metal) do not appear in the REE-added welded metals. The fractions of second-phase particles between 0.2 and 0.6 μm in Figures 6(a)–6(e) are 70.2%, 77.8%, 74.4%, 49.9%, and 46.4%, respectively. This shows the first increasing and then decreasing trend if the REE content increases. However, the fractions of particles

TABLE 6: Chemical composition of the welded metals (wt. %).

welded metal sample	C	Si	Mn	Mo	Al	Ti	Ni
REE0	0.033	0.30	1.59	0.23	0.013	0.030	2.39
REE1	0.12	0.25	0.97	0.11	0.011	0.028	1.07
REE2	0.16	0.29	1.44	0.17	0.014	0.031	1.77
REE3	0.056	0.42	1.47	0.23	0.014	0.029	2.41
REE4	0.036	0.33	1.42	0.25	0.013	0.032	2.34

TABLE 7: Mechanical properties of the welded metals with different amounts of REE.

welded metal	Tensile Strength/MPa	Yield Strength/MPa	Elongation /%	Average Impact Energy at -40˚C/J	Micro-hardness/HV$_{0.3}$
REE0	687	548	21.2	25.3	249
REE1	753	599	22.5	36.3	260
REE2	738	583	20	27.7	264
REE3	717	572	14.2	27	254
REE4	664	547	14.5	28	240

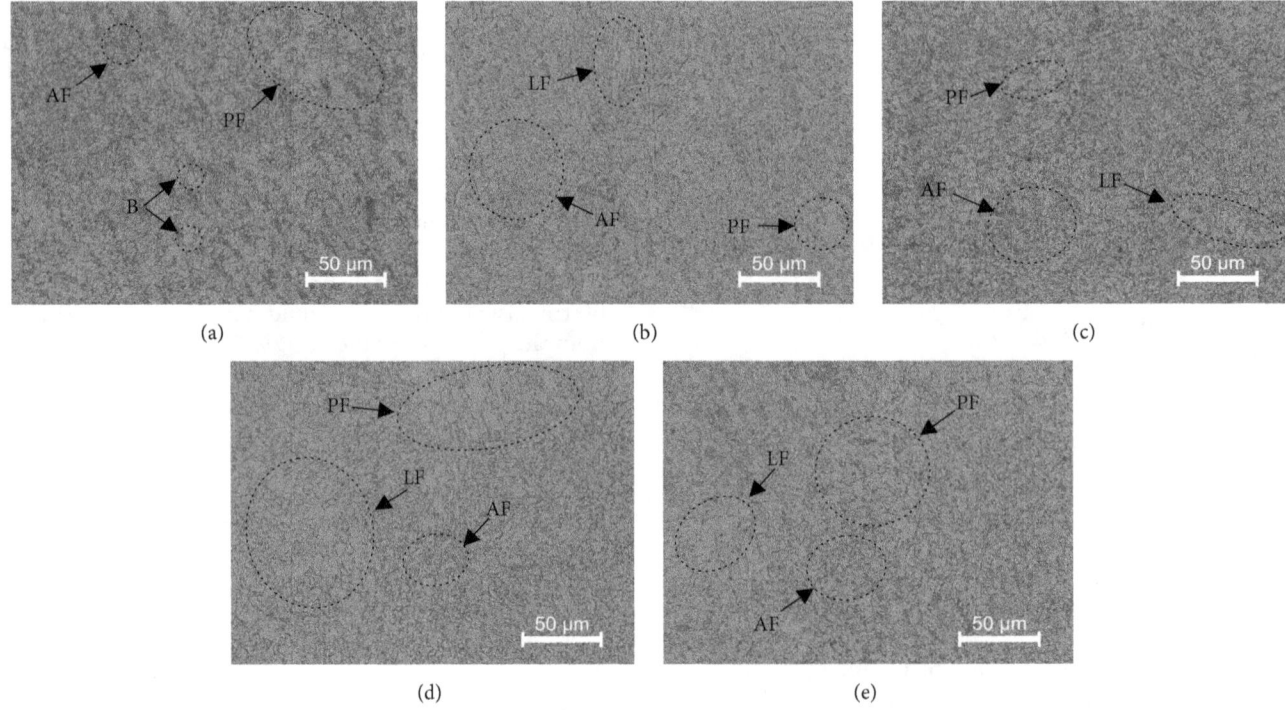

FIGURE 4: OM images of FCAW-welded metals for different REE contents:(a) REE-free, (b) 0.3 wt.%, (c) 0.5 wt.%, (d) 0.7 wt.%, and (e) 1.0 wt.%.

below 0.2 μm are 0%, 0%, 8.1%, 28.1%, and 44.9%, respectively, which represents an upward trend for increasing REE content. Hence, a very high REE content can reduce the size of the second-phase particles, and the slope describing the mechanical properties of welded metals becomes negative. When the REE content is optimal (0.3%), most of the second-phase particles form in the size range of 0.2 to 0.6 μm, which facilitates the formation of AF and results in the best mechanical performance of welded metals.

The EDS results for the second-phase particles are shown in Table 8. There are mainly three elements (Fe, Mn, and Ni) in the matrix, with a fluctuating variation of the components.

However, the types and compositions of the elements in the particles of different samples are somewhat different, whose main elements consisted of O, Al, Si, Ti, Mn, and Fe. Because the particles are located within the matrix, the EDS determination will include part component of the matrix into the results. Moreover, the content of Fe is about 45%, the highest of all elements; it can be considered that the Fe mainly stems from the matrix. There is a small amount of S in the second-phase particles in three groups of samples with 0.3% REE, 0.5% REE, and 0.7% REE, respectively. A certain amount of S element in the alloy steel is beneficial to the improvement of its machinability. However, for the welded

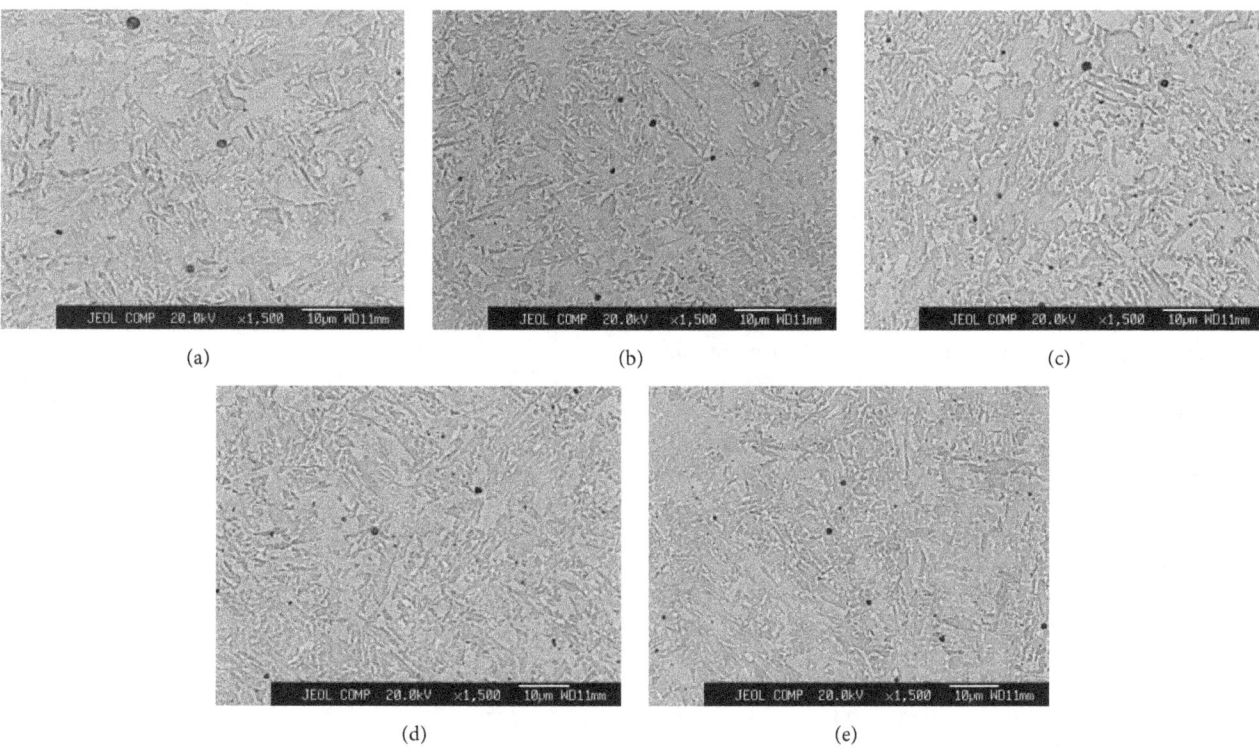

FIGURE 5: Typical SEM images of welded metals with different REE contents: (a) REE-free, (b) 0.3 wt.%, (c) 0.5 wt.%, (d) 0.7 wt.%, and (e) 1.0 wt.%.

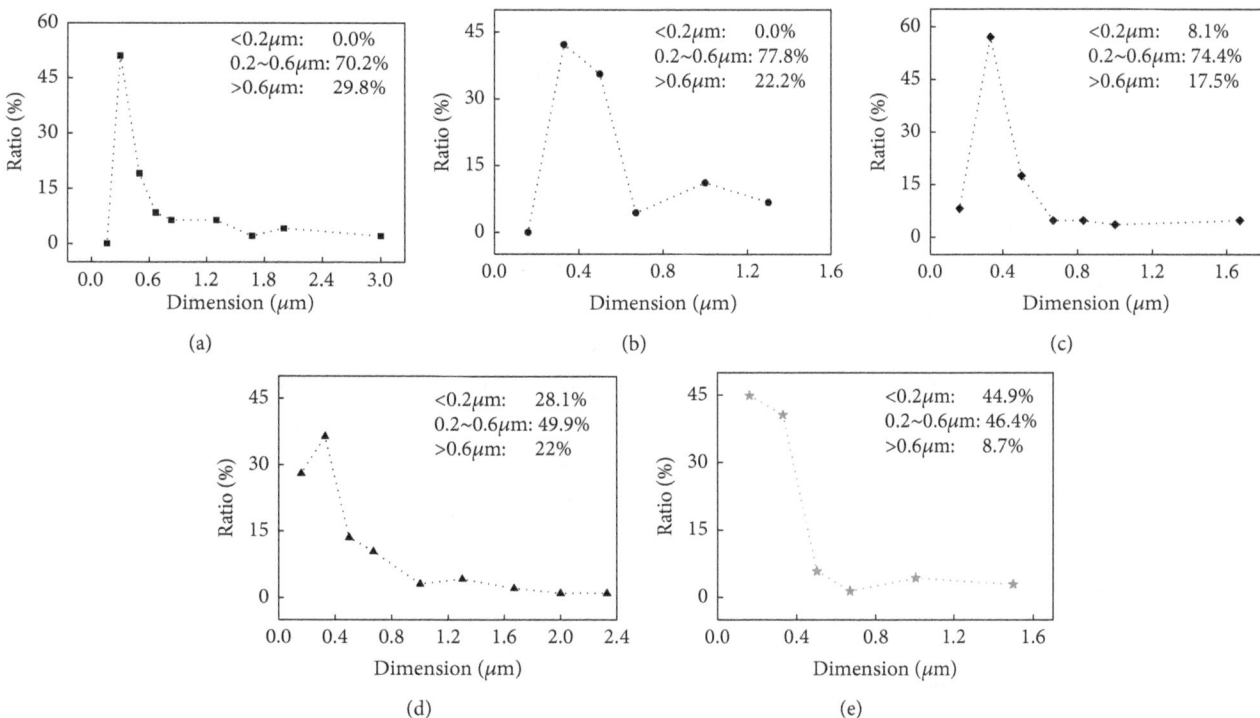

FIGURE 6: Dimension (size) and ratio of the second-phase particles in welded metals with different REE contents: (a) REE-free, (b) 0.3 wt.%, (c) 0.5 wt.%, (d) 0.7 wt.%, and (e) 1.0 wt.%.

TABLE 8: EDS results of the matrix or second-phase particles for each welded metal with varying REE content.

Items	Matrix			Second-phase particle							
Elements	Mn/%	Fe/%	Ni/%	O/%	Al/%	Si/%	Ti/%	Mn/%	Fe/%	S/%	Ni/%
REE0	1.42	96.27	2.31	16.86	4.26	2.57	14.12	18.55	42.91	/	0.73
REE1	1.72	96.13	2.15	16.92	3.91	6.47	6.73	15.34	49.85	0.14	/
REE2	2.39	94.99	2.62	14.38	5.25	5.10	9.29	17.32	46.65	2.01	/
REE3	1.43	95.61	2.96	20.61	6.09	6.00	12.25	13.30	39.25	2.50	/
REE4	1.23	96.39	2.38	16.68	4.53	6.66	6.48	14.99	50.67	/	/

metal, a tiny amount of S can greatly reduce its performance. In this work, it can be found that correct REE content can aid S to accumulate into the second-phase particles, which originally dissolved in the matrix. Thus, it will reduce the harmful effects on the matrix to increase the strength of weld metal. This conclusion is consistent with the above discussion of the mechanical properties in Figure 2. With the increasing REE addition, the relative contents of the nonmetallic elements O, Si, and S in the second-phase particles show a rising trend, and the content of Al and Fe increased slightly, while Ti and Mn decreased slightly. Mn and Ti tend to deoxidize and remove impurities to form refractory-phase impurities like MnS or TiO, combined with S or O. These discharged from the welded metal into the welding slag with the right content of nonmetallic elements O and S. Enrichment with oxygen in the welded metal tends to increase the possibility of the generation of weld blowholes, which can induce cracks and become the sources of microcracks under external load. This decreases both the strength and the low-temperature impact toughness of the weld metal. Sulphur could form banded FeS, when combined with Fe and weaken the consistency of the weld metal. The right amount of Si cannot only act as a deoxidant but also form Al-Mn silicate particles and facilitate nucleation. Therefore, adding REE causes the elements of O, S, and Si to accumulate in the second-phase particles instead of the matrix, which purifies the matrix and refines the grains or microstructure effectively. As a result, the strength and low-temperature impact toughness of the weld metal are improved using REE.

Phase compositions for each sample were determined using XRD, and the results are shown in Figure 7. The matrix consists of α-Fe (AF), and the second phase consists mainly of an Al-Ti phase and $(Al, Mn)_x SiO_4$ (Al-Mn silicate) phase. Furthermore, $(Al, Mn)_x SiO_4$ is composed of a variety of metallic or nonmetallic oxides including Al_2O_3, MnO, and SiO_2, which play a strong role in deoxygenation and greatly reduce oxygen in the welded metals. During the solidification process of the weld pool, the second-phase particles, which consist of Al_2O_3, MnO, and SiO_2, formed preferably with high surface-energy [25], where the crystal nuclei for ferrite form. In this way it can substantially reduce the potential barrier for nucleation during the phase transformation from γ to α phase. The second-phase particles in the weld metal are mostly a complex mixture of different phases. Therefore, each phase can be regarded as a high-energy region and form a nucleus for acicular ferrite. Multidimensional nucleation occurs at the end and causes acicular ferrite to overlap, which results in refined grains.

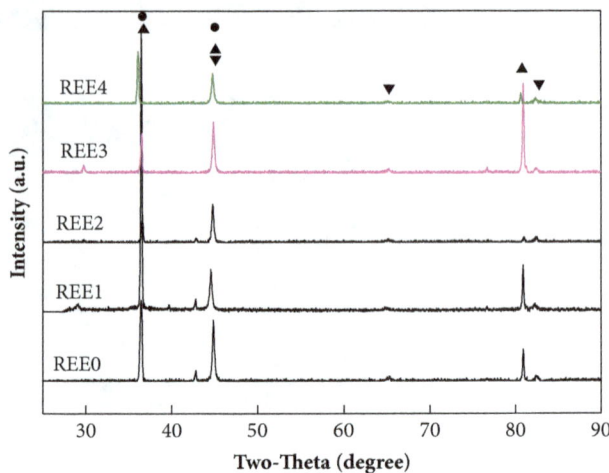

▼ α-Fe
▲ Al-Ti
● $(Al, Mn)_x SiO_4$

FIGURE 7: XRD analysis results of different FCAW-welded metals.

3.4. Electrochemical Properties. The EIS responses of the FCAW-welded metals with different REE contents in a 3.5 wt.% NaCl solution were measured to study effects on the electrochemical properties. Our results are shown as Nyquist, Bode, and Bode-phase plots in Figure 8. The Nyquist plots of all welded metals are imperfect semicircles (Figure 8(a)), i.e., capacitive arcs with similar capacitive response. Only a single depressed semicircle for each welded metal appears in the Nyquist plots, which indicates that only one time-constant is present in EIS. The imperfect semicircle-diameter for REE4 is the largest. It shows the largest capacitive arc and the best antidissolution properties. The imperfect semicircle-diameters for REE welded metals, with REE added (REE1, REE2 and REE3), are smaller than the REE-free welded metal. In the low-frequency range of the Bode plots (Figure 8(b)), the samples with higher electrical resistance possess stronger corrosion resistance. This is consistent with our analysis of the above Nyquist plots. In Figure 8(c), each spectrum shows one single sharp peak, with maximum phase-angles below 70°. This means the EIS capacitive response of the welded metals, with or without REE addition, is not pure or ideal. Hence, there is only one time-constant and low reactive resistance, which is similar to the results reported by R.M. Domene et al. [26].

To simulate the measured impedance data and explain the general corrosion process, the equivalent circuit shown

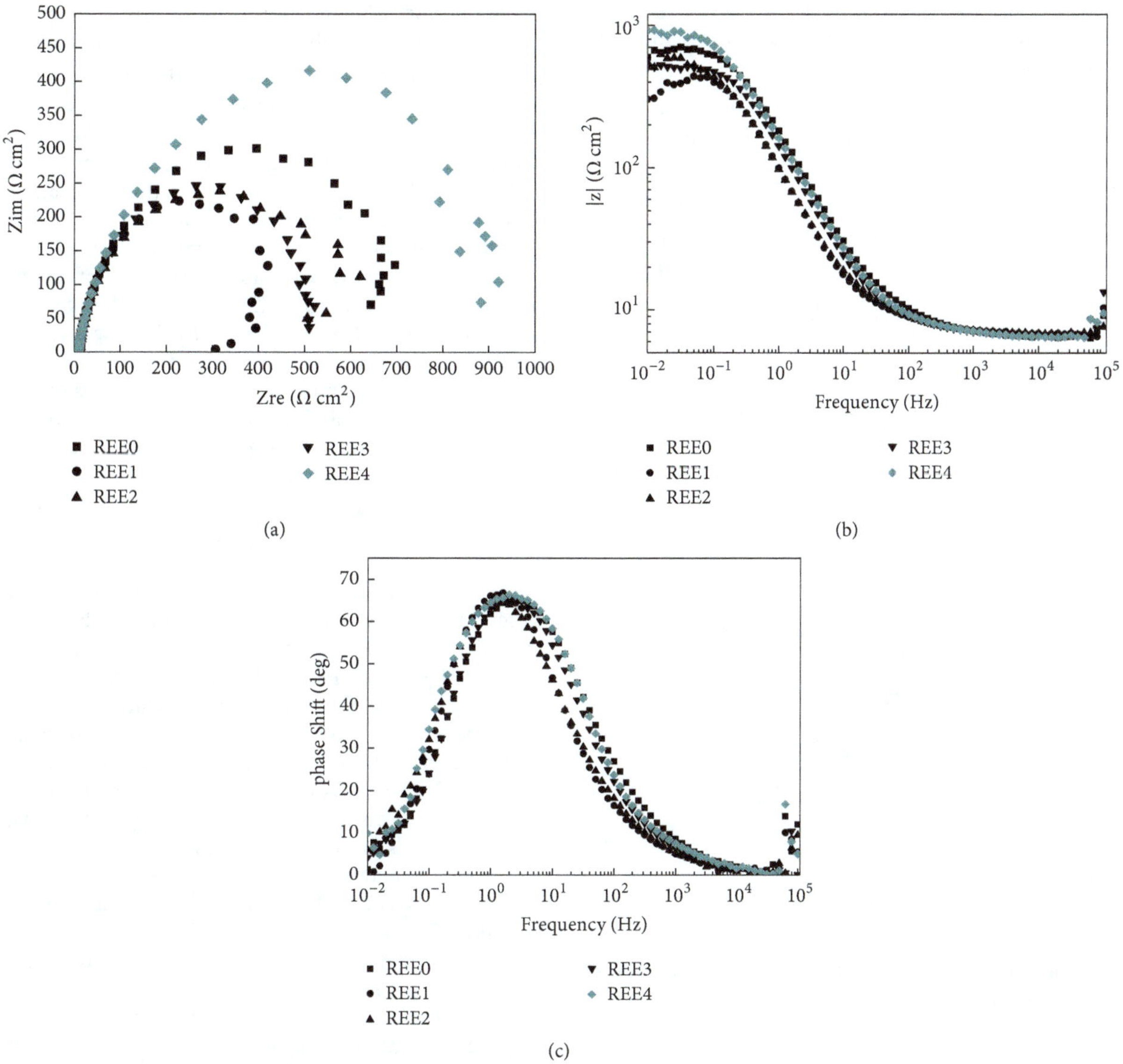

FIGURE 8: (a) Nyquist, (b) Bode, and (c) Bode-phase plots for the FCAW-welded metals with different REE contents versus OCP in a 3.5wt% NaCl solution.

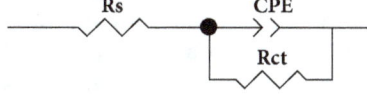

FIGURE 9: Equivalent circuit used to model the experimental EIS data.

TABLE 9: Impedance parameters for the EIS-tested welded metals.

welded metals	R_s ($\Omega \cdot cm^2$)	Capacitance (F)	R_{ct} ($\Omega \cdot cm^2$)
REE0	8.03	0.0006136	637.3
REE1	8.17	0.001366	401.7
REE2	8.19	0.001321	529.7
REE3	8.13	0.0008468	493.8
REE4	7.89	0.000742	843.7

in Figure 9 was adopted. The variations of the impedance parameters are shown in Table 9, and the respective errors are below 5%. Consistent with the active dissolution mechanism, R_s is the solution resistance, and R_{ct} represents the charge-transfer resistance from the metal to the electrolyte, which can be defined as the corrosion resistance for the sample in the electrolyte. Furthermore, C_{PE} accounts for the constant phase element corresponding to the metal in this equivalent model, as reported by Y. Chen et al. [27]. The welded metal

(with 1.00 wt.% REE addition) has the highest R_{ct} (983.0 $\Omega \cdot cm^2$), and the REE1 welded metal has the lowest R_{ct} (447.6 $\Omega \cdot cm^2$). Only the REE4 welded metal with 1.00 wt.% REE content shows a higher R_{ct} than its counterpart for the REE-free welded metal.

TABLE 10: Electrochemical data of the welded metals with different REE contents as obtained from potentiodynamic polarization measurements in 3.5 wt.% NaCl solution.

welded metals	E_{corr} (mV)	I_{corr} ($\mu A/cm^2$)	R_p (Ohms/cm^2)
REE0	-344	24.9	1047.2
REE1	-546	13.4	1945.7
REE2	-544	14.0	1862.5
REE3	-403	13.6	1918.3
REE4	-527	16.5	1579.5

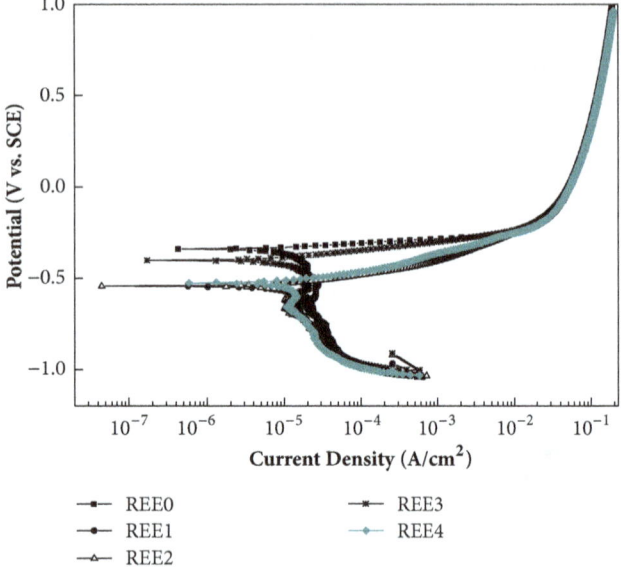

--■-- REE0
--●-- REE1
--△-- REE2
--✳-- REE3
--◆-- REE4

FIGURE 10: Potentiodynamic polarization curves for the welded metals with different REE contents.

After the EIS tests, potentiodynamic polarization measurements were carried out, in a 3.5 wt.% NaCl solution at room temperature. The results are shown in Figure 10 and Table 10. Corrosion potential (E_{corr}) is a static indicator for electrochemical corrosion resistance to describe corrosive susceptibility of a material. In this study, all welded metals with REE show a lower corrosion potential than the REE-free welded metal; see Table 10. The REE-free welded metal has the highest E_{corr} (-344 mV), while the REE1 with the minimum REE addition has the lowest E_{corr} (-546 mV). This indicates a small REE addition has already a significant effect on the corrosive susceptibility in welded metals. However, higher corrosive susceptibility does not mean the material corrodes easily. According to the Tafel rule, corrosion current density (I_{corr}) is a key factor and closely related to the corrosion-dissolution rate. Furthermore, the related polarization resistance (R_p) is used to determine the corrosion rate at any given time. In addition, lower I_{corr} and higher R_p indicate a higher corrosion resistance. The polarization curves for all welded metals with or without REE show typical signs of active dissolution in the 3.5 wt.%

NaCl solution. The active dissolution mechanism is further confirmed by the high I_{corr} compared to the calculated 10 $\mu A/cm^2$ for all welded metals; see Table 10. REE-free welded metal (REE0) has the highest I_{corr} (24.9 $\mu A/cm^2$) and the lowest R_p (1047.2 Ohms/cm^2). REE could retard the anodic process and cathode process to reduce the corrosion rate of welded metal. E_{corr} negatively moved, due to greater effect of REE on the equilibrium potential of cathode reaction [28]. The shapes of all polarization curves are basically the same, which shows that the corrosion mechanism of different REE has not changed. Both I_{corr} and R_p of the welded metals improved, due to the addition of REE. They are better than for Re-free welded metal. The REE addition could reduce the corrosion-dissolution rate and improve corrosion resistance of welded metals. REE1, with an REE addition of 0.2%, shows the strongest corrosion reduction effect, for the highest R_p (1945.7 Ohms/cm^2) and the lowest I_{corr} (13.4 $\mu A/cm^2$). In other words, the reduction effect weakens after adding REE, due to the increased I_{corr} and reduced R_p for REE2 to REE4.

It is interesting that R_{ct} and R_p of the welded metals with different REE contents show different trends. Considering the effect of REE, the difference between the R_{ct} and R_p values might depend on the second-phase particles. The correlations between second-phase particle size and the resistances are shown in Figure 11. It has been reported that a ferrite phase dissolves better during galvanic corrosion [29], and secondary phases act as a pathway in a corrosive environment [30]. The REE second-phase particles have no significant effect on the resistance of REE-free welded metals. REE R_{CT} changes inversely with the size of the second-phase particles in welded metals that contain REE; see Figures 11(b) and 11(d). The second-phase particles provide charge-transfer channels during the EIS tests. Large second-phase particles provide larger channels, which increase the active dissolution efficiency. By increasing the content of REE, the refining particles restrict the charge-transfer process between the welded metal and the electrolyte. This increases R_{ct} and leads to better corrosive resistance. The polarization test considers complex factors for the evaluation of corrosion properties, including the EIS response. For REE-containing welded metals, the R_p values are proportional to the second-phase particle sizes of the welded metals; see Figures 11(a) and 11(c). The R_p values are the opposite of the R_{ct} values. Welded metals with larger second-phase particles show higher R_p values. This can be due to the electric-potential increase of the second-phase particles containing REE. The REE content affects both sizes and ratios of the second-phase particles and the corrosion properties of HSLA welded metals as well as the mechanical properties.

The morphology of the welded metals after electrochemical testing is shown in Figure 12. Significant microcracks can be observed on the corrosive surface of the REE-free welded metal shown in Figure 12(a). The number of microcracks varies with the R_p values of the different REE-containing welded metals. No obvious microcracks were found on the corrosive surface of the REE1 sample; see Figure 12(b).

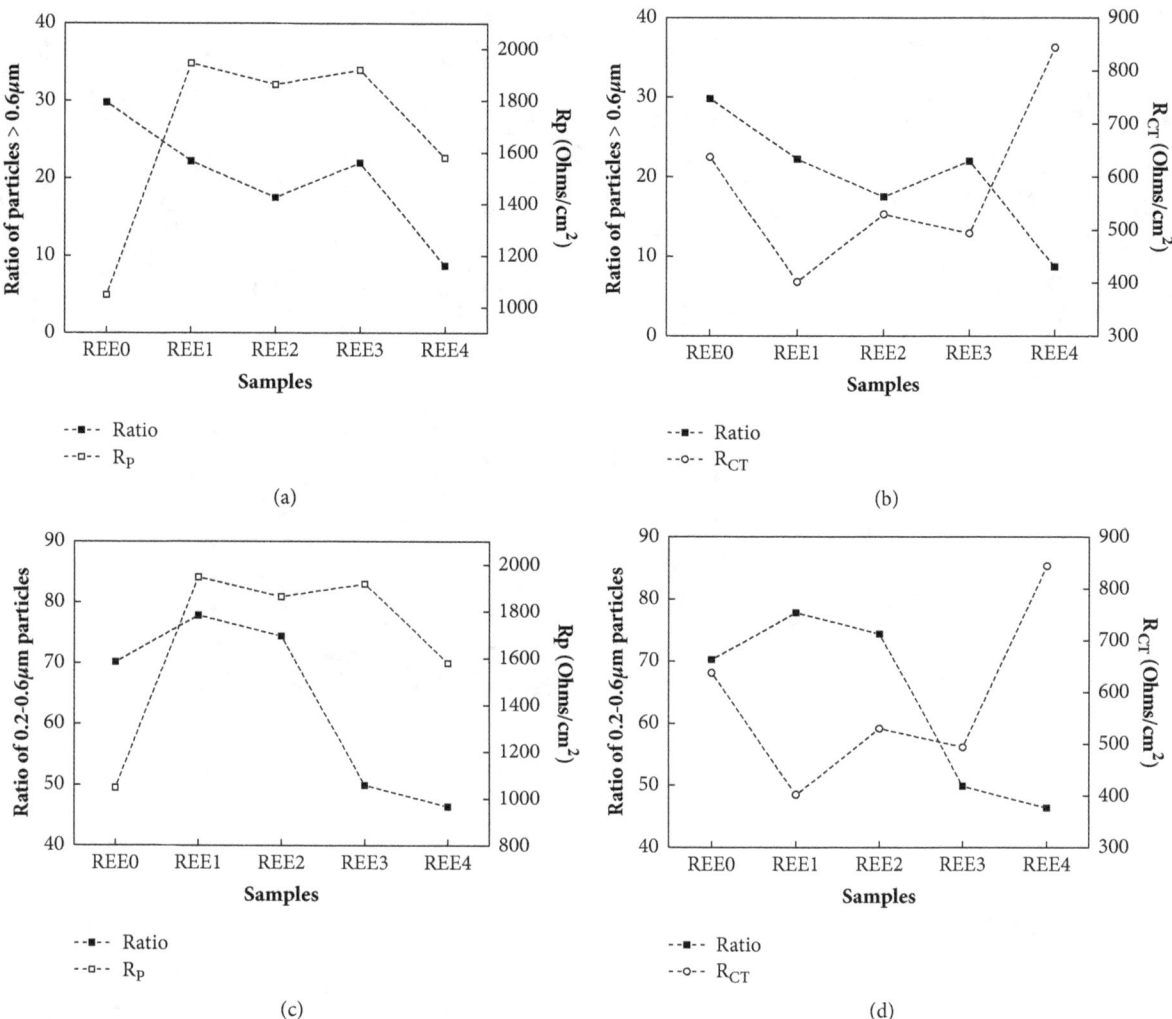

FIGURE 11: Correlation between second-phase particle sizes and the resistance in different welded metals.

However, microcracks REE can be seen clearly on the REE4 sample in Figure 12(e). During the electrochemical tests, the microcracks generated in the oxide films provide the channels for the electrolytes to infiltrate the oxide films. As a result, the charge-transfer process from the welded metal to the electrolyte accelerated. Both the REE-modified second-phase particles and microstructure of the welded metal facilitate the formation of dense oxide films thanks to the addition of REE.

For 10CrMo3NiV high-strength steel, the corrosion stability and the microstructure of the welded metal were significantly affected by the addition of REE, from the perspective of the effect on the size and ratio of second-phase particles. The result of the EIS tests and the potentiodynamic polarization curves indicates that the added REE in the flux-cored wire improves corrosion resistance in 10CrMo3NiV welded metals. To obtain detailed results of the second-phase particles affecting the electrochemical behavior of the welded metals, we plan to use scanning vibrating electrode technology (SVET) in further study. This will help reveal the underlying mechanism of

REE: how REE change for different corrosive times and electrolytes.

4. Conclusions

Rare-earth elements were added to the flux-cored wire to modify the performance of FCAW-welded metals with 10CrMo3NiV steel. Both OM and SEM studies indicate that the REE content changes the microstructure of welded metals by refining the second-phase particles. A content of 0.3 wt.% REE in the flux-cored wire is optimal to facilitate suitable second-phase particles in both the welded metal (0.2 μm to 0.6μm in diameter) and the main acicular ferrite microstructure. This amount of REE also helps avoid the accumulation of nonmetallic elements and improves the mechanical properties of the welded metals. Both potentiodynamic polarization and EIS tests show that REE addition reduces the charge-transfer channel effect of the second-phase particles. Overall, the corrosion properties as well as the mechanical properties of a welded metal can be improved substantially by adding 0.3 wt.% REE.

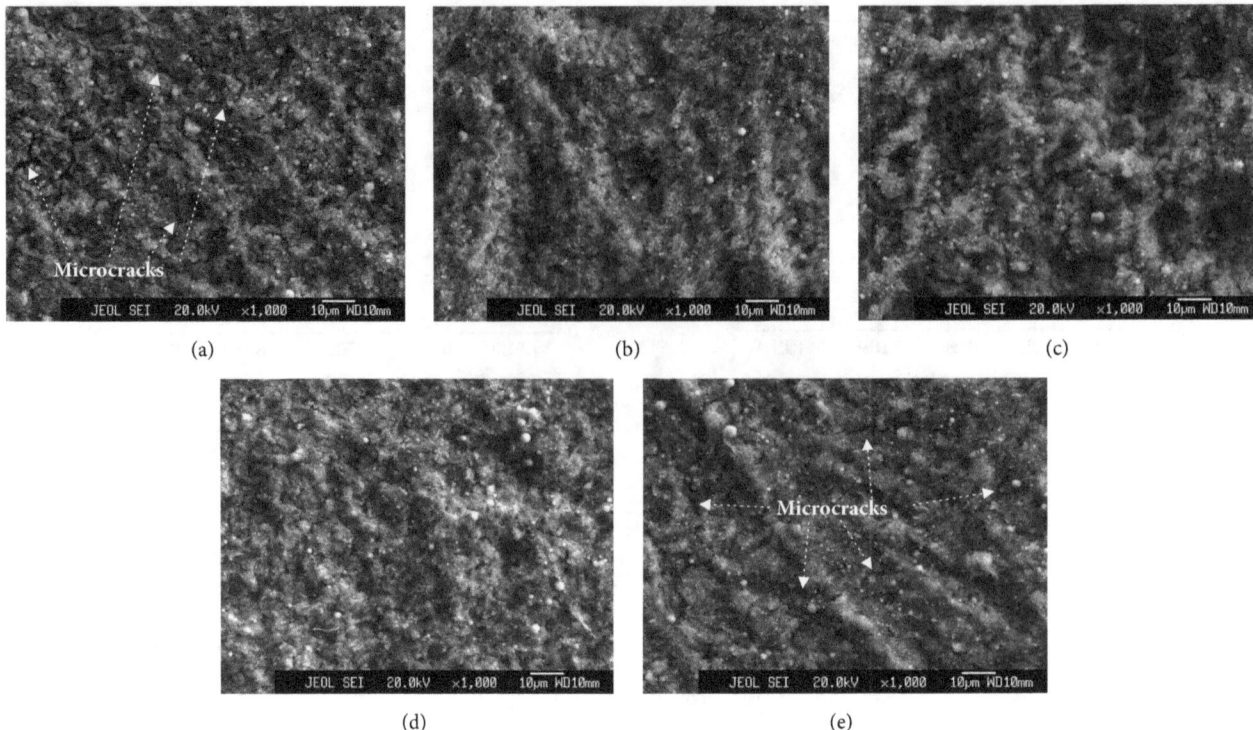

FIGURE 12: Morphology of the welded metals with different REE contents: (a) REE-free, (b) 0.3 wt.%, (c) 0.5 wt.%, (d) 0.7 wt.%, and (e) 1.0 wt.%, after electrochemical testing.

Conflicts of Interest

The authors declare that they have no conflicts of interest.

Acknowledgments

This research was supported by the National Natural Science Foundation of China (no. 51601043), the Scientific Program of GDAS (2016GDASPT-0205), the Technical Project of Guangdong and Guangzhou (201604046026, 2016B090918120, 201508030024, and 201704030112).

References

[1] *Committee on Effective Utilization of Weld Metal Yield Strength, Effective Use of Weld Metal Yield Strength for HY Steels*, National Materials Advisory Board, Washington DC, Wash, USA, 1983.

[2] J. H. Kim, J. S. Seo, H. J. Kim, H. S. Ryoo, K. H. Kim, and M. Y. Huh, "Effect of weld metal microstructures on cold crack susceptibility of FCAW weld metal," *Metals and Materials International*, vol. 14, no. 2, pp. 239–245, 2008.

[3] Y.-W. Xu, S.-H. Song, and J.-W. Wang, "Effect of rare earth cerium on the creep properties of modified 9Cr-1Mo heat-resistant steel," *Materials Letters*, vol. 161, pp. 616–619, 2015.

[4] J. Hufenbach, A. Helth, M.-H. Lee et al., "Effect of cerium addition on microstructure and mechanical properties of high-strength Fe85Cr4Mo8V2C1 cast steel," *Materials Science and Engineering: A Structural Materials: Properties, Microstructure and Processing*, vol. 674, pp. 366–374, 2016.

[5] W. M. Garrison Jr. and J. L. Maloney, "Lanthanum additions and the toughness of ultra-high strength steels and the determination of appropriate lanthanum additions," *Materials Science and Engineering: A Structural Materials: Properties, Microstructure and Processing*, vol. 403, no. 1-2, pp. 299–310, 2005.

[6] H.-L. Liu, C.-J. Liu, and M.-F. Jiang, "Effect of rare earths on impact toughness of a low-carbon steel," *Materials and Corrosion*, vol. 33, no. 1, pp. 306–312, 2012.

[7] G. Liu, G. Zhang, and R. Li, "Electronic theoretical study of the interaction between rare earth elements and impurities at grain boundaries in steel," *Journal of Rare Earths*, vol. 21, no. 3, pp. 372–374, 2003.

[8] D. Zhang, C. Wu, and R. Yang, "Interaction between cerium and phosphorus segregating in the grain boundaries in α-Fe studied by computer modelling and Auger electron spectroscopy," *Materials Science and Engineering: A Structural Materials: Properties, Microstructure and Processing*, vol. 131, no. 1, pp. 93–97, 1991.

[9] S. Song, H. Sun, and M. Wang, "Effect of rare earth cerium on brittleness of simulated welding heat-affected zones in a reactor pressure vessel steel," *Journal of Rare Earths*, vol. 33, no. 11, pp. 1204–1211, 2015.

[10] X. Jiang and S.-H. Song, "Enhanced hot ductility of a Cr-Mo low alloy steel by rare earth cerium," *Materials Science and Engineering: A Structural Materials: Properties, Microstructure and Processing*, vol. 613, pp. 171–177, 2014.

[11] L. Chen, X. Ma, M. Jin, J. Wang, H. Long, and T. Mao, "Beneficial Effect of Microalloyed Rare Earth on S Segregation in High-Purity Duplex Stainless Steel," *Metallurgical and Materials Transactions A: Physical Metallurgy and Materials Science*, vol. 47, no. 1, pp. 33–38, 2016.

[12] Y. Tomita, "Improved fracture toughness of ultrahigh strength steel through control of non-metallic inclusions," *Journal of Materials Science*, vol. 28, no. 4, pp. 853–859, 1993.

[13] J. Gao, P. Fu, H. Liu, and D. Li, "Effects of rare earth on the microstructure and impact toughness of H13 steel," *Metals*, vol. 5, no. 1, pp. 383–394, 2015.

[14] J. Pu, S. Yu, and Y. Li, "Role of inclusions in flux aided backing submerged arc welding," *Journal of Materials Processing Technology*, vol. 240, pp. 145–153, 2017.

[15] N. Yan, S. F. Yu, and Y. Chen, "Inclusions,microstructure and properties of flux copper backing submerged arc weld metal," *Science and Technology of Welding and Joining*, vol. 20, no. 5, pp. 418–424, 2015.

[16] B. Hinton, "Corrosion inhibition with rare earth metal salts," *Journal of Alloys Compounds*, vol. 180, pp. 15–25, 1992.

[17] C. Wang, F. Jiang, and F. Wang, "The characterization and corrosion resistance of cerium chemical conversion coatings for 304 stainless steel," *Corrosion Science*, vol. 46, no. 1, pp. 75–89, 2004.

[18] E. Gharibshahiyan, A. H. Raouf, N. Parvin, and M. Rahimian, "The effect of microstructure on hardness and toughness of low carbon welded steel using inert gas welding," *Materials and Corrosion*, vol. 32, no. 4, pp. 2042–2048, 2011.

[19] C. Zhang, X. Song, P. Lu, and X. Hu, "Effect of microstructure on mechanical properties in weld-repaired high strength low alloy steel," *Materials and Corrosion*, vol. 36, pp. 233–242, 2012.

[20] B. Basu and R. Raman, "Microstructural variations in a high-strength structural steel weld under isoheat input conditions," *Welding Journal (Miami, Fla)*, vol. 81, no. 11, 2002.

[21] G. Spanos, R. W. Fonda, R. A. Vandermeer, and A. Matuszeski, "Microstructural changes in HSLA-100 steel thermally cycled to simulate the heat-affected zone during welding," *Metallurgical and Materials Transactions A: Physical Metallurgy and Materials Science*, vol. 26, no. 12, pp. 3277–3293, 1995.

[22] Y. Cai, R. Liu, Y. Wei, and Z. Cheng, "Influence of Y on microstructures and mechanical properties of high strength steel weld metal," *Materials & Design (1980-2015)*, vol. 62, pp. 83–90, 2014.

[23] T. Koseki and G. Thewlis, "Inclusion assisted microstructure control in C-Mn and low alloy steel welds," *Materials Science and Technology*, vol. 21, no. 8, pp. 867–879, 2005.

[24] C. A. Williams, P. Unifantowicz, N. Baluc, G. D. W. Smith, and E. A. Marquis, "The formation and evolution of oxide particles in oxide-dispersion- strengthened ferritic steels during processing," *Acta Materialia*, vol. 61, no. 6, pp. 2219–2235, 2013.

[25] R. A. Ricks, P. R. Howell, and G. S. Barritte, "The nature of acicular ferrite in HSLA steel weld metals," *Journal of Materials Science*, vol. 17, no. 3, pp. 732–740, 1982.

[26] R. M. Fernández-Domene, E. Blasco-Tamarit, D. M. García-García, and J. García-Antón, "Effect of alloying elements on the electronic properties of thin passive films formed on carbon steel, ferritic and austenitic stainless steels in a highly concentrated LiBr solution," *Thin Solid Films*, vol. 558, pp. 252–258, 2014.

[27] Y. Chen, T. Hong, M. Gopal, and W. P. Jepson, "EIS studies of a corrosion inhibitor behavior under multiphase flow conditions," *Corrosion Science*, vol. 42, no. 6, pp. 979–990, 2000.

[28] G. A. Zhang and Y. F. Cheng, "Corrosion of X65 steel in CO_2-saturated oilfield formation water in the absence and presence of acetic acid," *Corrosion Science*, vol. 51, no. 8, pp. 1589–1595, 2009.

[29] H.-Y. Ha, M.-H. Jang, T.-H. Lee, and J. Moon, "Interpretation of the relation between ferrite fraction and pitting corrosion resistance of commercial 2205 duplex stainless steel," *Corrosion Science*, vol. 89, no. C, pp. 154–162, 2014.

[30] D. H. Kang and H. W. Lee, "Study of the correlation between pitting corrosion and the component ratio of the dual phase in duplex stainless steel welds," *Corrosion Science*, vol. 74, pp. 396–407, 2013.

14

Corrosion Behavior of Welded Joint of Q690 with CMT Twin

Peng Liu [ID],[1,2] **Shanguo Han,**[2] **Yaoyong Yi** [ID],[2] **and Cuixia Yan** [ID][1]

[1]*Faculty of Materials Science and Engineering, Kunming University of Science and Technology, 68 Wenchang Road, Kunming, Yunnan 650093, China*
[2]*Guangdong Welding Institute (China-Ukraine E. O. Paton Institute of Welding), Guangdong Key Laboratory of Modern Welding Technology, 363 Changxing Road, Guangzhou, Guangdong 510650, China*

Correspondence should be addressed to Yaoyong Yi; yiyaoyong@hotmail.com and Cuixia Yan; cuixiayan09@gmail.com

Academic Editor: Sergey Konovalov

Low alloy steel of Q690 was welded with the method of CMT Twin. The corrosion behavior of welded joint had been investigated using scanning vibrating electrode technique (SVET) in 3.5% NaCl solution. The research results showed that the appearance of the troostite increased the hardness of the heat affected zone. Furthermore, the corrosion products of different microstructure were identical, and the white products ($Fe(OH)_2$) of welded joint turned into products of rufous ($Fe(OH)_3$). The quantitative information provided by SVET was discussed, and the corrosion degree was measured by some parameters. In comparison with other areas, the corrosion rates of the overheated zone and the base metal were higher. Then, the corrosion resistance of the weld zone with CMT Twin was greatly improved, when compared with that of the base metal. Therefore, Ni has significant influence on corrosion resistance of weld zone. In summary, it can be discovered that the corrosion rates of various zones were related to the welding heat input.

1. Introduction

Q690 steel is known as a typical low carbon bainite steel with excellent mechanical properties, and it is widely used in offshore engineering industry [1, 2]. However, the complex chloride environment tends to be invalidated ahead of lifetime in welded parts, so it is significant to study the corrosion behavior of welded parts [3, 4]. Furthermore, the chloride environment contains more inorganic salts (such as NaCl and $MgCl_2$) [5], and Cl^- is the most important corrosive ion. Subsequently, Cl^- is able to destroy the protective effect of the corrosion scales, and it significantly increases the activity of the matrix. So Cl^- can accelerate the corrosion rate, and eventually lifetime of the welded joint will be reduced. Owing to the difference of microstructure in the welded joint, the corrosion situation is different. Various traditional corrosion detection methods [6, 7] obtain the corrosion condition of the whole electrode only and cannot get the corrosion behavior of the microregions. On microstructure and grain size of the welded joint, the electrochemical method of microregion has peculiar advantages over the classical electrochemical method.

Scanning vibrating electrode technique (SVET) [8, 9] is to study the corrosion resistance of samples by detecting the local corrosion potential (current) information of samples without touching the surface of the sample. However, welding defects can affect the electric potential of SVET, thus affecting test results [10, 11]. Therefore, SVET has been widely used in the research of various types of corrosion, such as pitting, intergranular corrosion, and stress corrosion [12, 13].

In previous researches the corrosion behavior of welded joint was studied with SVET. Wang et al. studied the corrosion behavior of HAZ in acidic soil solution with SVET [14]. The results demonstrated that microstructure of granular bainite mixed with ferrite was with the highest charge transfer resistance and the most positive current density value. Acicular ferrite of base metal displayed the lowest charge transfer resistance and the most negative current density. Liu et al. studied the corrosion properties of welded joints with different heat input in high strength low alloy steel [15]. Fuertes et al. used SVET to investigate oxide dissolution and initiation and propagation of corrosion on the welded zone [16]. Luo et al. used SVET to study stress corrosion cracking in heat affected zone [17]. The formation and development

TABLE 1: The chemical compositions of the tested steel (Q690) and ER 110S-G (wt%).

materials	C	Si	Mn	P	S	V	Ti	Cr	Ni	Cu
Q690	≤0.18	≤0.50	≤1.70	≤0.025	≤0.02	≤0.15	≤0.2	≤0.3	≤0.50	≤0.30
ER 110S-G	≤0.1	0.4-0.8	1.5-1.8	≤0.02	≤0.02	—	—	—	1.3-1.6	0.2-0.4

of stress corrosion cracking were explained. Ma et al. used SVET to investigate the localized corrosion behaviors of the welded joint [18]. They discovered that there were the maximum microhardness and the densest microstructure in the welded zone compared with the other zones. And the welded zone presented the lowest current density due to the presence of iron oxides and the densest microstructure, thus showing the excellent corrosion resistance. Indeed, the SVET results reveal that the corrosion rates of welded specimens are associated with the welding heat input and corrosion products.

The above studies indicate that the various microstructure of the welded joint is an important parameter of affecting corrosion resistance. Adopting local electrochemical measurement techniques for characterizing the localized corrosion of welded joint can better reveal the corrosion mechanism, and the corrosion behavior of welded joint with different microstructure is clearly evaluated. It is relatively rare to study the corrosion resistance of welded joints with SVET technology in chloride environment. Therefore, it is significant for the research of Q690 in 3.5% NaCl solution.

In this work, a SVET method provided further valuable information on different microstructure of the welded joints and evaluated the corrosion resistance of the welded joint. Combined with the microstructure and hardness analysis, the corrosion behavior of Q690 with CMT Twin in 3.5% NaCl solution was carried out.

2. Experimental

2.1. Materials.
The low alloy steel of Q690 was welded with CMT Twin, and the dimension of groove was shown in Figure 1. In this work, the welding wire (type: OK AristoRod 69/ER110S-G, produced by ESAB company) of 1.2 mm in diameter was utilized as the filling material in welding groove. Moreover, a welding machine (type: TransPuls Synergic CMT 5000, produced by Fronius Company) was utilized, and a mixture of 82% Ar + 18% CO_2 was selected. The chemical compositions of the base material Q690 and the wire were shown in Table 1.

2.2. The Test Method

2.2.1. Hardness Analysis.
Microhardness test was conducted through a nanoindentation measuring device (type: HVS 1000). The test force was 1 Kgf and the force duration was 10 s. The distance between each point was 0.5 mm. The connection between hardness value and microstructure is very close. The hardness is very important in mechanical properties of the materials.

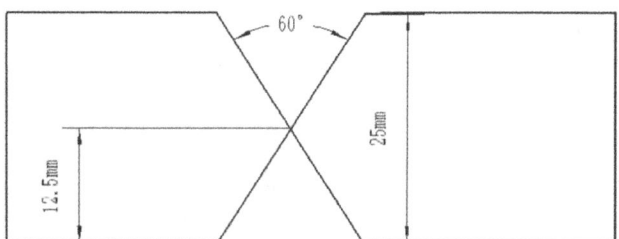

FIGURE 1: Schematic illustration of welded groove.

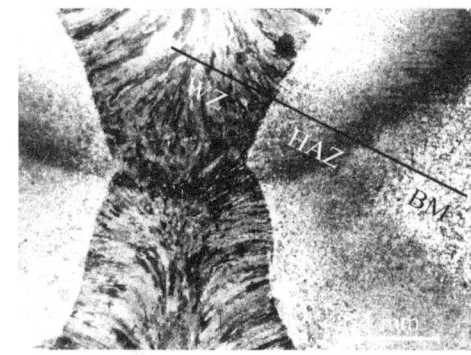

FIGURE 2: Cross section macro-images of the weld joint.

2.2.2. Microstructure Analysis.
The micro-images of cross section were shown in Figure 2, which exhibited that joint had good welding quality. Microstructure samples were machined with 10 mm×10 mm×5 mm, which ensured that the specimen contained base metal (BM), heat affected zone (HAZ), and weld zone (WZ). Prior to microstructure testing, the sample was polished sequentially with the number of 120, 240, 400, 600, 800, and 1000 emery papers and then polished to the mirror with 1.5 mm diamond paste. After drying, the specimen was etched by a mixture of 4% nitric acid and ethanol (weight fraction). The sample was dried with air compressor; microstructure of the sample was observed with metalloscope.

2.2.3. Corrosion Behavior Analysis.
All sides of the specimen were embedded in an epoxy resin except the exposure surface, and the working area was 14 mm×14 mm. Polish the face of the specimen, clean the surface with anhydrous ethanol, and dry it with air compressor. Subsequently, the SVET measurements were conducted through a PAR370 Scanning Electrochemical Workstation. The SVET test was adopted by the Versa SCAN electrochemical scanning system, and the scanning microprobe was a Pt-Ir microelectrode with a

diameter of 50 um. At the beginning of the test, the sample was placed in the electrolytic cell, and the surface of the sample remained horizontal.

The scanning probe vibrated in the perpendicular direction at the surface of the sample, and the distance between the probe and sample surface was 100 um. The vibrating amplitude of the microelectrode was 30 um and the vibrating frequency was 50 Hz in the direction to the surface. The scanning area was 5 mm × 3 mm, and it scanned from weld zone to base metal.

SVET measurements were carried out in electrochemical corrosion testing apparatus at room temperature in 3.5wt% NaCl solution [19–21]. And the probe was immersed into 3.5 wt% NaCl aqueous solution. The results of SVET were the ion potential difference of local oxidation-reduction reaction on the specimen surface, and the data were converted to local current density according to the specification of data processing in instrument.

The SVET current density maps and the statistical analysis of the data were performed with Origin software. The current densities were displayed in three-dimensional (3D) maps, showing the spatial distribution of the current density as a function of the (x,y) position in the scan region. The current values in the SVET map are positive for anodic currents and negative for cathodic currents. The contour map of the current densities is at the bottom of the 3D maps.

The integrated anodic current (I_{Int}) was used to characterize the corrosion resistance of different microstructure. I_{Int} was evaluated by integration of the overall anodic current (i_A) on SVET current density map. The whole scan area was split into 21×21 small squares, and we calculated the scan area (S) and number (n) of measurement points in each microstructure. The scan area of WZ, overheated zone (OZ), normalized zone (NZ), incomplete normalized zone (INZ), and BM, respectively, was 4.9, 2.2, 2.5, 0.4, and 5.0 mm², and n of WZ, OZ, NZ, INZ, and BM, respectively, was 142, 56, 74, 20, and 149. We calculated the anodic current on each microstructure, and calculation formula is shown by

$$I_{int} = \frac{S \sum i_A}{n} \qquad (1)$$

(see [22]) where i_A is the anodic current density ($i_A > 0$) measurement points in each microstructure.

3. Result and Discussion

3.1. The Hardness of the CMT Twin Welded Joint.
The distribution of hardness in welded joint is shown in Figure 3. It can be shown that the hardness values of the welded joint are symmetric distribution as a centre of the welded zone. It can be concluded that the welded joint shows relatively uniform fluctuations of hardness in BM, and the value is maintained at about 275 HV. In the vicinity of the heat affected zone, the hardness of BM decreases to a certain extent, which is due to the growth of grain on the edge of the HAZ.

The hardness of the HAZ in Figure 3 rises relatively fast, where the troostite has a high hardness (about 310 ~ 332 HV). Simultaneously, the decrease of troostite near the WZ leads to

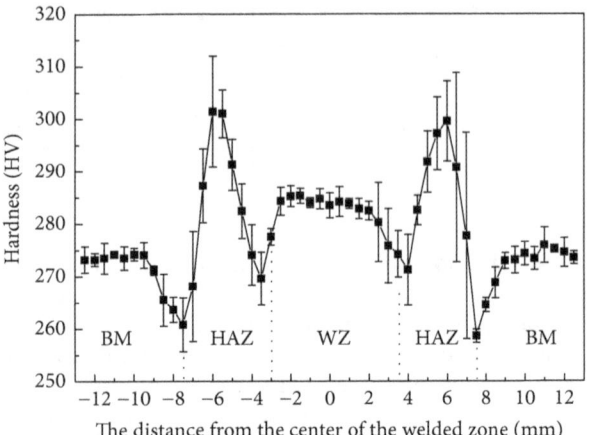

FIGURE 3: The distribution of the hardness in the welded joint.

further decrease of the hardness on the edge of the HAZ. And the amplitude of descent in hardness is little lower than that of the BM, as presented in Figure 3. It can be shown in Figure 3 that the welded joint shows obvious hardness fluctuation in HAZ, which reveals the inhomogeneous microstructure of HAZ. This leads to heavy corrosion for HAZ in 3.5% NaCl solution. However, the hardness of the HAZ is determined by the tendency of brittleness in the BM and the cooling speed of the HAZ. This means that the brittleness and hardness of the BM are determined by the chemical composition, while the cooling rate of the HAZ is mainly affected by the welding specification. Generally, the chemical composition of the BM is certain, and the hardness of the HAZ can only be improved by changing the cooling rate after welding. It can be confirmed that the microstructure of the WZ consists of the bainite and a small amount of acicular ferrite, and the hardness is kept at 285 HV, slightly higher than that of the BM.

3.2. Microstructure of the Welded Joint.
It can be shown in Figure 4(a) that the microstructure of BM consists of a mixture of ferrite with low carbon slabs and carbide. Meanwhile, the same size of the bainite is parallel with arrangement in direction, which formed the bainite-rich region. The microstructure of lower bainite (LB) is mostly featured with the lath ferrite (F), and from the austenite grain boundary to the intragranular one the parallel growth of ferrite is shown. Simultaneously, the short rod-like black product with intermittent is cementite, which exists in lath ferrite. Moreover, it can be shown in Figure 4 that the welded joint shows obvious crystal boundary of austenite in BM, and the distribution of ferrite is very homogeneous. Specifically, the grain orientation in the region is larger, and plenty of bainite regions are formed.

Figure 4(b) shows optical micrographs of the cross section of the different welding regions (incomplete normalized zone (INZ) and NZ) of the welded joint. It can be shown that the troostite (black substance) is precipitated out of the coarse bainite in HAZ, and it gradually diffuses into the BM. It obviously detects the diffusion process of the troostite in

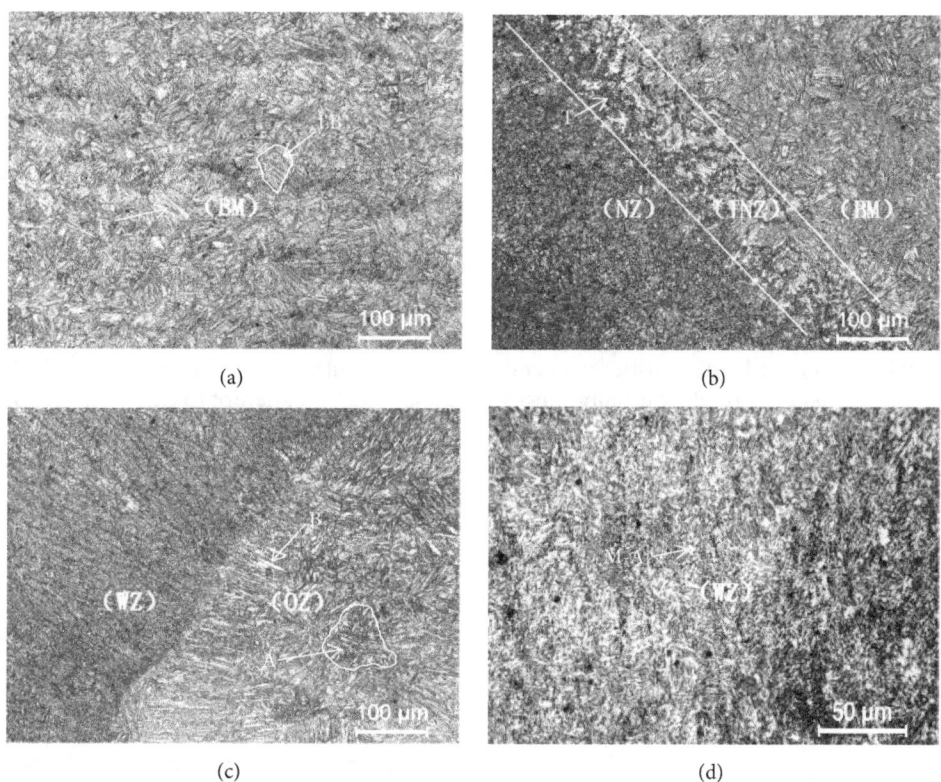

(a)

(b)

(c)

(d)

FIGURE 4: The microstructure of BM, INZ, NZ, OZ, and WZ.

INZ, while NZ is completely covered by the troostite (T). This means that this zone has high mechanical properties (hardness and strength).

Figure 4(c) shows the microstructure of OZ, in which the upper left corner is WZ. It can be seen that the microstructure adjacent to WZ is relatively coarse and the grain boundary is clear. Moreover, the microstructure adjacent to the fusion line shows that the microstructure of the welded zone is relatively finer compared with the BM, and there also exists black granular troostite in the grain boundary of austenite (A). In the area of BM adjacent to WZ, it can be clearly shown that the microstructure of the BM is basically perpendicular to the fusion line, and bainite (B) becomes longer. In HAZ, the grain boundary of austenite is obvious, and the microstructure of HAZ is obviously more coarse than that of BM, as presented in Figure 4. Owing to the lower heat input and high speed of cooling after welding, massive bainite with a high density dislocation is formed with dispersive distribution of carbide. The appearance of bainite presents a thin strip and a series of slabs are arranged in parallel. As a result, the same direction bainite grows to form a regional strip, and the slabs of the bainite intersect each other.

Figure 4(d) presents the microstructure of WZ; it can be shown in Figure 4(c) that the microstructure is finer. After magnification, the granular bainite (M-A constituent) is distributed homogeneously. It is well known that the cooling rate greatly affected the formation of granular bainite. The cooling rate influences the growth of factors such as the generation of a significantly quenched hardening structure. Moreover,

the low cooling rate increases the austenitic isothermal time, which is favorable for the short-range diffusion of carbon atoms. The quantity of the carbon concentrated in the primary austenite will be suppressed, being affected by the welding heat input, when the bainite ferrite is generated during the bainite transformation. The bainite ferrite formation is prevented by the primary austenite, while the carbon concentration reached a critical value. Therefore, the residual austenite transforms into the M-A island that is distributed within the ferrite matrix. Subsequently, the lath and granular bainite microstructures are generated. Therefore, a large number of small M-A islands are distributed evenly in ferrite matrix. In the process of granular bainite formation, ferrite with precipitating out of austenite firstly is featured with the shape of lath. Meanwhile, lath ferrite with high dislocation density leads to high strength of materials. Owing to the hard phase of M-A islands, the M-A islands are precipitated by way of small and dispersed precipitation. Indeed, the interaction between M-A islands and the dislocation hinders the dislocation movement. In other words, the way of dispersion strengthening improves the strength of the steel.

3.3. Corrosion Behavior of the CMT Twin Welded Joint. The corrosion of low carbon alloy steel in 3.5% NaCl solution is typical electrochemical corrosion. When low carbon alloy steel is immersed in 3.5% NaCl solution, the distribution of components, phase, and the surface stress of surface in material are not uniform. Indeed, the distribution of electrode potential between material surface and interface of water is inhomogeneous on the microscale. There are countless corrosion microcells on the surface of the metal, resulting in the corresponding anode and cathode areas. The anode is an active dissolving reaction, namely, $Fe-2e=Fe^{2+}$. In the meantime, the corrosive particles in the environment are mainly O_2 and Cl^-, in which the concentration of oxygen is consistent, and the cathode is the depolarization reaction of oxygen, namely, $O_2+2H_2O+4e=4OH^-$. Therefore, the corrosion rate of the materials may be accelerated by influencing the adsorption behavior of the corrosive Cl^-. Furthermore, Cl^- is a characteristic adsorption ion, which is able to form a chemical bond of coordination with metal and adsorb on the surface of the metal. This effect usually causes the change of charge on the metal surface.

3.4. SVET Current Density Maps. The SVET current density maps for various periods in 3.5% NaCl solution are demonstrated in Figure 5. Simultaneously, the boundary lines of different microstructure were displayed in X axis. The anodic current appeared at different sites, and the anodic area is mainly located at the BM in 0 hour (Figure 5(a)). The 0 hour is the time for SVET to scan the once.

At the beginning of time (0 h), the corrosion resistance of the BM and HAZ is worse. In addition, due to the influence of the heat input during the process of welding, there is obvious differentiation of microstructure between BM and WZ. There is obvious potential difference between the WZ and BM, resulting in galvanic corrosion. Following this, the cathode region appears in the WZ and thus is protected. In the meantime, the microstructure in the HAZ is relatively coarse. As a result, the corrosion resistance of the WZ is better than BM. The microstructure of WZ consists of the granular bainite (M-A constituent), as presented in Figure 4. Granular bainite consists of a mixture of ferrite and austenite, and ferrite is surrounded by austenite. In addition, phase transition temperature of bainite is low, and the diffusion ability of alloy elements is relatively worse. Therefore, the uniform bainite reduces the potential difference between the anode and cathode. Thus, it is indicated that the corrosion rate reduces to some extent, especially the tendency of localized corrosion. In addition, grain boundary of bainite is small angle grain boundary, and interfacial energy can be lower, and the impurity content of grain boundary is low, making it shows excellent corrosion resistance.

It has been established that, with the increase of time (2 h), the SVET current density in anode region and cathode region decreases. In the initial period of corrosion, the unstable product $Fe(OH)_2$ is formed on the surface of the sample, which to some extent inhibits the process of anodic dissolution of the corrosion reaction, thus reducing the corrosion rate [23]. The anodic current density peak appears in BM, and the area of higher anode current significantly reduces. In the meantime, the cathodic area appears in the HAZ, and the number significantly increases. When time is 4 h, the anodic current density peak appears in WZ. In other words, the anode area is moved from BM to the WZ. Owing to the porous structure of the new rust

layer, the rust layer cannot provide adequate protection, so the rust layer on the sample surface is continuously increasing, as shown in Figures 5(b)–5(d). When the corrosion time is 6 h, the peak current area and quantity of the anode decrease gradually. Thus, it means that the corrosion rate of the whole regional greatly reduces, and there are only scattered anode current peaks at the edge of the weld zone. The cathode region appears in HAZ and the BM, which improves the corrosion resistance of the two regions.

3.5. Current Peak of Anodic and Cathodic. It is possible to go further and attempt to extract more information from the SVET results. As shown in Figure 6, it can be observed that the welded joint exhibits the current peak of anodic ($i_{A,max}$), cathodic ($i_{C,max}$), and the average current density (i_{Ave}) in SVET maps. It can be seen from the graph that the peak of cathode current and anode current are decreasing gradually, while the average current is moving around zero. The test of corrosion was conducted without external polarization, so the anodic currents and cathodic currents are balanced and the net current should be zero. In general, the current peak of cathode and anode should be equal in value, and the average current density should be zero. However, the difference between the anode and cathode current peaks in SVET maps results in an average current deviation of zero. The reason for this deviation may be that the current density of the SVET maps is not measured at the same time. In the process of scanning, corrosion behavior and current distribution are constantly changing. As shown in Figures 5(a)–5(d), the current density of different microstructure varies considerably with the increase of corrosion time. This shows that the process of electrochemical is instantaneous, which changes at any time.

3.6. Corrosion Behavior of Different Zone. Integrated anodic current of different microstructure is as shown in Figure 7. It shows that the integrated anodic current (I_{Int}) of OZ and BM is larger than that of WZ in the whole corrosion process. Combined with SVET maps, the corrosion tendency of microstructure in WZ is relatively high and the corrosion rate is low. At the same time, the degree of corrosion in WZ is lower than that of BM. As the iron dissolves, the integrated anodic current in BM and OZ decreases gradually. Moreover, the corrosion products decrease the corrosion rate of WZ and BM. Due to the coarse microstructure in HAZ, the integrated anodic current increases (0-2 h). With the oxidation of $Fe(OH)_2$, the integrated anodic current of BM and OZ increases slightly (2-4 h). As the size of microstructure varies, I_{Int} in the BM decreases, while I_{Int} in OZ is relatively stable (4-6 h). Generally, the size of the integrated anodic current reflects the corrosion rate of different regions. With the formation of corrosion products, the corrosion rate of NZ gradually decreases (2-6 h). The weld zone is protected by the cathode and the degree of corrosion is little. Compared with the NZ, the corrosion resistance of the INZ is enhanced, which is due to the decrease of the amount of the troostite.

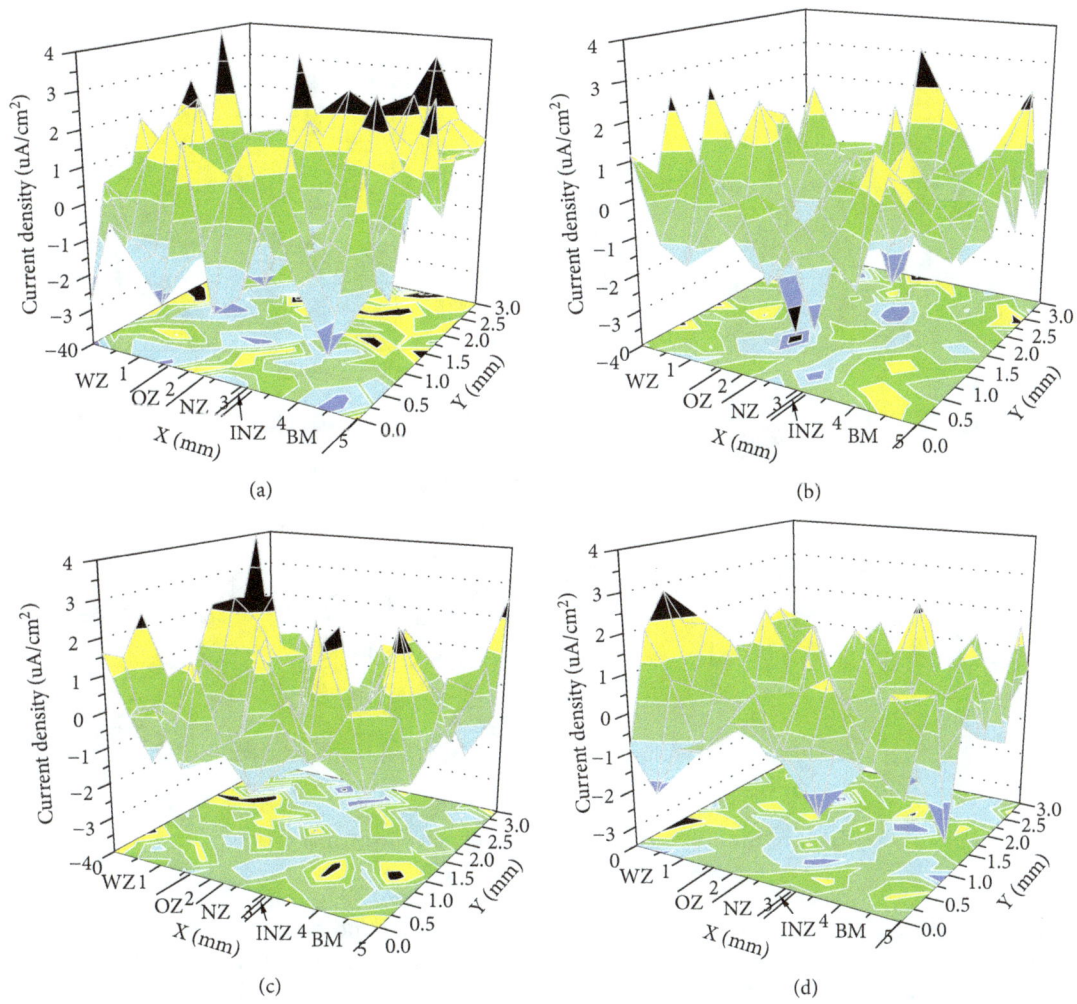

FIGURE 5: SVET current density maps of (a) 0 h, (b) 2 h, (c) 4 h, and (d) 6 h.

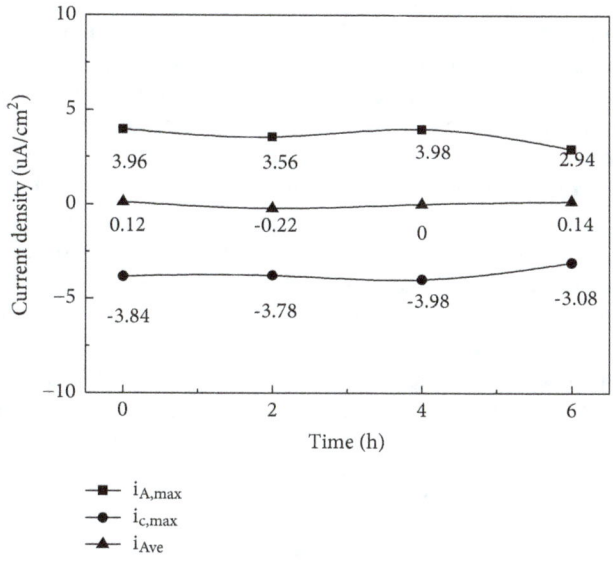

FIGURE 6: The current peak of anodic ($i_{A,max}$), cathodic ($i_{C,max}$), and the average current density (i_{Ave}) in SVET maps.

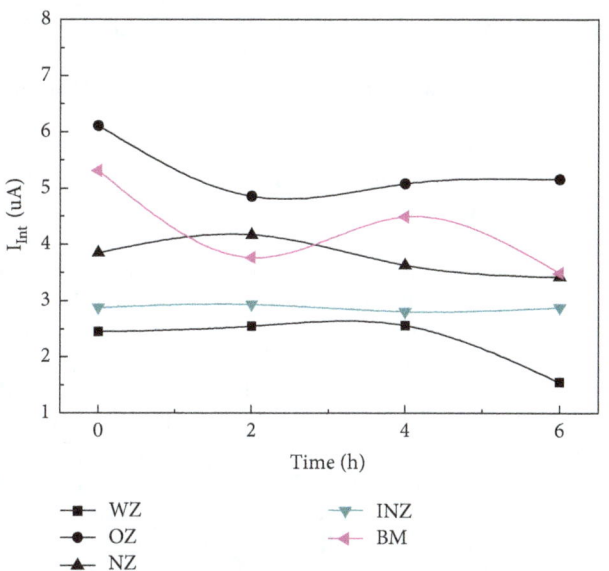

FIGURE 7: Integrated cathodic current (I_{Int}) of different microstructure.

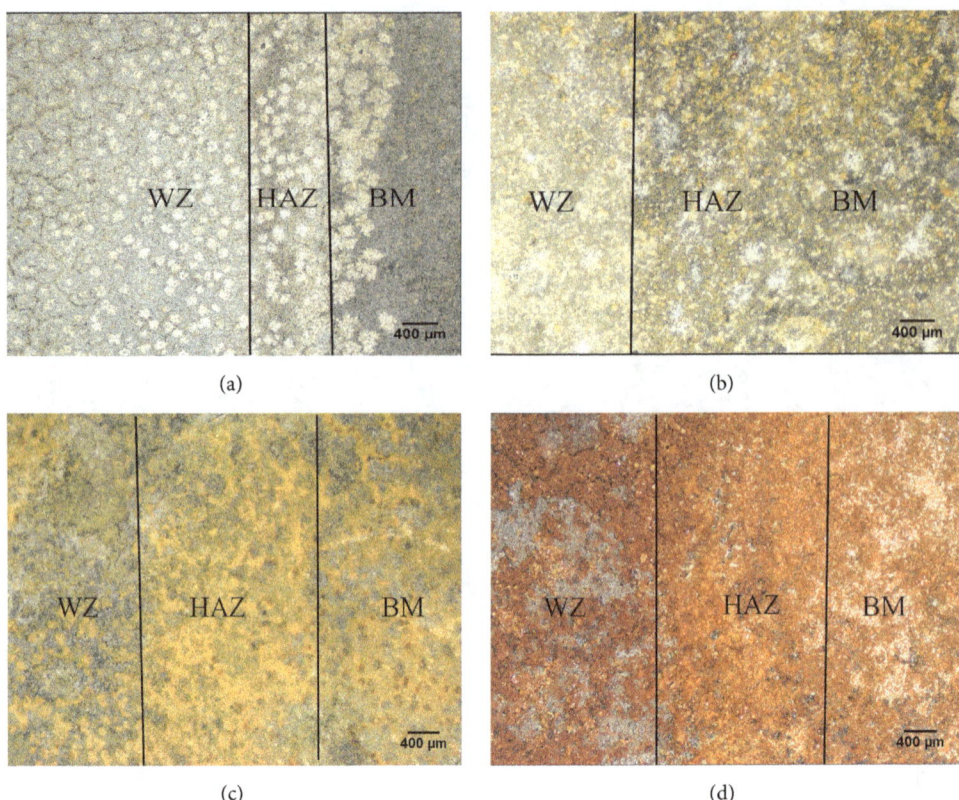

FIGURE 8: The macroscopic appearance of different corrosion time.

3.7. Macro Analysis of Welded Joint after Corrosion. The macroscopic appearance of different corrosion times in welded joint is demonstrated in Figure 8. Figure 8(a) reveals that different degrees of corrosion occurred in the BM, WZ, and HAZ. That is to say, the results are in agreement with the distribution of current density maps (Figure 5(a)). The white corrosion products, including $Fe(OH)_2$, are mainly distributed in the HAZ and the zone adjacent to the HAZ (BM and WZ). In the WZ, the white corrosion products, which are away from the area of HAZ, gradually decrease. Moreover, on the right of Figure 8(a), the coarse microstructure of HAZ leads to lesser corrosion product.

When corrosion time is 2 h, white corrosion products disappeared, and a few corrosion products ($Fe(OH)_2$) with rufous appear. Furthermore, it is apparent that the boundary of the HAZ and the BM basically disappeared. That is to say, the corrosion products of the BM and the HAZ are obviously more than the WZ. It is revealed that $Fe(OH)_2$ is easily oxidized to products of gray-green. The products of rufous are mainly concentrated in the HAZ, and the BM has fewer corrosion products and the least ones in WZ, as presented in Figure 8(c). When corrosion time is 6 h, there are obvious differences in WZ, HAZ, and BM. Figure 8(d) shows the corrosion degree of WZ is the minimum, and the corrosion resistance is excellent. Furthermore, the corrosion products in BM have not been completely oxidized, and some corrosion products of white ($Fe(OH)_2$) exist.

As Figure 8 shows, the corrosion degree of each region is not uniform at any time. The grains of granular bainite in

WZ are finer and evenly distributed, and the distribution of microstructure is uniform. It is well known that the content of granular bainite is high in Q690. In addition, the small cathode and large anode lead to reducing the corrosion rate of WZ in the process of corrosion. As shown in Figure 8, the WZ displays the excellent corrosion resistance. On the contrary, the HAZ with coarse microstructure exhibits the worst corrosion resistance. It may also be that nickel increases the corrosion resistance of the weld area. As shown in Table 1, there are more nickel elements in the weld zone, in comparison with BM. The study shows that the addition of nickel can form a protective rust layer on the surface, thus improving the corrosion resistance of the welded zone [24]. Ni has ability to resist the corrosive action of Cl^- ions. Accordingly, Ni exhibits high corrosion resistance in 3.5% NaCl solution. This indicates that, as a result of Ni in the welded zone, the aggressive action of Cl^- ions on the surface is suppressed. This behavior exhibits very important result that the Ni has influence on corrosion resistance in 3.5% NaCl solution.

In summary, it can be discovered that the corrosion rates of various zone are associated with the welding heat input. From this perspective, the corrosion degree of microstructure is mainly HAZ > BM > WZ.

4. Conclusions

The microstructure, hardness, and corrosion behavior in welded joint of Q690, which was welded with CMT Twin, were studied. The experimental findings support the following major conclusions:

(1) The hardness of the base metal was the minimum, while the hardness of weld zone with tiny microstructure was higher than that of the base metal. Owing to the coarse microstructure, the hardness of overheated zone decreased. The hardness of normalized zone increased with the appearance of the troostite.

(2) As the corrosion time increases, the corrosion products of different microstructure were identical, and the white products of the welded joint were turned into products of rufous. In the role of oxygen, $Fe(OH)_2$ was oxidized to $Fe(OH)_3$. Also, the overheated zone and the base metal presented a high corrosion rate. The degree of corrosion in weld zone, which is protected by the cathode, was mild. The corrosion resistance of the weld zone with CMT Twin was greatly improved compared with that of the base metal. This was related to the existence of nickel.

Conflicts of Interest

The authors declare that there are no conflicts of interest regarding the publication of this paper.

Acknowledgments

This research was finished with the fund of Science and Technology Project of Guangdong Province (2017A010106007, 2017A010102008, and 201101C0104901263) and Science and Technology Project of Guangzhou (201508030023). The authors are grateful to Guangdong Welding Institute (China-Ukraine E. O. Paton Institute of Welding).

References

[1] Z. S. Li, S. L. Li, X. H. Zhang, and J. Z. Liu, "Interaction study of carbide precipitation and impurity segregation under temper embrittlement conditions in a coarse-grained heat-affected zone in Q690 steel," *Advanced Materials Research*, vol. 1015, pp. 189–193, 2014.

[2] X. Chen and Y. Huang, "Hot deformation behavior of HSLA steel Q690 and phase transformation during compression," *Journal of Alloys and Compounds*, vol. 619, pp. 564–571, 2015.

[3] H. Wan, C. Du, Z. Liu, D. Song, and X. Li, "The effect of hydrogen on stress corrosion behavior of X65 steel welded joint in simulated deep sea environment," *Ocean Engineering*, vol. 114, pp. 216–223, 2016.

[4] R. Pérez-Mora, T. Palin-Luc, C. Bathias, and P. C. Paris, "Very high cycle fatigue of a high strength steel under sea water corrosion: A strong corrosion and mechanical damage coupling," *International Journal of Fatigue*, vol. 74, pp. 156–165, 2015.

[5] Y. Ye, C. Wang, H. Chen, Y. Wang, J. Li, and F. Ma, "An analysis of the tribological mechanism of GLC film in artificial seawater," *RSC Advances*, vol. 6, no. 39, pp. 32922–32931, 2016.

[6] F. Gan, G. Tian, Z. Wan, J. Liao, and W. Li, "Investigation of pitting corrosion monitoring using field signature method," *Measurement*, vol. 82, pp. 46–54, 2016.

[7] V. Verdingovas, M. S. Jellesen, and R. Ambat, "Colorimetric visualization of tin corrosion: A method for early stage corrosion detection on printed circuit boards," *Microelectronics Reliability*, vol. 73, pp. 158–166, 2017.

[8] D. Mata, M. Serdechnova, M. Mohedano et al., "Hierarchically organized Li-Al-LDH nano-flakes: A low-temperature approach to seal porous anodic oxide on aluminum alloys," *RSC Advances*, vol. 7, no. 56, pp. 35357–35367, 2017.

[9] A. C. Bastos, M. C. Quevedo, O. V. Karavai, and M. G. Ferreira, "Review—on the application of the scanning vibrating electrode technique (svet) to corrosion research," *Journal of The Electrochemical Society*, vol. 164, no. 14, pp. C973–C990, 2017.

[10] V. I. Danilov, L. B. Zuev, S. V. Konovalov et al., "On the effect of electric potential on resistance of metals surface to microindentation," *Journal of Surface Investigation X-ray Synchrotron and Neutron Techniques*, vol. 1, pp. 157–161, 2010.

[11] S. A. Nevskii, S. V. Konovalov, and V. E. Gromov, "Effect of the electric potential of the aluminum surface on stress relaxation," *Technical Physics*, vol. 56, no. 6, pp. 877–880, 2011.

[12] M. G. Taryba and S. V. Lamaka, "Plasticizer-free solid-contact pH-selective microelectrode for visualization of local corrosion," *Journal of Electroanalytical Chemistry*, vol. 725, pp. 32–38, 2014.

[13] A. S. Gnedenkov, S. L. Sinebryukhov, D. V. Mashtalyar, and S. V. Gnedenkov, "Localized corrosion of the Mg alloys with inhibitor-containing coatings: SVET and SIET studies," *Corrosion Science*, vol. 102, pp. 269–278, 2016.

[14] L. W. Wang, Z. Y. Liu, Z. Y. Cui, C. W. Du, X. H. Wang, and X. G. Li, "In situ corrosion characterization of simulated weld heat affected zone on api x80 pipeline steel," *Corrosion Science*, vol. 85, pp. 401–410, 2014.

[15] W. Liu, H. Pan, L. Li et al., "Corrosion behavior of the high strength low alloy steel joined by vertical electro-gas welding and submerged arc welding methods," *Journal of Manufacturing Processes*, vol. 25, pp. 418–425, 2017.

[16] N. Fuertes, V. Bengtsson, R. Pettersson, and M. Rohwerder, "Use of SVET to evaluate corrosion resistance of heat tinted stainless steel welds and effect of post-weld cleaning," *Materials and Corrosion*, vol. 68, no. 1, pp. 7–19, 2017.

[17] L. Luo, Y. Huang, and F.-Z. Xuan, "Deflection behaviour of corrosion crack growth in the heat affected zone of CrNiMoV steel welded joint," *Corrosion Science*, vol. 121, pp. 11–21, 2017.

[18] H. Ma, Y. Gu, H. Gao, X. Jiao, J. Che, and Q. Zeng, "Microstructure, chemical composition and local corrosion behavior of a friction stud welding joint," *Journal of Materials Engineering & Performance*, vol. 2, pp. 666–676, 2018.

[19] B. Yan, H. Chen, and D. Kong, "Effects of laser remelting on salt spray corrosion behaviors of arc-sprayed Al coatings in 3.5% NaCl sea environment," *Transactions of the Indian Institute of Metals*, vol. 6, pp. 1–9, 2017.

[20] S. Weng, Y. Huang, F. Xuan, and F. Yang, "Pit evolution around the fusion line of a NiCrMoV steel welded joint caused by galvanic and stress-assisted coupling corrosion," *RSC Advances*, vol. 8, no. 7, pp. 3399–3409, 2018.

[21] H. S. Gadow, M. M. Motawea, and H. M. Elabbasy, "Investigation of myrrh extract as a new corrosion inhibitor for α-brass in 3.5% NaCl solution polluted by 16 ppm sulfide," *RSC Advances*, vol. 7, no. 47, pp. 29883–29898, 2017.

[22] M. Yan, V. J. Gelling, B. R. Hinderliter, D. Battocchi, D. E. Tallman, and G. P. Bierwagen, "SVET method for characterizing anti-corrosion performance of metal-rich coatings," *Corrosion Science*, vol. 52, no. 8, pp. 2636–2642, 2010.

[23] Y. Ma, Y. Li, and F. Wang, "The effect of β-FeOOH on the corrosion behavior of low carbon steel exposed in tropic marine environment," *Materials Chemistry and Physics*, vol. 112, no. 3, pp. 844–852, 2008.

[24] H. M. Abd El-lateef, A. El-Sayed, and H. S. Mohran, "Role of Ni content in improvement of corrosion resistance of Zn–Ni alloy in 3.5% NaCl solution. Part I: Polarization and impedance studies," *Transactions of Nonferrous Metals Society of China*, vol. 25, no. 8, pp. 2807–2816, 2015.

15

Electrochemical Properties of Oxide Scale on Steel Exposed in Saturated Calcium Hydroxide Solutions with or without Chlorides

Johan Ahlström [ID],[1,2] **Johan Tidblad,**[1] **Luping Tang,**[2]
Bror Sederholm,[1] **and Simon Leijonmarck**[1]

[1]*Swerea KIMAB, Isafjordsgatan 28A, Kista 16440, Sweden*
[2]*Division of Building Technology, Chalmers University of Technology, Gothenburg 41296, Sweden*

Correspondence should be addressed to Johan Ahlström; johan.ahlstrom@swerea.se

Academic Editor: Flavio Deflorian

The electrochemical properties of various iron oxide scales on steel exposed in saturated calcium hydroxide solutions were investigated. The iron oxide scales were manufactured by different heat treatments and grinding processes and characterized using X-ray diffraction and scanning electron microscope. The electrochemical properties of the scales were assessed by measuring the corrosion potential and using electrochemical impedance spectroscopy and potentiodynamic polarization curves. It was found that wustite and magnetite are less noble compared to hematite but are more effective as cathodic surfaces. The results show that the electrochemical properties of the mill scale can be an important contributing factor in the corrosion of steel in concrete.

1. Introduction

Chloride induced corrosion of steel in concrete is an important deterioration mechanism for concrete structures. Many studies have been performed to determine a critical chloride threshold level and a wide range of critical chloride threshold levels has been reported [1]. One explanation of the wide range is that studies have been conducted with steel samples in an as-received condition where the mill scale from production is intact or with modified surfaces obtained by, e.g., grinding, pickling, where the steel surface is without a mill scale. Generally, it is known that the chloride threshold level is lower on surfaces with a mill scale compared to surfaces without a mill scale [2–7]. It has been reported that the mill scale has several effects on corrosion. One effect is that a remaining mill scale decreases the electrical resistance and inhibits the formation of a passive film on the steel surface in concrete [8, 9]. Another effect is that the cathodic current is higher on a surface with a mill scale compared to a surface without mill scale [10] which affects the overall corrosion properties.

The exact mechanism the mill scale has on corrosion of steel in concrete is not fully understood. Ghods et al. [11] suggested in a microscopic study that the mill scale contains cracks which form a crevice between the mill scale and steel surface. Corrosion is then initiated through a process similar to classical crevice corrosion. The effect of different oxides on the corrosion of steel in concrete was examined by Avila-Mendoza et al. [12]. The results showed that the corrosion rate was higher on steel with red oxides, mainly hematite, compared with polished and furnace produced oxide surfaces. It was proposed that the higher corrosion rate was due to an alternative cathodic reaction: self-reduction of Fe_2O_3 to Fe_3O_4.

The mill scale originating in the steel production process consists of mainly three types of iron oxides: wustite FeO, magnetite Fe_3O_4, and hematite Fe_2O_3 [13]. These iron oxides have different chemical and physical properties. Wustite is the least stable iron oxide and decomposes to magnetite and iron at temperatures below 570°C. Magnetite is stable in alkaline solutions under reducing conditions and the electric conductivity is high compared with the other iron

TABLE 1: Chemical composition of the steel bars (% mass).

C	Mn	P	S	N	Cu
0.17	1.4	0.035	0.035	0.12	0.55

TABLE 2: Manufacturing procedure of the different scales.

Sample name	Procedure
Steel	Mill scaled removed; ground and degreased
Wustite	850°C, 10 min ⟶ cooling to 600°C ⟶ quenched in water ⟶ grinding to black color
Magnetite	850°C, 10 min ⟶ cooling to 600°C ⟶ cooling to RT ⟶ 400°C, 50 min ⟶ light grinding
Hematite	850°C, 8 min ⟶ 550°C, 3 hours

oxides, 100-1000 $\Omega^{-1}.cm^{-1}$. Hematite is stable over a wide pH range in oxidizing conditions and the electric conductivity is low, roughly $10^{-9}\,\Omega^{-1}.cm^{-1}$ [14]. This means that mill scales with different chemical compositions can have various effects on the corrosion of steel in concrete. The aim of this study is to produce mill scales with different chemical compositions and assess their electrochemical properties to show if electrochemistry could be a part of the mechanism the mill scale has on corrosion of steel in concrete.

2. Materials and Methods

2.1. Manufacturing of Oxide Scales. During annealing of steel a mill scale is formed on the steel surface which consists of three types of iron oxides. The part of the scale closest to the steel surface is dominated by wustite, FeO, the middle part is dominated by magnetite, Fe_3O_4, and the outermost part is dominated by hematite, Fe_2O_3.

Steel samples were cut from a hot rolled plate, S235 JR (EN 10025-2:2004), to the dimensions 25 x 25 mm and 2 mm thick. The composition of the steel can be seen in Table 1.

The mill scale on the hot rolled samples was removed by grinding with 1200 grit paper and thereafter cleaning in an ultrasonic bath for 10 minutes in 50% ethanol and 50% acetone. After cleaning, the steel samples were heat-treated in order to produce an oxide scale dominated by a certain iron oxide and in the following the samples are therefore named after the intended dominating oxide: wustite, magnetite, and hematite. The manufacturing procedure of the different scales can be seen in Table 2.

Bare steel samples were made for comparison with the manufactured oxide scale samples. The original mill scale was removed by wet grinding with 600 grit paper and then rinsed in alcohol, dried, and stored in a desiccator until the start of the electrochemical experiments.

All samples, except steel which were used as a reference, were annealed at 850°C for 8-10 minutes and then cooled with different rates depending on the desired oxide layer. Wustite is stable in temperatures higher than 570°C and decomposes to magnetite and iron below 570°C according to the reaction:

$$4FeO \longrightarrow Fe_3O_4 + Fe \qquad (1)$$

The transformation rate is slow at room temperature and fast between 400-480°C [15]. To produce a wustite rich oxide layer

the samples were first annealed at 850°C and then cooled to 600°C in the oven leaving the lid open. This was followed by quenching the samples in water to obtain a fast cooling rate to room temperature. In this way the samples were exposed to temperatures between 400-480°C for a very short time and the transformation reaction was suppressed. When the samples were quenched in water from higher temperatures than 600°C the oxide scale spalled of the steel surface. After cooling, the oxide surface was wet ground with 600 grit paper. During wet grinding the water turned red/orange indicating the removal of hematite from the sample. With further grinding the water turned grey which was also the color of the sample surface indicating magnetite. At the end of the grinding the water color and color of the sample turned black indicating wustite. At this point the grinding was stopped and the sample was rinsed with alcohol and dried with a hair dryer. All samples were then stored in a desiccator.

To produce a magnetite-rich scale the samples were annealed and then cooled in the oven to 600°C leaving the lid open, followed by air cooling to room temperature. This was followed by a second heat treatment at 400°C for 50 minutes. 400°C was chosen to obtain a fast transformation rate of wustite to magnetite and iron. After the heat treatment the samples were wet ground until the red water color disappeared as described earlier indicating that magnetite was the dominating oxide in the oxide scale. The samples were then rinsed in alcohol, dried, and stored in a desiccator.

To produce a hematite rich layer, the samples were annealed at two temperatures. The first annealing temperature was at 850°C as for the other samples and then the temperature was lowered to 550°C and the second heat treatment was performed for 3 hours. The relatively long heat treatment at 550°C was chosen to oxidize the magnetite without obtaining thick scales. After the second heat treatment, the samples were air cooled and rinsed with alcohol and dried before being stored in a desiccator.

To control if the desired oxides had been formed on the steel samples, the samples were analyzed with X-ray diffraction (XRD) and scanning electron microscope in combination with electron backscatter diffraction (SEM/EBSD). The XRD measurements were performed on crushed oxide scales with a Bruker D8 using CuK_α radiation. The crushed scale was collected by hammering on the sample surface using a steel hammer.

TABLE 3: Calculated composition from XRD spectra of the samples [%].

Phase	Wustite	Magnetite	Hematite
Fe	-	4	-
FeO (wustite)	88	17	26
FeO(OH)	9	-	-
Fe_3O_4 (magnetite)	2	77	39
Fe_2O_3 (hematite)	1	2	35

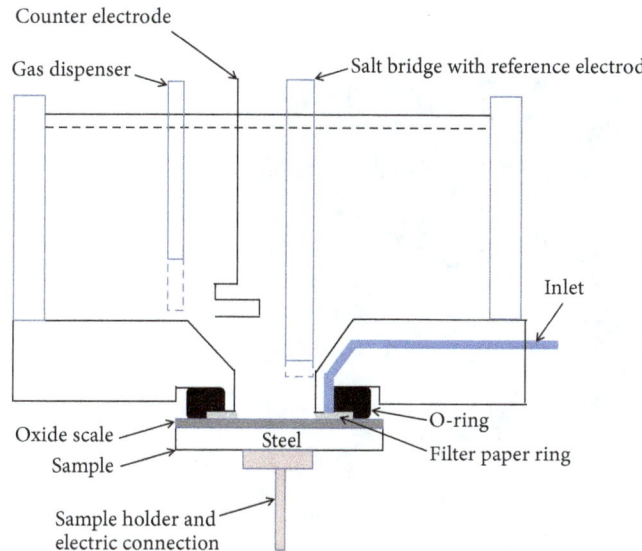

FIGURE 1: A schematic figure of the Avesta cell.

The sample preparation for the SEM/EBSD investigations were performed by first sputtering a thin gold layer onto the oxide scale surface, typically a few nm, and thereafter applying a nickel layer, typically 2 μm thick, by electrodeposition. The EBSD analyses were performed with an LEO 1530 FEG-SEM equipped with an Oxford EDS/EBSD system. A cross section of the manufactured oxide scale was analyzed where the cut samples were mounted in epoxy and ground with 1200 mesh paper and polished to 1 μm with diamond paste. The last step of the sample preparation was polishing in a colloidal silica suspension.

2.2. Electrochemical Measurements. The electrochemical experiments were performed using an Avesta cell in order to avoid crevice corrosion [16]; see Figure 1. The circular exposed sample area was 1 cm^2. The Avesta cell has an inlet where a saturated $Ca(OH)_2$ solution is pumped into the main cell via a filter paper that is put around the sample area. In this way there is no oxygen depletion at the edges of the exposed sample area and crevice corrosion can therefore be avoided.

The main cell was filled with 120 ml of either saturated $Ca(OH)_2$ solution or a saturated $Ca(OH)_2$ solution with 0.6M Cl^- added as NaCl. The pumping rate of the saturated $Ca(OH)_2$ into the cell was 3.3 ml/h which means that when the main cell was filled with the chloride containing

solution the chloride concentration was slowly diluted during the experiments. The reason why a chloride free solution was pumped into the cell was to eliminate problems with corrosion which would occur on steel samples below the filter paper when a chloride containing solution was used. To avoid chloride concentration gradients in the cell, air was bubbled into the cell through a gas dispenser.

The reference electrode used was a double junction Ag/AgCl sat KCl electrode immersed in a salt bridge with a saturated K_2SO_4 solution. A salt bridge was used as the Ag/AgCl electrode can become inaccurate in high pH solutions. A platinum wire was used as counter electrode. The measurements were performed using a Solartron 1286 potentiostat and a Solartron 1255 frequency analyzer.

The following procedure was performed for each sample. The open cell potential (OCP) was measured for 10 minutes. Electrochemical impedance spectroscopy (EIS) measurements were then performed in the frequency interval 10^5-10^{-3} Hz and amplitude ± 20 mV vs. OCP. Finally, a potentiodynamic polarization scan was made in either anodic or cathodic direction with the OCP as starting point. The scan rate was 0.2 mV per second and the samples were polarized to 0.8 V in the anodic direction and to -0.8 V in the cathodic direction. In total 16 samples (4 types of oxides exposed in saturated $Ca(OH)_2$ water with either 0 or 0.6M NaCl and polarized in either anodic or cathodic direction) were tested.

3. Results and Discussion

3.1. Analysis of Manufactured Oxide Scales

3.1.1. XRD. The manufactured oxide scales were analyzed with XRD and SEM/EBSD (SEM in combination with electron backscatter diffraction). The calculated composition from the XRD intensity spectra of crushed oxide scales can be seen in Table 3. According to the XRD results, the sample named wustite consisted of mostly wustite with minor content of hematite, magnetite, and FeO(OH). The FeO(OH) may have been formed due to oxidation of wustite in air before measurement. The sample named magnetite consisted of mostly magnetite, a small amount of wustite, and minor amount of iron and hematite. Iron is found in the mill scale since wustite transforms into magnetite and iron. According to the XRD results, the hematite sample consisted of approximately 35% hematite. Hematite is formed by oxidation of magnetite which is a relatively slow reaction compared to formation of wustite and magnetite at 850°C. If crushing of the oxide scale is incomplete then large flakes with hematite

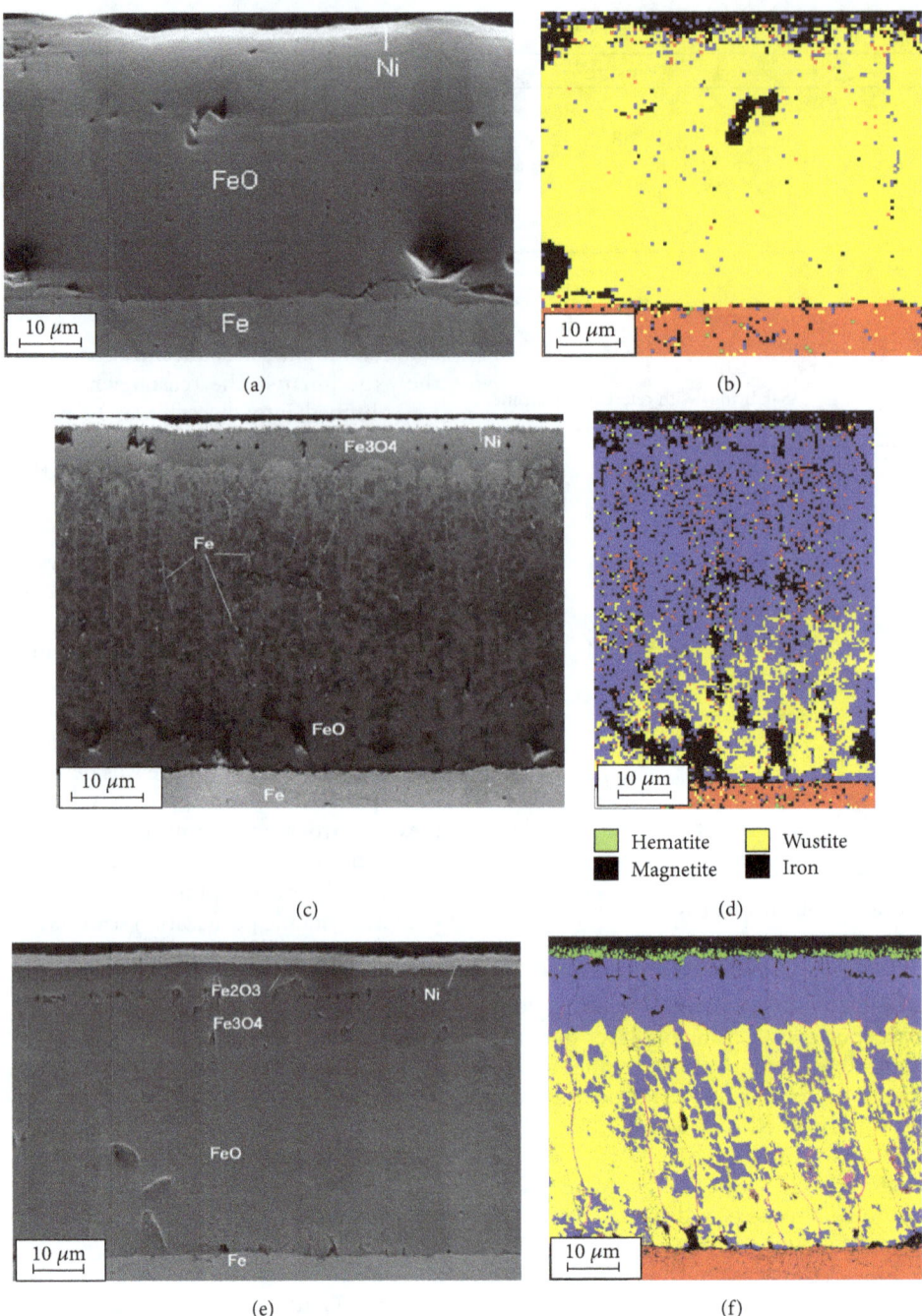

FIGURE 2: SEM images (black and white) and EBSD phase maps (color) on cross sections on the different oxide scale samples where a and b are on wustite, c and d are on magnetite, and e and f are on hematite.

on the surface will result in a positive bias in the XRD hematite signal. Therefore, it is possible that the hematite content is lower than 35%. However, what is important is that hematite is located on the surface (see SEM/EBSD below), influencing the electrochemical properties of the sample.

There is a fourth iron oxide, maghemite γ-Fe_2O_3, which is formed by oxidation of magnetite. Magnetite and maghemite have a very similar crystal structure and are therefore very difficult to distinguish from each other with diffraction based analysis. In a study by Cook, the mill scale was assessed

with Mössbauer spectroscopy which is a technique that can measure maghemite and found that the mill scale did not contain maghemite [17].

3.1.2. SEM/EBSD. One cross section sample of each oxide scale type was analyzed with SEM (in black and white) and EBSD (in color); see Figure 2. Generally, a nickel layer is seen on top of the oxide scales in the SEM figures which are not shown in the EBSD figures. The underlying steel is seen at the bottom of the figures. Figure 2(a) shows a SEM

TABLE 4: Corrosion potential vs. Ag/AgCl sat. KCl [mV]: average value of two samples during 10 minutes.

Solution	Steel	Wustite	Magnetite	Hematite
sat. Ca(OH)$_2$, 0M Cl	-180	-186	-101	80
sat. Ca(OH)$_2$, 0.6M Cl	-440	-183	-217	2

image of a cross section of the wustite sample where the oxide layer was measured to be approximately 50 μm thick consisting of columnar grains. Figure 2(b) is an EBSD phase map over a cross section of the wustite sample where wustite is the dominating oxide from the bulk steel to the outermost surface, which is in agreement with the XRD results.

SEM and EBSD images of a cross section of the magnetite sample can be seen in Figures 2(c) and 2(d), respectively. The thickness of the scale was measured to be approximately 70 μm. Small light dots can be seen in the SEM image and red dots in the EBSD figure within the oxide scale which is iron formed when wustite has transformed to magnetite and iron. Magnetite is the dominating oxide in the scale which is in agreement with the XRD results. Wustite is seen close to the steel surface. Traces of hematite at the top of the oxide scale indicate that hematite has not been completely removed in the grinding process. However, magnetite is the dominating oxide at the outermost surface of the sample.

SEM and EBSD images of a cross section of the hematite sample can be seen in Figures 2(e) and 2(f), respectively. The thickness of the scale was measured to be approximately 60 μm (between the top nickel layer and the bottom steel surface). In the EBSD figure, a mixture of wustite and magnetite is seen above the steel and a magnetite layer on top of the mixture. The hematite layer at the outermost surface was measured to be approximately 2 μm thick. The relatively thin hematite layer confirms the relatively low proportion of Fe$_2$O$_3$ in the XRD result which was taken from the powder of mixed layers. Because it is the outermost part of the scale that will have the most exposure to solution during the electrochemical experiments, this layer of hematite, even though very thin, is regarded as a representative sample of hematite.

3.2. Electrochemical Measurements

3.2.1. Open Cell Potential (OCP).
The OCP was measured for 10 minutes to investigate the corrosion potentials of the samples and the average potential of two samples can be seen in Table 4. The scatter of values between each sample type is approximately in relation to the start potential of each sample type for the test of anodic and cathodic potentiodynamic potential (PDP) as will be shown later in Figures 7 and 8. The steel and wustite samples had the least noble potential measured in the calcium hydroxide solution followed by magnetite. The hematite sample had the noblest potential.

In the solution containing chlorides the steel sample had a relatively large potential drop of approximately 250 mV lower than in the solution without chlorides. The potential difference of the iron oxides measured in the two solutions differed by approximately 100 mV. The steel sample had the least noble potential followed by magnetite and the

wustite. The hematite sample had the noblest potential. The magnetite-rich sample contained iron particles which may lower the corrosion potential in the chloride containing solution if the iron particles are exposed to the solution.

Theoretically, if a steel surface with mill scale is connected to a steel surface without mill scale, the steel surface would act as an anode and the mill scale as a cathode. It is likely that this galvanic cell would affect the passivation behavior of the steel, especially in a chloride containing solution where the potential difference in this study is measured to be about 440 mV between steel and hematite. The distinctly different potentials of the oxide, as compared to the raw steel surface, also indicate that the oxides are relatively pore-free.

Avila-Mendoza et al. [12] compared the corrosion potential for two types of steel samples embedded in mortar, one oxidized at 800°C for 20 seconds and another mirror polished. They found that the corrosion potential for the oxide samples was generally more positive than that for the oxide free samples when exposed in various solutions. This is in agreement with the results in this study. In a number of papers [2, 10, 18, 19] the corrosion potentials were compared between "as-received" samples and oxide free samples exposed in alkaline solution or embedded in concrete. Akhoondan and Sagues [10] found that the measured corrosion potentials were nobler for samples in an as-received condition than those for oxide free samples whereas the others [2, 18, 19] found that the oxide free samples had nobler potentials than the "as-received" samples. The different results may be explained by different exposure conditions and that the oxide scale in as-received condition differs between studies. It is likely that the oxide scale is relatively free from cracks in studies where the corrosion potential was nobler than oxide free samples and that the oxide scale contained many cracks in studies where the potential was less noble. If the scale contains many cracks it is possible that corrosion is initiated due to a galvanic effect by the oxide layer which lowers the corrosion potential. Unfortunately, the oxide layer is seldom characterized in terms of number of cracks, thickness and composition, which makes it more difficult to explain any effects of the oxide layer and differences between studies.

3.2.2. Electrochemical Impedance Spectroscopy.
A representative Bode plot of each sample type, obtained from electrochemical impedance spectroscopy (EIS) measurements, can be seen in Figure 3, where samples were exposed in saturated Ca(OH)$_2$ solution, and in Figure 4 where samples were exposed to a chloride containing Ca(OH)$_2$ solution. In the solution without chlorides wustite had the lowest impedance at 1 mHz, since wustite is the least stable oxide and it is possible that oxidation of wustite was ongoing. Steel, magnetite, and hematite all had high impedances at 1

TABLE 5: Average total impedance of two samples at 1 mHz [10^6 Ohm.cm^2].

Solution	Steel	Wustite	Magnetite	Hematite
sat. Ca(OH)$_2$, 0M Cl	0.9	0.15	1.8	1.8
sat. Ca(OH)$_2$, 0.6M Cl	0.015	0.3	0.7	1.9

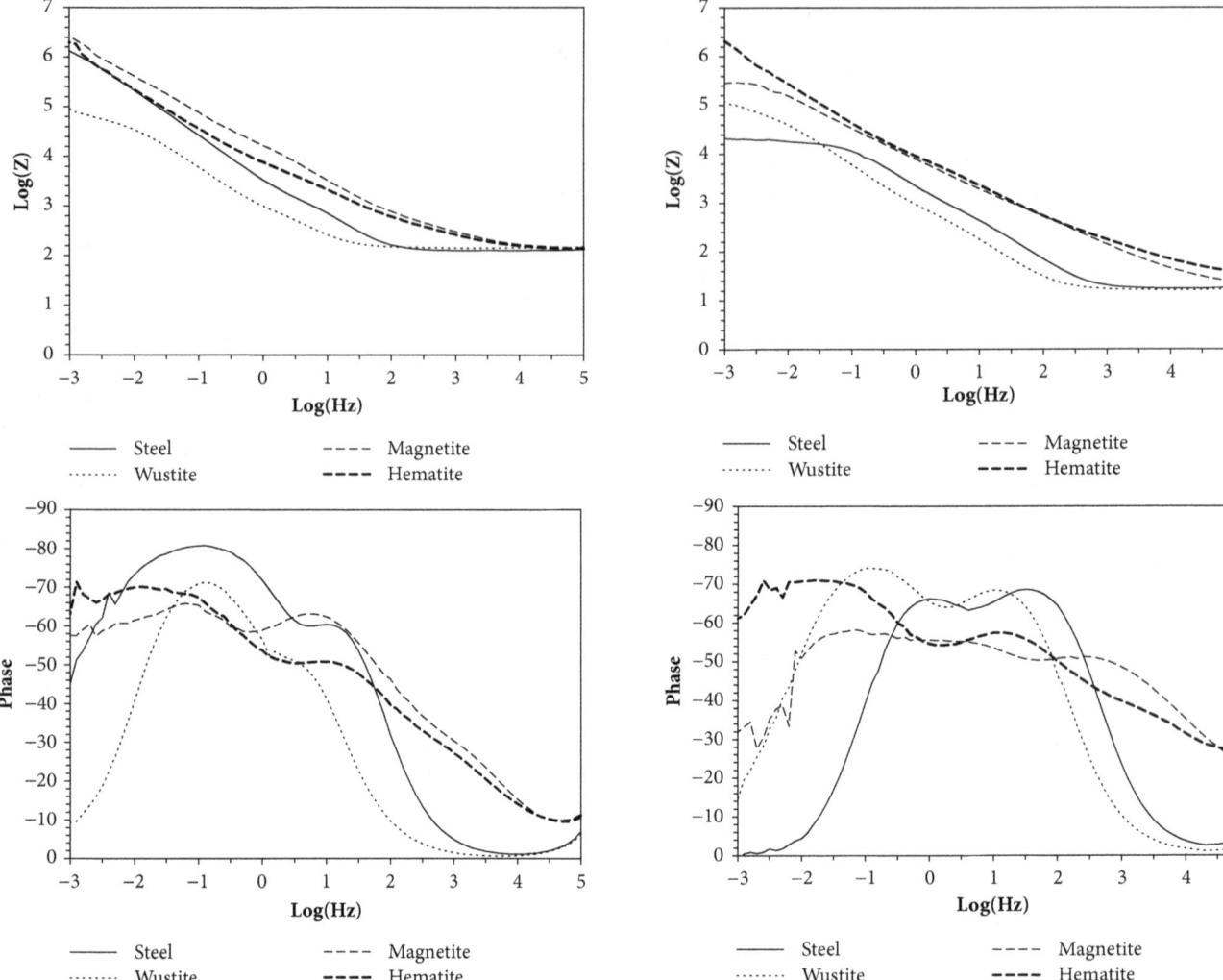

FIGURE 3: Example of Bode plots obtained in a sat. Calcium hydroxide solution.

FIGURE 4: Example of Bode plots obtained in sat. Calcium hydroxide solution with 0.6 M Cl$^-$.

mHz due to stable behavior in chloride free solution. The impedance at 1 mHz measured in solutions with and without chlorides can be seen in Table 5. The iron oxides had at least two peaks in the Bode phase plot which may correlate with the properties of the oxide scale in the midfrequency range and the properties of the double layer in the lower frequency range. The impedance of the steel sample at 1 mHz was relatively high and the phase angle in the Bode plot was high in a broad frequency range which is attributed to a protective passive layer. This is in accordance with [20], in which it was stated that a protective passive layer is formed during the first 10-20 minutes in concrete pore solutions.

In the solution containing chlorides, the iron oxides had similar Bode plots relative to the plots obtained in the solution without chlorides. The iron oxides are not strongly affected by chlorides. The impedance remained high for magnetite and hematite and relatively low for wustite. The steel sample had significantly lower impedance which is attributed to a destabilized passive layer. This is in agreement with reported studies in the literature where the impedance modulus is lower for steel exposed in chloride containing alkaline solutions compared to solutions without chlorides [5, 20, 21].

The EIS results were further analyzed by fitting the data to the equivalent circuit in Figure 5. This equivalent circuit

TABLE 6: Parameter data obtained from the equivalent circuit fitting for samples exposed in sat. $Ca(OH)_2$. The data values are an average of two samples.

Parameter	Steel	Wustite	Magnetite	Hematite
CPE_{Oxide} [$\mu F.cm^{-2}$]	30	110	9	20
R_{oxide} [$M\Omega.cm^2$]	0.004	0.002	0.02	0.01
CPE_{DL} [$\mu F.cm^{-2}$]	29	111	15	18
R_{DL} [$M\Omega.cm^2$]	1.22	0.14	14	59

TABLE 7: Parameter data obtained from the equivalent circuit fitting for samples exposed in sat. $Ca(OH)_2$ with 0.6 M Cl^-. The data values are an average of two samples.

Parameter		Steel	Wustite	Magnetite	Hematite
CPE_{Oxide}	[$\mu F.cm^{-2}$]	53	101	21	21
R_{oxide}	[$M\Omega.cm^2$]	$4*10^{-11}$	0.002	0.002	0.01
CPE_{DL}	[$\mu F.cm^{-2}$]	15	130	11	17
R_{DL}	[$M\Omega.cm^2$]	0.012	0.25	2	124

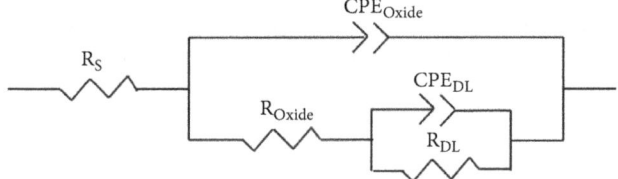

FIGURE 5: Proposed equivalent electrical circuit for fitting EIS data.

showed good fitting results with the experimental data and has been used in other studies on rebar corrosion, e.g., [22]. The capacitance of the oxide and double layer was each described by a constant phase element (CPE) which represents the inhomogeneity of the steel surface and oxide layer.

In the equivalent circuit, R_S is the solution resistance, CPE_{Oxide} is the constant phase element for the oxide, R_{oxide} is the resistance of the oxide, CPE_{DL} is the constant phase element for the double layer, and R_{DL} is the resistance of the double layer.

The results from the equivalent circuit simulation can be seen in Table 6, for samples exposed to the solution without chlorides and in Table 7 for samples in the solution containing chlorides. Generally, the simulation results confirm that chlorides have a small effect on the iron oxides. The fitted data for magnetite and hematite are similar in both solutions. This is surprising since the tabulated electrical conductivity in [14] is much higher for magnetite than that for hematite and this would result in a much smaller R_{oxide} for magnetite compared to hematite. Theoretically, if the electrical conductivity is 10^{-9} $\Omega^{-1}.cm^{-1}$ [14] for hematite and the hematite layer thickness is roughly 2 μm, this gives a resistance of $2*10^5$ Ω, and for magnetite the resistance would be approximately $6*10^{-5}$ Ω using the electrical conductivity 100 $\Omega^{-1}.cm^{-1}$ [14] and layer thickness 70 μm. It is therefore possible that the magnetite sample was not ground sufficiently and had residual hematite on the sample which is confirmed by small hematite dots on top of the magnetite

scale in Figure 2(d). No significant difference of the fitted values could be seen for the wustite samples exposed in either solution. For the steel samples it is clearly seen that the chloride containing solution destabilizes the passive layer, which is manifested as decreased values for R_{oxide} and R_{DL} in the chloride containing solution. The capacitance increased for steel samples in the chloride containing solution, which is attributed to roughening of the corroding surface, formation of corrosion products, and formation of charged iron species [23].

Not many studies have been reported in the literature with EIS data comparing steel with mill scale and steel without mill scale. Shi et al. [9] fitted EIS data to an equivalent circuit and found that the charge transfer resistance was higher for sand-blasted samples than that for as-received samples exposed to an alkaline solution without chlorides. In chloride containing solutions the charge transfer resistance for as-received samples was higher than that for sand-blasted samples when exposed in a 0.5 M NaCl solution but lower than that for sand-blasted samples when exposed in a 1 M NaCl solution. Ghods et al. [5] determined chloride threshold levels based on EIS data and found that the chloride threshold level was lower for as-received samples than that for turned and polished samples when exposed to alkaline solutions. As-received samples are inhomogeneous and may or may not contain cracks and defects which can explain the scatter in the reported results. The impedance modulus of the oxide scales produced in the present study is much higher compared to steel exposed to alkaline solution with chlorides since the oxide scales are protective and relatively free from cracks.

3.2.3. Potentiodynamic Polarization (PDP). All oxide scales and steel samples were assessed with PDP exposed to saturated calcium hydroxide solution with and without chlorides. Figure 6 shows the PDP curves for ground steel samples exposed to the solutions with and without chlorides. Each anodic and cathodic polarization curve started from OCP and was obtained for one sample. From the curves it can be seen that for the steel exposed to the solution without

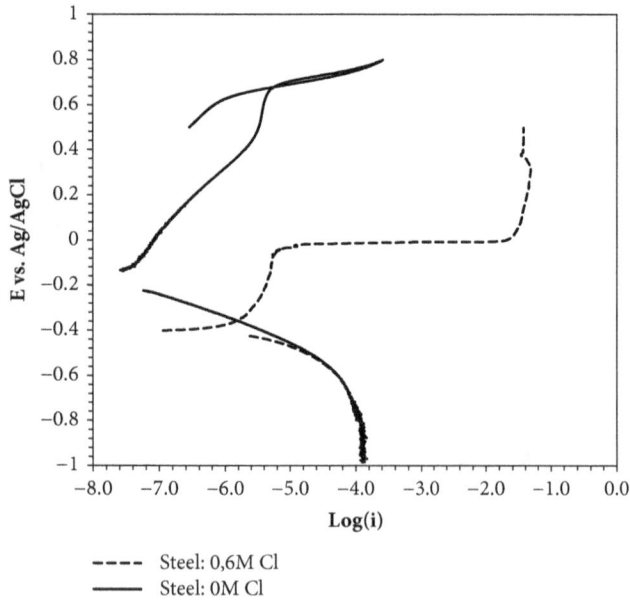

FIGURE 6: Anodic and cathodic potentiodynamic polarization curves for a steel exposed in a saturated Ca(OH)$_2$ solution and a steel sample in saturated Ca(OH)$_2$ solution with 0.6 M Cl$^-$.

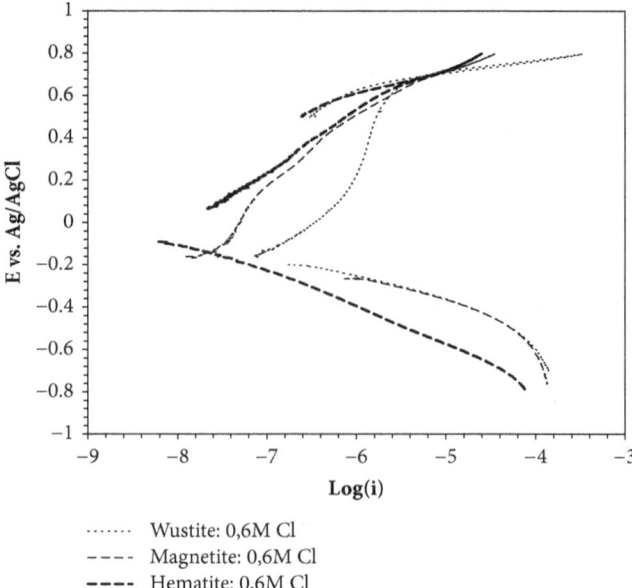

FIGURE 8: Anodic and cathodic potentiodynamic potential curves for the iron oxides exposed in saturated Ca(OH)$_2$ solution with 0.6 M Cl$^-$.

current for hematite and magnetite is generally lower than that for wustite. This can be explained by the fact that these oxides are more stable than wustite. The higher anodic current for the wustite sample is also in accordance with the previous XRD results showing that the wustite sample contained FeO(OH), which means that the wustite had probably started to oxidize during the wet grinding and is easily oxidized. The cathodic current curve is a result of dissolved oxygen in the solution and possibly to some extent reduction of the oxide. Both reactions may occur simultaneously. Determining which reaction dominates is beyond the scope of this work.

The anodic polarization reversed at 0.8 V vs Ag/AgCl to the cathodic direction. The reversed current was at the same level or lower than the current in the anodic direction. This means that the underlying steel is not corroding and that the increased current at 0.7 V is due to electrolysis of water.

Figure 8 shows the PDP curves for the iron oxides exposed in saturated calcium hydroxide with chlorides. Each anodic and cathodic polarization curve started from OCP and was obtained for one sample. The general behavior of the potential curves is similar to the behavior of the potential curves obtained in the solution without chlorides. This is also in agreement with the previous OCP and EIS results; i.e., the oxide samples do not change very much in solution with or without chlorides. The anodic current for magnetite and hematite is lower than that for wustite since wustite is oxidized at a higher rate. The cathodic current for wustite and magnetite was higher than that for hematite. This suggests that hematite has a higher resistivity and a less effective cathodic surface compared to the other oxides.

The anodic polarization reversed at 0.8 V vs Ag/AgCl to the cathodic direction. The reversed current was in the

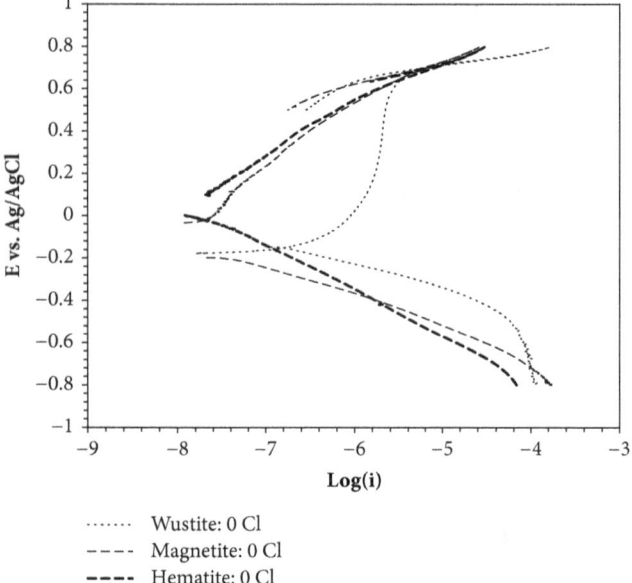

FIGURE 7: Anodic and cathodic potentiodynamic potential curves for the iron oxides exposed in saturated Ca(OH)$_2$ solution.

chlorides the anodic current is relatively low indicating a passive behavior, whilst for the steel exposed in the solution with chlorides the anodic current increases significantly at about 0 mV rel. Ag/AgCl, indicating pitting corrosion. These results are similar to results found in the literature, e.g., [2, 21, 24].

The PDP curves for wustite, magnetite, and hematite exposed in saturated calcium hydroxide can be seen in Figure 7. Each anodic and cathodic polarization curve started from OCP and was obtained for one sample. The anodic

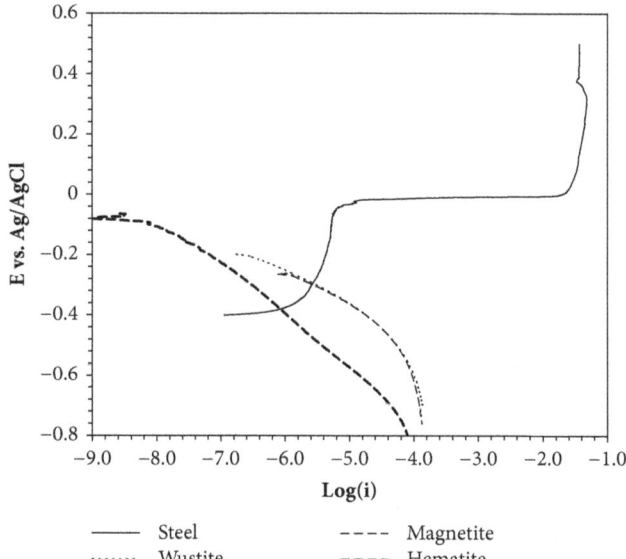

FIGURE 9: Cathodic curve for the iron oxides and an anodic curve for the steel sample.

same level or lower than the current in anodic direction. This means that the underlying steel is not corroding and is protected by the mill scale.

Figure 9 shows an overlay of previously shown selected polarization curves and the anodic polarization curve for steel and the cathodic polarization curves for the iron oxides exposed in saturated calcium hydroxide with chlorides. All iron oxides samples had a higher OCP compared to steel. The cathodic polarization curves for the iron oxides intersect with the anodic polarization curve for steel. Theoretically, this means that the steel would act as an anode and the iron oxides as cathode and the anodic current for steel would be higher if the steel was in contact with a mill scale. This can be one of the reasons why steel in an as-received condition generally has lower corrosion resistance compared to oxide free steel samples. If a small steel surface at a crack within the mill scale is exposed to a chloride containing solution, the small steel area would act as an anode and the mill scale surface would act as a cathode. The mill scale would polarize the underlying steel in the anodic direction and chlorides would migrate into the crack, continue to the steel surface, and initiate pitting corrosion.

Only a few studies have been reported in the literature where PDP curves from samples in an as-received condition have been compared with those from oxide free samples and there is a lack of quantitative information about the compositions of the mill scale in those studies. Mahalatti et al. [8] measured a higher current during anodic polarization on as-received samples compared to oxide free samples embedded in concrete with or without chlorides. The largest difference in current density was measured in chloride free concrete. One explanation may be that these as-received samples contained a large amount of wustite and therefore high anodic currents were measured. Shi et al. [9] measured a somewhat higher current during anodic polarization for

as-received samples compared to oxide free samples exposed to alkaline solutions. The anodic polarization curves in the present study show that the mill scales protect the underlying steel in chloride containing solutions since these scales are relatively free from cracks. It is likely that the results from samples in an as-received condition found in the literature contain cracks where corrosion initiated on the underlying steel due to a galvanic effect as proposed in the current study.

4. Conclusions

In this study a combination of heat treatment and grinding processes was used to manufacture steel samples with three different synthetic mill scales: one dominated by wustite (FeO); one dominated by magnetite (Fe_3O_4); and one with hematite on the surface. Based on the results from electrochemical measurements the following differences can be observed:

(i) In chloride containing solutions the untreated steel sample had the least noble potential followed by samples with scales dominated by magnetite and wustite. The noblest potential was observed for the sample with hematite on the surface. The difference in potential between samples could be as high as 440 mV.

(ii) When polarizing samples in the cathodic direction, the measured currents for the samples with scales dominated by magnetite and wustite were significantly higher than those for the sample with hematite.

In conclusion, wustite and magnetite are less noble but more effective as cathodic surfaces than hematite. The results show that the electrochemical properties of the mill scale can be an important contributing factor in the corrosion of steel in concrete.

Data Availability

The XRD data used to support the findings of this study are included within the supplementary information file. The SEM/EBSD images used to support the findings of this study are included within the article. The OPC data used to support the findings of this study are included within the article. The EIS data used to support the findings of this study are included within the article. The simulated EIS parameters used to support the findings of this study are included within the article. The PDP curves used to support the findings of this study are included within the article.

Conflicts of Interest

The authors declare that they have no conflicts of interest.

References

[1] U. Angst, B. Elsener, C. K. Larsen, and Ø. Vennesland, "Critical chloride content in reinforced concrete—a review," *Cement and Concrete Research*, vol. 39, no. 12, pp. 1122–1138, 2009.

[2] L. Li, "Chloride corrosion threshold of reinforcing steel in alkaline solutions - Open-circuit immersion tests," *Corrosion*, vol. 57, no. 1, pp. 19–28, 2001.

[3] R. G. Pillai and D. Trejo, "Surface condition effects on critical chloride threshold of steel reinforcement," *ACI Materials Journal*, vol. 102, no. 2, pp. 103–109, 2005.

[4] M. Manera, Ø. Vennesland, and L. Bertolini, "Chloride threshold for rebar corrosion in concrete with addition of silica fume," *Corrosion Science*, vol. 50, no. 2, pp. 554–560, 2008.

[5] P. Ghods, O. B. Isgor, G. A. McRae, and G. P. Gu, "Electrochemical investigation of chloride-induced depassivation of black steel rebar under simulated service conditions," *Corrosion Science*, vol. 52, no. 5, pp. 1649–1659, 2010.

[6] T. U. Mohammed and H. Hamada, "Corrosion of steel bars in concrete with various steel surface conditions," *ACI Materials Journal*, vol. 103, no. 4, pp. 233–242, 2006.

[7] D. Boubitsas and L. Tang, "The influence of reinforcement steel surface condition on initiation of chloride induced corrosion," *Materials and Structures*, vol. 48, no. 8, pp. 2641–2658, 2015.

[8] E. Mahallati and M. Saremi, "An assessment on the mill scale effects on the electrochemical characteristics of steel bars in concrete under DC-polarization," *Cement and Concrete Research*, vol. 36, no. 7, pp. 1324–1329, 2006.

[9] J. Shi and J. Ming, "Influence of mill scale and rust layer on the corrosion resistance of low-alloy steel in simulated concrete pore solution," *International Journal of Minerals, Metallurgy and Materials*, vol. 24, no. 1, pp. 64–74, 2017.

[10] M. Akhoondan and A. A. Sagüés, "Comparative cathodic behavior of ~9% Cr and plain steel reinforcement in concrete," *Corrosion*, vol. 68, no. 4, 2012.

[11] P. Ghods, O. B. Isgor, G. A. McRae, J. Li, and G. P. Gu, "Microscopic investigation of mill scale and its proposed effect on the variability of chloride-induced depassivation of carbon steel rebar," *Corrosion Science*, vol. 53, no. 3, pp. 946–954, 2011.

[12] J. Avila-Mendoza, J. M. Flores, and U. C. Castillo, "Effect of superficial oxides on corrosion of steel reinforcement embedded in concrete," *Corrosion*, vol. 50, no. 11, pp. 879–885, 1994.

[13] X. Hu, B. Zhang, S. Chen, F. Fang, and J. Jiang, "Oxide Scale Growth on High Carbon Steel at High Temperatures," *Journal of Iron and Steel Research, International*, vol. 20, no. 1, pp. 47–52, 2013.

[14] R. M. Cornell and U. Schwertmann, *The Iron Oxides*, Wiley-VCH, 1996.

[15] N. Otsuka, T. Doi, Y. Hidaka et al., "In-situ measurements of isothermal wüstite transformation of thermally grown feo scale formed on 0.048 mass% fe by synchrotron radiation in air," *ISIJ International*, vol. 53, no. 2, pp. 286–293, 2013.

[16] R. Ovarfort, "New electrochemical cell for pitting corrosion testing," *Corrosion Science*, vol. 28, no. 2, pp. 135–140, 1988.

[17] D. C. Cook, "Application of Mössbauer spectroscopy to the study of corrosion," *Hyperfine Interactions*, vol. 153, no. 1-4, pp. 61–82, 2004.

[18] A. Poursaee and C. M. Hansson, "Reinforcing steel passivation in mortar and pore solution," *Cement and Concrete Research*, vol. 37, no. 7, pp. 1127–1133, 2007.

[19] D. Boubitsas and L. Tang, "The influence of reinforcement steel surface condition on initiation of chloride induced corrosion," *Materials and Structures/Materiaux et Constructions*, vol. 48, no. 8, pp. 2641–2658, 2015.

[20] H. B. Gunay, O. B. Isgor, and P. Ghods, "Kinetics of Passivation and Chloride-Induced Depassivation of Iron in Simulated Concrete Pore Solutions Using Electrochemical Quartz Crystal Nanobalance," *Corrosion*, vol. 71, no. 5, pp. 615–627, 2015.

[21] F. Zhang, J. Pan, and C. Lin, "Localized corrosion behaviour of reinforcement steel in simulated concrete pore solution," *Corrosion Science*, vol. 51, no. 9, pp. 2130–2138, 2009.

[22] D. A. Koleva, K. Van Breugel, J. H. W. De Wit, E. Van Westing, N. Boshkov, and A. L. A. Fraaij, "Electrochemical behavior, microstructural analysis, and morphological observations in reinforced mortar subjected to chloride ingress," *Journal of The Electrochemical Society*, vol. 154, no. 3, pp. E45–E56, 2007.

[23] J. Flis, H. W. Pickering, and K. Osseo-Asare, "Interpretation of impedance data for reinforcing steel in alkaline solution containing chlorides and acetates," *Electrochimica Acta*, vol. 43, no. 12-13, pp. 1921–1929, 1998.

[24] M. Liu, X. Cheng, X. Li, and T. J. Lu, "Corrosion behavior of low-Cr steel rebars in alkaline solutions with different pH in the presence of chlorides," *Journal of Electroanalytical Chemistry*, vol. 803, pp. 40–50, 2017.

16

The Red Sea as a Corrosive Environment: Corrosion Rates and Corrosion Mechanism of Aluminum Alloys 7075, 2024, and 6061

Aisha H. Al-Moubaraki ⑩ **and Hind H. Al-Rushud**

Chemistry Department, Faculty of Sciences, King Abdulaziz University, Al Faisaliah Campus, Jeddah, Saudi Arabia

Correspondence should be addressed to Aisha H. Al-Moubaraki; ahm13988@hotmail.com

Academic Editor: Jerzy A. Szpunar

Corrosion behavior of Al 7075, Al 2024, and Al 6061 in the Red Sea water was studied using weight loss (WL) measurements and potentiodynamic polarization (PDP) technique. The corrosion patterns and corrosion products formed on Al alloys were characterized using optical photography (OP), scanning electron microscopy (SEM), and energy-dispersive spectroscopy (EDS). The results showed that WL data were consistent with bimodal model rather than the power law function and the corrosion rates exhibit a continuous decrease with exposure time. The increasing order of the Red Sea corrosivity on the studied Al alloys can be given as follows: Al 6061 < Al 2024 < Al 7075. The results of temperature effect revealed that an increase in temperature resulted in an increase in both anodic and cathodic current density and a decrease in corrosion potential. Al 7075 was less influenced by temperature than the other alloys. Pitting corrosion was the predominant corrosion pattern detected on all Al alloy surfaces after prolonged immersion in the Red Sea water. The appearance of S peak in EDS spectra of Al 7075 after corrosion gives an indication of the contribution of bacteria in the corrosion process.

1. Introduction

Desirable properties of aluminum such as low density (only 2.7 g/cm^3), recyclability, thermal and electrical conductivity, and, in some degree, corrosion resistance, make aluminum the most consumed nonferrous metal in the world [1]. Aluminum is often alloyed with other elements to improve its mechanical properties and increase its usefulness [2]. Commercially wrought aluminum alloys are classified into different series depending on their principle alloying elements such as 2xxx (Al-Cu alloys), 5xxx (Al-Mg alloys), 6xxx (Al-Mg-Si alloys), and 7xxx series (Al-Zn-Cu alloys). Wrought aluminum alloys are used in marine engineering applications, specifically 5xxx and 6xxx series, as in shipbuilding, coastal equipment, and desalination of seawater [3–5]. In fact, aluminum alloys have less corrosion resistance than pure metal; so, they are often used with an appropriate protection method [2, 3].

Corrosion is defined by ISO as "physicochemical interaction (usually of an electrochemical nature) between a metal and its environment which results in changes in the properties of the metal and which often leads to impairment of the function of the metal, the environment, or the technical system of which these form a part" [6]. From a thermodynamic view, a layer of aluminum oxide is formed on aluminum surface in pH ranges of 4 to 9, acting as a physical barrier that separates aluminum from its environment and prevents further dissolution; but, the stability of the oxide layer can be influenced by the presence of aggressive anions in the environment such as chloride in seawater [7, 8]. The properties of the protective oxide film can also be affected by alloying elements. Those alloying elements are present not only in aluminum solid solution but also in intermetallic (IM) particles. The microstructures of aluminum alloys are complex and highly heterogeneous compared to other alloys [9]. IM particles are randomly distributed in the alloys matrix and have different microstructures and composition than the solid solution. As a result, they have different electrochemical potentials which produce an opportunity to create microgalvanic cells between these particles and the

TABLE 1: Percentage composition of the studied Al alloys.

Alloy	Cr	Cu	Fe	Mg	Mn	Si	Ti	Zn	Zr	Al
7075	0.18–0.28	1.2–2.0	0.5	2.1–2.9	0.3	0.4	0.2	5.1–6.1	0.25	Remainder
2024	0.10	3.8–4.9	0.5	1.2–1.8	0.3–0.9	0.5	0.15	0.25	-	Remainder
6061	0.04–0.35	0.15–0.4	0.7	0.8–1.2	0.15	0.4–0.8	0.15	0.25	-	Remainder

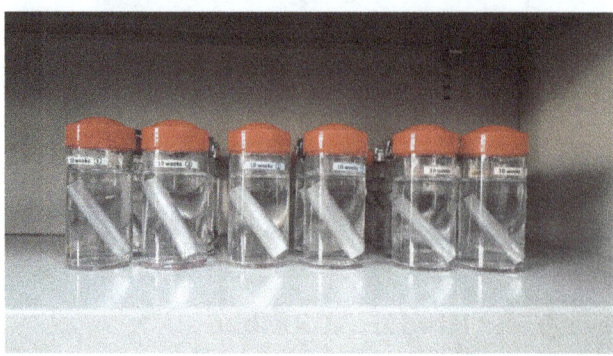

FIGURE 1: The final arrangement for weight loss experiments.

aluminum matrix [10, 11]. Both the heterogeneous nature of alloys and presence of aggressive anions in the environments can cause loss of passivity at specific sites of aluminum alloy surface and lead to localized corrosion [7, 12, 13]. Muller and Galvele [14] studied the role of alloying elements in pitting corrosion of Al-Zn, Al-Cu, and Al-Mg binary alloys in 1 M NaCl. It was observed that pitting potential decreased with increase in Zn content up to 3%, addition of copper increased pitting potential dramatically up to 5%, and Mg did not affect the pitting potential.

Various environmental factors such as dissolved gases, temperature, pH, and micro- and macroorganisms can affect corrosion in seawater [15, 16]. Most of these factors are interrelated with each other as an increase in temperature will reduce the dissolved oxygen and result in a change in biological activities; thus, evaluation of the corrosion process in seawater is complicated. Dissolved oxygen is an important factor because oxygen reduction is a primary cathodic reduction for corrosion of Al alloys in seawater [17]. Aluminum and its alloys have been reported to experience microbial influenced corrosion (MIC). Sulfate reducing bacteria (SRB) present in seawater is the most widely known and studied as the cause of MIC. SRB form tubercles (discrete hemispherical mounds) on aluminum surface and enhance pitting corrosion as a result of low pH of electrolyte under tubercles [18, 19]. In corrosion studies, seawater is often duplicated by 3.5 NaCl but salt concentration is not efficient to duplicate the properties of natural seawater. However, many researchers have studied the corrosion behavior of aluminum alloys in natural seawater and their results revealed that aluminum and its alloys mainly suffered from pitting corrosion in seawater [20–22]. In previous studies on aluminum corrosion by pitting, researchers distinguished between crystallographic and alkaline pitting. When cathodic intermetallic particles enhance the dissolution of the surrounding matrix, it is called alkaline pits. This type of pitting occurs below the pitting potential, while crystallographic pits are formed when an alloy reaches its pitting potential [23].

Using three different aluminum alloys (Al 7075, Al 2024, and Al 6061), the present research studied the Red Sea as a corrosive environment. Factors such as immersion time and temperatures on the corrosion behavior of aluminum alloys in seawater were studied using weight loss (WL) and potentiodynamic polarization (PDP) measurements. Change in aluminum alloy surfaces was detected at different conditions using optical photography (OP), scanning electron microscope (SEM), and electron-dispersive energy spectroscopy (EDS).

2. Materials and Methods

2.1. Materials. Three different types of aluminum alloys were used in this study. The chemical composition of aluminum alloys is given in Table 1. Specimens, 10 to 15 mm in diameter and 40 to 50 mm in length, were cut from the respective metal rods. Prior to each experiment, the exposed surfaces were scratched with a series of emery paper from 80 to 1200 grades. Then, they were washed several times with deionized water and then with ethanol and finally dried using a stream of air. The specimens were then weighed accurately and immersed in the test solution. The test solution was natural seawater and it was collected directly from the Red Sea in the western region (Obhur, Jeddah, Saudi Arabia). Several properties of the sample such as pH, electrical conductivity, total alkalinity, and content of different ions were demonstrated in previous work [24].

2.2. Methods

2.2.1. Weight Loss (WL). Polished and preweighed Al specimens were placed in airtight glass containers containing 60 mL of seawater for different immersion periods (1, 2, 3, 5, 7, 10, 22, and 32 weeks) at an ambient temperature (laboratory temperature was 21 ± 1°C) under stagnant conditions. To facilitate the identification of specimens immersed in the seawater at different conditions, numbered stickers were used in each case (Figure 1). At the end of each immersion period,

TABLE 2: Weight loss and corrosion rates of Al alloys at different immersion periods in the Red Sea water.

Immersion period (weeks)	Weight loss (g m^{-2})			Corrosion rate, CR$_{WL}$ (g m^{-2} year^{-1})		
	Al 7075	Al 2024	Al 6061	Al 7075	Al 2024	Al 6061
1st	0.733	0.463	0.312	38.184	24.111	16.247
2nd	1.227	0.781	0.539	31.956	20.328	14.034
3rd	1.685	0.820	0.648	29.313	14.269	11.268
5th	2.284	1.721	0.997	23.813	17.950	10.392
7th	2.366	1.891	1.094	17.630	14.088	8.150
10th	2.396	1.994	1.147	12.490	10.394	5.981
22nd	3.826	2.567	1.663	9.069	6.084	3.942
32nd	4.173	3.241	1.880	6.796	5.279	3.062

the specimens were assessed by weight loss measurement. Layers of corrosion products were scraped off the Al surfaces with a bristle brush and immediately immersed in the pickling solution containing concentrated nitric acid (86%) for five minutes (ASTM G1) [25]. After corrosion products were completely removed, the specimens were rinsed with deionized water and ethanol, dried with a stream of air, and reweighed to determine weight loss. Duplicate experiments were conducted for all the weight loss experiments. The corrosion rate (CR$_{WL}$) was calculated in g/m^2 year according to the following equation:

$$CR_{WL} \left(g/m^2 \ year \right) = \frac{\Delta W}{At}, \tag{1}$$

where ΔW is weight loss in grams (g), A is area in square meters (m^2), and t is the exposure time in year.

2.2.2. Potentiodynamic Polarization (PDP).
Potentiodynamic polarization measurements were performed in a three-electrode cell. A cylinder specimen of Al alloys was used as working electrode and was embedded in a Teflon holder using epoxy resin, giving an exposed area of 0.699 to 1.269 cm^2. Platinum mesh was used as the counter electrode and silver/silver chloride (Ag/AgCl(s)/KCl saturated (aq)) was used as the reference electrode. Potentiodynamic polarization measurements were done using ACM Gill AC Potentiostat/Galvanostat model 1649 connected to a personal computer. Prior to each experiment, the working electrode was treated as described in the weight loss method and then dipped in the test solution. After reaching a steady-state potential, the potentiodynamic curves were carried out by changing, linearly, the electrode potential from the starting potential (−1200 mV) with respect to the reference electrode towards a less negative direction with the required scan rate (1 mV/s) till the end of the experiment at −200 mV. All the electrochemical measurements were done twice at different conditions. Corrosion current densities were determined by extrapolation of cathodic Tafel line to corrosion potential using the ACM Gill software.

2.3. Surface Examination Study

2.3.1. Optical Photograph.
Optical photographs of Al alloy specimens were taken after the chemical experiments, to evaluate gross changes in the metal surface and to perform a cursory evaluation of the forms of corrosion (e.g., general, pitting) at different conditions. Optical photographs were taken using VMS-004 USB microscope.

2.3.2. SEM-EDS Measurement.
The scanning electron microscopy (SEM) associated with the energy-dispersive X-ray spectroscopy (EDS) was used to investigate the surface morphology of aluminum alloys and analyze the elements on the surfaces of alloys before and after immersion in seawater for five weeks.

3. Results and Discussion

3.1. Weight Loss

3.1.1. Time Dependence of Corrosion Rate.
Table 2 and Figure 2 give the weight loss and corrosion rates (CR$_{WL}$) of Al alloys at different immersion periods in the Red Sea water. The results showed that weight loss increased (Figure 2(a)), while the corrosion rates decreased (Figure 2(b)) with increase in time of exposure. Decrease in corrosion rates with time is frequently observed in many types of environments [26–28] and is caused by the gradual build-up of a thicker and more protective layer. This layer is not truly passivating as it acts as a membrane, especially in the presence of aggressive ions such as the Cl$^-$ ions. If Cl$^-$ ions are present in the aqueous phase, they can penetrate the corrosion product layer and corrode the Al alloys (increase in weight loss value), although at a reduced rate [24]. The results indicate that the CRs depend on the composition of the studied Al alloys. Sinyavskii and Kalinin [21] reported that if cathodic additives and impurities in aluminum alloys such as Cu and Fe increase, the susceptibility to localized corrosion increases. This explains the poor corrosion resistance of Al 7075 and Al 2024 and high corrosion resistance of Al 6061 alloy. Zinc

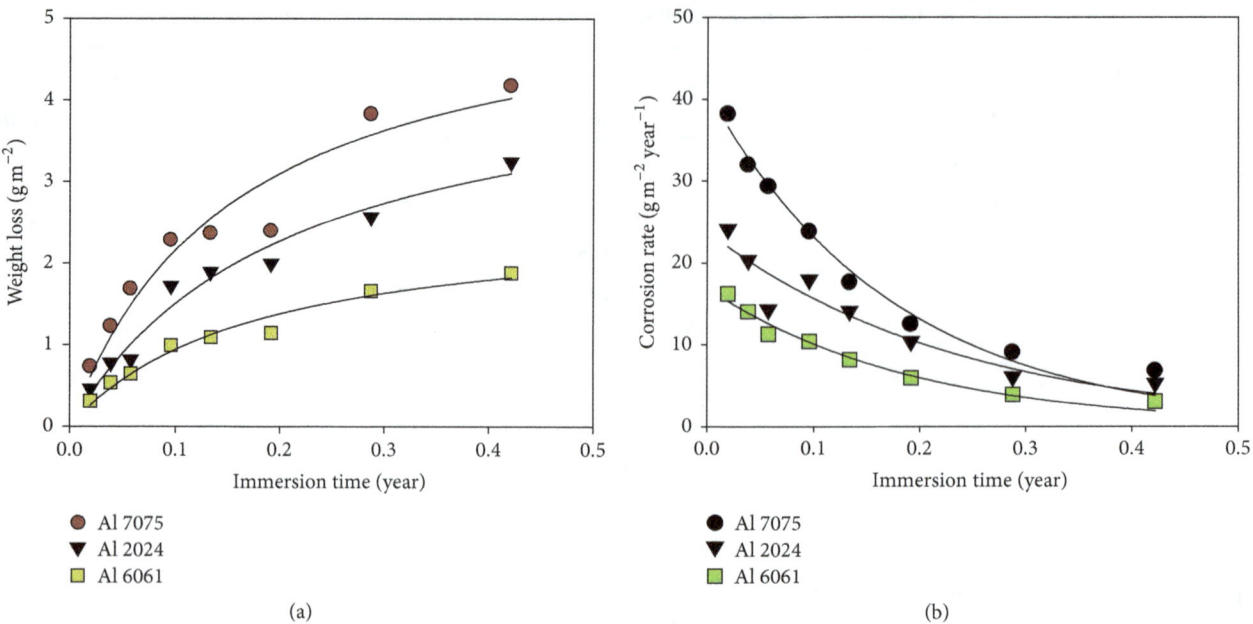

FIGURE 2: Dependence of weight loss and corrosion rates of Al alloys in the Red Sea water on immersion time.

content in Al 7075 reduced its corrosion resistance and led to higher corrosion rate. According to the corrosion rates values, the increasing order of the Red Sea water corrosivity on the studied Al alloys can be given as follows.

$$\xrightarrow{\text{Corrosion rate}}$$

$$\text{Al } 6061 < \text{Al } 2024 < \text{Al } 7075 \qquad (2)$$

$$\xleftarrow{\text{Corrosion resistance}}$$

Corrosion pattern can be monitored at different periods by visual inspection as shown in Figures 3–5. It is obvious that the alloys suffered from localized corrosion, especially pitting form. Deep and shallow pits appeared on the Al alloy surface. The samples with pitting corrosion had white sediment of corrosion products above every pit. Around the pits, there was a brighter area, which is the cathodic area where the reduction occurs. The color of the surface, the number of pits, and the spread of the pits were different for each alloy and different immersion periods. The Red Sea has a high concentration of chloride ions and according to Frankel [28], pitting corrosion will only occur in the presence of aggressive anionic species, and chloride ions are usually the cause. The severity of pitting tends to vary with the logarithm of the bulk chloride concentration [7]. The aggressiveness of chloride anion is attributed to its ability to interfere with passivation as it is a small anion with high diffusivity

3.1.2. Kinetic Study. Most of the previous data sets in corrosion studies appeared to fit the power law function (see (3)) which is derived based on the oxygen diffusion mathematics by Tammann [29]:

$$C = At^n, \qquad (3)$$

where C is corrosion loss or pit depth, t is exposure time, and A and n are constants obtained from fitting the function to data. However, a power model involving logarithmic transformation of the immersion time and corrosion loss must be applied in order to obtain the linear form of the above equation as follows:

$$\log C = \log A + n \log t. \qquad (4)$$

Figure 6 shows log-log plots for weight loss (g m^{-2}) of Al 7075, Al 2024, and Al 6061, respectively, with immersion periods. The data trends of the studied alloys did not show a good fit to a straight line, so they are inconsistent with the power law function model. Melchers [30, 31] reviewed the data sets for long-term corrosion of aluminum and reported that the bimodal model appears to be applicable to long-term corrosion and the power law function is only limited to the short-term corrosion loss. The bilinear graph obtained on log-log coordinates as an alternative obeys an equation of the following type [32, 33]:

$$C = C_1 t_1^{n_1 - n_2} t^{n_2} \qquad (t \geq t_1), \qquad (5)$$

where C is the weight loss after t years, C_1 is the first year weight loss, t_1 is the duration (in years) of the first exposure period, whose slope is n_1, and n_2 is the slope of the second period. One possible reason for this extraordinary behavior may lie in the formation with time of more compact layers which impede the diffusion of the reactive species that participate in the corrosion reactions. However, even though the immersion time in the current experiments did not exceed 32 weeks, the data showed that two distinct lines provide a better fit for the data of Al alloys than a single trend line and this is consistent with the bimodal model. The values of C_1, n_1, n_2, and correlation coefficients $(r_1^2 \text{ and } r_2^2)$ are estimated and recorded in Table 3.

TABLE 3: Corrosion kinetic parameters for Al alloys in the Red Sea water.

Kinetic parameters	Al 7075	Al 2024	Al 6061
C_1 (g m^{-2})	12.505	9.029	5.136
n_1	0.71	0.763	0.706
n_2	0.420	0.326	0.355
r_1^2	0.997	0.920	0.992
r_2^2	0.952	0.964	0.978

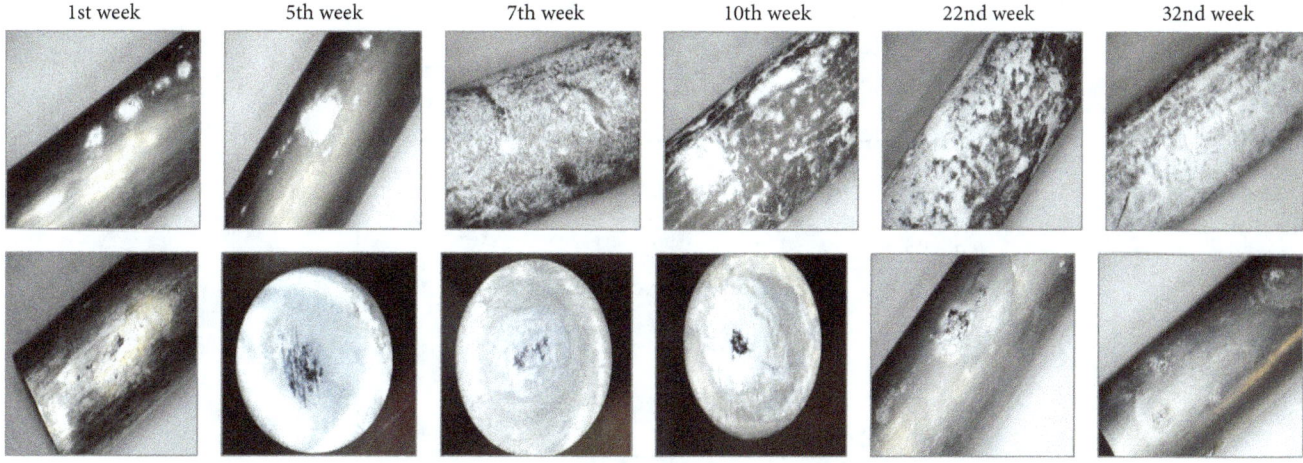

FIGURE 3: Visual images of Al 7075 specimens after immersion in the Red Sea water at different time intervals.

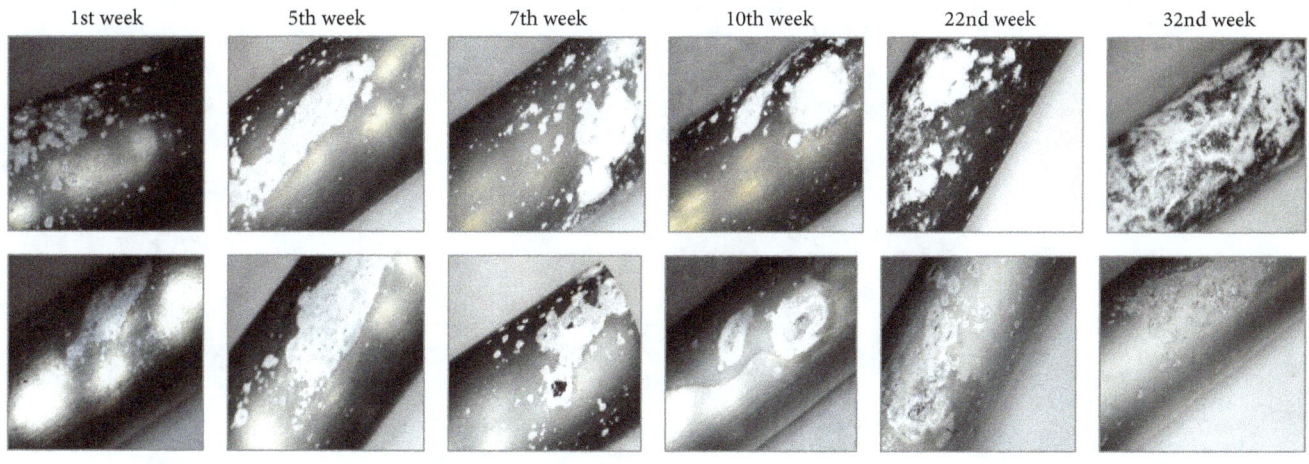

FIGURE 4: Visual images of Al 2024 specimens after immersion in the Red Sea water at different time intervals.

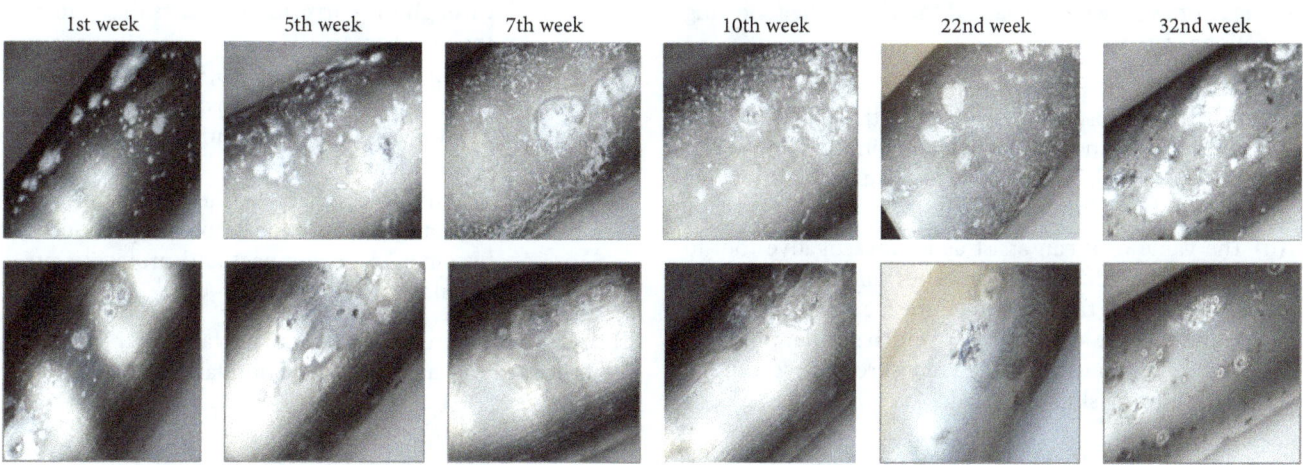

FIGURE 5: Visual images of Al 6061 specimens after immersion in the Red Sea water at different time intervals.

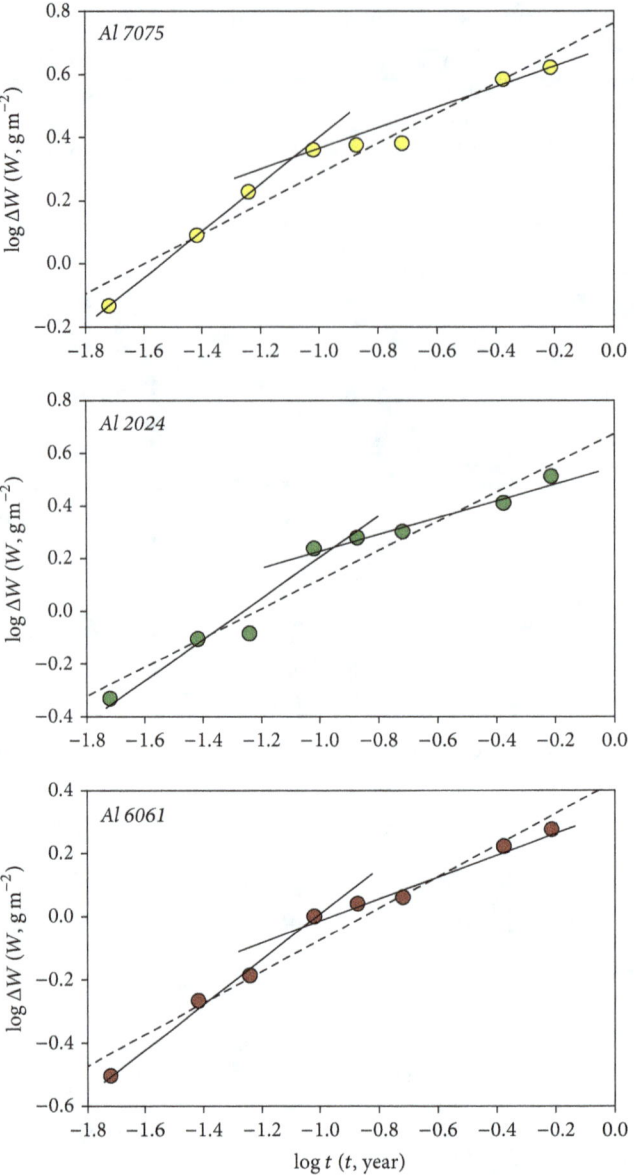

FIGURE 6: $\log \Delta W$ versus $\log t$ for Al alloys immersed in the Red Sea water.

By examining the data obtained in Table 3, the following observations were obtained:

(a) The values of weight loss after one year (C_1) indicate that the corrosion of Al alloys in the Red Sea water continues to increase significantly in the same order; Al 7075 is the highest mass loss more frequently, while Al 6061 is the lowest mass loss.

(b) The value of n can assist as an investigative tool to indicate the diffusion properties through the corrosion products that formed on the metal surface. It was found that the values of n_2 are much smaller than that of n_1. The values of n can be interpreted according to the following situations [33]:

(i) When n is roughly equal to 0.5, the corrosion process is controlled by ideal diffusion process through the corrosion products that remain on the metal surface.

(ii) When n is greater than 0.5, the diffusion process is accelerated due to corrosion products detachment by cracking, dissolution, soluble complex formation, and so on.

(iii) When n is less than 0.5, the diffusion process is decreased because the corrosion products layer becomes more compact and protective on the metal surface with time.

The reason for this observation can be stated after considering the proposed mechanism of the bimodal behavior.

$$4Al + 3O_2 + 6H_2O \longrightarrow 4Al(OH)_3 \qquad (6)$$

$$2Al + 3H_2O \longrightarrow Al_2O_3 + 3H_2 \uparrow \qquad (7)$$

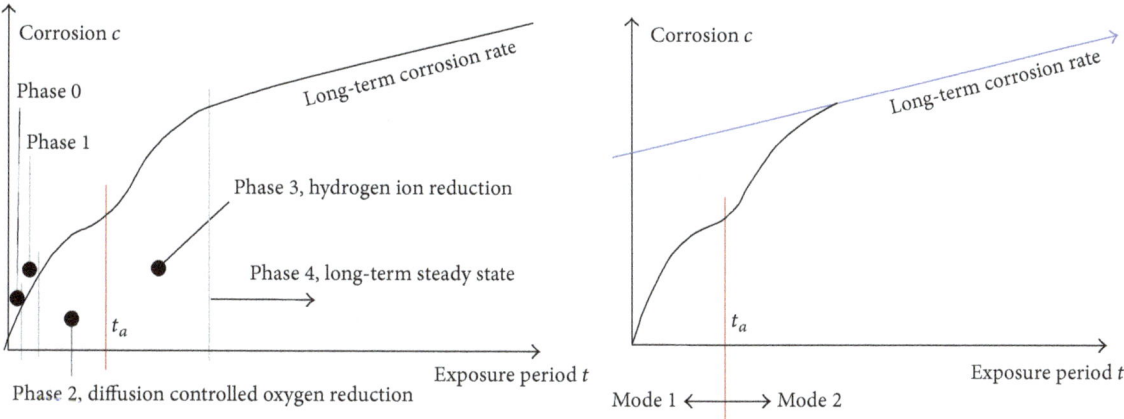

FIGURE 7: Schematic bimodal model for long-term corrosion loss and pit depth in marine environments. The change from mode 1 with predominantly oxic corrosion conditions to mode 2 with predominantly anoxic conditions occurs at t_a.

According to the above equations, reduction of oxygen and hydrogen ion is the two possible cathodic reactions. In fact, they cannot occur at the same environment conditions. For oxygen reduction, sufficient amount of oxygen must be supplied to the corroded surfaces, whereas hydrogen ion reduction can occur under anoxic conditions. Melchers Centre [30, 31] claimed that continuous formation of corrosion products and nonuniform topography of the corroded surface creates localized anoxic environments and permits hydrogen ion reduction. As the corrosion process progresses with time, the primary cathodic reaction changes to be hydrogen ion reduction instead of oxygen reduction. This change in cathodic reaction and, as a result, in the rate determining step, is the cause of the bimodal trend. Figure 7 shows the schematic bimodal model for corrosion loss as a function of exposure time [31]. As the first trend, a gradual decrease in corrosion rate was observed and this behavior is also represented by the power law function. The nonuniform and nonhomogeneous topography of corroding surfaces will develop as corrosion proceeds, though the thickness and properties of corrosion products on the corroding surfaces and over the pits also become nonuniform. As a result, local regions with low oxygen concentration will be produced and differential aeration cells will be established. These are sufficient conditions for the initiation and progression of pitting corrosion. Oxygen reduction is likely to occur in the shallower regions and corrosion is controlled by inward oxygen diffusion through corrosion products. As the second trend, further build-up of corrosion products will develop highly localized anoxic conditions, especially at the deeper corroded parts (pits). Hydrogen ion reduction is permitted thermodynamically under anoxic condition. In this case, corrosion rate is controlled by the outward diffusion of hydrogen gas. Hydrogen gas diffuses faster than oxygen through corrosion products (because hydrogen molecules are smaller than oxygen molecules) and an increase in corrosion rate is expected if the hydrogen ion reduction becomes the cathodic reaction in sufficient number of areas; this change in the rate trend causes the inflection in the bimodal model. In steady state, hydrogen diffusion cannot remain the rate limiting step for a long time. Other aspects have been considered as reason for the polarized equation (7). This causes a slow decline in instantaneous corrosion rate for longer term exposure such as the inward diffusion of water through the corrosion product, the outward diffusion of soluble corrosion products, or, less likely, the kinetics of the reactions. A review of the previous data revealed that a long period of time (Figure 7 line (t_a)) was required to permit deviation from the first trend; so, the short-term corrosion loss of data did not reach this time and was consistent with the power law function. This time can be affected by temperature, oxygen concentrations, or experimental conditions. In the current study, the weight loss experiments were carried out under stagnant conditions at laboratory temperature; this procedure kept corrosion products adherent on the alloy surfaces and high amount of corrosion products accumulated quickly on the corroded surfaces. As a result, the required time to reach the inflection point was reduced to a few weeks. This may be the reason for the observation of the bimodal model with short-immersion corrosion loss.

3.2. Potentiodynamic Polarization (PDP)

3.2.1. Effect of Immersion Time. Anodic curves can present useful information on passivation or depassivation of a metal dipped in solutions containing aggressive ions [34, 35]. Figure 8 shows the polarization curves for Al 7075, Al 2024, and Al 6061, respectively, in the Red Sea water after different immersion periods. Al 6061 and Al 2024 show well-defined breakdown potential at all immersion times. Passivity region or difference between corrosion potential and breakdown potential broaden as immersion time increased which may be as a result of accumulation of corrosion products during immersion and build-up of insoluble products. Polarization curves of Al 7075 (2, 24, and 72 h) show a sharp increase in the current density directly above corrosion potential. In this case, the corrosion potential of Al 7075 was often remarked as the first breakdown potential [36]. As the immersion time

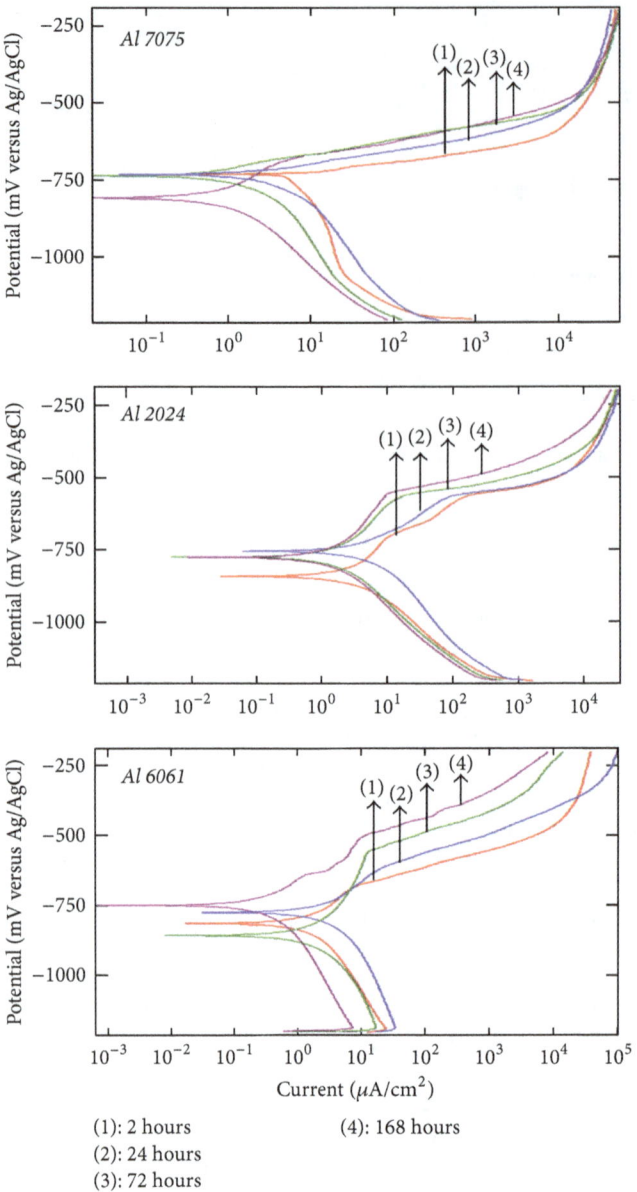

FIGURE 8: Potentiodynamic polarization curves of Al alloys in the Red Sea water at different time interval.

TABLE 4: Potentiodynamic polarization parameters of Al alloys in the Red Sea water at different immersion periods.

Time (hours)	Al 7075		Al 2024		Al 6061	
	$-E_{corr}$ (mV)	i_{corr} ($\mu A/cm^2$)	$-E_{corr}$ (mV)	i_{corr} ($\mu A/cm^2$)	$-E_{corr}$ (mV)	i_{corr} ($\mu A/cm^2$)
2	732	5.78	842	3.20	814	1.38
24	731	6.29	757	4.31	773	2.28
72	735	2.14	777	1.56	855	1.28
168	806	1.64	777	1.19	749	0.27

increased (168 h), the anodic curve of Al 7075 exhibited small passive window and the breakdown potential was distinguished from the corrosion potential. Cathodic curves of the studied alloys shifted to less current at 72 and 168 h of immersion, but the shape of the cathodic curves did not change; thus, it is expected that the cathodic reaction mechanism remains the same over all immersion times. The

electrochemical parameters such as E_{corr} and i_{corr} were estimated and listed in Table 4. The influence of immersion time on the above electrochemical parameters is discussed in the following.

(i) The values of i_{corr} tend to increase as immersion time increases, up to a maximum time of 24 h and then

TABLE 5: Corrosion potential E_{corr} (mV) and breakdown potential E_b (mV) of aluminum alloys at different temperature.

Temperature (°C)	Al 7075		Al 2024		Al 6061	
	$-E_{corr}$ (mV)	$-E_b$ (mV)	$-E_{corr}$ (mV)	$-E_b$ (mV)	$-E_{corr}$ (mV)	$-E_b$ (mV)
10	715	-	797	708	781	700
25	732	-	842	708	814	695
45	793	691	909	704	978	689
60	956	685	968	677	1035	628

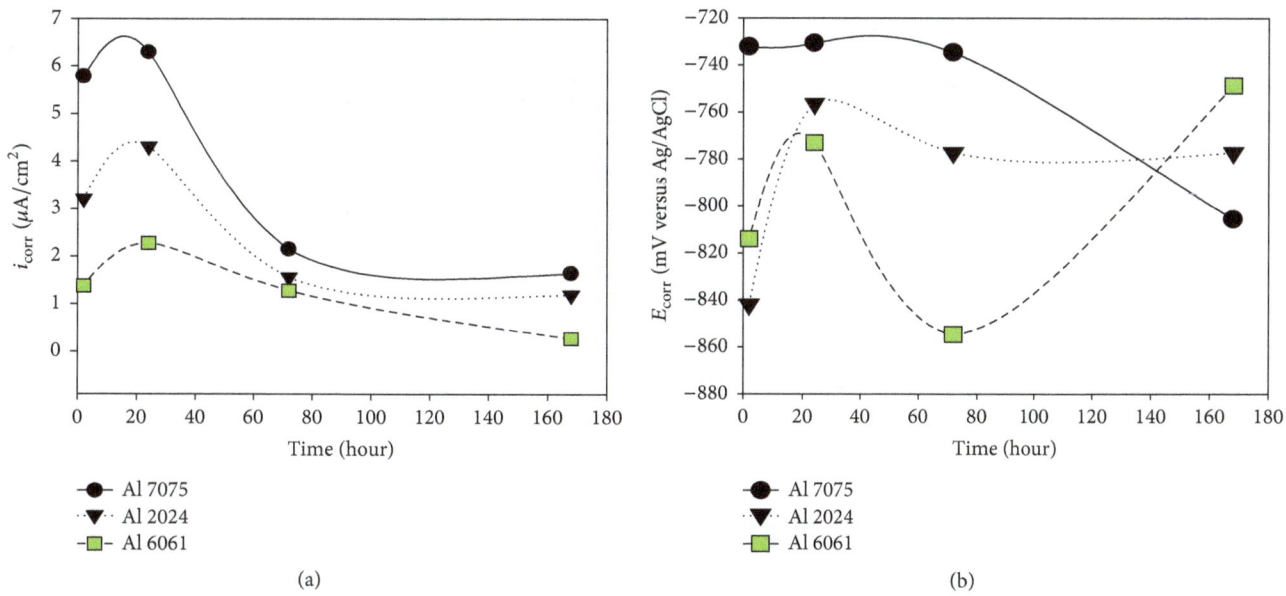

(a)

(b)

FIGURE 9: Variation of (a) i_{corr} and (b) E_{corr} with immersion periods.

tend to decrease at immersion time more than 24 h, as shown in Figure 9(a). The decrease in i_{corr} with immersion time may be attributed to the formation of a protective layer of corrosion products, composed of mainly less soluble aluminum compounds. According to i_{corr} values, Al 7075 has the highest corrosion current, while Al 6061 has the lowest corrosion current density at all immersion times. They resist corrosion in an order like that obtained previously from chemical studies:

$$Al\ 6061 > Al\ 2024 > Al\ 7075 \qquad (8)$$

(ii) Figure 9(b) shows no definite trend for E_{corr} values. Although the corrosion products are quickly formed on the Al surface, they take a considerably longer time to reach a steady state which is not seen in PDP measurements. The times of immersion included in the PDP study are not sufficient to reach steady state.

3.2.2. Effect of Temperature.

Polarization curves of Al 7075, Al 2024, and Al 6061 immersed in the Red Sea water for 2 h were recorded as a function of temperature and are presented in Figure 10. Temperature affects chemical kinetics

of the corrosion reaction and enhances cathodic reaction by increasing oxygen diffusion. The examination of polarization responses of aluminum alloys revealed a continuous increase in both anodic and cathodic current density and a decrease in corrosion potential with increase in temperature. Table 5 shows corrosion potential E_{corr} and breakdown potential E_b of the studied alloys at different temperatures.

The obtained results can be interpreted as follows:

(i) E_{corr} values shifted to more negative values with increase in temperatures, indicating that, with increase in the Red Sea water temperatures, the corrosion process comes under cathodic control. Increasing the temperature will increase the rate of oxygen diffusion to the metal surface, thus increasing corrosion current density (Figure 11(a)) because more oxygen is available for the cathodic reduction process. In the current study, this fact did not conflict with the fact that increasing temperature reduces oxygen solubility, because the studied temperatures did not exceed 80°C and the studied systems are closed, whereby oxygen was unable to escape; so, corrosion continued to increase with temperature [37].

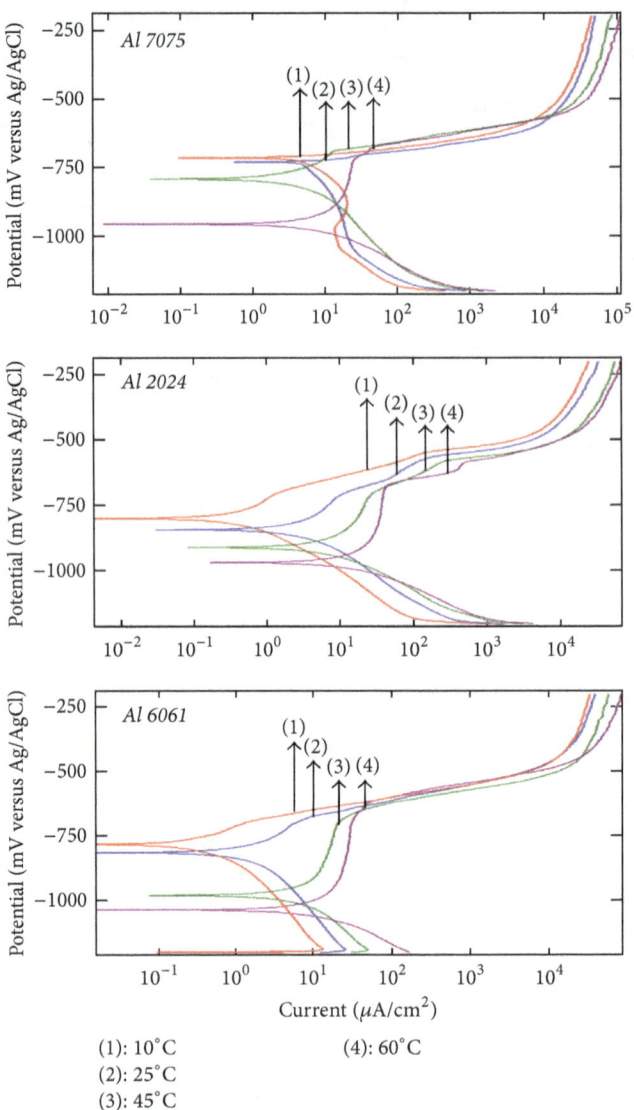

(1): 10°C   (4): 60°C
(2): 25°C
(3): 45°C

FIGURE 10: Potentiodynamic polarization curves of Al alloys in the Red Sea water at different temperatures.

(ii) It is shown that lowering corrosion potentials was accompanied by raising breakdown potential (E_b) which resulted in expanding the passive region ΔE of all alloys (Figure 11(b)), especially at high temperatures (45°C and 60°C). Increasing passivation with temperatures was also reported in a previous work [38]. However, increase in passivity region did not result in decrease in the current density. Immersion of alloys in the Red Sea water for 2 h at high temperature resulted in high reaction rate and formation of more amounts of corrosion products. Since a thick barrier of corrosion products is considered as a type of passivation [18], the presence of more corrosion products on alloys surfaces may be the reason for promoting the passivity. One other possible reason is that increased temperature leads to rapid dissolution of active elements (Mg and Al) from S-phase particles (Al_2CuMg), leaving Cu-rich regions on the alloys surfaces that may increase breakdown potentials with temperature [39, 40].

(iii) Al 7075 experienced the highest current density followed by Al 2024 and then Al 6061 at all temperatures with the exception of 60°C; Al 2024 exhibited higher corrosion current than Al 7075. Figure 12 reveals how the order of corrosion current density at Tafel region of the studied Al alloys changed at high temperature 60°C compared to low temperature 25°C. Other researchers [17] also found that Al 2024 exhibited higher cathodic current density in chloride solution than Al 7075 and they related this order by the number of cathodic active IM particles. Numbers of catholically active Cu-containing IM particles increased in this order: Al 6061 < Al 7075 < Al 2024. Corrosion current rate of Al 7075 was less influenced by temperature than Al 6061 and Al 2024; Cavanaugh

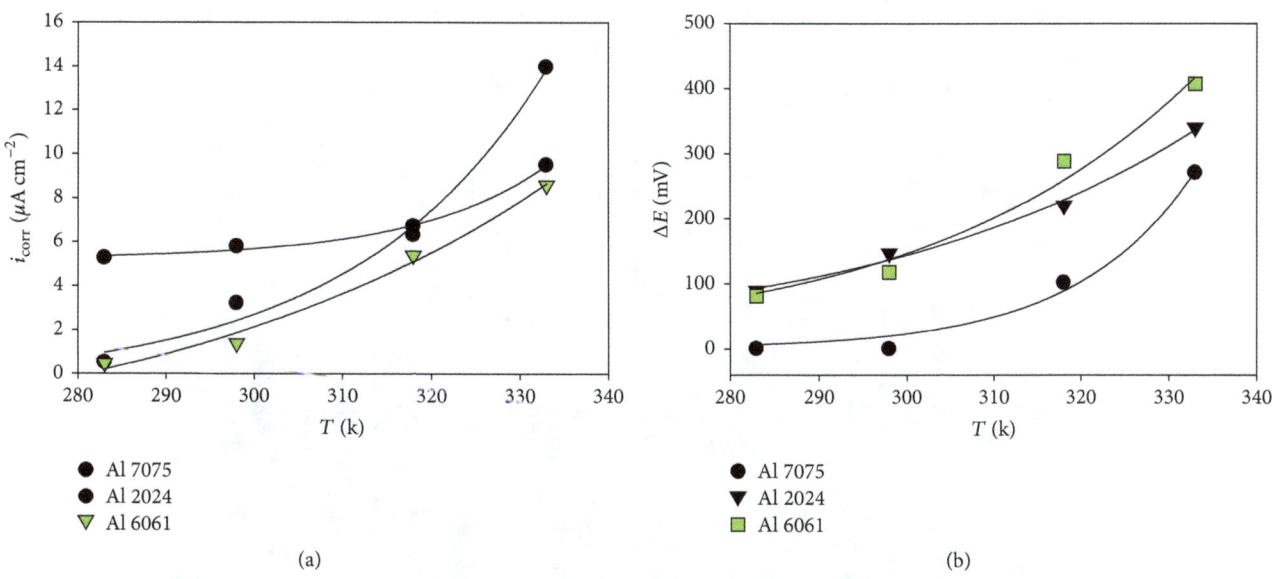

(a)   (b)

FIGURE 11: Dependence of (a) i_{corr} and (b) ΔE of Al alloys in the Red Sea water on temperature.

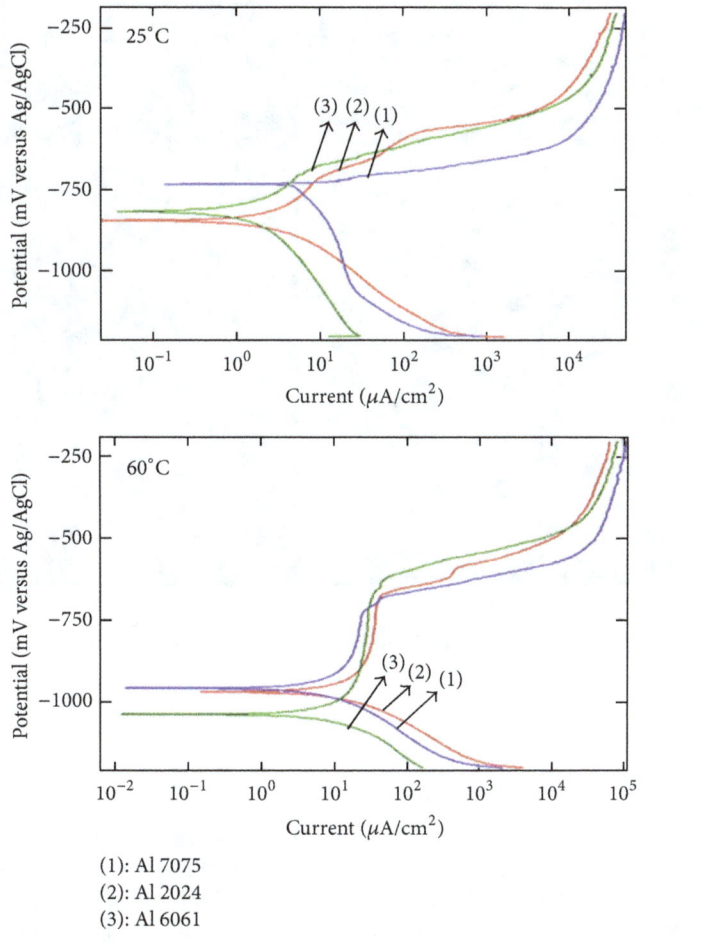

(1): Al 7075
(2): Al 2024
(3): Al 6061

FIGURE 12: Potentiodynamic polarization curves of Al alloys in the Red Sea water at 25°C and 60°C temperature.

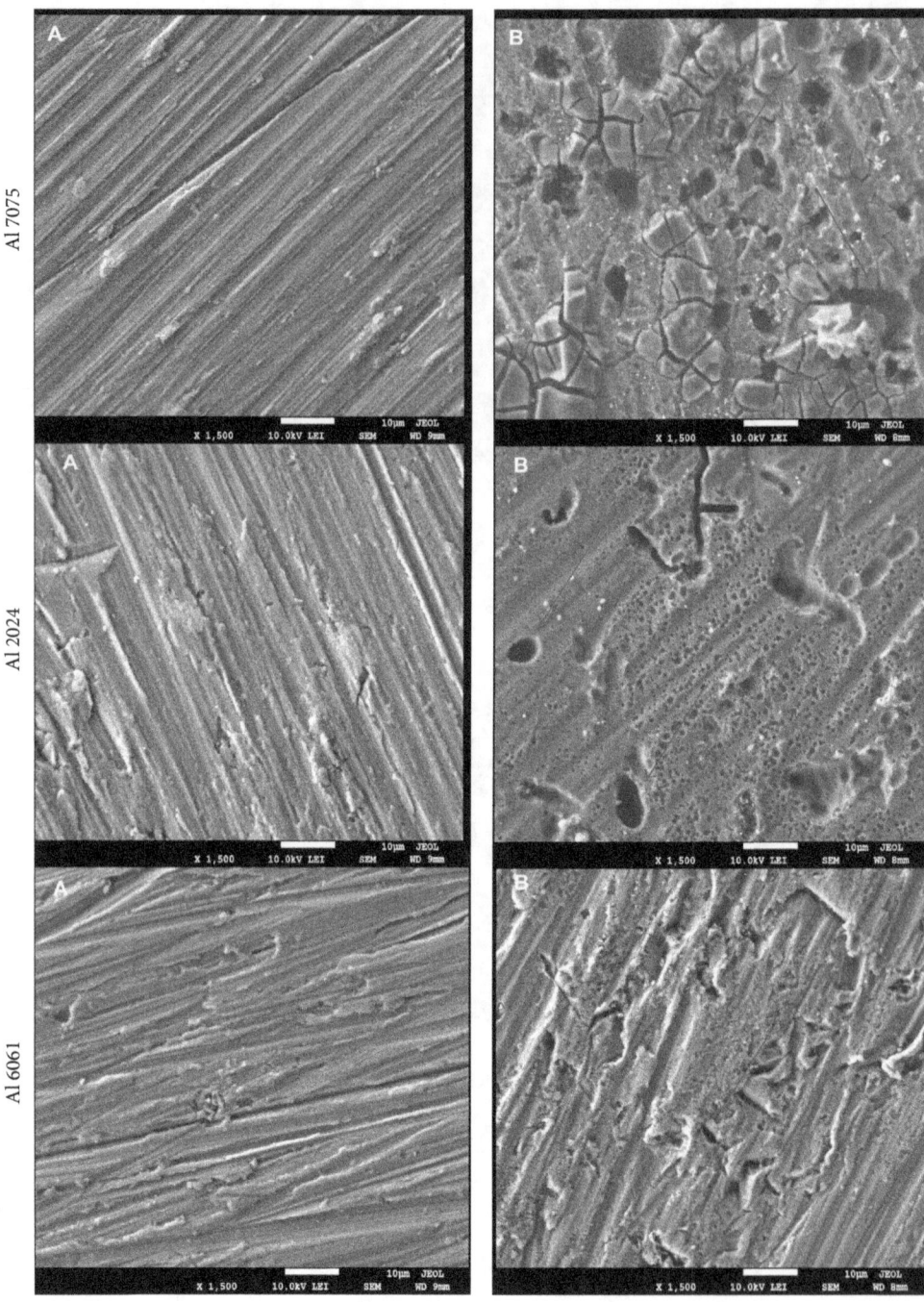

FIGURE 13: SEM images of Al 7075, Al 2024, and Al 6061 "A" before and "B" after being immersed in the Red Sea water for 5 weeks. The arrows indicate the pits.

et al. [40] reported nearly the same behavior of $MgZn_2$ (one of the constituent particles population in Al 7075) with temperature in chloride solution; the electrochemical response of $MgZn_2$ did not show great variation with temperature from 25°C to 60°C, so we predicted that $MgZn_2$ particle has a very important effect on the overall behavior of Al 7075.

3.3. Surface Characterization (SEM and EDS). Figure 13 depicts SEM images of the studied alloys (Al 7075, Al 2024, and Al 6061, resp.) "A" before and "B" after immersion in the Red Sea water for five weeks. Before taking images, sample A was abraded and treated as referred in Section 2.1 and sample B was pickled in nitric acid to remove corrosion products. SEM images "A" display noncorroded surfaces with scratches

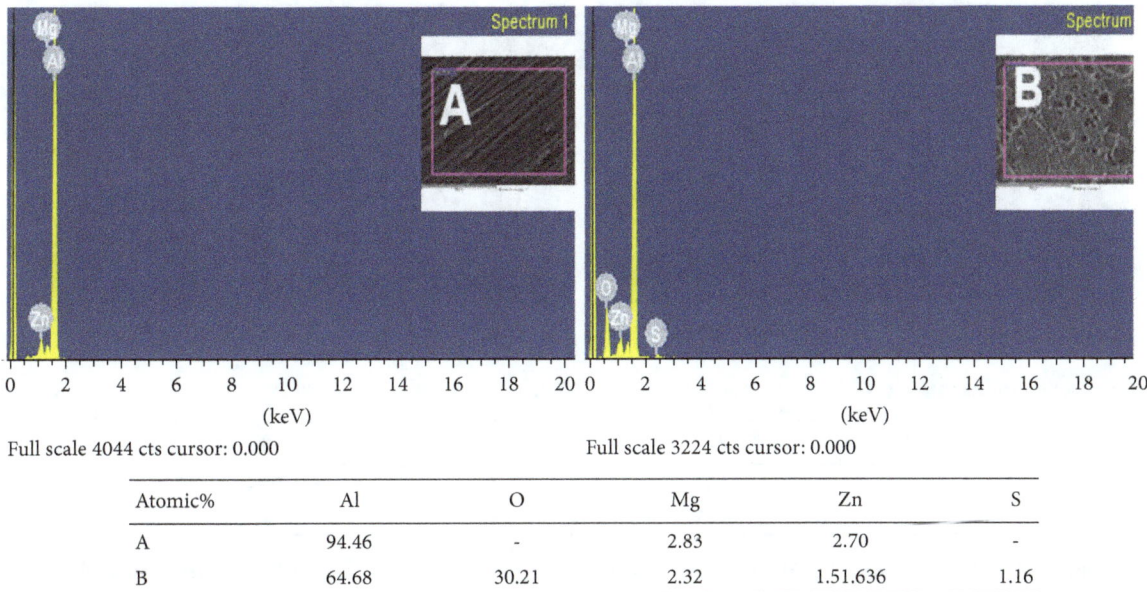

Atomic%	Al	O	Mg	Zn	S
A	94.46	-	2.83	2.70	-
B	64.68	30.21	2.32	1.51.636	1.16

FIGURE 14: EDS results of Al 7075 "A" before and "B" after being immersed in the Red Sea water for 5 weeks.

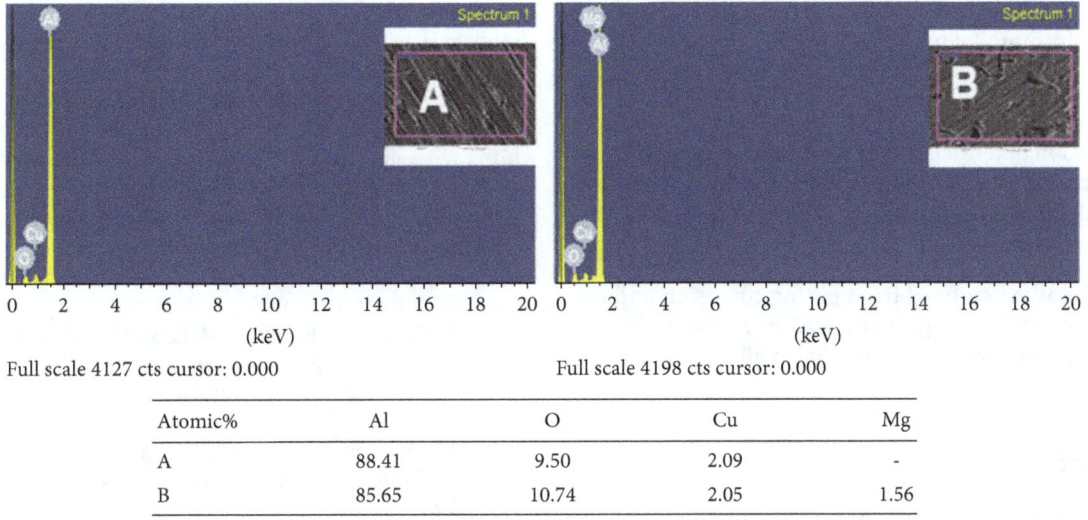

Atomic%	Al	O	Cu	Mg
A	88.41	9.50	2.09	-
B	85.65	10.74	2.05	1.56

FIGURE 15: EDS results of Al 2024 "A" before and "B" after being immersed in the Red Sea water for 5 weeks.

due to polishing. Heterogeneous nature of aluminum alloys surfaces is evident by images "A". SEM images "B" show damaged surfaces due to localized attack by the Red Sea water. Al 6061 is less damaged than other alloys. Pits can be observed in Al 7075 and Al 2024 images. Flakes of corrosion products are shown in Al 7075.

For EDS results represented in Figures 14–16, two different samples of the same alloy type were detected by EDS to analyze the elements present on the whole surface (A, noncorroded sample, and B, corroded sample); so, it is thought that EDS analysis lost some degree of precision. In general, Al percent decreased after immersion in seawater and oxygen increased. Copper content in Al 2024 was almost constant before and after corrosion because of the cathodic nature of the copper and Cu-rich particles with respect to

alloy matrix. S peak was observed in Al 7075 spectra after corrosion. Appearance of S on Al 7075 after corrosion gives an indication of the contribution of bacteria in corrosion process as SRB. Sulfate reducing bacteria exist in seawater and obtain their energy by reducing sulfate ions to highly corrosive sulfide ions such as S^{2-}, HS^-, or H_2S or metal sulfides which can accelerate the electrochemical reaction of corrosion. Previous literature proved that SRB induces pitting corrosion for Al 7075 and Al 2024 [18].

4. Conclusion

The corrosion behavior of Al 7075, Al 2024, and Al 6061 was investigated in this research and the results revealed the following:

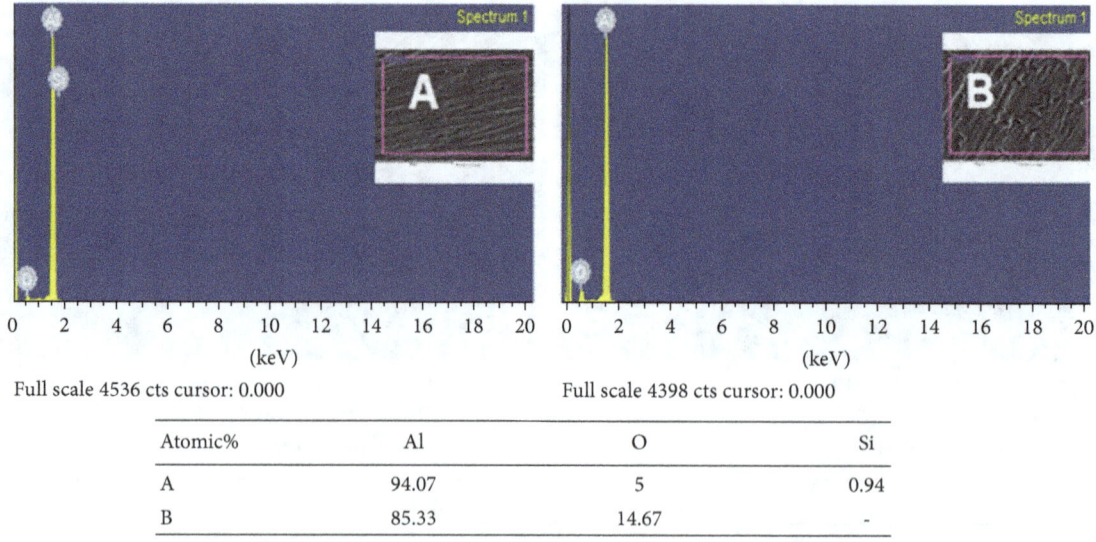

Atomic%	Al	O	Si
A	94.07	5	0.94
B	85.33	14.67	-

FIGURE 16: EDS results of Al 6061 "A" before and "B" after being immersed in the Red Sea water for 5 weeks.

(i) The corrosion rates of Al alloys decreased with time and the weight loss data for 32 weeks were consistent with the bimodal model.

(ii) Increase in temperature led to an increase in both anodic and cathodic current density and a decrease in corrosion potential. Lowering corrosion potential was associated with an increase in breakdown potential which resulted in expanding the passivity region with temperature

(iii) All the alloys suffered from pitting corrosion and the most corroded alloy was Al 7075. Al 6061 showed the highest corrosion resistance at all the immersion times.

Conflicts of Interest

The authors declare that there are no conflicts of interest regarding the publication of this article.

Acknowledgments

The authors are immensely grateful to Professor Robert E. Melchers, University of Newcastle, for his comments on WL data. He agreed with us that the data are consistent with bimodal model and explained the reason for the observation of the bimodal model with short-immersion time as we wrote in kinetic study section.

References

[1] A. Sverdlin, "Physical metallurgy and processes," in *Handbook of Aluminum*, G. E. Totten and D. S. MacKenzie, Eds., vol. 1, pp. 1–32, CRC Press, New York, NY, USA, 2003.

[2] K. A. Chandler, *Marine and Offshore Corrosion*, Marine Engineering Series, Elsevier, Oxford, UK, 2014.

[3] R. L. Reuben, *Materials in Marine Technology*, Springer-Verlag, London, UK, 1994.

[4] J. E. Hatch, Ed., *Aluminium: Properties and Physical Metallurgy*, ASM International, Ohio, USA, 1984.

[5] J. R. Davis, Ed., *Alloying: Understanding the Basics*, ASM International, Ohio, USA, 2001.

[6] R. Javaherdashti, *Micrbiologically Influenced Corrosion*, Springer-Verlag, London, UK, 2008.

[7] E. McCafferty, *Introduction to Corrosion Science*, Springer-Verlag, NewYork, NY, USA, 2010.

[8] J. R. Davis, Ed., *Corrsion of Aluminum and Aluminum Alloys*, ASM International, Ohio, USA, 1999.

[9] A. E. Hughes, N. Birbilis, J. M. C. Mol, S. J. Garcia, X. Zhou, and G. E. Thompson Z, *Recent Trends in Processing and Degradation of Aluminium Alloys*, Z. Ahmad, Ed., In Tech, Rijeka, Croatia, 2011.

[10] T. H. Muster, A. E. Hughes, and E. G. Thompson, *Copper Distributions in Aluminium Alloys*, Nova, New York, NY, USA, 2009.

[11] N. L. Sukiman, X. Zhou, N. Birbilis et al., "Aluminum Alloys: New Trends in Fabrication and Application," Z. Ahmad, Ed., pp. 47–97, In Tech, Rijeka, Croatia, 2012.

[12] J. Soltis, "Passivity breakdown, pit initiation and propagation of pits in metallic materials–review," *Corrosion Science*, vol. 90, pp. 5–22, 2015.

[13] Z. Szklarska-Smialowska, "Pitting corrosion of aluminum," *Corrosion Science*, vol. 41, no. 9, pp. 1743–1767, 1999.

[14] I. L. Muller and J. R. Galvele, "Pitting potential of high purity binary aluminium alloys-I. AlCu alloys. Pitting and intergranular corrosion," *Corrosion Science*, vol. 17, no. 3, pp. 179–193, 1977.

[15] R. Baboian, Ed., *Corrosion Tests and Standards*, ASTM International, West Conshohocken, PA, USA, 2005.

[16] D. A. Shifler, T. Tsuru, P. M. Natishan, and S. Ito, Eds., *Corrosion in Marine And Saltwater Environment II*, The Electrochemical Society, Nueva Jersey, USA, 2005.

[17] I.-W. Huang, B. L. Hurley, F. Yang, and R. G. Buchheit, "Dependence on temperature, pH, and Cl- in the uniform corrosion of aluminum alloys 2024-T3, 6061-T6, and 7075-T6," *International Journal of Corrosion*, vol. 199, pp. 242–253, 2016.

[18] E. Ghali, *Corrosion Resistance of Aluminum and Magnesium Alloys*, John Wiley & Sons, Hoboken, NJ, USA, 2010.

[19] D. Ornek, A. Jayaraman, T. K. Wood, Z. Sun, C. H. Hsu, and F. Mansfeld, "Pitting corrosion control using regenerative biofilms on aluminium 2024 in artificial seawater," *Corrosion Science*, vol. 43, no. 11, pp. 2121–2133, 2001.

[20] R. Rosliza, H. B. Senin, and W. B. W. Nik, "Electrochemical properties and corrosion inhibition of AA6061 in tropical seawater," *Colloids and Surfaces A: Physicochemical and Engineering Aspects*, vol. 312, no. 2-3, pp. 185–189, 2008.

[21] V. S. Sinyavskii and V. D. Kalinin, "Marine corrosion and protection of aluminum alloys according to their composition and structure," *Protection of Metals*, vol. 41, no. 4, pp. 317–328, 2005.

[22] H. Ezuber, A. El-Houd, and F. El-Shawesh, "A study on the corrosion behavior of aluminum alloys in seawater," *Materials and Corrosion*, vol. 29, no. 4, pp. 801–805, 2008.

[23] A. Aballe, M. Bethencourt, F. J. Botana, M. J. Cano, and M. Marcos, "Localized alkaline corrosion of alloy AA5083 in neutral 3.5% NaCl solution," *Corrosion Science*, vol. 43, no. 9, pp. 1657–1674, 2001.

[24] A. H. Al-Moubaraki, A. Al-Judaibi, and M. Asiri, "Corrosion of C-steel in the red sea: effect of immersion time and inhibitor concentration," *International Journal of Electrochemical Science*, vol. 10, no. 5, pp. 4252–4278, 2014.

[25] ASTM, *G1-03 (2017)e1, Standard Practice for Preparing, Cleaning, and Evaluating Corrosion Test Specimens*, ASTM International, West Conshohocken, PA, USA, 2017.

[26] A. U. Malik, S. Ahmad, I. Andijani, and S. Al-Fouzan, "Corrosion behavior of steels in Gulf seawater environment," *Desalination*, vol. 123, no. 2-3, pp. 205–213, 1999.

[27] I. O. Wallinder and C. Leygraf, "Seasonal variations in corrosion rate and runoff rate of copper roofs in an urban and a rural atmospheric environment," *Corrosion Science*, vol. 43, no. 12, pp. 2379–2396, 2001.

[28] G. S. Frankel, "Pitting corrosion of metals: a review of the critical factors," *Journal of The Electrochemical Society*, vol. 145, no. 6, pp. 2186–2198, 1998.

[29] G. Tammann, *Lehrbuch der Metallographie*, Leopold Voss, Leipzig, Germany, 2nd edition, 1923.

[30] R. E. Melchers, "Bi-modal trend in the long-term corrosion of aluminium alloys," *Corrosion Science*, vol. 82, pp. 239–247, 2014.

[31] R. E. Melchers, "Time dependent development of aluminium pitting corrosion," *Advances in Materials Science and Engineering*, vol. 2015, Article ID 215712, 2015.

[32] M. Morcillo, S. Feliu, and J. Simancas, "Deviation from bilogarithmic law for atmospheric corrosion of steel," *British Corrosion Journal*, vol. 28, no. 1, pp. 50–52, 1993.

[33] J. H. Wang, F. I. Wei, Y. S. Chang, and H. C. Shih, "The corrosion mechanisms of carbon steel and weathering steel in SO2 polluted atmospheres," *Materials Chemistry and Physics*, vol. 47, no. 1, pp. 1–8, 1997.

[34] B. Yi, D. Lin, Y. Chen, N. Dai, X. Du, and J. Zhang, "Effect of Cl- and SO42- on corrosion behavior of reinforcing steel in simulated concrete pore solutions," *Corrosion Science and Protection Technology*, vol. 28, no. 2, pp. 97–102, 2016.

[35] G. Sahoo and R. Balasubramaniam, "On the corrosion behaviour of phosphoric irons in simulated concrete pore solution," *Corrosion Science*, vol. 50, no. 1, pp. 131–143, 2008.

[36] Z. Zhao and G. S. Frankel, "On the first breakdown in AA7075-T6," *Corrosion Science*, vol. 49, no. 7, pp. 3064–3088, 2007.

[37] L. Garverick, Ed., *Corrosion in the Petrochemical Industry*, ASM intrernationa, Ohio, USA, 1994.

[38] M. Belkhaouda, L. Bazzi, R. Salghi et al., "Effect of the heat treatment on the behaviour of the corrosion and passivation of 3003 aluminium alloy in synthetic solution," *Journal of Materials and Environmental Science*, vol. 1, no. 1, pp. 25–33, 2010.

[39] Q. Meng and G. S. Frankel, "Effect of Cu content on corrosion behavior of 7xxx series aluminum alloys," *Journal of The Electrochemical Society*, vol. 151, no. 5, pp. B271–B283, 2004.

[40] M. K. Cavanaugh, J.-C. Li, N. Birbilis, and R. G. Buchheit, "Electrochemical characterization of intermetallic phases common to aluminum alloys as a function of solution temperature," *Journal of The Electrochemical Society*, vol. 161, no. 12, pp. C535–C543, 2014.

17

Synergistic Effect of Carbamide and Sulfate Reducing Bacteria on Corrosion Behavior of Carbon Steel in Soil

Ximing Li ⓘ[1] **and Cheng Sun**[2]

[1]*Chemical and Biomolecular Engineering, The University of Akron, Akron, OH 44325, USA*
[2]*State Key Laboratory for Corrosion and Protection, Institute of Metals Research, Chinese Academy of Sciences, Shenyang 110016, China*

Correspondence should be addressed to Ximing Li; xmli2012@gmail.com

Academic Editor: Ramazan Solmaz

Synergistic effect of carbamide and sulfate reducing bacteria (SRB) on corrosion behavior of carbon steel was studied in soils with moisture of 20% and 30%, by soil properties measurement, weight loss, polarization curve, and electrochemical impedance spectroscopy. The results show that carbamide decreased the soil redox potential and increased soil pH. In soil without SRB, carbamide made corrosion potential of Q235 steel much more positive and then inhibited corrosion. Meanwhile, in soil with SRB, 0.5 wt% carbamide restrained SRB growth and inhibited biocorrosion of Q235 steel. Corrosion rate of carbon steel decreased in soil with 30% moisture compared with that with 20% moisture.

1. Introduction

Soil corrosion is one of most common corrosion behaviors of metals underground. There are numerous factors influencing the soil aggressiveness such as soil humidity, acidity, soil salinity, and composition of microbes [1–4]. These factors interchange over time, making soil corrosion study much more complicated. The idea that soil corrosion depends on such factors as soil resistivity [2], pH [5], redox potential [6], water content [7, 8], and sulfate reducing bacteria (SRB) [9, 10] is widely accepted and studied. pH, soil acidity, is a comprehensive reflection of soil chemical properties, especially the salt content. It affects both anode and cathode polarization process, which are related to corrosion dynamics [5]. It is well known that metals suffer serious corrosion in strongly acidic environment. Since in strongly acidic soil, with decreasing pH, the depolarized potential of hydrogen increases, consequently corrosion rate increases. Meanwhile in soils with a high content of organic matters and organic acid, even with pH being neutral, metals can still suffer serious corrosion [11].

The soil redox potential reflects various kinds of oxidation-reduction equilibrium, normally changing from +700 mV to -300 mV. It is mainly affected by soil salt composition and content, aeration condition, organic matter, and so on. A soil with strong reducing property is commonly considered as a reference index of biocorrosion [12]. The main kinds of microbe related to soil corrosion include aerobic corrosion bacteria, such as sulfur-oxidizing bacteria and iron bacteria, and anaerobic corrosion bacteria such as SRB, referred to as Microbiological Influenced Corrosion (MIC) [13, 14]. Of various types of microorganisms, anaerobic SRB are the most common ones that induced MIC of underground pipeline steel [15]. Many studies have investigated SRB induced corrosion of carbon steel in soil simulation solution [3, 16].

The soil corrosiveness increases with soil moisture [12, 17]. But when saturated the diffusion of oxygen would be suppressed. Besides, soil moisture depended on local climate. In Shenyang, north part of China, the average soil moisture is 20%. Few studies have been done on the corrosion behavior of carbon steel in saturated soil [18, 19]. In addition, most of the studies on soil corrosion were conducted in soil simulated solution [20], ignoring some soil factors. For example, large amount of nitrogenous fertilizer has been used on farmland for agriculture development, of which carbamide is the most

TABLE 1: Compositions of the soil (mg/100 g soil).

chemical composition								organic content	whole nitrogen content	total salt content	pH
NO_3^-	Cl^-	SO_4^{2-}	HCO_3^-	Ca^{2+}	Mg^{2+}	K^+	Na^+				
4.6	3.1	4.8	23.4	5.7	3.2	0.2	1.4	2260	91	46.4	7.75

TABLE 2: Experimental matrix.

Experiments	Soil A	Soil B	Soil C	Soil D
Soil moisture (wt.%)	20%	20%	30%	30%
Carbamide (wt.%)	0.5 wt%	0.5 wt%	0.5 wt%	0.5 wt%
Activated SRB	No	Yes	No	Yes

common type. It would be of importance to study the effect of carbamide on metals underground. However, little work has been done. Carbamide is able to be adsorbed on the steel surface to inhibit its corrosion [21]. There are also reports that carbamide can be used as a noncorroding alternative to rock salt for road deicing. In addition, it has been applied as corrosion inhibitor in metal pickling.

Besides, nitrogen is one of the essential elements to constitute protein and nucleic acid for bacteria in soil. Hence carbamide can serve as nitrogen source for SRB growth. The synergistic effect of carbamide and sulfate reducing bacteria (SRB) on corrosion behavior of carbon steel in soils will be interesting. In our previous work [19], the low content of carbamide promoted the growth of SRB and prevented the microbiological corrosion of Q235 steel. In this work, further studies in different soil moisture of 20% and 30% are done to see the effect of excess carbamide on SRB growth and on biocorrosion behavior of carbon steel.

2. Experimental Design

2.1. Materials and Preparation

2.1.1. Coupon Preparation. Carbon steel Q235, with a composition (wt.%) of 0.30 C, 0.019 P, 0.029 S, 0.01 Si, 0.42 Mn, and balance Fe, was cut into two varied sizes, 10×10×3 mm and 20×20×3 mm, which will be used as working and counter electrodes and weight-loss test samples. The coupons were abraded with a series of grit papers (200, 400, 600, 800, and 1000) followed by cleaning in acetone and alcohol and then dried. The working and counter electrodes were embedded in epoxy resin to give working areas of 10 mm × 10 mm for electrochemical measurements in a three-electrode cell. The working surface was cleaned with acetone and distilled water. All the prepared coupons were sterilized under ultraviolet rays prior to each experiment.

2.1.2. Soil Preparation. The chemical compositions of soil used in this work are given in Table 1. The soils were sterilized by heating at 121.8°C at high pressure. Then the sterilized soil and distilled water were mixed proportionally in experimental containers, to make different soils with moisture content of 20% and 30%.

2.1.3. Carbamide and SRB. Carbamide was purchased from Sinopharm, with the nitrogen content of 46.4 wt.% according to the national standard GB2440-2001. Before experiment set-up, different amounts of carbamide were measured by electronic analytical balance up to 0.0001 g, sterilized under ultraviolet rays, and mixed into different soils prepared above.

The sulfate reducing bacteria (SRB) used in this study are Desulfovibrio desulfuricans, same SRB strain as described in previous paper [20, 22]. SRB cultures were incubated in an anaerobic environment in the API RP-38 medium (g/l), containing $MgSO_4·7H_2O$ 0.2, KH_2PO_4 0.5, NaCl 10.0, ascorbic acid 1.0, sodium lactate 4.0, yeast extract 1.0, and $Fe(NH_4)_2(SO_4)_2$ 0.02. The pH value of the culture solution was between 7.0 and 7.1. SRB species were activated in an incubator for 24 hours and then were added to and mixed carefully with the prepared soil under bubbling N_2 in the experiment container. Then, the experiment containers were weighted and sealed over experiment period to keep the water moisture constant. The experimental matrix is listed in Table 2.

2.2. Testing Environments and Methods.
All the experiments were performed at room temperature for 65 days.

2.2.1. Soil Redox Potential and pH. Soil redox potential was measured by a multimeter, 5 Pt electrodes, and a saturated calomel electrode (SCE). The redox potential was calculated from

$$E_h = E_m + 250 + 60 * (pH - 7) \ (mV) \tag{1}$$

where Eh is redox potential at pH=7 (mV, standard hydrogen scale); Em is the mean of the potential measured from the five platinum electrodes (mV) [23]. Soil pH was also measured from time to time, using the same method as in published article [22].

2.2.2. Immersion Tests. The specimens of 20×20×3 mm were used for weight-loss tests, which were buried at a depth of 10 cm below the surface of the soils in the same experimental containers. The prepared specimens were weighed to a precision of 0.1 mg before tests. After the experiment, the extracted specimen was pickled in a mixture containing hydrogen

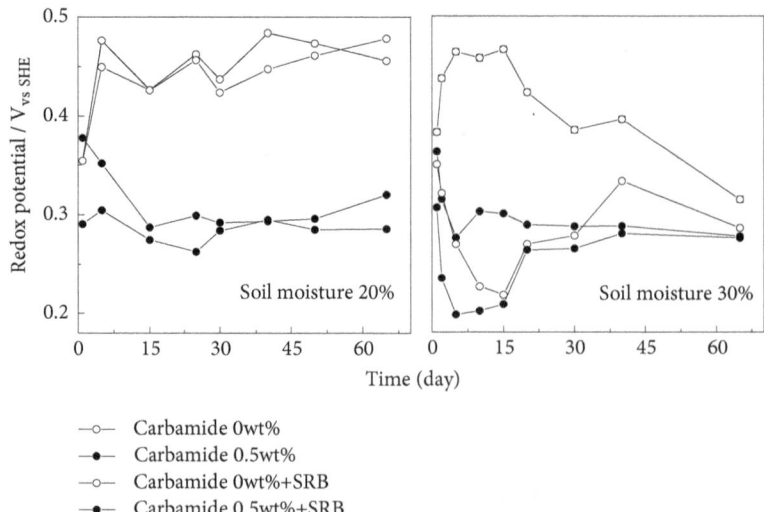

FIGURE 1: The soil redox potential changes over time.

chloride (HCl) 500 ml, urotropine 20 g, and water (H$_2$O) 500 ml for 10 min at room temperature, then cleaned with water, dried at 105. 8°C for 30 min in a furnace, cooled, and weighed. The same process was repeated until the difference between the last two results was less than 0.0001 g. Weight losses were calculated and converted into uniform corrosion rates (μm/y). Each experiment used triplicate specimens to guarantee the reliability of the results. After the tests, surface appearance of tested specimen was observed using scanning electron microscopy (SEM).

2.2.3. EIS and Polarization Tests. Electrochemical tests were performed in a three-electrode system, with a graphite electrode used as the counter electrode, a saturated copper/bluestone (Cu/CuSO$_4$) electrode as the reference electrode, and carbon steel as working electrode. The tests were conducted using the PARSTAT 2273 electrochemical measurement system. In EIS, an alternating current signal with a frequency range from 10 kHz to 1 mHz and an amplitude of 10 mV was applied to the working electrode at the corrosion potential. Potentiodynamic polarization was tested at the potential scanning velocity 0.5 mV/s and with the scope being ±0.25 V.

3. Results and Discussion

3.1. Soil Properties. The changes of soil redox potential over time are shown in Figure 1. Obviously in soils with and without SRB, the addition of carbamide made the soil redox potential decrease dramatically. That is because carbamide increased the soil organic matter content. The redox potential was more negative in soil with SRB, in agreement with published literature [22]. It is commonly accepted that the more negative the soil redox potential, the more the aggressiveness the carbon steel suffered [12]. However, this may not be the case in neutral or alkaline soil: the more positive the soil redox potential is, the more serious the steel would suffer. Hence soil

pH needs to be measured over time. Effect of soil moisture on soil redox potential was also compared in Figure 1. The redox potential in soil moisture of 30% was lower than that in soil moisture of 20%. That is because the more the water content, the lower the oxygen level in soil.

As mentioned above, pH is an important parameter to measure soil corrosivity. Hence, the changes of soil pH were recorded and drawn in Figure 2. pH value in soils with carbamide is much higher than in soils without carbamide, where the fact that some carbamide was decomposed into carbon dioxide (CO$_2$) and ammonia (NH$_3$) makes sense. When NH$_3$ volatilized, pH declined slightly after 35 days, as shown in Figure 2. Considering the effect of SRB, pH value was a little bit lower in soil with SRB. This is because the ultimate metabolic product of SRB was low-carbon chain fatty acid, which formed relative acid environment [24, 25]. Some studies have shown that metabolic products of bacteria can promote corrosion of carbon steel in soil. In soil with moisture of 30%, compared to 20%, pH value decreased in soil with carbamide and increased in soil without carbamide.

So, the addition of carbamide decreased soil redox potential and made the soil pH higher than 7, which theoretically decreases the soil corrosivity and then inhibits corrosion behavior of carbon steel. Meanwhile, the addition of SRB in soil decreased soil redox potential, and soil pH was lower than 7 in soil moisture of 20% and neutral in 30% moisture, generally increasing the soil corrosivity and accelerating corrosion of carbon steel. The synergistic effect of carbamide and SRB on corrosion behavior of carbon steel in soil will be very interesting, considering that carbamide can be the growth resource of SRB.

3.2. SRB Growth Analysis. To check the effect of carbamide on SRB growth, different addition levels of carbamide were studied in soil, shown in Figure 3. The method of how to detect the number of live SRB has been published in our previous work [20, 22], which follows the national standard

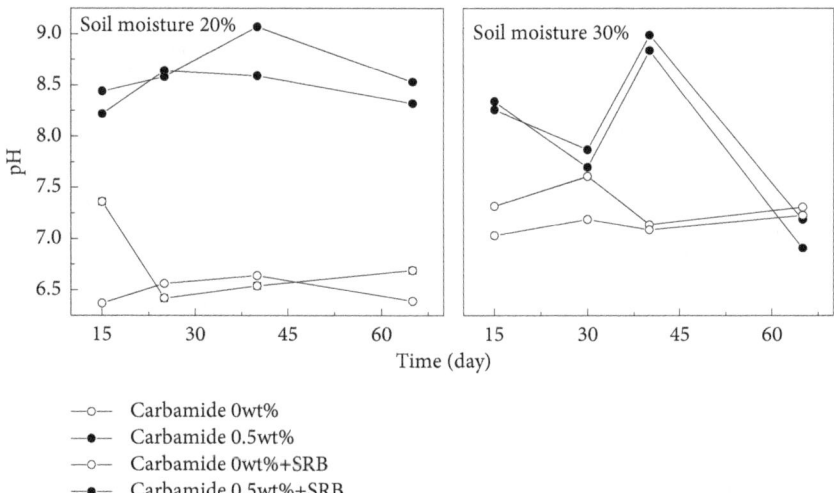

FIGURE 2: Soil pH changes over time.

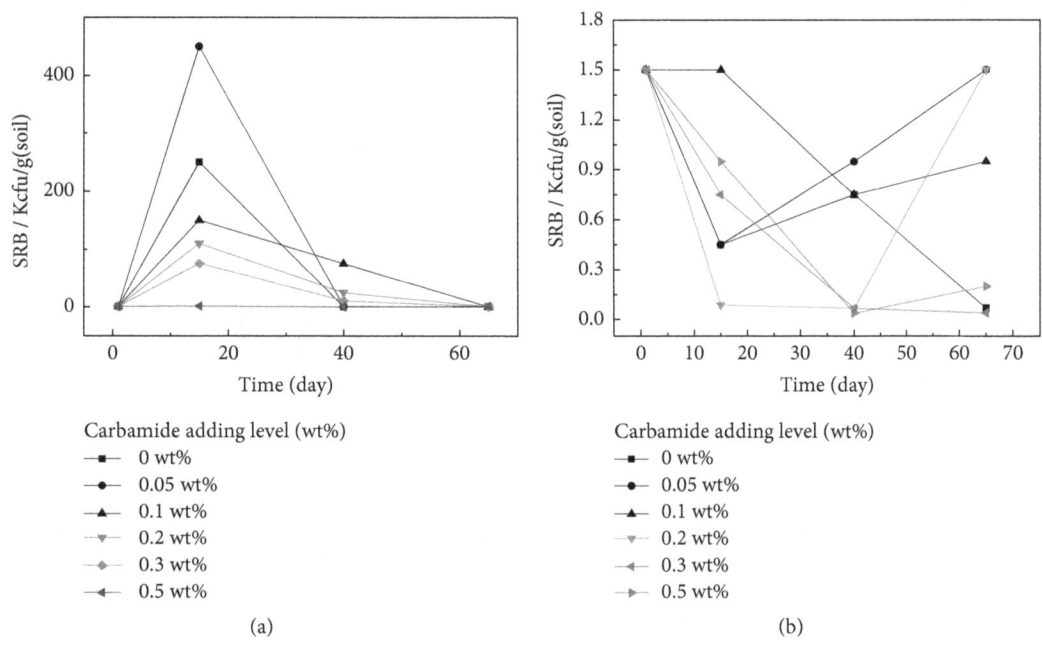

FIGURE 3: Number of SRB over time in soil with moisture of 20% (a) and 30% (b).

[26]. In the early stage of the experiment, the amount of SRB in soils with low content of carbamide is more compared with higher carbamide. Over time, due to the decomposition of the carbamide, the soils with higher level of carbamide have more live SRB, but significantly lower than the initial number. This indicated that carbamide promoted the growth of SRB but in a certain limited level. With excess amount of carbamide, SRB growth was compressed. With soil moisture of 30%, the amount of SRB decreased first and then increased, and the changes of SRB were different in soils with different concentrations of carbamide. In general, low content of carbamide promoted SRB growth, while elevated levels of carbamide (0.5 wt%) inhibit the growth of SRB, same as the result in 20% moisture.

3.3. *Corrosion Potential of Carbon Steel.* The corrosion potential of carbon steel fluctuated at an initial period due to the unstable soil environment (Figure 4). After 10 days, it kept constant. In soils without SRB, corrosion potential of carbon steel was more positive in soil containing carbamide than that in soil without carbamide. Carbamide was easily adsorbed on the steel surface to form a protective film [6]. However, when the soil environment was unstable, the adsorbed film was incomplete and corrosion potential tended to be more negative. When the soil environment was stable, corrosion potential dramatically increased in soil with carbamide, which indicated that carbamide inhibited steel corrosion. Meanwhile, in soil with SRB, corrosion potential tended to be negative in soil with carbamide, the reason for which

TABLE 3: Fitting results of polarization curves.

Fitting results of polarization curves		0 wt%		0.5 wt%		0 wt%+SRB		0.5 wt%+SRB	
		I_{corr} ($\mu A/cm^2$)	E_{corr} (mV)	I_{corr} ($\mu A/cm^2$)	E_{corr} (mV)	I_{corr} ($\mu A/cm^2$)	E_{corr} (mV)	I_{corr} ($\mu A/cm^2$)	E_{corr} (mV)
Soil moisture 20%	1 day	1.552	-556.6	35.42	-689.9	0.6235	-781.7	2.561	-798.6
	40 days	9.063	-486.5	2.227	-324.0	1.280	-890.3	0.7698	-891.2
	65 days	6.745	-489.7	0.5043	-277.0	3.899	-897.9	2.178	-822.8
Soil moisture 30%	1 day	9.068	-786.6	51.64	-821.1	5.083	-823.9	2.129	-823.9
	40 days	2.440	-638.8	0.3075	-214.4	5.222	-869.3	3.089	-319.9
	65 days	3.252	-780.0	1.284	-242.9	0.5128	-918.7	4.096	-224.4

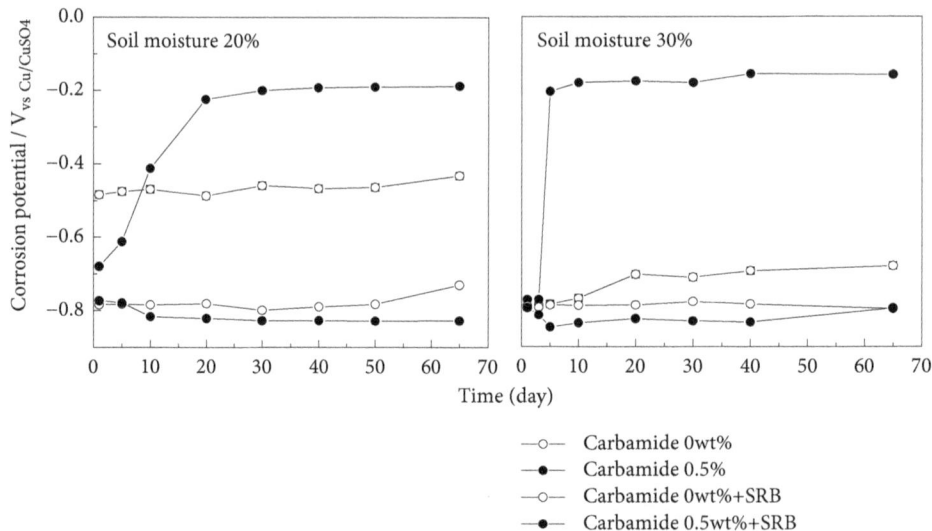

FIGURE 4: Corrosion potential of Q235 steel over buried time.

was discussed before. Proper amount of carbamide provided energy for the growth of SRB and promoted soil corrosion of carbon steel. Hence the effects of carbamide on soil corrosion turned to its effects on SRB growth. But excessive carbamide decreased the number of SRB, which in turn inhibited soil corrosion of Q235 steel. To compare corrosion potential in different soil moisture, there were minor changes except that in soil without carbamide. In that experiment group, corrosion potential of Q235 steel in soil with moisture of 30% was more negative than that in soil with moisture of 20%. Probably the increased water molecule reduced the corrosion potential, which played no role in soil with carbamide.

3.4. Polarization Curves of Carbon Steel. The polarization curves of carbon steel Q235 in different soils after 1 day, 40 days, and 65 days are shown in Figures 5 and 6, respectively, and the fitting results are listed in Table 3. At day 1, in contrast with the groups in soils without carbamide, corrosion current I_{corr} was larger and corrosion potential E_{corr} was more negative in soil with 0.5 wt% carbamide. That is likely due to the incomplete protective film which would cause localized corrosion on steel surface. In soil with SRB, I_{corr} of Q235 steel was smaller than that in soils without SRB, since initially the

biofilm of SRB was formed on the sample surface to protect it from corrosion [20]. In soil with moisture of 30%, I_{corr} increased over all.

Over time, E_{corr} of Q235 steel in soil without SRB shifted to be positive, and I_{corr} decreased. Because initially oxygen depolarization was the leading cathodic reaction, over time oxygen was consumed and cathode reaction was prevented. Besides the slope of anodic polarization became larger, which shows that there were corrosion products on the steel surface, and the diffusion process of cation became more difficult. Comparing with the case in soils without carbamide, E_{corr} of carbon steel in soil with carbamide was more positive and I_{corr} was much lower, indicating that carbamide inhibited corrosion reaction of carbon steel. In soil with moisture of 30%, the slope of cathode polarization curve was even smaller, showing the cathode reaction was inhibited, and I_{corr} was decreased. Besides, E_{corr} shifted to be more negative, which agrees with published literature [8]. In soil with SRB, E_{corr} was much more negative than that in soils without SRB. I_{corr} increased after 40 days except the group in soil with moisture of 30% and containing 0.5 wt% carbamide. I_{corr} increased because of the death of SRB and the degradation of SRB metabolite products. It has been reported that corrosion

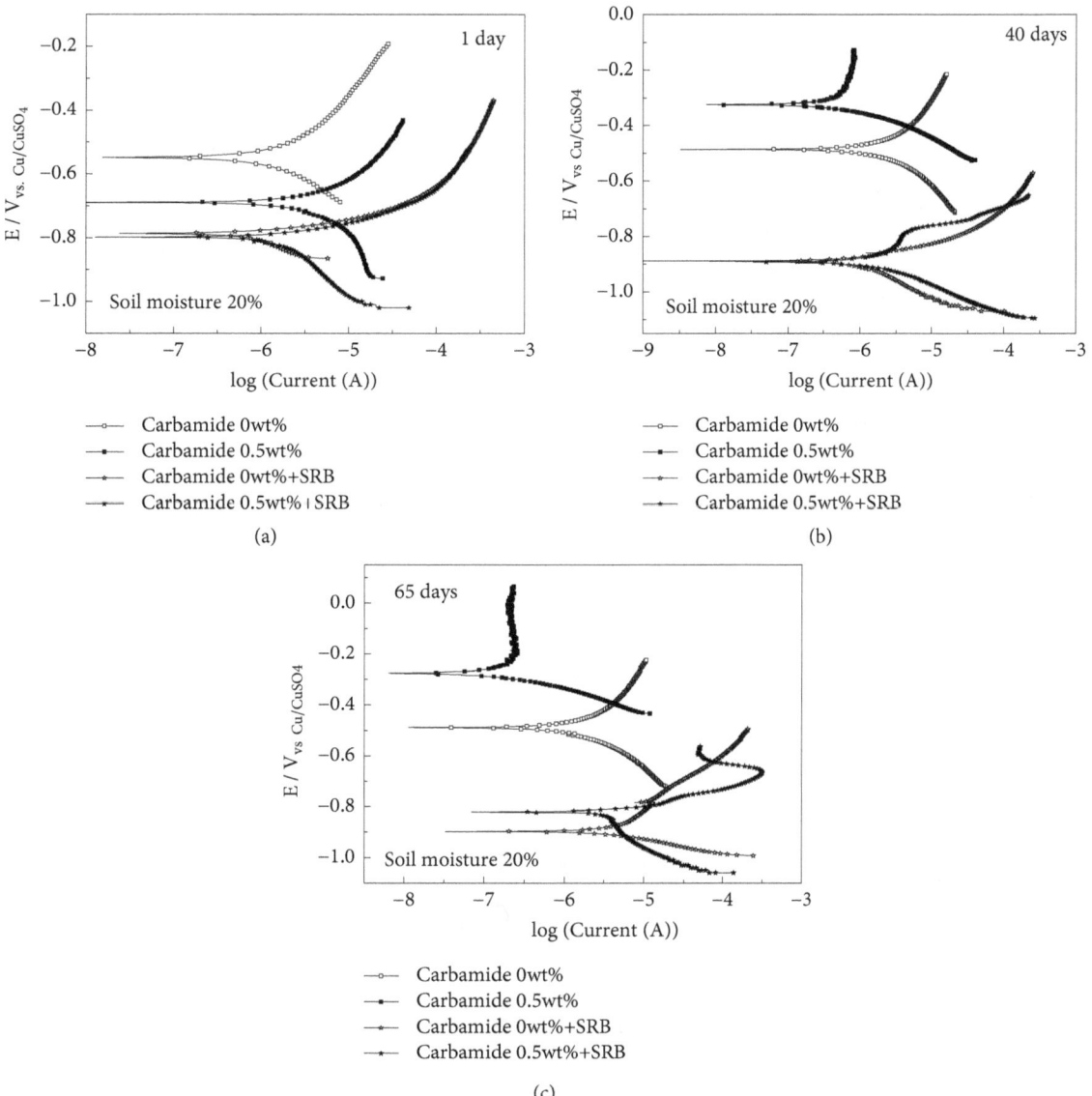

FIGURE 5: Polarization curves of Q235 steel in soil with moisture of 20% after 1 day (a), 40 days (b), and 65 days (c), respectively.

rate of Q235 steel was enhanced in the death phase of SRB [27]. In soil containing 0.5 wt% carbamide, there appeared passivation but soon activation on the anodic polarization plot, which was due to the formation and fracture of mixed products on the steel surface. The cathodic curves became flat, showing that the reduction of corrosion products was prevented.

After long term experiments (see 65 days), in soil without SRB, the anode Tafel slope was dramatically larger than the cathode Tafel slope, which indicates that the corrosion rate was controlled by the anode process [28]. This is probably because the corrosion products prevent the dissolution of the iron. In soil moisture of 20% current rate was declined, and in soil moisture of 30% current rate instead increased slightly possibly because of the rupture of corrosion products. The current rate in soil containing 0.5 wt% carbamide was

dramatically smaller than that in soil without carbamide. In soils with SRB, the anode process was passivated due to its corrosion products on the surface. But still corrosion rate was larger than that in soil without SRB, which indicates that SRB and its metabolite promoted corrosion of Q235 steel.

3.5. Analysis of the Electrochemical Impedance Spectroscopy (EIS). EIS was one of the effective means to investigate the electrochemical properties of the corroded surface of carbon steel in soils. However, soil environment is very complicated, and corrosion products of carbon steel usually combine with soil particles closely, which make EIS plots hard to analyze and various analysis methods have been proposed [29, 30]. To be sample, in the Bode plots, the impedance modulus at the high/medium frequencies reflects the resistance of all products on the steel surface, and that at the low frequencies

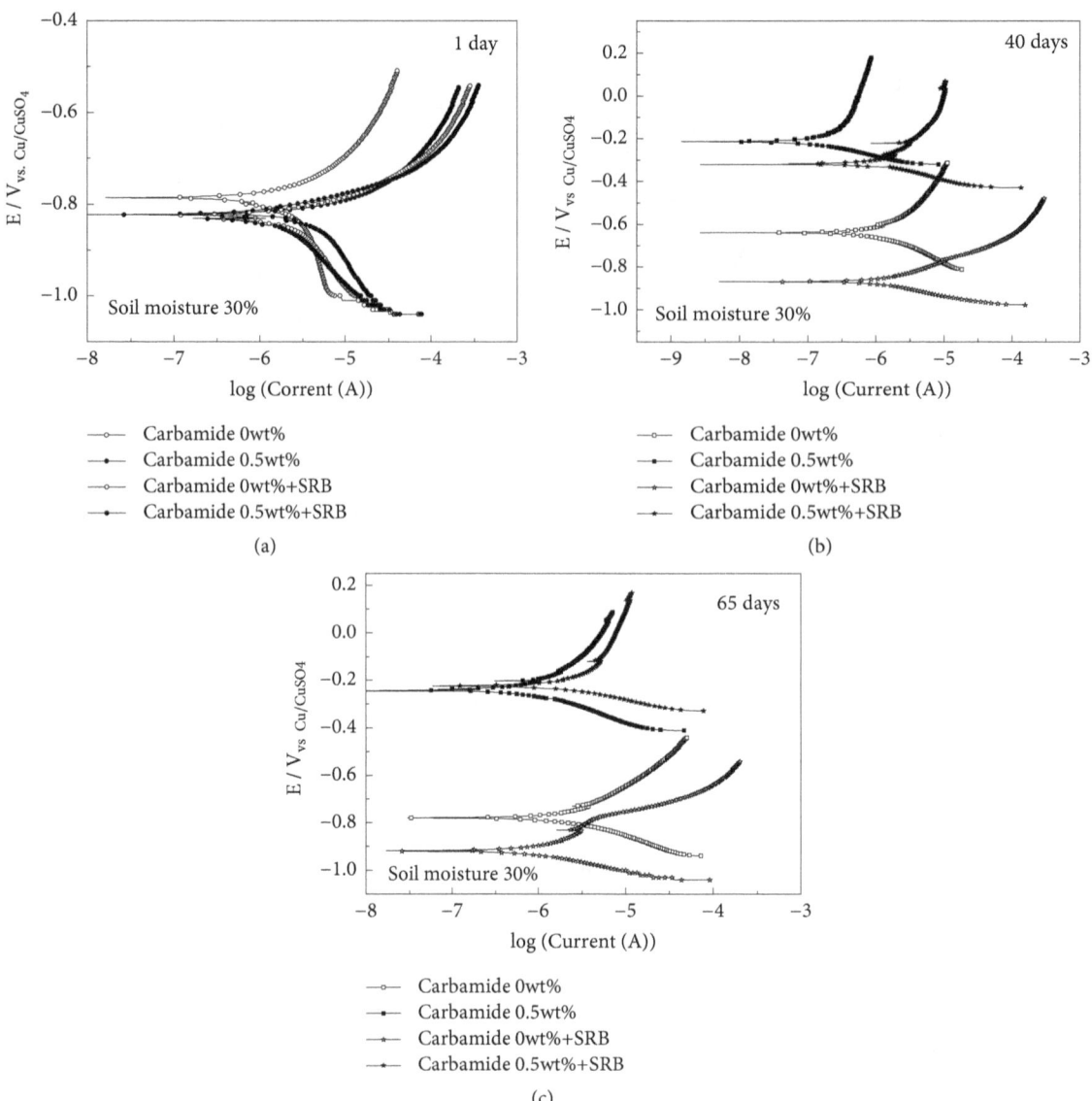

FIGURE 6: Polarization curves of Q235 steel in soil with moisture of 30% after 1 day (a), 40 days (b), and 65 days (c), respectively.

reflects the charge transfer resistance related to the iron dissolution reaction. Figures 7 and 8 show EIS Bode plots of the carbon steel Q235 in soils with moisture 20% and 30% after 3 days, 30 days, and 65 days, respectively.

After 3 days, there was few corrosion products on the steel surface, so the resistance at high/medium frequency was generally low in all experiment groups. In soil with moisture of 20%, the resistance at low frequency in soils containing 0.5 wt% carbamide was smaller than that in soil without carbamide, which was due to the same reason as before. Carbamide was adsorbed on the steel surface to form an incomplete film, which promoted the charge transfer. In soils with SRB, at the initial period carbamide and SRB were both adsorbed on the steel surface to form a protective film, so the resistance at the low frequency in soil added carbamide was larger than that in soil without carbamide, in accordance with the results of polarization curves. In soil with moisture of

30%, whether in soils with and without SRB, charge transfer resistance in soil containing carbamide was lower than that in soil without carbamide, indicating carbamide inhibited the corrosion reaction. Besides, the impedance at low frequency was smaller than that in soil with moisture of 20%.

After 30 days, there were corrosion products on the steel surface; thus the resistance at the high/medium frequencies increased. One or two time constants were found in the plots, which was dependent on the combination of corrosion products on the sample surface, which can be confirmed by SEM and EDS data, shown in Figure 11. But the corrosion products combined with soil grains were easily broken off, which shows that the resistance at the high frequency cannot reflect the real resistance of corrosion process. The resistance at the low frequency instead shows the corrosion situation more accurately. In soil with moisture of 20%, the resistance at the low frequency increased in soils without SRB and

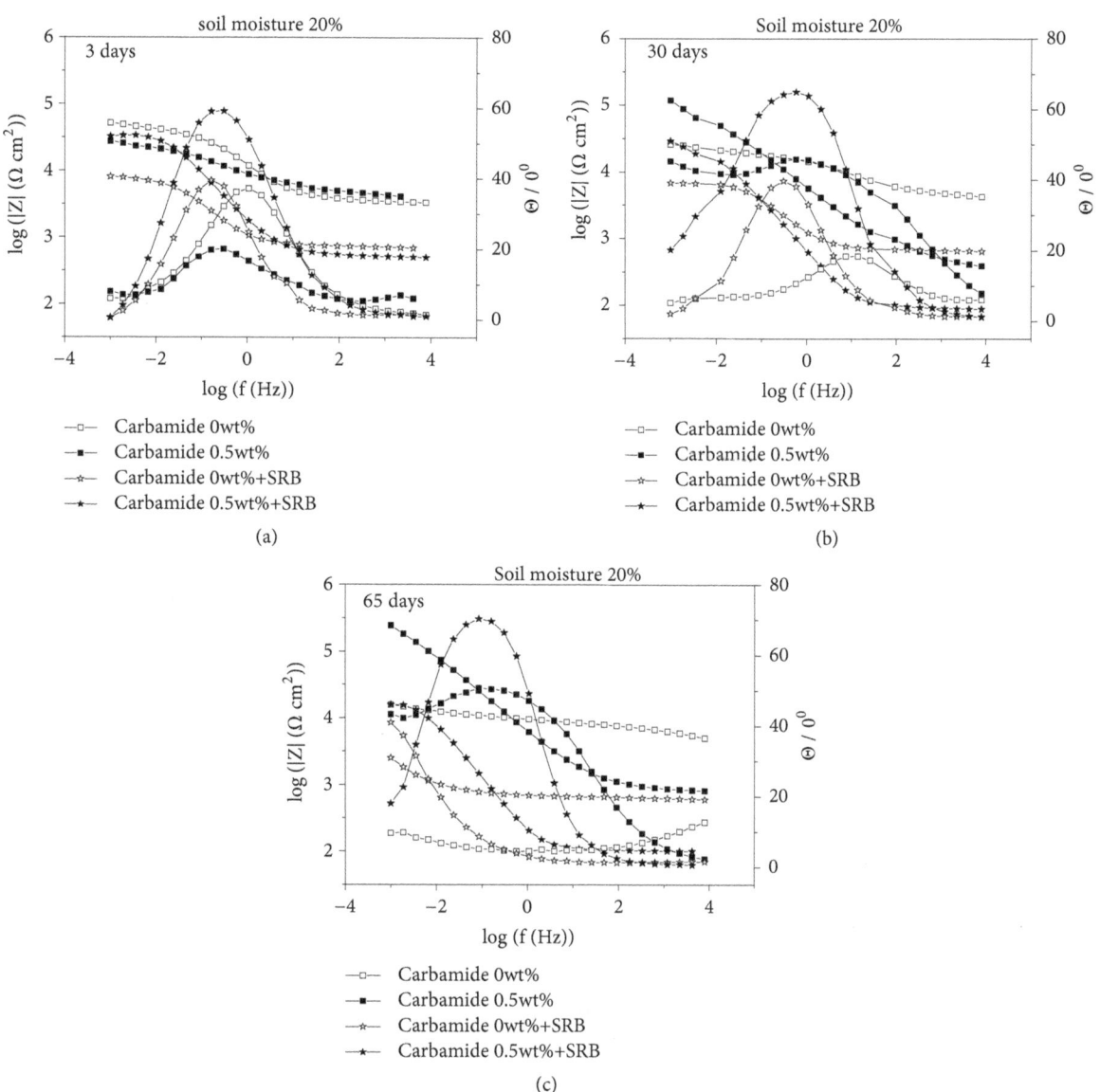

FIGURE 7: EIS plots of the carbon steel Q235 in soils with moisture of 20 wt% after 3 days (a), 30 days (b), and 65 days (c).

decreased in soils with SRB which was in accordance with the results of polarization curves. In soil moisture of 30%, the resistance at the low frequency increased due to the corrosion products on the steel surface. The charge transfer resistance in soil added carbamide was larger than that in soil with no carbamide, which indicates that carbamide could inhibit the further corrosion in soil. Over time, there were more and more corrosion products on the surface of carbon steel and then the diffusion of corrosion products became more difficult. Hence, after 65 days, there was Warburg impedance in EIS plot, which indicates that the process was controlled by concentration polarization [31].

In addition, electric equivalent circuits (EEC) models were selected to fit EIS data of carbon steel in soils with moisture 20 wt%, shown in Figure 9. R_s represents the soil electrolyte resistance, and R_f represents all the resistance

of products on the surface of sample. R_t and Q_{dl} represent the charge transfer resistance and the double layer capacitance, and W represents the Warburg impedance. Then the corrosion resistance R_{cr} was calculated to be the sum of R_f, R_t, and diffusion resistance W_R, to check the evolution of carbon steel in different soils over time. The SEM and EDS data were drawn to show the surface morphologies of carbon steel in soils with moisture 20 wt% (Figure 11). It can be seen from Figure 10 that initially R_{cr} was similar in all soils except in soils with no carbamide but with SRB. In soils with no carbamide but with SRB, R_{cr} was higher, in agreement with the polarization test. Over time, the R_{cr} value of carbon steel in soils with SRB decreased, and localized corrosion was overserved on surface (Figure 11(a)). At 15 days, in soils with SRB, localized corrosion occurred, while the element S was detected at the location where

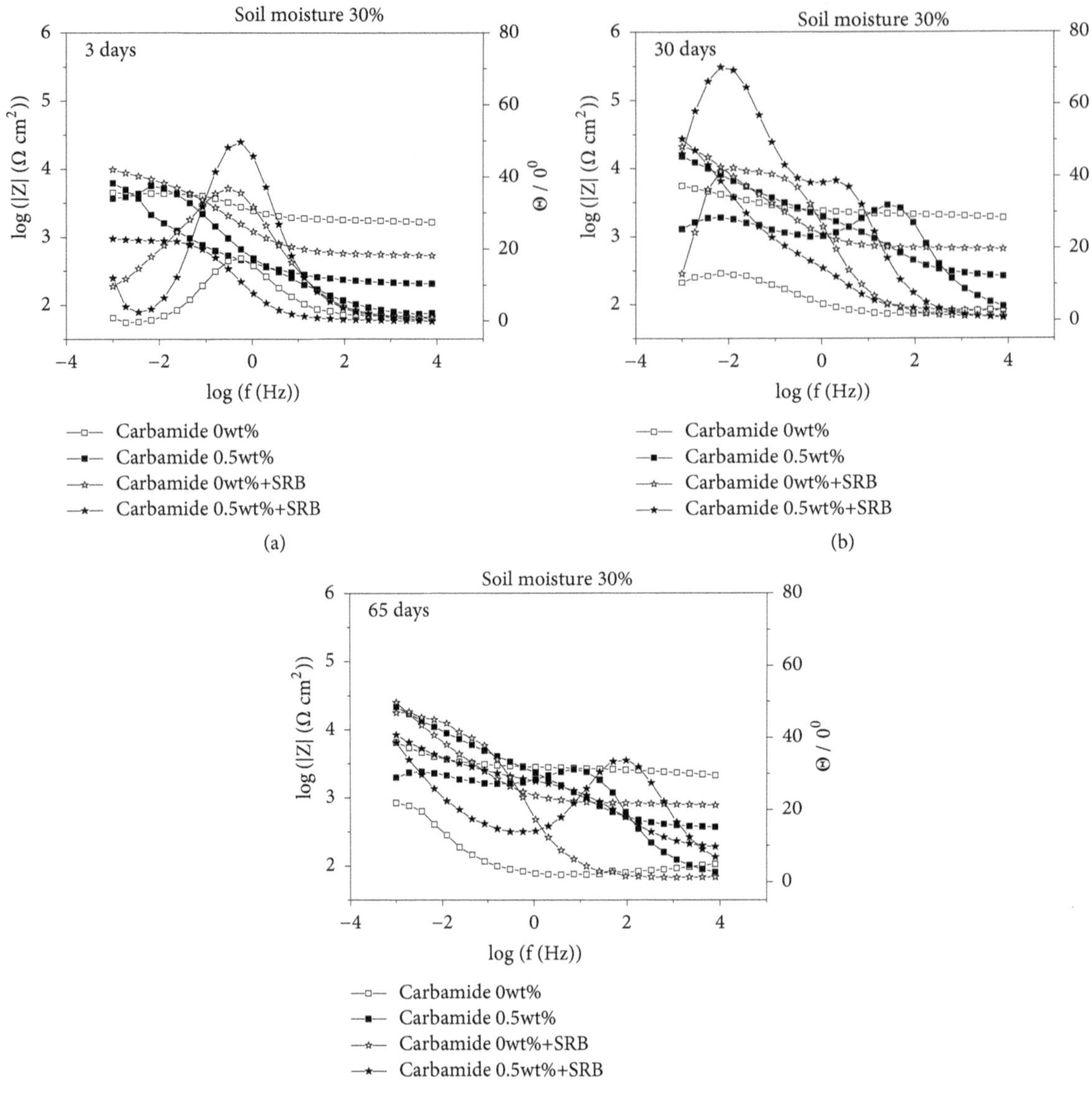

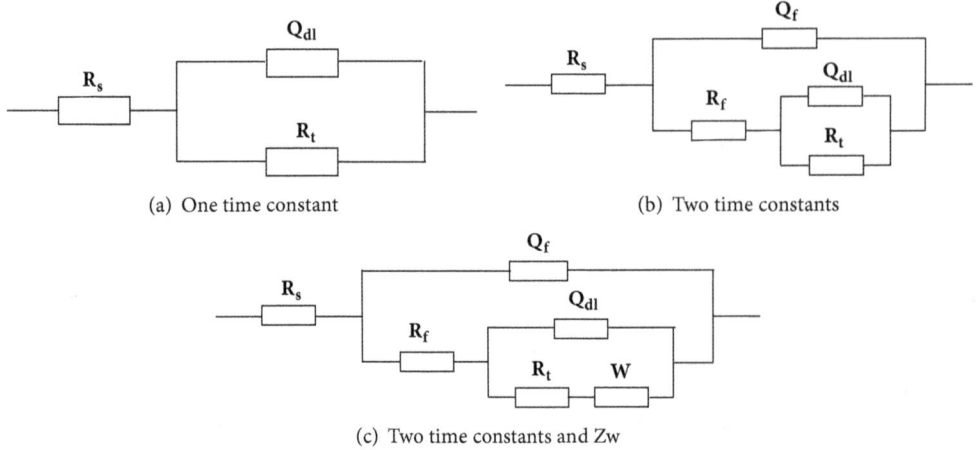

FIGURE 8: EIS plots of the carbon steel Q235 in soils with moisture of 30 wt% after 3 days (a), 30 days (b), and 65 days (c).

(a) One time constant

(b) Two time constants

(c) Two time constants and Zw

FIGURE 9: Equivalent circuits of the different time constants.

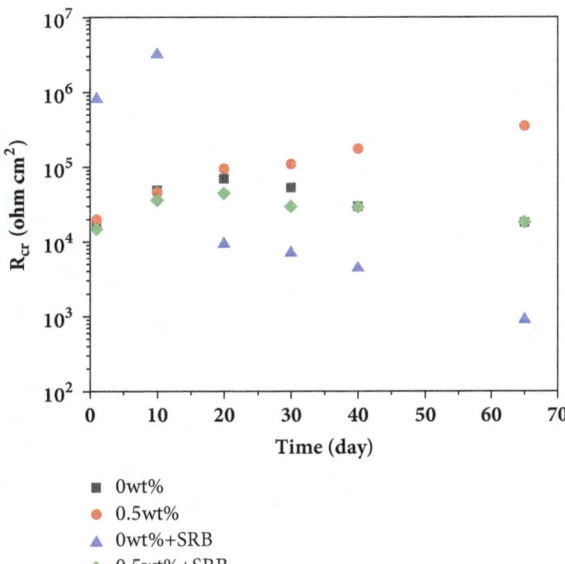

FIGURE 10: EIS fitting parameter R_t changes over time in different systems.

SRB was detected (Figure 11(b)). After 65 days, localized corrosion was also detected on the sample surface and SRB metabolic products combined with corrosion products and soil particles were observed (Figures 11(c) and 11(d)). R_{cr} of carbon steel in soils with carbamide was higher than that without carbamide, indicating carbamide inhibited corrosion reaction of iron. After 65 days, a thick layer of corrosion products combined with steel oxides was observed (Figure 11(e)).

3.6. Average Corrosion Rate. Average corrosion rates (Figure 12) were calculated from the weight losses of the samples. In soils without SRB, 0.5 wt% carbamide remarkably inhibited soil corrosion, while in soil with SRB, the addition of carbamide also decreased average corrosion rate of Q235 steel but the reason is different from the previous case. Besides, corrosion rate in soil with SRB was much larger than that in soils without SRB, which further proved that SRB accelerated soil corrosion. The effect of water content in soil on corrosion rate was also shown. In soil without SRB, corrosion rate was lower in moisture of 30%, as the saturated water content allowed even less dissolved oxygen and then prevented cathodic polarization, with corrosion rate down. In soil with SRB, corrosion rate was also smaller in moisture of 30% because SRB are able to survive or even take advantage of the presence of molecule oxygen [32]. The surface morphologies are shown in Figures 13 and 14. To check this effect, different concentrations of carbamide were added in soil with and without SRB and weight-loss tests were done to measure the corrosion rate. The results (Figure 15) show that in soil without SRB carbamide inhibits corrosion and the inhibition effect increases with increasing carbamide concentration. In soils with SRB, 10% humidity promotes the corrosion; the effect increases with the increase of the carbamide concentration [23]. Meanwhile,

in soils with 20% moisture, carbamide promotes corrosion but the effect decreased with the carbamide concentration increasing. With 0.5 wt% of carbamide, it inhibited biocorrosion of carbon steel. In soils with 30% moisture, the inhibition of corrosion increases with increasing carbamide concentration. Hence, carbamide can be used as an effect inhibitor in soils for both common corrosion and biocorrosion.

4. Conclusions

Carbamide decreased soil redox potential, and so did SRB. Some carbamide was decomposed into CO_2 and NH_3, which increased the pH value. SRB decreased soil pH due to its metabolic products. But the effect of carbamide took a bigger part. In soil without SRB, carbamide induced the corrosion potential of Q235 steel which was much more positive and inhibited its corrosion. In inoculated soil, excessive carbamide retrained the growth of SRB and inhibited biocorrosion of Q235 steel. In different water content, there was less oxygen in soil with the moisture of 30%, so corrosion rate declined in both soils with and without SRB. Carbamide can be a potential inhibitor for soil corrosion and biocorrosion.

Conflicts of Interest

The authors declare that they have no conflicts of interest.

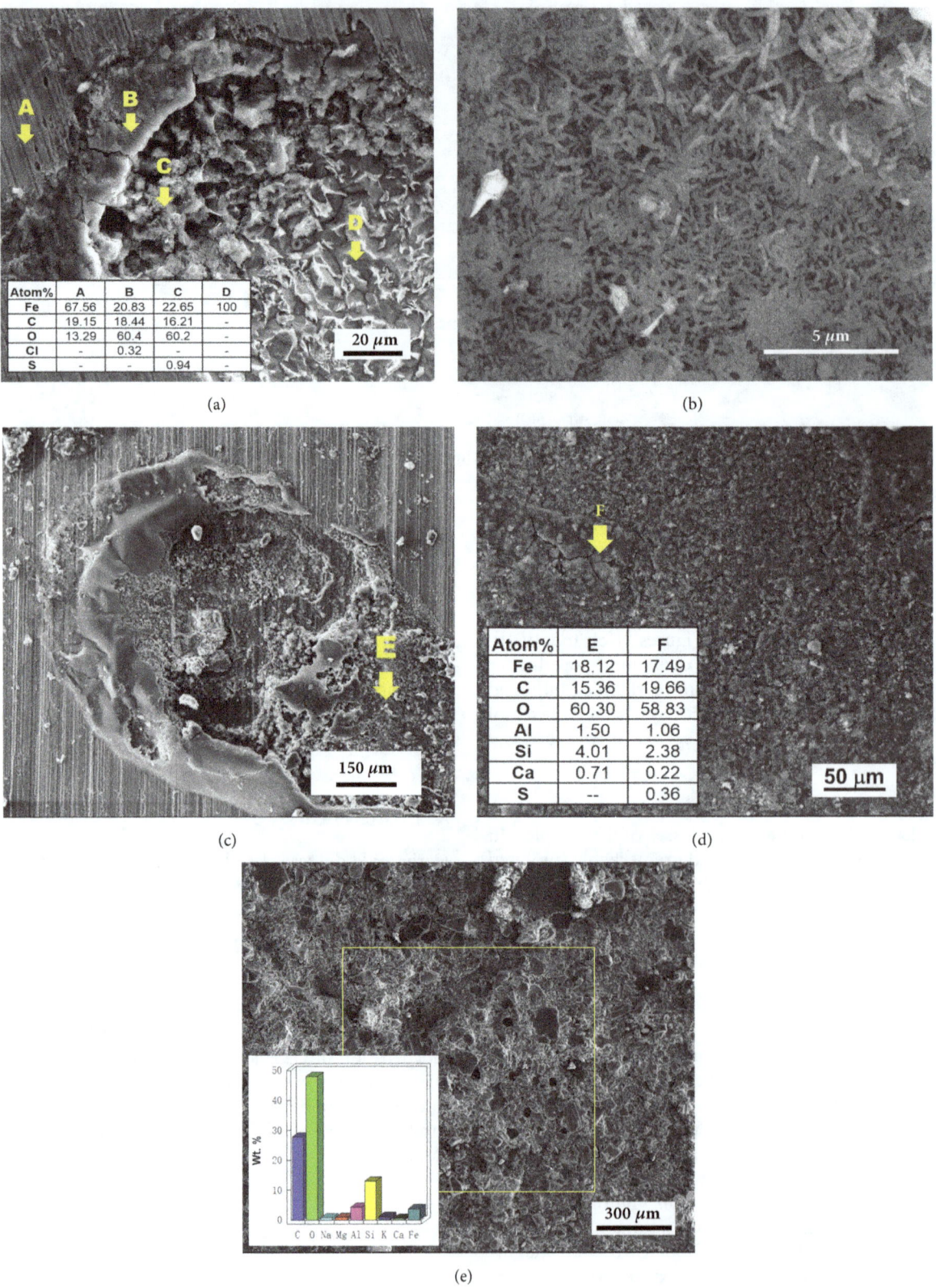

FIGURE 11: SEM and EDS data of specimen in soils with moisture of 20 wt%: (a and b) 0 wt% carbamide+SRB @15 days; (c and d) 0 wt% carbamide+SRB @ 65 days; (e) 0.5 wt% carbamide+SRB @ 65 days.

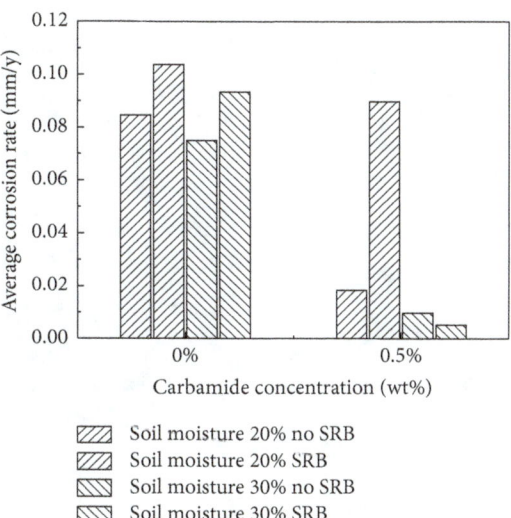

FIGURE 12: Average corrosion rate (mm/y) of Q235 steel in soil after 65 days.

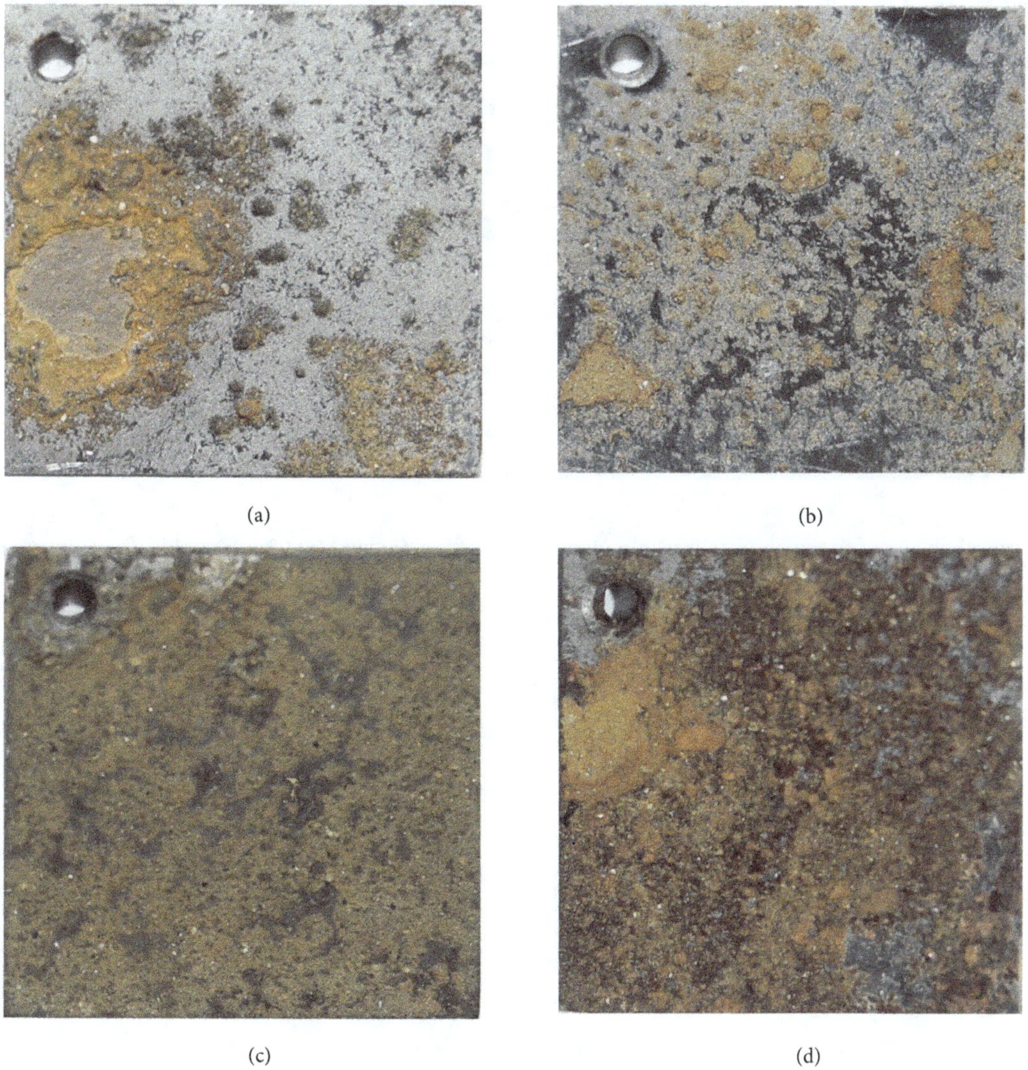

(a)

(b)

(c)

(d)

FIGURE 13: Surface morphologies of specimen in soils with moisture of 20% after 65 days. (a) 0% carbamide, (b) 0.5% carbamide, (c) 0% carbamide+SRB, and (d) 0.5% carbamide+SRB.

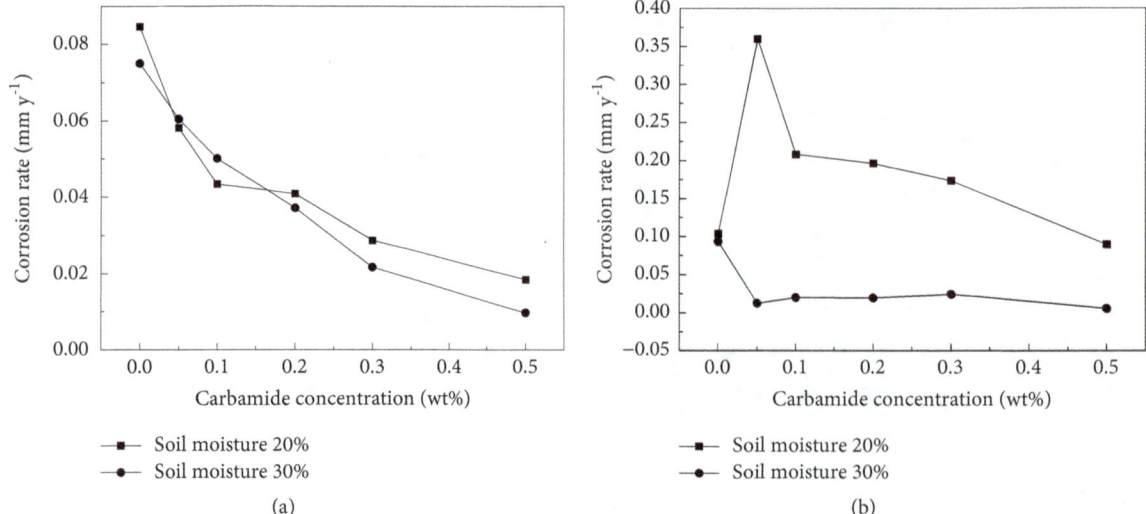

FIGURE 14: Surface morphologies of specimen in soils with moisture of 30% after 65 days. (a) 0% carbamide, (b) 0.5% carbamide, (c) 0% carbamide+SRB, and (d) 0.5% carbamide+SRB.

FIGURE 15: Average corrosion rate of carbon steel in soil without SRB (a) and with SRB (b) with carbamide concentration.

References

[1] G. H. Booth, A. W. Cooper, P. M. Cooper, and D. S. Wakerley, "Criteria of soil aggressiveness towards buried metals. I. experimental methods," *British Corrosion Journal*, vol. 2, no. 3, pp. 104–108, 1967.

[2] I. Matsushima, "Carbon steel-corrosion by soils," in *Uhlig's Corrosion Handbook*, pp. 609–613, 3rd edition, 2011.

[3] F. M. Sani, A. Afshar, and M. Mohammadi, "Evaluation of the simultaneous effects of sulfate reducing bacteria, soil type and moisture content on corrosion behavior of buried carbon steel API 5L X65," *International Journal of Electrochemical Science*, vol. 11, no. 5, pp. 3887–3907, 2016.

[4] X. Li and H. Castaneda, "Influence of soil parameters on coating damage evolution of X52 pipeline steel under cathodic protection conditions," in *Corrosion*, NACE International, 2014.

[5] M. Yan, C. Sun, J. Xu, and W. Ke, "Electrochemical behavior of API X80 steel in acidic soils from Southeast China," *International Journal of Electrochemical Science*, vol. 10, no. 2, pp. 1762–1776, 2015.

[6] S. Cheng, "Effect of urea on microbiologically induced corrosion of carbon steel in soil," *Acta Physico-Chimica Sinca*, vol. 28, no. 11, pp. 2659–2668, 2012.

[7] R. Akkouche, C. Rémazeilles, M. Jeannin, M. Barbalat, R. Sabot, and P. Refait, "Influence of soil moisture on the corrosion processes of carbon steel in artificial soil: active area and differential aeration cells," *Electrochimica Acta*, vol. 213, pp. 698–708, 2016.

[8] F. Qin, "Effect of soil moisture content on corrosion behavior of× 70 steel," *International Journal of Electrochemical Science*, vol. 13, pp. 1603–1613, 2018.

[9] H. H. Collins, "Protection of spun ductile iron pipes against corrosion by soils," in *Proceedings of the 5th International Conference on the Internal External Protection of Pipes*, 1983.

[10] H. Liu and Y. Frank Cheng, "Mechanism of microbiologically influenced corrosion of X52 pipeline steel in a wet soil containing sulfate-reduced bacteria," *Electrochimica Acta*, vol. 253, pp. 368–378, 2017.

[11] R. Javaherdashti, *Microbiologically Influenced Corrosion: An Engineering Insight*, Springer International Publishing, 2016.

[12] A. I. M. Ismail and A. M. El-Shamy, "Engineering behaviour of soil materials on the corrosion of mild steel," *Applied Clay Science*, vol. 42, no. 3-4, pp. 356–362, 2009.

[13] B. Little, P. Wagner, and F. Mansfeld, "An overview of microbiologically influenced corrosion," *Electrochimica Acta*, vol. 37, no. 12, pp. 2185–2194, 1992.

[14] S. Y. Li, Y. G. Kim, K. S. Jeon, Y. T. Kho, and T. Kang, "Microbiologically influenced corrosion of carbon steel exposed to anaerobic soil," *Corrosion*, vol. 57, no. 9, pp. 815–828, 2001.

[15] D. Wang, F. Xie, X. Li, X. Wang, J. Liu, and M. Wu, "Effect of interfacial film on the corrosion behaviour of X80 pipeline steel in a neutral soil environment containing sulphate-reducing bacteria," *Corrosion Reviews*, vol. 35, no. 6, pp. 445–453, 2017.

[16] H. Liu and Y. F. Cheng, "Microbial corrosion of X52 pipeline steel under soil with varied thicknesses soaked with a simulated soil solution containing sulfate-reducing bacteria and the associated galvanic coupling effect," *Electrochimica Acta*, vol. 266, pp. 312–325, 2018.

[17] N. N. Glazov, S. M. Ukhlovtsev, I. I. Reformatskaya, A. N. Podobaev, and I. I. Ashcheulova, "Corrosion of carbon steel in soils of varying moisture content," *Protection of Metals*, vol. 42, no. 6, pp. 601–608, 2006.

[18] X. M. Li, O. Rosas, and H. Castaneda, "Deterministic modeling of API5L X52 steel in a coal-tar-coating/cathodic-protection system in soil," *International Journal of Pressure Vessels and Piping*, vol. 146, pp. 161–170, 2016.

[19] C. Sun, L. Xi-Ming, and X. Jin, "Effect of urea on microbiologically induced corrosion of carbon steel in soil," *Acta Physico-Chemica Sinica*, vol. 28, no. 11, pp. 2659–2668, 2012.

[20] J. Xu, K. Wang, C. Sun et al., "The effects of sulfate reducing bacteria on corrosion of carbon steel Q235 under simulated disbonded coating by using electrochemical impedance spectroscopy," *Corrosion Science*, vol. 53, no. 4, pp. 1554–1562, 2011.

[21] F. F. Mozheiko, T. N. Potkina, and I. I. Goncharik, "Effect of inhibitors on corrosion resistance of carbon steel in suspensed liquid combined fertilizer," *Russian Journal of Applied Chemistry*, vol. 81, no. 9, pp. 1705–1709, 2008.

[22] S. Cheng et al., "Effect of urea on microbiologically induced corrosion of carbon steel in soil," *Acta Physico-Chemica Sinica*, vol. 28, no. 11, pp. 2659–2668, 2012.

[23] N. S. C. T. N. Stations, Ed., *Test Methods of Soil Corrosion of Materials*, Science Press, Beijing, China, 1990.

[24] H. Castaneda and X. D. Benetton, "SRB-biofilm influence in active corrosion sites formed at the steel-electrolyte interface when exposed to artificial seawater conditions," *Corrosion Science*, vol. 50, no. 4, pp. 1169–1183, 2008.

[25] C. Sun, J. Xu, F. H. Wang, and C. K. Yu, "Effect of sulfate reducing bacteria on corrosion of stainless steel 1Cr18Ni9Ti in soils containing chloride ions," *Materials Chemistry and Physics*, vol. 126, no. 1-2, pp. 330–336, 2011.

[26] C. N. Standards, "Examination of bacteria and algae in industrial circulating cooling water," in *Examination of Soil Fangi - Standard of Plate Count*, part 4, 2009.

[27] W. P. Iverson, "Direct evidence for the cathodic depolarization theory of bacterial corrosion," *Science*, vol. 151, no. 3713, pp. 986–988, 1966.

[28] B. M. Wei, Ed., *The Corrosion Theory and Application of Metals*, Chemical Industry Press, Beijing, China, 2004.

[29] X. Li and H. Castaneda, "Coating studies of buried pipe in soil by novel approach of electrochemical impedance spectroscopy at wide frequency domain," *Corrosion Engineering, Science and Technology*, vol. 50, no. 3, pp. 218–225, 2015.

[30] X. Li and H. Castaneda, "Damage evolution of coated steel pipe under cathodic-protection in soil," *Anti-Corrosion Methods and Materials*, vol. 64, no. 1, pp. 118–126, 2017.

[31] C. N. Cao, Ed., *Principles of Electrochemistry of Corrosion*, Chemical Industry Press, Beijing, China, 2008.

[32] W. Dilling and H. Cypionka, "Aerobic respiration in sulfate-reducing bacteria," *FEMS Microbiology Letters*, vol. 71, no. 1-2, pp. 123–128, 1990.

18

Experimental Study on the Influence of Sulfate Reducing Bacteria on the Metallic Corrosion Behavior under Disbonded Coating

Qingmiao Ding,[1] Liping Fang,[2] Yanyu Cui,[1] and Yujun Wang[1]

[1]*Airport School, Civil Aviation University of China, Tianjin, China*
[2]*Guangxi Colleges and Universities Key Laboratory of Beibu Gulf Oil and Natural Gas Resource Effective Utilization, Qinzhou University, Qinzhou 535011, China*

Correspondence should be addressed to Qingmiao Ding; qmding@cauc.edu.cn

Academic Editor: Flavio Deflorian

A rectangle disbonded coating simulation device was used to research the effect of sulfate reducing bacteria (SRB) on the metallic corrosion behavior under disbonded coating by the electrochemical method. The results showed that the metal self-corrosion potential at the same test point had little change in the initial experiment stage, whether the solution was without or with SRB. The potential amplitude in the solution with SRB was larger than that without SRB in the later corrosion period. The corrosion current density of the metal at the same test point increased gradually over time in the solution with or without SRB, and SRB could accelerate the corrosion of the metal in the disbonded crevice. The metal self-corrosion potential in the crevice had little change in the SRB solution environment after adding the fungicide, but the corrosion current density decreased significantly. That meant the growth and reproduction of SRB were inhibited after adding the fungicide, so the metal corrosion rate slowed down. Among the three kinds of solution environment, increasing the coating disbonded thickness could accelerate the corrosion of the metal in the crevice, and it was the largest in the solution with SRB.

1. Introduction

With the rapid development of buried pipeline construction, more and more attention had been paid to the problem of pipeline corrosion [1–5]. The buried pipelines were arranged in a crisscross pattern, and this would cause fire and explosion once they start to corrode and leak; that would threaten the personal safety and cause environmental pollution. The anticorrosion coating of the buried pipeline often had disbonded area with broken open holiday due to the mechanical damage, aging and degradation, cathodic disbondment, and other reasons in the process of pipeline installation and using [6–8]. It formed the special corrosive environment between the anticorrosion coating and pipeline surface, and the moisture, the soluble ion like CO_2 and O_2 and other corrosive media in the soil got into the disbonded area through the holiday; that caused the corrosion of the metal

under the disbonded coating [9–11]. A large number of studies showed that the content of moisture and bacteria in the backfill soil was more than that in the original uncultivated soil, and that would cause microbial corrosion (MIC) due to microbial activity under disbonded coating [12–14]. The sulfate reducing bacteria (SRB) as a kind of microorganism widely existed in soil, sea, river water, underground pipeline, oil gas well, and so on, and they made the pipeline more prone to perforation in the solution with SRB. So the metal corrosion caused by SRB in recent years was taking more and more attention; there have been massive researches about SRB corrosion on emergence [15–18] and some reporting about metal corrosion under disbanded coating [19, 20], but up to now, there was no study on the SRB corrosion behavior under disbonded coating. Therefore, we built a rectangle crevice device for the metal corrosion under disbonded coatings and researched the regularity of SRB corrosion on

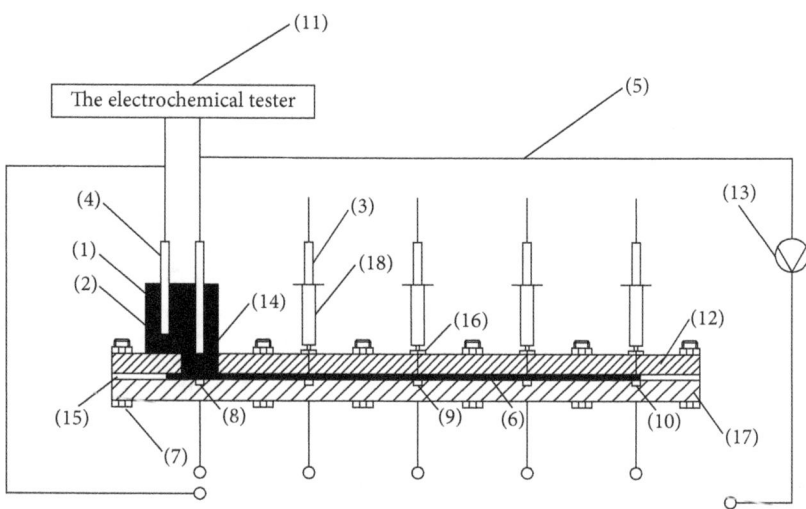

FIGURE 1: The experimental device. (1) Organic glass box; (2) soil simulated solution; (3) the saturated calomel electrode with salt-bridge; (4) auxiliary anode; (5) wire; (6) gap space; (7) bolt; (8) working electrode at test point 1; (9) working electrode at test point 2; (10) working electrode at test point 3; (11) electrochemical tester; (12) organic glass plate; (13) voltmeter; (14) the point of coating damage; (15) the spacer; (16) rubber plug; (17) epoxy resin board; (18) needle tubing.

the metal under disbonded coatings by the electrochemical method. It is meaningful to the safe operation and long-term service of the pipeline in the soil environment.

2. Experiment Content

2.1. Experimental Device. Experiment device was shown in Figure 1; we use the epoxy resin plate and organic glass plate of 30 cm × 10 cm × 0.5 cm while simulating the situation of coating disbonded, also we put a gasket of 0.5 mm thickness between them for regulating the thickness of simulation stripping gap. We drill a round hole of 1 cm diameter as point 1 to simulate coating damage point on organic glass plate near the end of 7 cm and drill the round holes of 0.5 cm diameter at 80 mm and 160 mm from the damaged point, respectively, as the electrochemical parameters test point 2 and point 3. We set an organic glass box on point 1 of the organic glass plate to contain simulated soil solution and set the working electrodes, respectively, in the epoxy resin plate with points 1~3 in the organic glass plate while simulating the pipeline metal under the disbonded coating.

2.2. The Experimental Material. The working electrode was made from X80 steel, and its chemical composition was shown in Table 1. The sample was made of the small cylindrical specimens, 10 mm diameter, and the electrode surface was polished successively with 60#~2000# sandpaper by MP-2 metallographic sample pregrinding machine until the metal surface was smooth and there were no obvious signs; then we washed away the surface oil with acetone and washed away the surface residues water and acetone with anhydrous ethanol and then put it into the drying box for use. The reference electrode was saturated calomel electrode (in this paper, if there were no special instructions, the potential was defined relatively to the saturated calomel reference electrode). The auxiliary electrode was platinum electrode.

The composition of the solution was as follows: 0.1712 g Na_2SO_4, 0.1600 g Na_2CO_3, 0.0865 g $NaHCO_3$ 0.5125 g NaCl, and 1 L distilled water.

SRB was selected as the experimental bacterial strain. The medium composition of SRB strains was as follows: 0.5 g KH_2PO_4, 0.06 g $CaCl_2$, 0.06 g $MgSO_4 \cdot 7H_2O$, 1.0 g NH_4Cl, 0.3 g $C_6H_5Na_3O_7 \cdot 2H_2O$, 6.0 g $CH_3CH(OH)COONa$, 1.0 g yeast extract powder, and 1 L distilled water. We put the ready-prepared medium in the high temperature steam sterilization pot (120°C) to sterilize 30 min before starting the experiment. We use glutaraldehyde ($C_5H_8O_2$) as the fungicide.

2.3. The Experiment Content. The rectangle crevice device was used to study the SRB corrosion behavior of the metal in the crevice under the disbonded thickness of 0.5 mm and 1.0 mm, respectively. The experimental period was 14 days, and we test the electrochemical impedance spectroscopy (EIS) and polarization curves by the electrochemical workstation in the test period.

3. The Experimental Results and Discussion

3.1. The SRB Corrosion Behavior of the Metal in Different Solution Environments. As shown in Figure 2 and Table 2, the self-corrosion potential of the metal in the crevice at the same test point had a little change in the solution without or with SRB in the initial stage when the disbonded thickness was 0.5 mm, and it tended to be stable. But there was obvious potential amplitude of self-corrosion potential in the later stage, and the amplitude affected with SRB was much larger than that without SRB. The self-corrosion potential of the metal without SRB was more negative than that with SRB. With the extension of time, the corrosion current density of the metal in the crevice gradually increased at the same test point in two solutions, and the corrosion current density of the metal affected with SRB was significantly greater than that

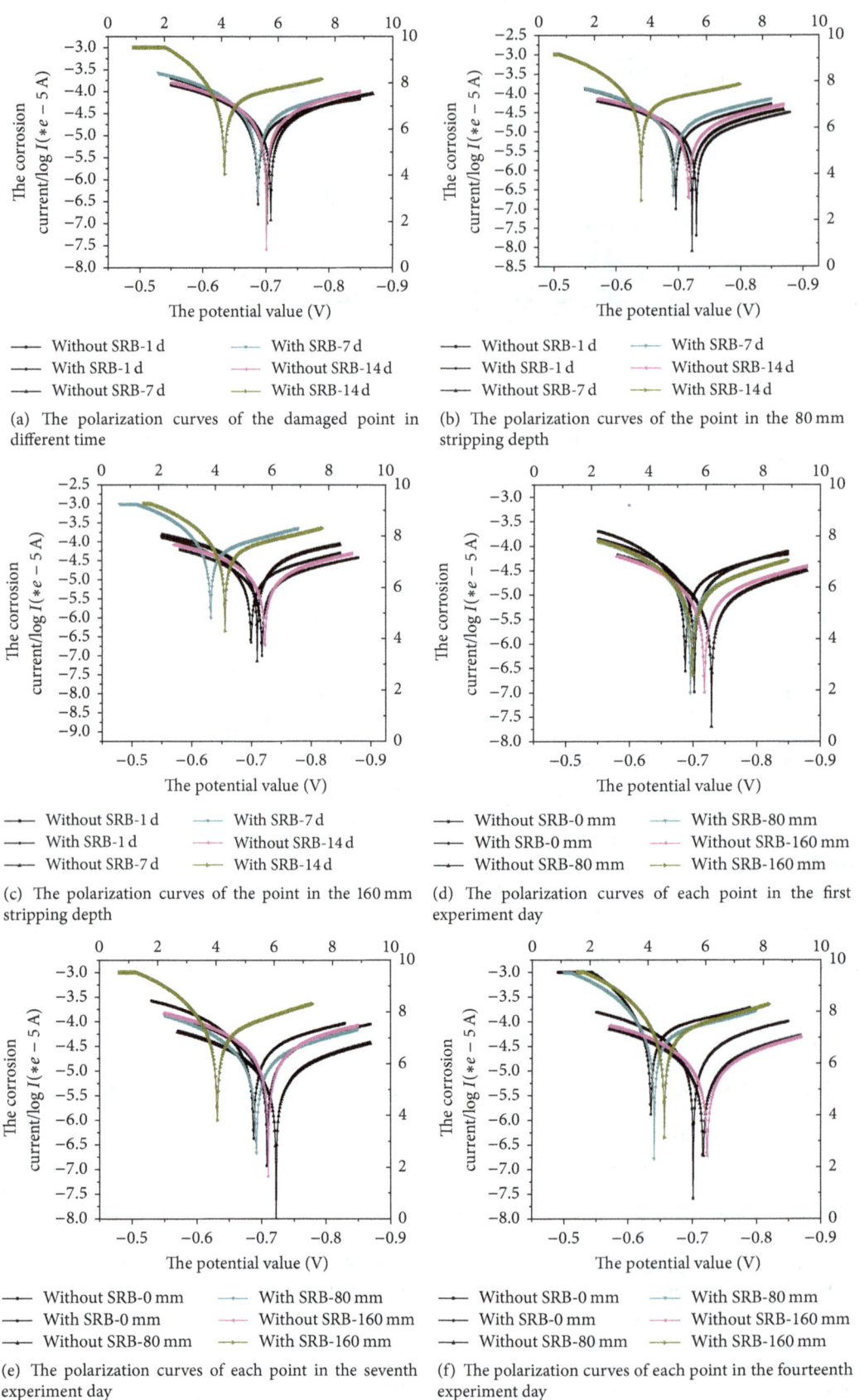

(a) The polarization curves of the damaged point in different time

(b) The polarization curves of the point in the 80 mm stripping depth

(c) The polarization curves of the point in the 160 mm stripping depth

(d) The polarization curves of each point in the first experiment day

(e) The polarization curves of each point in the seventh experiment day

(f) The polarization curves of each point in the fourteenth experiment day

FIGURE 2: The polarization curves of the metal; the influence on the metal corrosion process with or without SRB.

TABLE 1: The chemical composition of X80 steel (%).

C	Si	Mn	Cr	Mo	Ni	Al	Cu	Nb	Ti	Pb	Fe
0.042	0.189	1.560	0.028	0.243	0.230	0.034	0.153	0.060	0.019	0.005	97.464

TABLE 2: The electrochemical parameters.

Simulated solution	Experiment time	Stripping depths/mm	The corrosion current density/μA/cm^2	The self-corrosion potential/mV	The ratio of anode Tafel slope and cathode Tafel slope
Soil simulated solution (No. 1)	1 d	0	23.1029	−702	1.7328
		80	19.2514	−722	1.6538
		160	11.5943	−718	1.5963
	7 d	0	43.7829	−708	1.5726
		80	30.3086	−722	1.5250
		160	26.0457	−725	1.4844
	14 d	0	45.2457	−701	1.4111
		80	32.3429	−717	1.4049
		160	27.7543	−723	1.4831
Soil simulated solution + SRB (No. 2)	1 d	0	25.8171	−688	1.8134
		80	20.2457	−696	1.6897
		160	19.5143	−700	1.6727
	7 d	0	30.9143	−688	1.7509
		80	28.4971	−692	1.7250
		160	21.7143	−695	1.6395
	14 d	0	78.5143	−675	1.7515
		80	87.6457	−680	1.2216
		160	95.4914	−676	1.0870
Soil simulated solution + SRB + fungicide (No. 3)	1 d	0	16.3886	−694	1.9859
		80	19.0400	−692	1.4779
		160	21.2457	−696	1.3274
	7 d	0	32.4343	−698	1.6243
		80	40.1143	−732	1.4063
		160	43.4286	−691	1.4588
	14 d	0	45.6114	−684	1.4157
		80	58.8800	−689	1.4475
		160	64.1714	−684	1.4206

without SRB; that meant SRB could accelerate the corrosion rate of the metal in the crevice. The amplification of the corrosion current density in the initial stage was less than that in the later stage in the solution environment with SRB. This was because the metal surface in the crevice would generate a complete microbial membrane in the solution with SRB in the initial stage of corrosion, and it provided protection for the metal in the crevice and inhibited the corrosion. But as time goes on, the corrosive material produced by SRB metabolism damaged the microbial membrane and made it rupture and fall off. At the same time, the depolarization role of SRB was gradually weakened; that made the corrosion rate of the metal in the crevice speed up. In a word, the corrosion current density of the metal in the crevice still gradually increased in the solution with SRB. At the same experimental time, the self-corrosion potential of the metal

in the crevice had negative trend whether the solution is with or without SRB as the deep crack extension. The corrosion current density decreased with the increase of the stripping depths in the simulated solution without SRB. But for the solution with SRB, the corrosion current density decreased with the increase of the stripping depths in the initial stage and midstage and increased with the increase of the stripping depths in the later stage of the experiment. In addition, because the ratio of anode Tafel slope and cathode Tafel slope was bigger than 1, we could see that when the disbonded thickness was 0.5 mm, the control factor was anode control.

As shown in Figure 3 and Table 2, when the disbonded thickness was 0.5 mm, the simulated solution with SRB was numbered as No. 2 and the simulated solution with SRB and fungicide was numbered as No. 3. In the No. 3 solution, the self-corrosion potential did not change much compared with

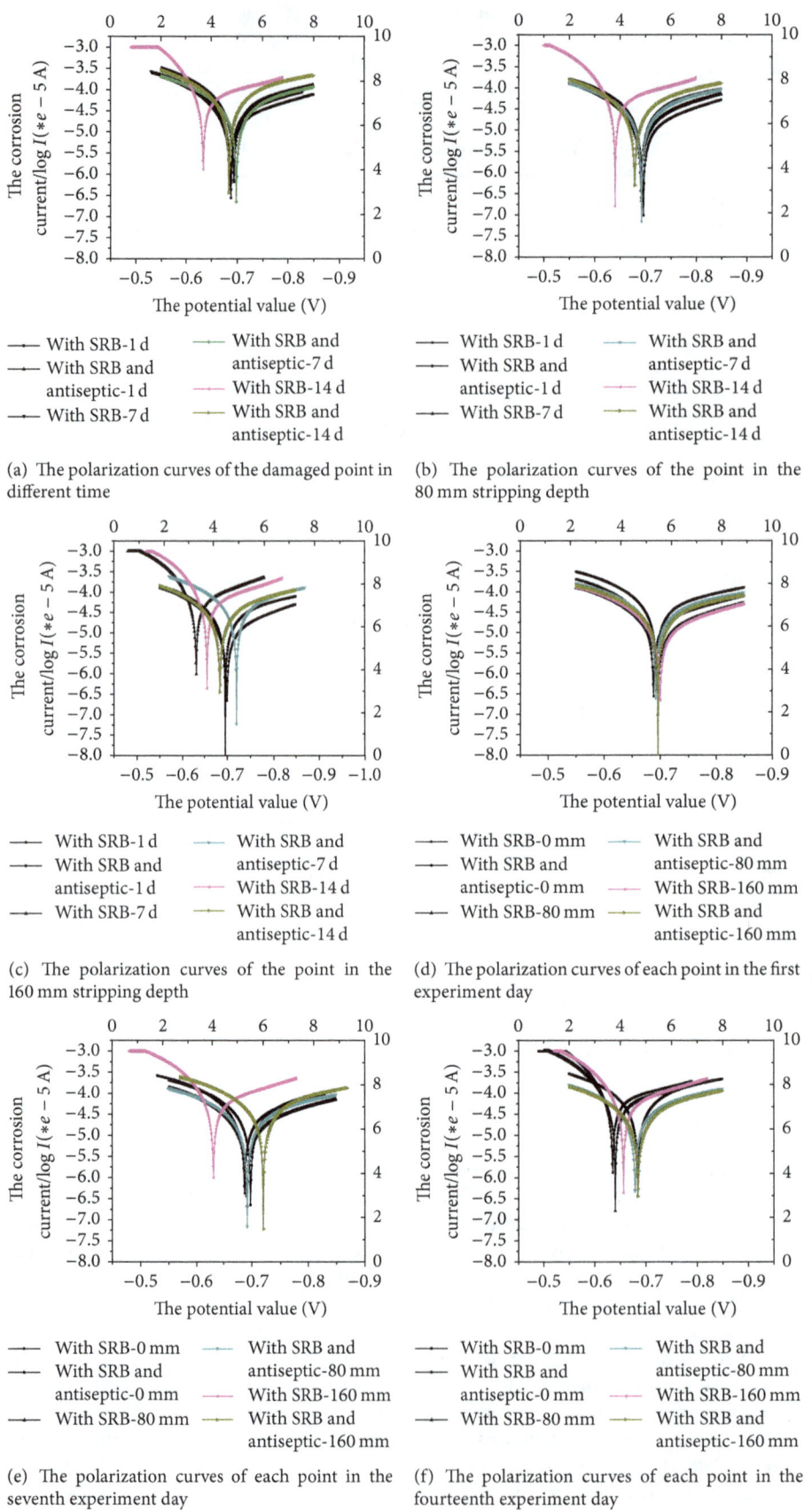

(a) The polarization curves of the damaged point in different time

(b) The polarization curves of the point in the 80 mm stripping depth

(c) The polarization curves of the point in the 160 mm stripping depth

(d) The polarization curves of each point in the first experiment day

(e) The polarization curves of each point in the seventh experiment day

(f) The polarization curves of each point in the fourteenth experiment day

FIGURE 3: The polarization curves of the influence on the metal corrosion process with or without fungicide.

TABLE 3: The polarization resistance under the disbonded coatings in different solution environment in the seventh experiment day/Ω.

Simulated solution	At the damage point/mm	At the point of 80 mm stripping depth/mm	At the point of 160 mm stripping depth/mm
Soil simulated solution	1426.4	3246.3	3231.6
Soil simulated solution + SRB	955.1	1632.7	1803.1
Soil simulated solution + SRB + Fungicide	919.3	1290.7	1408.1

that in the No. 2 solution, but the corrosion current density significantly decreased. That meant the growth and reproduction of SRB would be inhibited after adding the fungicide and then slowed down the corrosion rate of the metal in the crevice. As the experiment time goes on, the self-corrosion potential at the same test point shifted toward negative before shifting toward positive, and the corrosion current density increased. This was because fungicide, although inhibiting the growth and reproduction of SRB, meanwhile damaged the integrity of metal surface microbiofilm and made the microbial film protection reduce, so the corrosion speed was still very fast. At the same test time, the self-corrosion potential did not change much with the increase of the stripping depths, but the corrosion current density increased; that was different from the solution only with SRB. In addition, we also could see that when the disbonded thickness was 0.5 mm, the control factor of the corrosion in No. 3 solution was anode control in the crevice at different test point. During the whole experimental period, the change of this ratio presented the irregularity obviously; it was indicated that the motion of charged particles in the SRB solution was irregular after adding the fungicide, and the speed change was larger than before, so the conversion trend of anode control in the whole corrosion process was not obvious.

In order to accurately analyze the electrochemical impedance spectroscopy which was shown in Figure 4, the electrochemical impedance spectroscopy under different conditions was fitted through ZSimpWin software in this article. We use electrochemical equivalent circuit C (CR (CR)) comprised of resistance R and capacitance C to represent the electrode process. The equivalent circuit diagram was shown in Figure 5.

R_s was the solution resistance between the reference electrode and the working electrode in the crevice. C_f was the capacitor of adsorption film formed on the corrosion metal surface. R_f was the adsorption film resistance formed on corrosion metal surface due to the microbial attachment. C_d was the double layer capacitance between the metal surface and the electrolyte solution. R_p was the polarization resistance; it was associated with Faraday process and anodic reaction; thus it could reflect the corrosion well, so we would use R_p to characterize the corrosion rate.

As shown in Figure 5, when the disbonded thickness was 0.5 mm, the EIS of every point in the crevice in different simulated solution environment in the seventh experiment day was composed of single high-frequency capacitance and single low-frequency capacitance. The high-frequency section corresponded to the impedance signal of the corrosion product, and low-frequency section was the corrosion reaction of metal substrate. At the same stripping depths, the

capacitance arc radius of the metal in solution No. 2 was the minimum, and the capacitance in solution No. 3 came second; the capacitance of the solution without SRB was the maximum. That meant the corrosion resistance of the metal under the disbonded coating in the solution with SRB was the worst, and the corrosion was the most serious. In the solution environment with SRB, the fungicide is added, the corrosion resistance of the metal under the disbonded coating was enhanced, but the corrosion was still very serious, and the capacitance arc radius was far less than that in the solution without SRB. In addition, according to the polarization resistance of various points in the crevice in different solution environment from Table 3, we could see that the polarization resistance of the metal in the crevice in the solution with SRB was the minimum, the solution with SRB and the fungicide added came second, and they were far less than that without SRB. Therefore, the presence of SRB in the solution would reduce the resistance of the metal corrosion process in the disbonded crevice, which meant that SRB could accelerate the corrosion rate of the metal in the disbonded crevice. The results above were consistent with the conclusions obtained by the method of polarization curves analysis.

3.2. The Effect of SRB on Corrosion Behavior of the Metal under Different Disbonded Thickness.

According to the electrochemical parameters in Table 4, in the above three kinds of simulated solution, with the increase of the coating disbonded thickness, the self-corrosion potential of the metal in the crevice shifted negatively and the corrosion current density increased obviously. As shown in Figure 6, when the coating disbonded thickness increased in the simulated solution without SRB, the crevice space became larger, and the dissolved oxygen content increased, and that accelerated the corrosion rate of the metal in the crevice. At the same experiment time, the increase amplitude of the metal corrosion current density in the crevice at the distance of 160 mm was significantly higher than that of 80 mm and that at damaged point; that meant the crevice thickness increased, of the metal which at the longer distance in the crevice had also been supplemental oxygen. In the simulated solution with SRB, as shown in Figure 7, increasing the coating disbonded thickness, the amplitude of the metal corrosion current density in the crevice was significantly greater than the other two kinds; that meant larger aperture space was suitable for the growth and reproduction of SRB, improving the activity of SRB in the crevice, thus greatly accelerating the corrosion rate of metal in the crevice. As shown in Figure 8, when the fungicide was added to the simulated SRB solution, the amplitude of the metal corrosion current density in the initial experiment

TABLE 4: The electrochemical parameters.

Simulated solution	Experiment time	Stripping depths/mm	Disbonded thickness (mm)	The corrosion current density/μA/cm^2	The self-corrosion potential/mV	The ratio of anode Tafel slope and cathode Tafel slope
Soil simulated solution	1 d	0	0.5	23.1029	−702	1.7328
			1.0	35.1429	−719	1.6204
		80	0.5	19.2514	−722	1.6538
			1.0	35.3486	−724	1.7384
		160	0.5	11.5943	−718	1.5963
			1.0	34.3171	−718	1.5686
	7 d	0	0.5	43.7829	−708	1.5726
			1.0	38.2800	−719	1.4733
		80	0.5	30.3086	−722	1.5250
			1.0	38.1543	−717	1.3973
		160	0.5	26.0457	−725	1.4844
			1.0	36.8343	−716	1.3790
	14 d	0	0.5	45.2457	−701	1.4111
			1.0	45.8971	−718	1.4050
		80	0.5	32.3429	−717	1.4049
			1.0	48.0971	−720	1.3542
		160	0.5	27.7543	−723	1.4831
			1.0	46.9549	−713	1.3135
Soil simulated solution + SRB	1 d	0	0.5	25.8171	−688	1.8134
			1.0	35.6800	−707	1.8220
		80	0.5	20.2457	−696	1.6897
			1.0	31.9029	−710	1.6832
		160	0.5	19.5143	−700	1.6727
			1.0	31.7543	−705	1.6649
	7 d	0	0.5	30.9143	−688	1.7509
			1.0	66.9429	−692	1.7047
		80	0.5	28.4971	−692	1.7250
			1.0	67.3714	−706	1.6529
		160	0.5	21.7143	−695	1.6395
			1.0	65.6229	−701	1.5372
	14 d	0	0.5	78.5143	−675	1.7515
			1.0	100.4343	−692	1.5221
		80	0.5	87.6457	−680	1.2216
			1.0	96.6857	−710	1.4924
		160	0.5	95.4914	−676	1.0870
			1.0	98.8457	−701	1.4101
Soil simulated solution + SRB + fungicide	1 d	0	0.5	16.3886	−694	1.9859
			1.0	33.6514	−708	1.9256
		80	0.5	19.0400	−692	1.4779
			1.0	34.5714	−719	1.5530
		160	0.5	21.2457	−696	1.3274
			1.0	32.9943	−713	1.6431
	7 d	0	0.5	33.4433	−698	1.6243
			1.0	54.1429	−695	1.7249
		80	0.5	43.1143	−732	1.4063
			1.0	51.7086	−703	1.7372
		160	0.5	47.4286	−691	1.4588
			1.0	50.7714	−696	1.5878
	14 d	0	0.5	48.6114	−684	1.4157
			1.0	79.0457	−690	1.6985
		80	0.5	63.8800	−689	1.4475
			1.0	77.1771	−712	1.4383
		160	0.5	75.1714	−684	1.4206
			1.0	78.0971	−700	1.4900

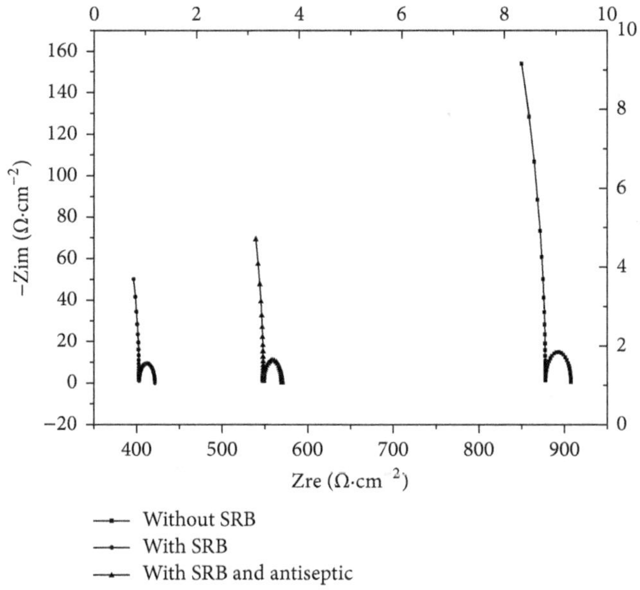

(a) The EIS of the damaged point

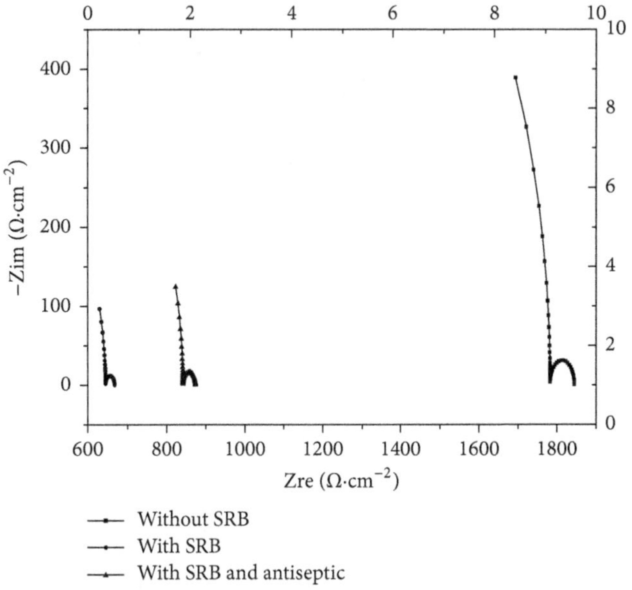

(b) The EIS of the point in the 80 mm stripping depth

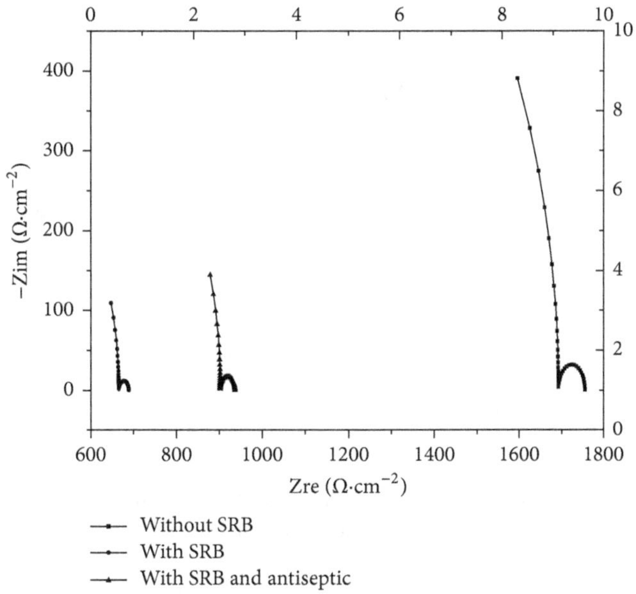

(c) The EIS of the point in the 160 mm stripping depth

FIGURE 4: The EIS of the base metal under stripping coating in the seventh experiment day.

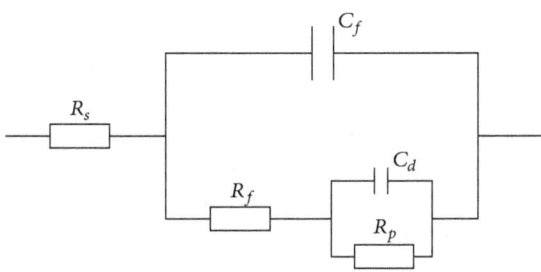

FIGURE 5: The equivalent circuit diagram.

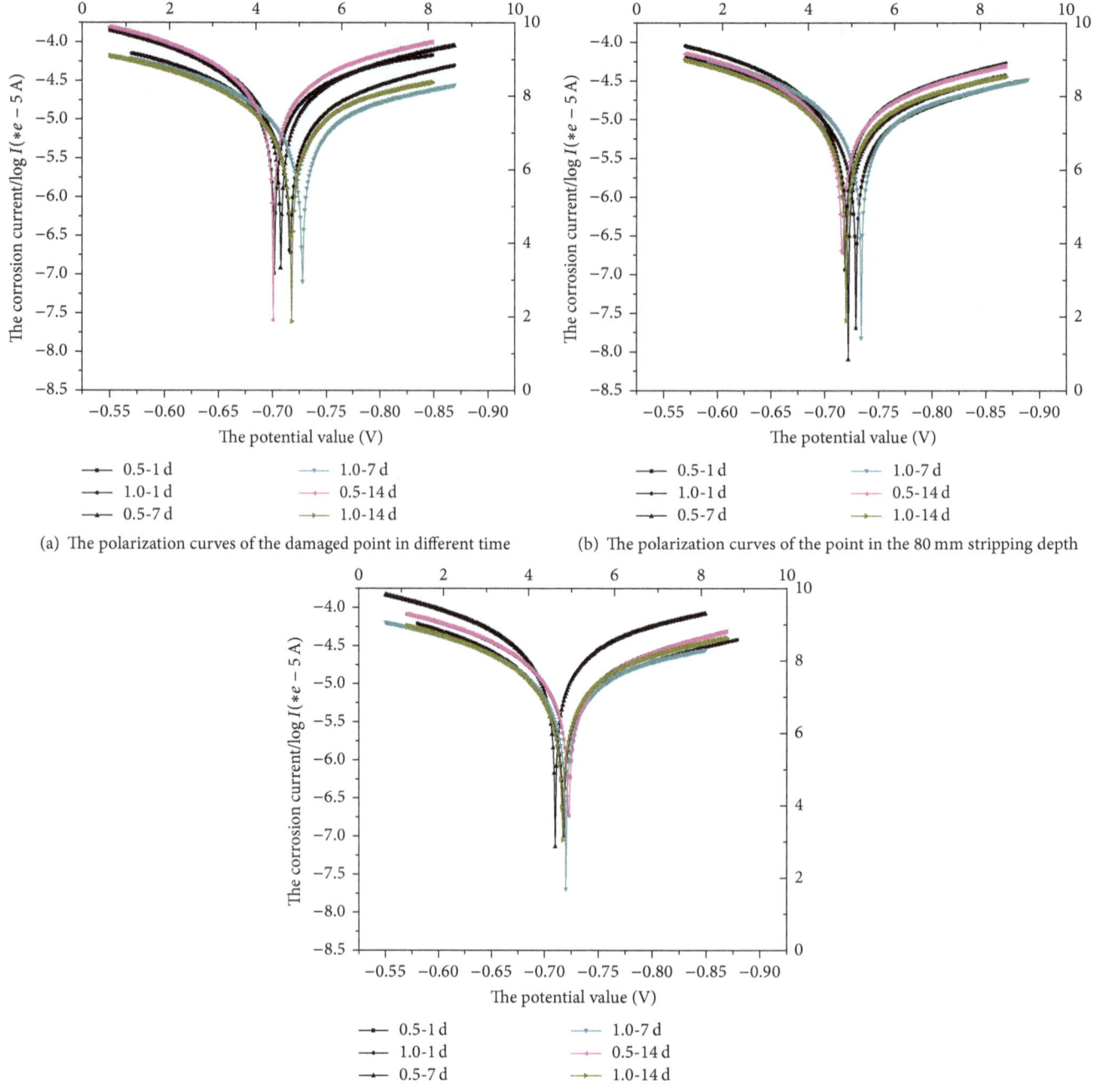

(a) The polarization curves of the damaged point in different time

(b) The polarization curves of the point in the 80 mm stripping depth

(c) The polarization curves of the point in the 160 mm stripping depth

Figure 6: The polarization curves of various points of the metal substrate in the disbonded crevice under different disbonded thickness in the simulated solution without SRB.

stage was significantly higher than that in the later stage of corrosion with the increase of the coating disbonded thickness. In addition, the ratio of anode Tafel slope and cathode Tafel slope was still more than 1 after increasing the coating disbonded thickness in these three kinds of simulated solution, so the control factor of the corrosion was anode control, but as the coating disbonded thickness increases this ratio had a decreasing trend with and without SRB. The anode control trend was gradually weakened, but this ratio had an increasing trend in the solution environment with SRB and

added fungicide, and the anode control trend was gradually enhanced.

The EIS used the ZSimpWin software for fitting. The electrochemical equivalent circuit C (CR (CR)) was composed of a resistance R and a capacitance C to represent the electrode process, the equivalent circuit diagram was the same as Figure 5, and there was no longer a detailed introduction.

As shown in Figure 9, when the stripping depth was 80 mm, the capacitance arc radius of the disbonded thickness of 0.5 mm was bigger than that of 1.0 mm in the three kinds of

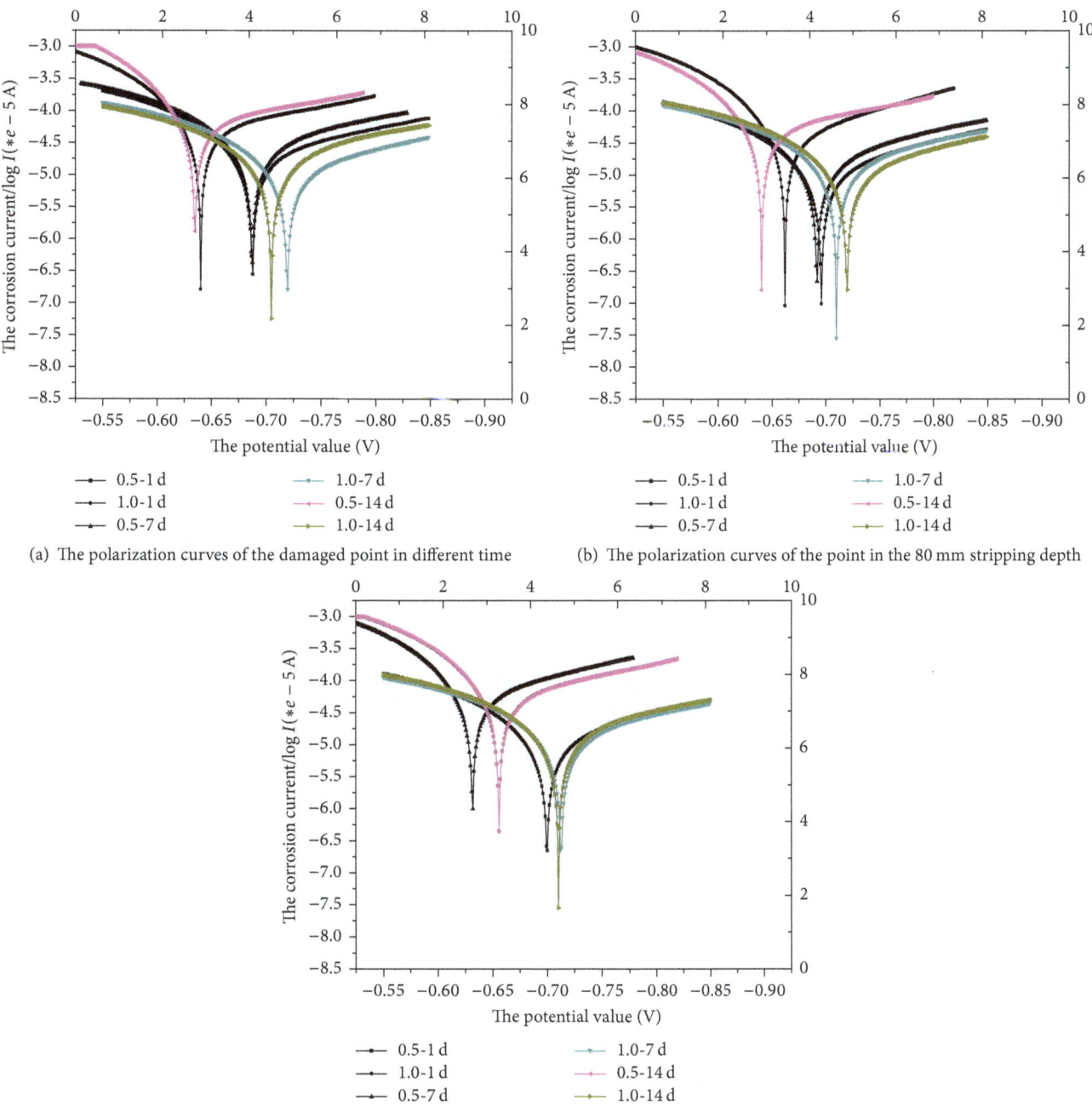

(a) The polarization curves of the damaged point in different time

(b) The polarization curves of the point in the 80 mm stripping depth

(c) The polarization curves of the point in the 160 mm stripping depth

FIGURE 7: The polarization curves of various points of the metal substrate in the disbonded crevice under different disbonded thickness in the simulated solution with SRB.

simulated solution in the seventh day. That meant the corrosion resistance of the metal in the crevice with the disbonded thickness 1.0 mm was worse; the corrosion was more serious. In addition, according to the polarization resistance of the metal in Table 5, we could see that the polarization resistance of the metal in the crevice at the thickness of 0.5 mm was bigger than that of 1.0 mm in these three kinds of simulated solution, and the difference of the polarization resistance of the metal in the solution with SRB was the maximum. Therefore, increasing the coating disbonded thickness would enhance the resistance of the metal corrosion process, and

it also could accelerate the corrosion rate of the metal in the disbonded crevice. The change of the coating disbonded thickness had a far greater impact on the corrosion rate of the metal in the simulated solution with SRB. The results above were consistent with the conclusions obtained by the method of polarization curve analysis.

4. Conclusion

(1) Whether the simulated solution is without or with SRB, the self-corrosion potential of the metal in the crevice under

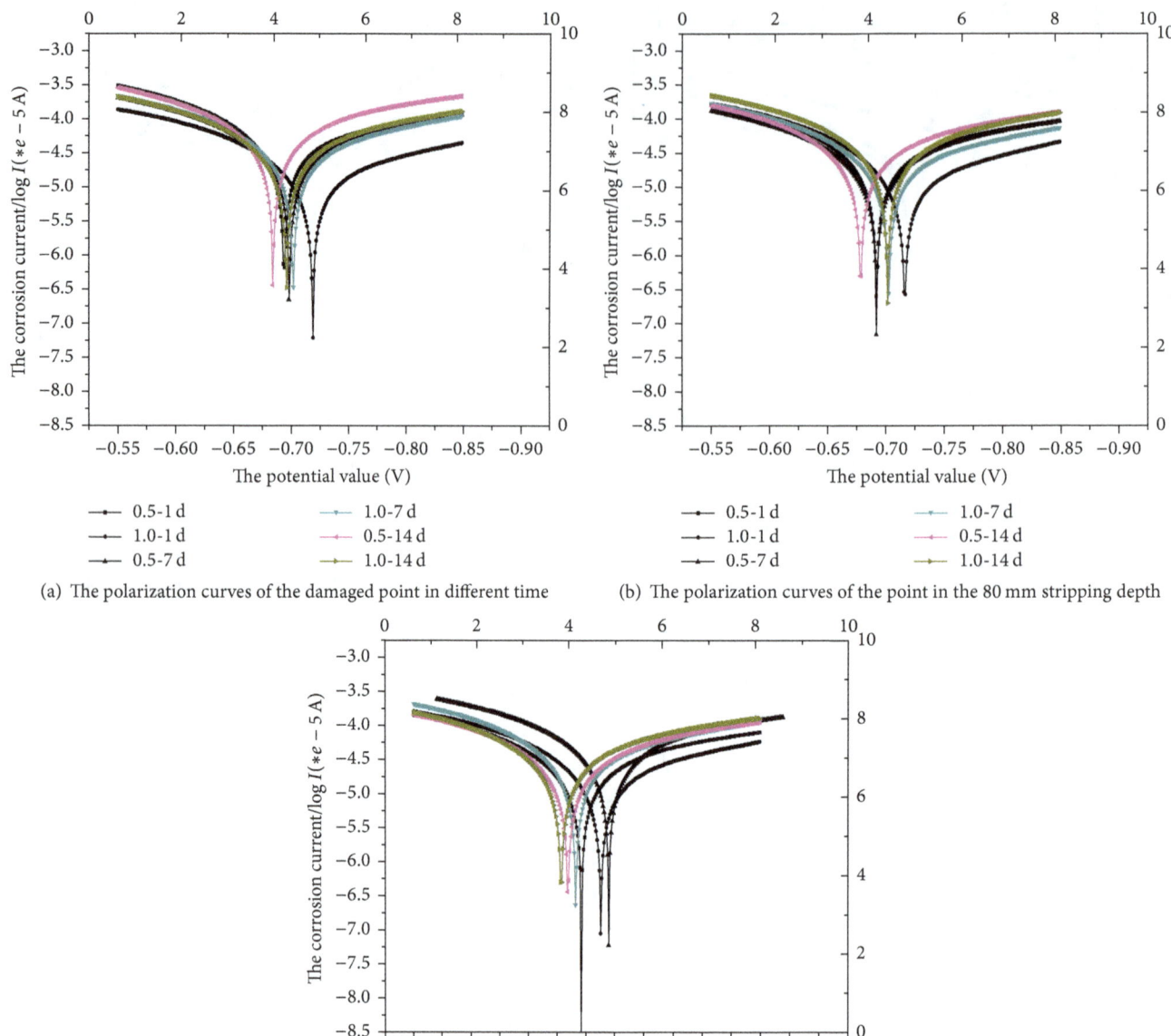

(a) The polarization curves of the damaged point in different time

(b) The polarization curves of the point in the 80 mm stripping depth

(c) The polarization curves of the point in the 160 mm stripping depth

FIGURE 8: The polarization curves of various points of the metal substrate in the disbonded crevice under different disbonded thickness in the simulated solution with SRB and fungicide.

TABLE 5: The polarization resistance of the metal at the distance of 80 mm under different disbonded thickness/Ω.

The disbonded distance/mm	The experiment time	The simulated solution	The disbonded thickness/mm	R_p/Ω
80	7 d	Soil simulated solution	0.5	3246.3
			1.0	3090.4
		Soil simulated solution + SRB	0.5	1632.7
			1.0	1152.4
		Soil simulated solution + SRB + fungicide	0.5	1321.2
			1.0	1290.7

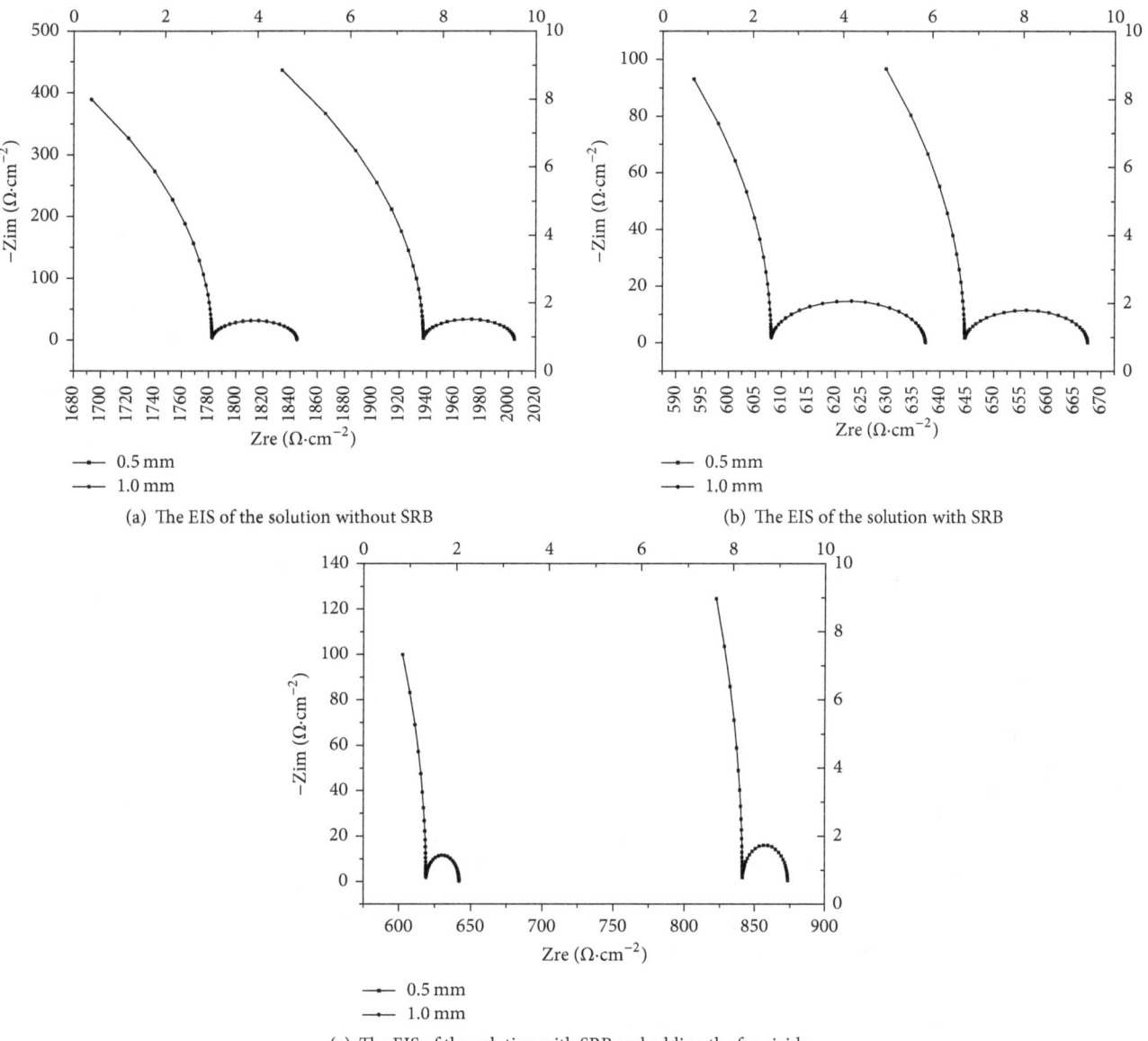

(a) The EIS of the solution without SRB

(b) The EIS of the solution with SRB

(c) The EIS of the solution with SRB and adding the fungicide

FIGURE 9: The EIS of the metal at the distance of 80 mm under different disbonded thickness in the seventh experiment day.

the same disbonded distance had little change in the initial experiment stage. In the later stage, the self-corrosion potential of the metal in the crevice had amplitude obviously, and the amplitude in the SRB solution environment was greater than that without SRB. With the extension of the experiment time, whether the solution is without or with SRB, the corrosion current density of the metal in the crevice under the same disbonded distance increased gradually, and the corrosion current density of the metal in the crevice in the simulated solution with SRB was significantly greater than that without SRB; that meant SRB could accelerate the corrosion of the metal in the disbonded crevice.

(2) In the SRB solution where the fungicide was added at the same time, the self-corrosion potential of the metal in the disbonded crevice did not change much, but the corrosion current density significantly decreased; that meant the

growth and reproduction of SRB were inhibited after adding the fungicide, so as to slow down the corrosion rate of metal in the crevice. With the experiment time extension, the self-corrosion potential of the metal in the crevice shifted negatively before moving to positive direction at the same depths, but the corrosion current density was increasing. When the experimental time was the same, with the increase of the coating disbonded depths, the self-corrosion potential of the metal in the crevice did not change much, but the corrosion current density increased.

(3) In the three kinds of simulated solution, with the increase of the coating disbonded thickness, the self-corrosion potential of the metal in the disbonded crevice shifted negatively, and the corrosion current density increased significantly. It indicated that increasing the coating disbonded thickness could accelerate the corrosion of the metal in the

crevice, and in the simulated solution with SRB, the amplitude of the corrosion current density of the metal in the crevice was greater than that in other two kinds of solution.

Conflicts of Interest

The authors declare that there are no conflicts of interest regarding the publication of this article.

Acknowledgments

The research work was supported by the Central College Foundation of CAUC (3122017038), the Opening Project of Guangxi Colleges and Universities Key Laboratory of Beibu Gulf Oil and Natural Gas Resource Effective Utilization (2016KLOG21), and the National Science Foundation (U1633111).

References

[1] I. U. L. Xu and W. Xiaoyu, "The research of oil and gas pipeline corrosion and protection technology," in *Advances in Petroleum Exploration and Development*, vol. 7, pp. 102–105, 2014.

[2] D. B. Yadav, K. S. Jha, and R. Kumar, "Pipeline Corrosion in a 24" dia Crude Oil Pipeline due to Interference from High Voltage AC Transmission Line: A Case Study," in *Proceedings of the 10th Pipeline Technology Conference 2015*, EITEP Institute, 2015.

[3] G. Cui, Z.-L. Li, C. Yang, and M. Wang, "The influence of DC stray current on pipeline corrosion," *Petroleum Science*, vol. 13, no. 1, pp. 135–145, 2016.

[4] Y. Liu, B. Zhang, Y. Zhang, L. Ma, and P. Yang, "Electrochemical polarization study on crude oil pipeline corrosion by the produced water with high salinity," *Engineering Failure Analysis*, vol. 60, pp. 307–315, 2016.

[5] H. Tian and Y. Frank Cheng, "Novel inhibitors containing multi-functional groups for pipeline corrosion inhibition in oilfield formation water," *Corrosion*, vol. 72, no. 4, pp. 472–485, 2016.

[6] S. Lyon, R. Bingham, and D. Mills, "Advances in corrosion protection by organic coatings: what we know and what we would like to know," *Progress in Organic Coatings*, vol. 102, pp. 2–7, 2017.

[7] B. T. Richards, M. R. Begley, and H. N. G. Wadley, "Mechanisms of ytterbium monosilicate/mullite/silicon coating failure during thermal cycling in water vapor," *Journal of the American Ceramic Society*, vol. 98, no. 12, pp. 4066–4075, 2015.

[8] A. A. Vereschaka, B. Y. Mokritskii, N. N. Sitnikov, G. V. Oganyan, and A. Y. Aksenenko, "Study of mechanism of failure and wear of multi-layered composite nano-structured coating based on system Ti-TiN-(ZrNbTi)N deposited on carbide substrates," *Journal of Nano Research*, vol. 45, pp. 110–123, 2017.

[9] F. Varela, M. Forsyth, and M. Y. J. Tan, "Electrochemically monitoring localized corrosion patterns and CP effectiveness under disbonded coatings," *NACE - International Corrosion Conference Series*, vol. 2015-, 2015.

[10] Y. A. N. Maocheng, X. U. Jin, Y. U. Libao, W. U. Tangqing, S. U. N. Cheng, and K. E. Wei, "EIS analysis on stress corrosion initiation of pipeline steel under disbonded coating in near-neutral pH simulated soil electrolyte," *Corrosion Science*, vol. 110, pp. 23–34, 2016.

[11] Y. Chen, "Electrochemical impedance spectroscopy study for cathodic disbonding test technology on three layer polyethylene anticorrosive coating under full immersion and alternating dry–wet environments," *International Journal of Electrochemical Science*, pp. 10884–10894, 2016.

[12] D. Xu, Y. Li, and T. Gu, "Mechanistic modeling of biocorrosion caused by biofilms of sulfate reducing bacteria and acid producing bacteria," *Bioelectrochemistry*, vol. 110, pp. 52–58, 2016.

[13] D. Xu and T. Gu, "Carbon source starvation triggered more aggressive corrosion against carbon steel by the Desulfovibrio vulgaris biofilm," *International Biodeterioration and Biodegradation*, vol. 91, pp. 74–81, 2014.

[14] D. Xu, Y. Li, F. Song, and T. Gu, "Laboratory investigation of microbiologically influenced corrosion of C1018 carbon steel by nitrate reducing bacterium Bacillus licheniformis," *Corrosion Science*, vol. 77, pp. 385–390, 2013.

[15] H. Venzlaff, D. Enning, J. Srinivasan et al., "Accelerated cathodic reaction in microbial corrosion of iron due to direct electron uptake by sulfate-reducing bacteria," *Corrosion Science*, vol. 66, pp. 88–96, 2013.

[16] D. Enning and J. Garrelfs, "Corrosion of iron by sulfate-reducing bacteria: New views of an old problem," *Applied and Environmental Microbiology*, vol. 80, no. 4, pp. 1226–1236, 2014.

[17] S. Chen, P. Wang, and D. Zhang, "Corrosion behavior of copper under biofilm of sulfate-reducing bacteria," *Corrosion Science*, vol. 87, pp. 407–415, 2014.

[18] M. G. Chesnokova, V. V. Shalaj, Y. A. Kraus et al., "Analysis of corrosion defects on oil pipeline surface using scanning electron microscopy and soil thionic and sulfate-reducing bacteria quantification," *Procedia Engineering*, vol. 152, pp. 247–250, 2016.

[19] W. Wang, Q. Wang, C. Wang, and J. Yi, "Experimental studies of crevice corrosion for buried pipeline with disbonded coatings under cathodic protection," *Journal of Loss Prevention in the Process Industries*, vol. 29, no. 1, pp. 163–169, 2014.

[20] F. N. Varela, M. Y. Tan, and M. Forsyth, "A novel approach for monitoring pipeline corrosion under disbonded coatings," in *Proceedings of the 2014 10th International Pipeline Conference*, p. V002T06A068, Calgary, Alberta, Canada.

19

Effect of Additional Sulfide and Thiosulfate on Corrosion of Q235 Carbon Steel in Alkaline Solutions

Bian Li Quan,[1,2] **Jun Qi Li,**[1,2] **and Chao Yi Chen**[1,2]

[1]*College of Materials and Metallurgy, Guizhou University, Guiyang 550025, China*
[2]*Guizhou Province Key Laboratory of Metallurgical Engineering and Process Energy Saving, Guiyang 550025, China*

Correspondence should be addressed to Jun Qi Li; jqli@gzu.edu.cn

Academic Editor: Flavio Deflorian

This paper investigated the effect of additional sulfide and thiosulfate on Q235 carbon steel corrosion in alkaline solutions. Weight loss method, scanning electron microscopy (SEM) equipped with EDS, X-ray photoelectron spectroscopy (XPS), and electrochemical measurements were used in this study to show the corrosion behavior and electrochemistry of Q235 carbon steel. Results indicate that the synergistic corrosion rate of Q235 carbon steel in alkaline solution containing sulfide and thiosulfate is larger than that of sulfide and thiosulfate alone, which could be due to redox reaction of sulfide and thiosulfate. The surface cracks and pitting characteristics of the specimens after corrosion were carefully examined and the corrosion products film is flake grains and defective. The main corrosion products of specimen induced by S^{2-} and $S_2O_3^{2-}$ are FeS, FeS_2, Fe_3O_4, and FeOOH. The present study shows that the corrosion mechanism of S^{2-} and $S_2O_3^{2-}$ is different for the corrosion of Q235 carbon steel.

1. Introduction

The use of high-sulfur bauxite makes Bayer solutions contain high content of S^{2-} in recent years. Therefore, in Bayer liquor of alumina extraction besides $Al(OH)_3$ and caustic, other substances are impurities, including Na_2SO_3, Na_2SO_4, $Na_2S_2O_3$, Na_2S and other trace impurities, and so forth [1]. The content of four kinds of sulfide is $0.4\,g\cdot L^{-1}$, $0.8\,g\cdot L^{-1}$, $1.2\,g\cdot L^{-1}$, and $1.6\,g\cdot L^{-1}$, respectively. Among them, $Na_2S_2O_3$ and Na_2S are corrosion activators to steel corrosion, which remain the most serious forms of metal failure [2, 3]. However, as an important equipment material, carbon steel has been widely used and plays a significant role in alumina industry which is used to dissolution equipment, dilution tank, Bayer mother liquor evaporation equipment, and so forth. With the progress of production, the corrosion of sulfides has been recognized as a serious problem, which results in a great economic loss and corrosion cracking and pitting of steel equipment has been experienced in different alumina industry. So, the study on corrosion of steel equipment and control is a very critical issue.

The mechanism of steel corrosion and formation of the passive film have been always studied by many scholars [4–6]. The role of sulfide in the highest concentrations of alkaline solution like white liquor has been studied on the corrosion behavior of carbon steel [7–9]. But in the higher concentrations of caustic solution sulfur-contained species like alumina industry the research of the role of sulfide on steel corrosion is too sparse and corrosion behavior of 16MnR low alloy steel in sulfide-containing Bayer solutions was researched by Xie and Chen [3]. The effect of S^{2-}, NaOH, and Al_2O_3 on the corrosion of 16MnR low alloy steel had been studied by polarization curves and electrochemical impedance spectroscopy (EIS). But the effect of the synergy effect of S^{2-} and $S_2O_3^{2-}$ on Q235 carbon steel corrosion has not been discussed in detail.

Moreover, these corrosion experiments in literature have been performed using mostly conventional static stainless steel autoclave, dynamic and static immersion, and so on. The author will use the salt fog box to discuss corrosion of Q235 carbon steel in alkaline solutions containing sulfide and thiosulfate [10]. The aim is to use the accelerated corrosion of the salt fog box.

TABLE 1: Chemical composition of Q235 carbon steel (wt.%).

Element	C	Si	Mn	Cr	P	S	Fe
Content (%)	0.375	0.026	0.239	0.098	0.052	0.015	Bal

The purpose of the study is to evaluate corrosion behavior of Q235 carbon steel through additional sulfide and thiosulfate to alkaline solutions and corrosion experiments are made in the salt fog box. The effect of S^{2-} and $S_2O_3^{2-}$ on corrosion behavior of steel and mechanisms will be discussed in detail by employing weight loss measurements, scanning electron microscopy (SEM) equipped with an energy dispersive spectroscopy (EDS), X-ray photoelectron spectroscopy (XPS), and electrochemistry. The fundamental purposes are to gain the corrosion characteristics of Q235 carton steel in sulfur-containing alkaline solutions and to provide fundamental information for steel corrosion in sulfur-containing Bayer mother liquor and hence to provide insight into the understanding of the corrosion behavior and mechanism of Q235 carbon steel in alumina industry of the Southwest China.

2. Experimental

2.1. Specimens and Test Solutions. The chemical compositions (wt%) of Q235 carbon steel used in this study are listed in Table 1. The steel coupon was machined from steel with the 15 mm × 15 mm × 2.5 mm square type. Prior to experiment, the coupons were rinsed with distilled water, polished orderly with number 180, 240, 360, and 600 silicon carbide paper and degreased with acetone, and finally dried with cold air.

The sulfur-containing alkaline solutions contained 115 g·L^{-1} NaOH, 0, 5 g·L^{-1} S^{2-}, and 0, 4, 5 g·L^{-1} S$_2$O$_3$$^{2-}$ at 328 K. The solutions were prepared by dissolving NaOH, Na$_2$S·9H$_2$O, and Na$_2$S$_2$O$_3$·5H$_2$O into 1000 mL of deionized water, respectively. The pH of the solutions is 14 by laboratory test, so H$^+$ can barely exist in such strong alkaline solution. All reagents are analytically pure.

2.2. The Corrosion Experiment. All of the corrosion experiments were carried out for 120 h at 328 K (maximum temperature of the salt fog box) in the salt fog box (GB/T 24195-2009). The samples were taken out one by one at regular intervals to analyze the corrosion behavior. The test solutions were supplemented daily. After being taken out from the salt fog box, these coupons were rinsed using distilled water and dried with cold air.

Prior to each corrosion experiment, the coupon was weighed to sensitivity level of 0.0001 g using an analytical balance (AR1140/C) for the original weight (W_1) and its area (S) was measured. Before weight loss measurements, the corrosion products were removed using the chemical products cleanup method (the 500 mL of hydrochloric acid + 500 mL deionized water + 10 g six methyl tetramine solutions) (GB/T 6074-1992) [11]. Finally, the sample was weighed again in order to obtain the final weight (W_2).

The corrosion rate was calculated using weight loss measurement and the formula to calculate is given as

$$R = 8.76 \times 10^4 \times (W_1 - W_2) \times (S \times t \times D)^{-1}, \quad (1)$$

where R is corrosion rate (mm/a), W_1 is the quality of the sample before the corrosion test (g) with an accuracy of ±0.0001 g, W_2 is corrosion sample after the removal of corrosion products quality (g) with an accuracy of ±0.0001 g, S is the sample surface area (cm^2) with an accuracy of 0.01 cm^2, t is corrosion time (h) with an accuracy of 1 h, and D is the density of test steel (g/cm^3).

2.3. Surface Analysis. The morphology and composition of Q235 carbon steel surface were observed with scanning electron microscopy (SEM) (SUPRA 40) and energy dispersive spectroscopy (EDS) (AZ tec.).

The composition of the passive films was analyzed by means of PHI Quantum 2000 Scanning ESCA Microprobe X-rays photoelectron spectrometer (XPS). Monochromatized Al Ka radiation (1486.6 eV) was used as the excitation source. After being exposed to the solutions for different concentrations of S^{2-} and S$_2$O$_3$$^{2-}$, the sample was dried and transferred to the XPS instrument for analysis. The binding energy values were calibrated with reference to the C 1s peak at 284.8 eV. The data processing and Photoelectron analytical peak would be parsed using Multipeak 8.0; the narrow scan spectra were fitted with XPSPeak 4.1 software.

2.4. Electrochemical Measurement. Electrochemical testing was carried out in a conventional three-electrode cell at 338 K. The temperature of 338 K is to simulate the alumina dilution process. Using a saturated calomel electrode (SCE) as the reference electrode, platinum electrodes are auxiliary electrode and Q235 steel as working electrode. Before electrochemical experiment the sample was immersed for 120 h in the solutions. The work area of Q235 steel is 2.25 cm^2, nonworking parts sealed with epoxy resin. The cell was placed in a water bath to achieve the test temperature. Potentiodynamic polarization curves measurements were performed at a potential scan rate of 1 mV/s. The potential range was from −1.5 V to 0.5 V. The EIS measurements were carried out at OCP over the frequency from 100 kHz to 10 mHz by 5 points per decade. All potentials reported in this paper were measured with respect to the SCE. All the tests were repeated to obtain reproducibility of results.

3. Results and Discussion

3.1. Corrosion Rate. Weight loss measurement is a more reliable way to get the corrosion rate compared with other methods [12]. Table 2 shows the corrosion rate (mm/a) of Q235 carbon steel based on weightlessness corrosion experiment in the alkaline solutions with different concentrations of S^{2-} and S$_2$O$_3$$^{2-}$ at 328 K. The corrosion rate of steel increased significantly in the presence of S$_2$O$_3$$^{2-}$ or S^{2-}, and the corrosion rate much increased when it is synergy corrosion. Thus it can be seen that S^{2-} and S$_2$O$_3$$^{2-}$ were corrosion

TABLE 2: Corrosion rate of Q235 carbon steel in alkaline solutions.

S^{2-} concentration $(g \cdot L^{-1})$	$S_2O_3^{2-}$ concentration $(g \cdot L^{-1})$	Corrosion rate (mm/a)
0	0	0.01
0	4	0.03
5	0	0.08
5	4	0.18
5	5	0.10

TABLE 3: Element distributions of corrosion products with different concentrations of sulfide and thiosulfate at 328 K (wt.%).

Concentrations of sulfide and thiosulfate	O	S	Fe
$0 \, g \cdot L^{-1} \, S^{2-}$ and $4 \, g \cdot L^{-1} \, S_2O_3^{2-}$	31.43	4.55	64.02
$5 \, g \cdot L^{-1} \, S^{2-}$ and $0 \, g \cdot L^{-1} \, S_2O_3^{2-}$	7.67	30.28	62.05
$5 \, g \cdot L^{-1} \, S^{2-}$ and $4 \, g \cdot L^{-1} \, S_2O_3^{2-}$	8.60	17.61	73.79
$5 \, g \cdot L^{-1} \, S^{2-}$ and $5 \, g \cdot L^{-1} \, S_2O_3^{2-}$	2.78	7.89	89.32

activators for Q235 carbon steel when they are separate in alkaline solution; however, when S^{2-} and $S_2O_3^{2-}$ exist altogether, the reduction reaction of $S_2O_3^{2-}$ accelerates iron loss of electrons and the hydrolysis of S^{2-} was also inhibited, so the corrosion is accelerated. The reduction reaction of $S_2O_3^{2-}$ could be described as follows:

$$S_2O_3^{2-} + 5H_2O + 8e^- = 2HS^- + 8OH^- \qquad (2)$$

$$S^{2-} + H_2O = HS^- + OH^- \qquad (3)$$

When the concentration of $S_2O_3^{2-}$ is $5 \, g \cdot L^{-1}$, the loss of electrons of S^{2-} and $S_2O_3^{2-}$ inhibits the iron loss of electrons. So, the corrosion rate decreases as follows:

$$2S^{2-} + S_2O_3^{2-} + 6OH^-$$
$$= S_2^{2-} + 2SO_3^{2-} + 3H_2O + 6e^- \qquad (4)$$

This is consistent with what was previously reported that thiosulfate is corrosive activator [3] and inconsistent with the conclusions of Xie et al. [13]. So, whether or not the effect of thiosulfate on corrosion of Q235 carbon steel is to accelerate or hinder which mainly depends on the concentration of sulfide when they are synergy corrosion [14]. Besides, in strong alkaline medium, the solutions have a lot of OH^- and hinder the secondary hydrolysis reaction of S^{2-} ion; therefore, the concentration of HS^- is higher, is helpful to generate the sulfide film, and thus decreases the corrosion rate of Q235 carbon steel in the initial stage of corrosion [15].

3.2. Surface Morphology and Composition. Figure 1 shows SEM images of corrosion products film on the specimen surfaces and the corresponding EDS in alkaline solutions with different concentrations S^{2-} and $S_2O_3^{2-}$ at 328 K after 120 hours of corrosion. It can be seen that the corrosion products cover unevenly the surface.

As shown in Figure 1, a large amount of corrosion is formed on steel surface with the additives of different concentrations S^{2-} and $S_2O_3^{2-}$; the corrosion is the most serious in the presence of S^{2-} (Figure 1(c)) and the presence of S^{2-} and $S_2O_3^{2-}$ (Figures 1(d) and 1(e)) [16]. As can be seen from Figures 1(c), 1(d), and 1(e), there are selective corrosion and pitting with carbon steel. The corrosion surface with a loose and porous structure is easy to fall off, which cannot protect the metal matrix. The corrosion of Q235 carbon steel is slight in pure alkaline solution and steel surface has some

defects and scratches (Figure 1(a)). Therefore, it is thought that the corrosion of Q235 carbon steel preferentially takes place in defects and scratches, mainly caused by polishing [17], whose surface roughness and surface residual have resulted in the faster corrosion rate.

Table 3 lists the distribution of elements with Q235 carbon steel for different concentrations of S^{2-} and $S_2O_3^{2-}$ in solutions at 328 K. From Table 3 it can be seen that the oxygen content is largest in the absence of S^{2-} and the sulfur content is largest in the absence of $S_2O_3^{2-}$, which decreases later with the concentration of $S_2O_3^{2-}$. The results show that the oxidation film is generated at the steel surface mainly in the absence of S^{2-}, and the corrosion products are sulfide in the absence of $S_2O_3^{2-}$. The compactness and stability of oxidation film are higher than those of sulfide. Results of EDS indicate that the corrosion mechanism of Q235 carbon steel S^{2-} and $S_2O_3^{2-}$ is different.

3.3. Potentiodynamic Polarization Curves. Figure 2 is the potentiodynamic polarization curves of Q235 carbon steel in alkaline solutions with different concentrations of S^{2-} and $S_2O_3^{2-}$ at 338 K. As can be seen from Figure 2, the anodic process exhibits active-passive-transpassive behaviors, which indicates that the surface of specimen formed compact corrosion products film at later stage and the anodic iron dissolution and the transformation of iron compounds. This could be due to the fact that the corrosion process involved transformation of corrosion products, always maintaining the surface active [18]. In alkaline solutions regardless of the concentrations of S^{2-} and $S_2O_3^{2-}$, Q235 carbon steel exhibits corrosion potential (E_{corr}) near -1.20 V versus SCE in the active state. At the potentials -1.0 V versus SCE, the curves exhibit a humped shape, revealing at least two reactions including one for the exposed steel anodically dissolved into the electrolyte to form FeO_2^- ions and another one attributed to the oxidation of $S_2O_3^{2-}$ to higher forms, as described by Zou and Chin [19].

The determination of corrosion parameters (E_{corr}, β_a, β_c, i_{corr}, i_c, and R_p) could provide more information about the overall corrosion process. Table 4 shows the corrosion potentials (E_{corr}), Tafel slopes (β_a and β_c represent anodic and cathodic, resp.), corrosion currents (i_{corr}), critical current density (i_c), and polarization resistance (R_p) obtained from Tafel fitting of the polarization curves data. The higher critical current density suggests that corrosion of steel increases. When S^{2-} concentration is $5 \, g \cdot L^{-1}$, the presence of $S_2O_3^{2-}$ caused a considerable increase in the critical current density

TABLE 4: Electrochemical parameters fitted from the potentiodynamic polarization curves of Q235 carbon steel.

Concentration (g·L^{-1})	E_{corr} (mV)	i_{corr} (μA/cm^2)	β_a (mV)	β_c (mV)	i_c (mA/cm^2)	R_p (Ohm)
5 g·L^{-1} S^{2-} + 5 g·L^{-1} S$_2$O$_3$$^{2-}$	−1155	365	228.0	122.0	2.4	31.1
5 g·L^{-1} S^{2-} + 4 g·L^{-1} S$_2$O$_3$$^{2-}$	−1179	792	112.6	119.0	10.4	14.6
5 g·L^{-1} S^{2-} + 0 g·L^{-1} S$_2$O$_3$$^{2-}$	−1242	372	172.8	116.9	2.1	31.7
0 g·L^{-1} S^{2-} + 4 g·L^{-1} S$_2$O$_3$$^{2-}$	−1167	14.5	132.2	135.3	0.8	1104

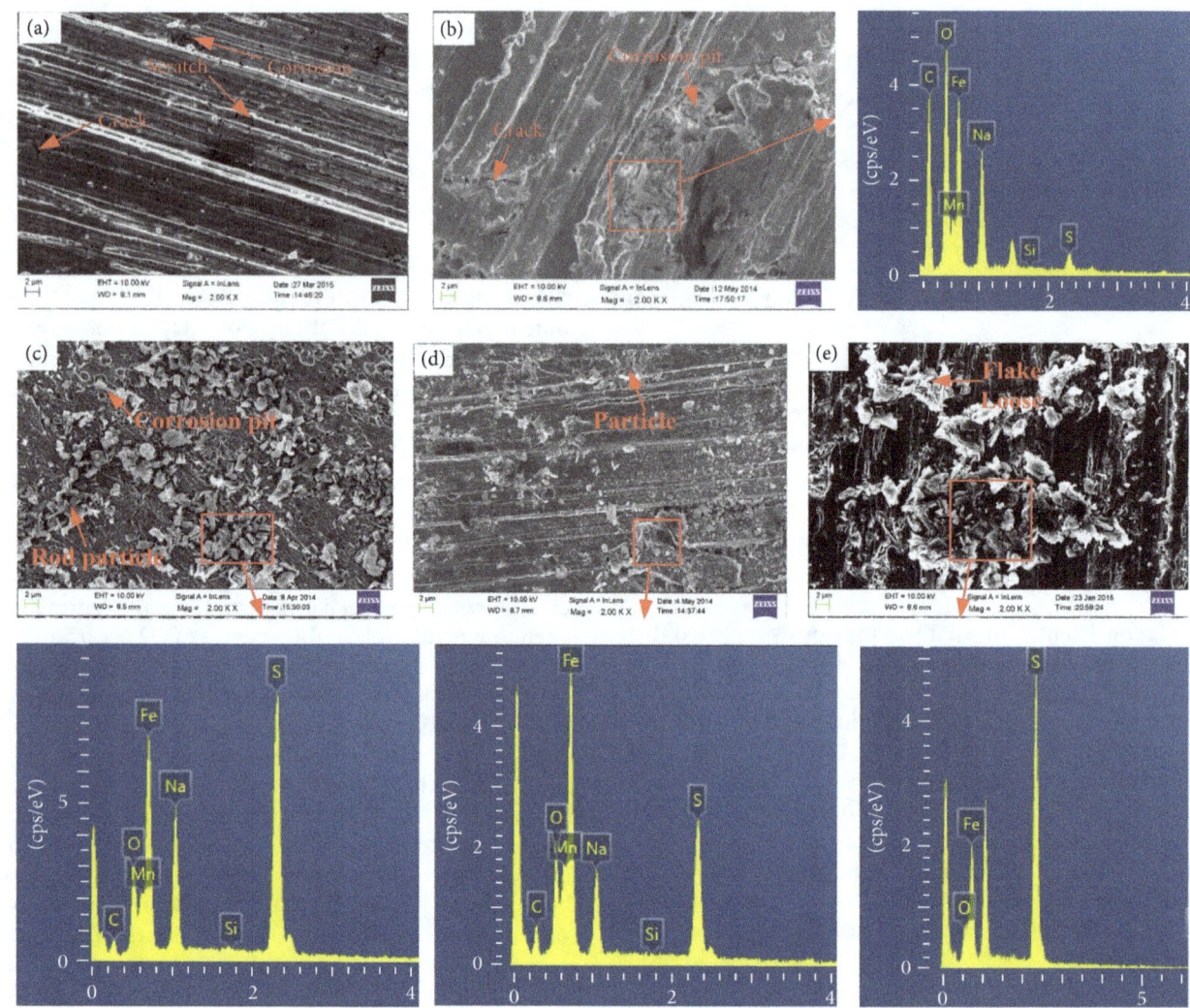

FIGURE 1: Morphologies (SEM) of Q235 carbon steel in alkaline solutions with different concentrations of S^{2-} and S$_2$O$_3$$^{2-}$ at 328 K. (a) 0 g·L^{-1} S^{2-} and 0 g·L^{-1} S$_2$O$_3$$^{2-}$, (b) 0 g·L^{-1} S^{2-} and 4 g·L^{-1} S$_2$O$_3$$^{2-}$, (c) 5 g·L^{-1} S^{2-} and 0 g·L^{-1} S$_2$O$_3$$^{2-}$, (d) 5 g·L^{-1} S^{2-} and 4 g·L^{-1} S$_2$O$_3$$^{2-}$, (e) 5 g·L^{-1} S^{2-} and 5 g·L^{-1} S$_2$O$_3$$^{2-}$.

(Table 4) which shows that S$_2$O$_3$$^{2-}$ is activator; however, the increase of S$_2$O$_3$$^{2-}$ concentration decelerated the corrosion of Q235 carbon steel when the concentration of S^{2-} is 5 g·L^{-1} [14], reflecting that S^{2-} played a more important role in the corrosion of Q235 carbon steel. Anodic Tafel slopes, β_c (Table 4), in S^{2-}-containing solutions are very similar (about 120 mV), regardless of the concentration of S$_2$O$_3$$^{2-}$.

However, the slopes of the anodic branch, β_a (Table 4), are considerably dependent on the concentration of S$_2$O$_3$$^{2-}$. This fact, along with higher values of these slopes compared with those from the anodic branch, indicates the complex nature of the oxidation process. The change of i_c is the opposite trend to the change of R_p (Table 4). The shape of cathodic polarization curve of S$_2$O$_3$$^{2-}$ is different from S^{2-}, which shows that the

TABLE 5: EIS fitting results of Q235 steel in alkaline solutions with different concentrations of S^{2-} and $S_2O_3^{2-}$.

Concentration/g·L^{-1}	$R_s/\Omega\cdot cm^2$	$R_{ct}/\Omega\cdot cm^2$	$CPE_{dl}(Y_0)/\Omega^{-1}\cdot cm^{-2}\cdot s^{-n}$	n_1	$R_f/\Omega\cdot cm^2$	$CPE_f(Y_0)/\Omega^{-1}\cdot cm^{-2}\cdot s^{-n}$	n_2	R^2
$5\,g\cdot L^{-1}\,S^{2-} + 5\,g\cdot L^{-1}\,S_2O_3^{2-}$	0.441	67.48	0.00132	0.88	12134	0.0074	0.52	$1e-20$
$5\,g\cdot L^{-1}\,S^{2-} + 4\,g\cdot L^{-1}\,S_2O_3^{2-}$	0.288	1.235	0.093	0.89	48.46	0.173	0.61	$1e-20$
$5\,g\cdot L^{-1}\,S^{2-} + 0\,g\cdot L^{-1}\,S_2O_3^{2-}$	0.743	78.21	0.26	0.99	105.4	0.149	0.92	$1e-20$
$0\,g\cdot L^{-1}\,S^{2-} + 4\,g\cdot L^{-1}\,S_2O_3^{2-}$	1.969	4234	0.0002	0.97	77719	0.00002	0.62	$1e-20$

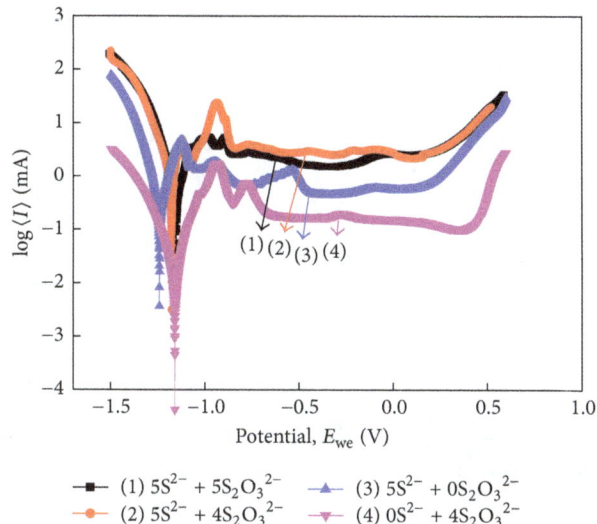

FIGURE 2: Potentiodynamic polarization curves of Q235 carbon steel in solutions with different concentrations of S^{2-} and $S_2O_3^{2-}$ at 338 K.

cathodic reactions of S^{2-} and $S_2O_3^{2-}$ are different for the corrosion of Q235 carbon steel, consistent with the EDS findings.

3.4. Electrochemical Impedance Spectroscopy.

Typical Nyquist and Bode plots of Q235 carbon steel after passivation for 120 h in alkaline solutions with different concentrations of S^{2-} and $S_2O_3^{2-}$ at 338 K at the open-circuit potential (OCP) are illustrated in Figure 3. Nyquist plots in Figure 3 indicate that the overall impedance of steel in alkaline solution without S^{2-} is the largest, but then it decreases in the presence of S^{2-}, suggesting an activation of the corrosion processes, as expected in the presence of sulfides [2]. In the presence of S^{2-}, the overall impedance reduces when the $S_2O_3^{2-}$ concentration increases from 0 to 4 g·L^{-1}, after increasing the $S_2O_3^{2-}$ concentration being 5 g·L^{-1}, and makes the surface more sensitive to S^{2-} and $S_2O_3^{2-}$ attack. These data of EIS testes in Figure 3 further support the results from potentiodynamic polarization curves in Figure 2. Deconvolution of the phase angle curves of Bode plots in Figure 3 indicate that two overlapped time constants are found for Q235 carbon steel in solutions in the presence of S^{2-} and $S_2O_3^{2-}$.

The experimental data was then fitted by Z-View, and the results are shown in Figure 3, which is in accordance with some references [20, 21]. As shown, the fitted curves are in good agreement with corresponding experimental data. To further understand the processes and mechanisms of Q235 carbon steel corrosion, electrical equivalent circuit was constructed (see Figure 4). In the model, R_s denotes the resistance between the working electrode and reference electrode, R_f and Q_f represent, respectively, the resistance and capacitance of the oxidation film or corrosion products film, and R_{ct} and Q_{dl} are the charge transfer resistance and constant phase element (CPE), respectively. The reason of Q for capacitors in analysis of impedance spectra is that the passive film is not considered as a homogeneous layer but rather as a defective layer. The impedance of Q is defined as

$$Z_Q = Y_0^{-1}\left(j\omega\right)^{-n}, \qquad (5)$$

where Y_0 is the admittance, ω is angular frequency in rad/s, $j^2 = (-1)$, and n is an exponential term which always lies between 0.5 and 1.0. When $n = 1$, the CPE describes an ideal capacitor. When $n = 0.5$, the CPE represents a Warburg impedance with diffusional character. For $0.5 < n < 1$, the CPE describes a distribution of dielectric relaxation times in frequency space. From the fitted values of electrical equivalent circuits (listed in Table 5), one can find that after the corrosion product film is generated, n_1 and n_2 decreases with the concentration of $S_2O_3^{2-}$, suggesting that the corrosion of the steel became more and more serious [22]. The change of R_f also corresponds to the change of polarization results (Table 4). Equivalent circuit mode shows that passive film does not completely cover the metal surface, which is due to surface roughness and other reasons that can cause the dispersion effect. The passive film is not a uniform layer, but a porous film of characteristic. Results are consistent with SEM-EDS analysis.

3.5. Surface Analysis of Specimens by XPS.

Figure 5 shows the decomposition of peaks for Fe2p3, S2p, and O1s of the passive films formed on Q235 carbon steel. Peak decompositions of Fe2p3, S2p, and O1s core level spectra are in agreement with published results [23–25]. In the O1s spectrum, a broad peak and two narrow peaks are observed at about 528–536 eV in the solutions. The O1s spectrum can be curve-fitted with two peaks. The peak at 529.92 eV is attributed to O^{2-} in oxides. The peak at 531.5 eV can be attributed to oxygen in FeOOH. The Fe2p3 spectrum can be curve-fitted with four peaks. The peaks at 706.91, 708.19, 710.49, and 712.82 eV correspond to FeS_2, Fe_3O_4, and FeS. The formation of a stable layer (Fe_3O_4), which can inhibit the diffusion of corrosive species, will be

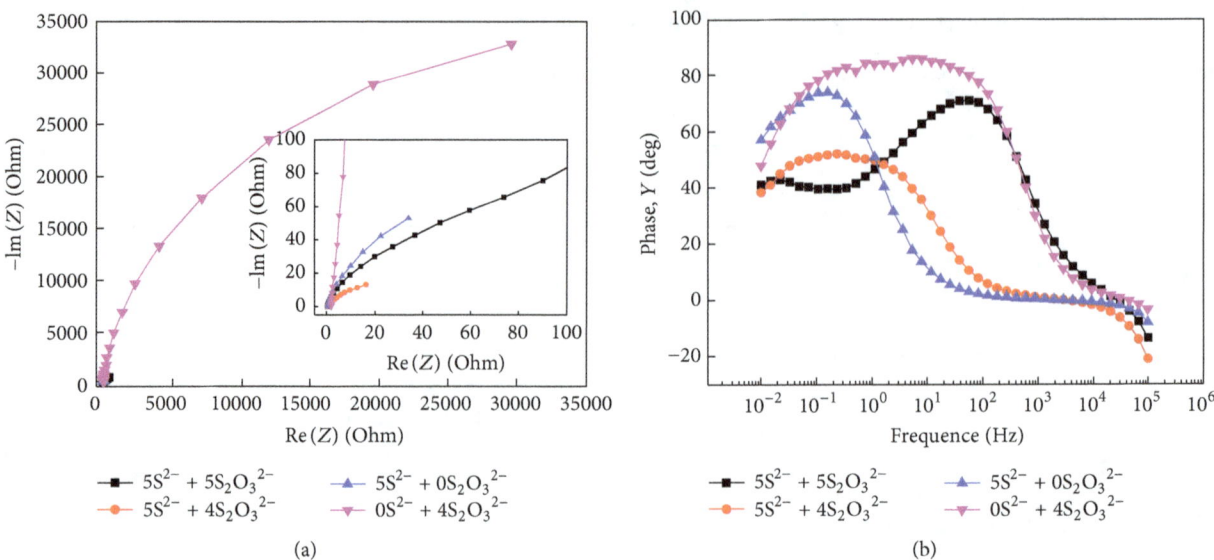

FIGURE 3: EIS of Q235 carbon steel in alkaline solutions with different concentrations of S^{2-} and $S_2O_3^{2-}$ solutions. (a) Nyquist plot and amplificatory plot and (b) Bode plot (phase angle versus frequency).

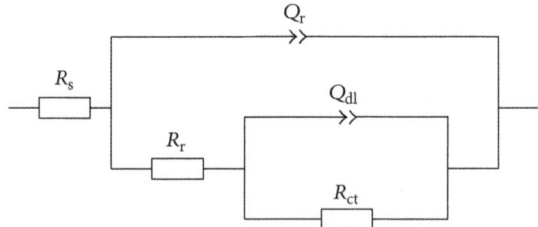

FIGURE 4: Equivalent circuit for EIS of Q235 steels in alkaline.

TABLE 6: XPS peak position and peak result of the corrosion scale of Q235 carbon steel.

Elements		Binging energy (eV)	Peaks	Peaks intensity (%)
Fe	2p3	706.91	FeS_2, III	18.72
		708.19	Fe_3O_4, II	12.54
		710.49	FeS, I	42.62
		712.82	FeS, I	26.12
S	2p1	163.02	FeS, I	22.63
		164.17	FeS_2, III	10.70
	2p3	161.84	FeS, I	45.26
		162.99	FeS_2, III	21.40
O	1s	529.92	Fe_3O_4, II	21.64
		531.50	FeOOH, IV	67.65
		535.16		10.51

beneficial to improve the corrosion resistance of specimen. The S2p spectrum can be curve-fitted with four peaks. The peaks at 163.02, 164.17, 161.84, and 162.99 eV correspond to FeS_2 and FeS. Table 6 shows XPS peak position and peak result analysis of the corrosion scale of Q235 carbon steel, which indicates that the corrosion scale of Q235 carbon steel mainly consisted of FeS, FeS_2, Fe_3O_4, and FeOOH.

4. Conclusions

The corrosion behavior of Q235 carbon steel was investigated using weight loss measurements, SEM-EDS, XPS, and electrochemistry in alkaline solutions adding sulfide and thiosulfate at 328 K. S^{2-} and $S_2O_3^{2-}$ are corrosion activators for Q235 carbon steel when they are separate in alkaline solution. A large number of granular corrosion products accumulated in the defects or the scratches, and severe selective corrosion and pitting corrosion on Q235 carbon steel surfaces were observed. The XPS analysis results indicated that the main corrosion products of specimen induced by S^{2-} and $S_2O_3^{2-}$ were FeS, FeS_2, Fe_3O_4, and FeOOH. The corrosion products

film of Q235 carbon steel is not a uniform layer, and the corrosion mechanism of S^{2-} and $S_2O_3^{2-}$ is different for the corrosion of Q235 carbon steel. The R_f values strongly depended on the stability and compactness of the corrosion products film of steel surface in sulfur-containing alkaline solutions. This paper studied only a production process of alumina industry, and the corrosion of the other aspects still needs further discussion.

Competing Interests

The authors declare that they have no competing interests.

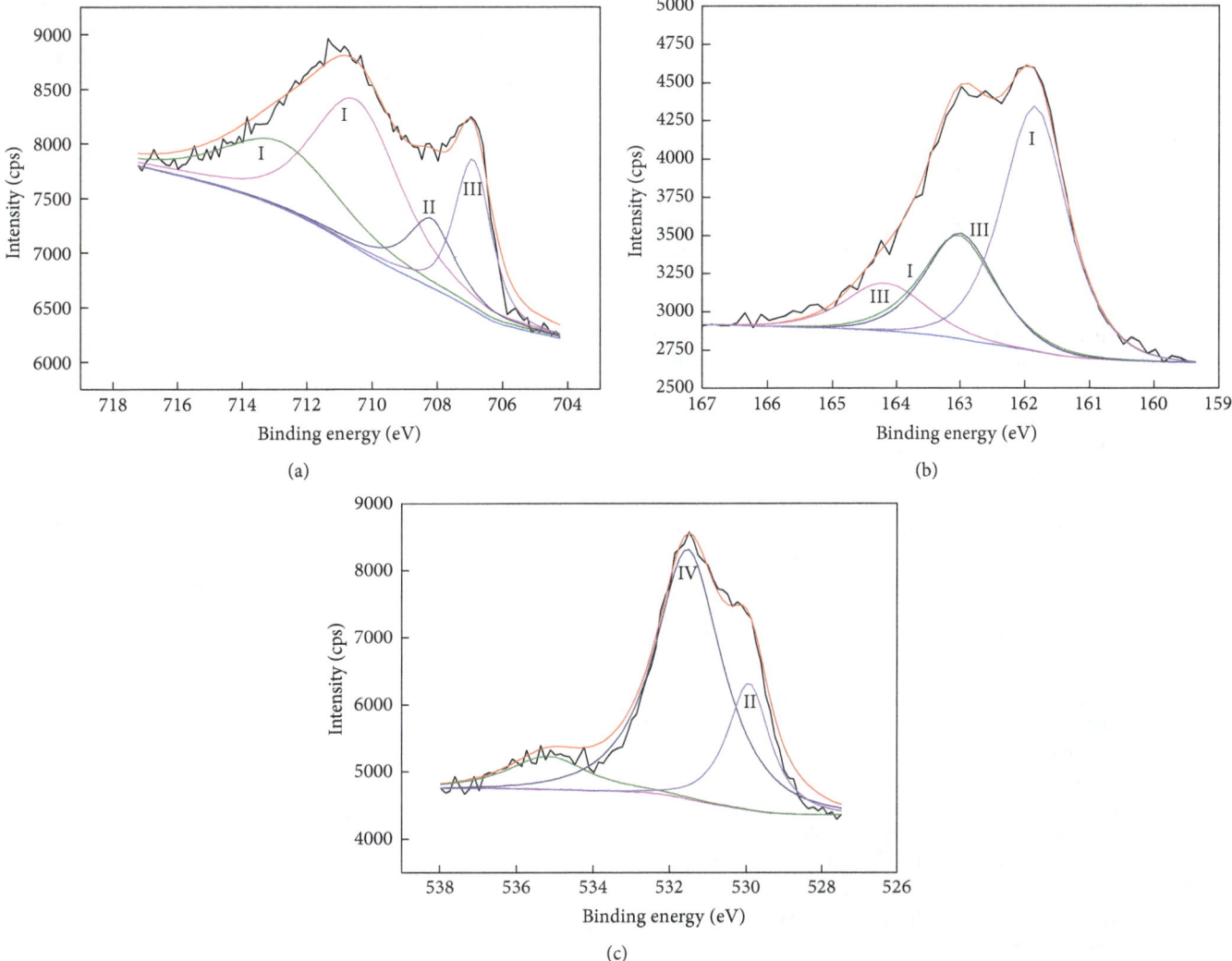

FIGURE 5: XPS patterns Fe 2p (a), S2p (b), and O1s (c) of Q235 steel surface corrosion: (I) FeS; (II) Fe_3O_4; (III) FeS_2; (IV) FeOOH.

Acknowledgments

This work was supported by the National Natured Science Foundation of China (Grant nos. 51264006, 51474079, and 51574095), Guizhou Natural Science Fund ([2014]7609 and [2015]7652), Project supported by Guizhou Education Department (KY [2015]334), NSFC (51264006, 51474079, and 51574095), Guizhou Natural Science Fund ([2014]7609 and [2015]7652), and 125 Key Fund of Education Department of Guizhou Province (2012)0002.

References

[1] S. W. Bi, *Alumina Production Process*, Chemical Industry Press, Beijing, China, 2006.

[2] A. Bhattacharya and P. M. Singh, "Electrochemical behaviour of duplex stainless steels in caustic environment," *Corrosion Science*, vol. 53, no. 1, pp. 71–81, 2011.

[3] Q. Xie and W. Chen, "Corrosion behavior of 16Mn low alloy steel in sulfide-containing Bayer solutions," *Corrosion Science*, vol. 86, pp. 252–260, 2014.

[4] R. E. Melchers, "Mathematical modelling of the diffusion controlled phase in marine immersion corrosion of mild steel," *Corrosion Science*, vol. 45, no. 5, pp. 923–940, 2003.

[5] R. E. Melchers and R. Jeffrey, "Early corrosion of mild steel in seawater," *Corrosion Science*, vol. 47, no. 7, pp. 1678–1693, 2005.

[6] C. Pillay and J. Lin, "The impact of additional nitrates in mild steel corrosion in a seawater/sediment system," *Corrosion Science*, vol. 80, pp. 416–426, 2014.

[7] T. Hemmingsen, F. Fusek, and E. Skavås, "Monitoring of the corrosion process on sulphide film formation with electrochemical and optical measurements," *Electrochimica Acta*, vol. 51, no. 14, pp. 2919–2925, 2006.

[8] E. B. Hansson, M. S. Odziemkowski, and R. W. Gillham, "Formation of poorly crystalline iron monosulfides: surface redox reactions on high purity iron, spectroelectrochemical studies," *Corrosion Science*, vol. 48, no. 11, pp. 3767–3783, 2006.

[9] S. M. Abd El Haleem and E. E. Abd El Aal, "Electrochemical behaviour of iron in alkaline sulphide solutions," *Corrosion Engineering Science and Technology*, vol. 43, no. 2, pp. 173–178, 2008.

[10] W. Z. Shi, L. W. Tong, Y. Y. Chen, Z. Li, and K. Shen, "Experimental study on influence of corrosion on behavior of steel material and steel beams," *Journal of Building Structures China*, vol. 33, no. 7, pp. 53–60, 2012.

[11] L. Yue, *The Research of Behavior and Effect and Mechanism of Rare Earths in Cu-P-RE Weathering Steel*, vol. 26, Northeastern University, Boston, Mass, USA, 2006.

[12] E. Poorqasemi, O. Abootalebi, M. Peikari, and F. Haqdar, "Investigating accuracy of the Tafel extrapolation method in HCl solutions," *Corrosion Science*, vol. 51, no. 5, pp. 1043–1054, 2009.

[13] Q. L. Xie, W. M. Chen, and Q. Yang, "Influence of sulfur anions on corrosion of 16mn low-alloy steel in sulfide-containing bayer solutions," *Corrosion*, vol. 70, no. 8, pp. 842–849, 2014.

[14] P. E. Hazlewood, P. M. Singh, and J. S. Hsieh, "Corrosion behavior of carbon steels in sulfide-containing caustic solutions," *Industrial and Engineering Chemistry Research*, vol. 45, no. 23, pp. 7789–7794, 2006.

[15] H.-Z. Cao, J.-Y. Zhang, G.-Q. Zheng, and J.-G. Yuan, "Polarization behavior of carbon steel in sulfur-bearing solution," *Corrosion & Protection*, vol. 23, no. 10, pp. 427–429, 2002.

[16] Z. M. Zhang, J. Q. Wang, E. H. Han et al., "Effects of surface condition on corrosion and stress corrosion cracking of alloy 690TT," *Journal of Chinese Society for Corrosion and Protection*, vol. 31, no. 6, pp. 441–445, 2011.

[17] X. Su, Z. X. Yin, and Y. F. Cheng, "Corrosion of 16Mn line pipe steel in a simulated soil solution and the implication on its long-term corrosion behavior," *Journal of Materials Engineering and Performance*, vol. 22, no. 2, pp. 498–504, 2013.

[18] M. A. Veloz and I. González, "Electrochemical study of carbon steel corrosion in buffered acetic acid solutions with chlorides and H_2S," *Electrochimica Acta*, vol. 48, no. 2, pp. 135–144, 2002.

[19] J. Y. Zou and D. T. Chin, "Anodic behavior of carbon steel in concentrated NaOH solutions," *Electrochimica Acta*, vol. 33, no. 4, pp. 477–485, 1998.

[20] B. He, C.-H. Lu, P.-J. Han, and X.-H. Bai, "Short-term electrochemical corrosion behavior of pipeline steel in saline sandy environments," *Engineering Failure Analysis*, vol. 59, pp. 410–418, 2016.

[21] H. Liu, D. Xu, A. Q. Dao, G. Zhang, Y. Lv, and H. Liu, "Study of corrosion behavior and mechanism of carbon steel in the presence of *Chlorella vulgaris*," *Corrosion Science*, vol. 101, pp. 84–93, 2015.

[22] J. Kang, J. Li, K.-Y. Zhao, X. Bai, Q.-L. Yong, and J. Su, "Passivation behaviors of super martensitic stainless steel in weak acidic and weak alkaline NaCl solutions," *Journal of Iron and Steel Research International*, vol. 22, no. 12, pp. 1156–1163, 2015.

[23] J. B. Tan, X. Q. Wu, E.-H. Han, X. Q. Liu, X. L. Xu, and H. T. Sun, "The effect of dissolved oxygen on fatigue behavior of Alloy 690 steam generator tubes in borated and lithiated high temperature water," *Corrosion Science*, vol. 102, pp. 394–404, 2016.

[24] L. Jinlong, L. Tongxiang, W. Chen, and G. Ting, "Comparison of corrosion behavior between coarse grained and nano/ultrafine grained alloy 690," *Applied Surface Science*, vol. 360, pp. 403–408, 2016.

[25] H. Luo, H. Su, C. Dong, K. Xiao, and X. Li, "Electrochemical and passivation behavior investigation of ferritic stainless steel in alkaline environment," *Construction and Building Materials*, vol. 96, pp. 502–507, 2015.

20

Study of the Corrosion Process of AZ91D Magnesium Alloy during the First Hours of Immersion in 3.5 wt.% NaCl Solution

Vanessa Mandarano Pinela, Leandro Antônio de Oliveira,
Mara Cristina Lopes de Oliveira ⓘ, and Renato Altobelli Antunes ⓘ

Universidade Federal do ABC (UFABC), Centro de Engenharia, Modelagem e Ciências Sociais Aplicadas (CECS),
09210-580 Santo André, SP, Brazil

Correspondence should be addressed to Renato Altobelli Antunes; renato.antunes@ufabc.edu.br

Academic Editor: Ramazan Solmaz

The AZ91D magnesium alloy was immersed in 3.5 wt.% NaCl solution at room temperature for times ranging from 1 minute up to 72 hours. The aim was to investigate the evolution of the corrosion process using confocal laser scanning microscopy (CLSM), electrochemical impedance spectroscopy, and X-ray photoelectron spectroscopy. The microstructure of the as-received alloy was initially characterized by optical microscopy and scanning electron microscopy (SEM). The crystalline phases were identified by X-ray diffractometry. The main phases were primary-α, eutectic-α, and β ($Mg_{17}Al_{12}$). Vickers microhardness markings were made on the surface of one etched sample to facilitate the identification of the same region at each different immersion time, thus enabling the observation of the corrosion process evolution. Corrosion initiates at the grain boundaries of the eutectic microconstituent and, then, propagates through primary α-grains. The β-phase was less severely attacked.

1. Introduction

Weight reduction is a serious concern in the automotive and aerospace industries. Magnesium alloys are the state-of-the-art materials when high strength-to-weight ratio is pursued. In this respect, they are gaining increasing interest for structural engineering applications [1–3]. Recyclability and good machinability are additional attributes that make them attractive materials to manufacture low weight parts [4]. In spite of these attractive attributes, the well-known chemical instability in aqueous environments is a core issue for magnesium alloys, limiting their applicability [5–7].

AZ91D is one of the most extensively used magnesium-based alloys with consolidated applications in the automotive industry [8, 9]. It is part of the AZ series, being alloyed with Al and Zn. Typically, its microstructure is comprised of a mixture of magnesium-α, an intermetallic β-phase rich in aluminum ($Mg_{17}Al_{12}$), and an eutectic phase consisting of alternating lamellae of the α and β phases [10, 11]. The corrosion behavior of the alloy is markedly affected by its microstructural features. The relative complex mixture of different phases can give rise to regions with distinct chemical activities, thereby triggering the formation of local electrochemical microgalvanic cells, with preferential dissolution of the least noble phases [12–14]. The β-phase is reported to play a core role in this scenario. Its morphology and distribution within the magnesium matrix are key factors for the corrosion resistance of the AZ91D alloy. Large volume fractions and continuous distribution across the matrix are considered to be beneficial to the general corrosion resistance of the alloy by forming a protective layer of aluminum-rich surface oxide. By contrast, if it is discontinuous and concentrated in small areas due to its relative low volume fraction, it can accelerate galvanic corrosion effects, acting as local cathodes [15–18]. In spite of the knowledge accumulated so far and the consensus on the critical role of the β-phase, several authors have recently highlighted the need for further studies on the corrosion mechanism of Mg-Al alloys in order to expand and consolidate their engineering applications [19–22].

Confocal laser scanning microscopy (CLSM) is a powerful tool for investigating the onset of corrosion processes at

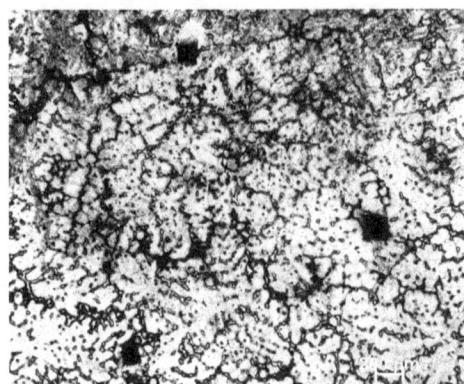

FIGURE 1: CLSM micrograph showing the Vickers impressions on the surface of the etched AZ91D alloy.

either in situ or ex situ conditions [23]. The high magnification and relative short acquisition times can be advantageously exploited in the corrosion field [24]. CLSM was successfully employed to study local corrosion process in aluminum and nickel alloys [25, 26]. Investigations devoted to corrosion of welded joints of high strength steels and pitting corrosion of stainless steels have been reported [27–30]. Notwithstanding, in spite of the increasing interest of magnesium alloys for structural applications, CLSM studies are hardly reported for elucidating the onset of corrosion processes of magnesium alloys in aqueous environments.

In this work, CLSM was employed to investigate the onset of corrosion processes of the AZ91D during short immersion times in 3.5 wt.% NaCl solution at room temperature. The alloy microstructure was initially characterized by optical and scanning electron microscopy (SEM). Corrosion evolution was monitored up to 72 h of immersion.

2. Materials and Methods

2.1. Material and Sample Preparation. A die-cast ingot of the AZ91D magnesium alloy (nominal composition in wt.% 8.3–9.7% Al, 0.35–1.00% Zn, Mn > 0.15%, Si < 0.10%, Fe < 0.005%, Cu < 0.03%, Ni < 0.002%, Mg balance) was kindly provided by Rima Industrial Magnésio S.A. (Brazil). Small square-shaped samples (0.50 cm^2) were cut from the as-cast ingot using a conventional metallographic cut-off machine. The samples were embedded in phenolic resin. Next, surface preparation was carried out by sequential grinding with waterproof SiC paper up to grit #1000, followed by polishing with diamond paste (6 μm). After washing with distilled water and ethanol the samples were chemically etched in a solution comprised of a mixture of glacial acetic acid, ethanol, and distilled water for 10 s. Next, the samples were washed with distilled water and dried in a warm air stream provided by a conventional air blowing gun.

Vickers microhardness indentation tester (EQUILAM HVS-1000) was employed to produce surface marking on the etched samples in order to allow for identifying the same region of the microstructure when subjecting the samples to CLSM analysis. Hence, three impressions were made using 1 kg load as shown by a representative micrograph displayed

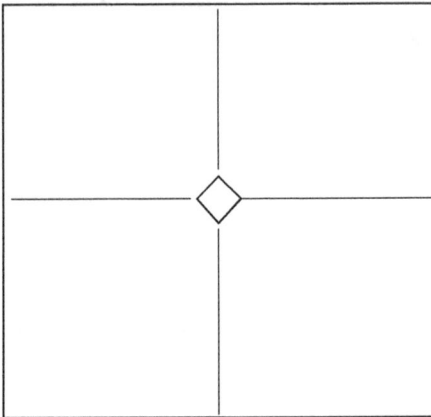

FIGURE 2: Schematic representation of the different quadrants evaluated around each Vickers impression mark.

in Figure 1. The marked regions were, thereafter, employed as a guide to register micrographs of the same region for the samples subjected to corrosion for different times. For each impression mark four different quadrants were evaluated to investigate the evolution of the corrosion process, as schematically shown in Figure 2. The impression mark is at the center of the figure and the evolution of the corrosion process of the AZ91D alloy was prospected at each one of the four quadrants. Thus, twelve different regions were evaluated for each immersion time.

2.2. Microstructural Characterization. X-ray diffraction (XRD) analysis was carried out to identify the crystalline phases of the AZ91D alloy using a Rigaku diffractometer (Multiflex, 40 kV, 20 mA) operating with Cu-kα radiation in the θ-2θ geometry. The analysis was performed in the range from 20 to 90°. SEM micrographs were obtained to examine to morphology of the different phases (JEOL JMS-6010LA). Elemental composition was determined by energy dispersive X-ray analysis (EDS).

2.3. Immersion Test. Samples prepared as described in Section 2.1 were immersed in 3.5 wt.% NaCl solution at room temperature for the following times: 1 minute, 5 minutes,

Study of the Corrosion Process of AZ91D Magnesium Alloy during the First Hours of Immersion...

193

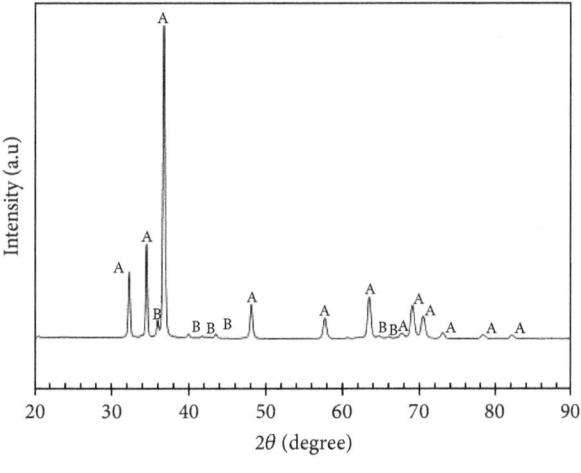

A: α-Mg
B: β-Mg$_{17}$Al$_{12}$

FIGURE 3: XRD pattern of the AZ91D magnesium alloy.

10 minutes, 20 minutes, 30 minutes, 50 minutes, 3 hours, 6 hours, 15 hours, 40 hours, and 72 hours. Only the results obtained for the samples immersed for 1 minute, 30 minutes, 3 hours, 6 hours, 15 hours, 40 hours, and 72 hours are presented. The surface morphologies of samples immersed for 5 minutes, 10 minutes, and 20 minutes are very similar to that observed after 1 minute. In the same respect, the surface morphology of the sample immersed for 50 minutes was very similar to that observed after 30 minutes. As consequence, these micrographs of samples immersed for 5 minutes, 10 minutes, 20 minutes, and 50 minutes are not presented throughout the text. After immersion, the sample was removed from the testing solution, washed with distilled water, dried in a warm air stream, and promptly subjected to CLSM analysis (Olympus OLS4100). The evolution of the corrosion process was accompanied by observing the same region of the microstructure at each time, using the hardness impressions as guides to locate the region of interest.

2.4. Electrochemical Impedance Spectroscopy (EIS). EIS measurements were carried out to evaluate the electrochemical behavior of the AZ91D upon immersion in 3.5 wt.% NaCl solution at room temperature. The data were acquired after the same immersion times evaluated during the immersion test (Section 2.3) in order to complement the characterization of the evolution of the AZ91D corrosion process. A conventional three-electrode cell setup was used for the measurements with a platinum wire as the counter-electrode, Ag/AgCl as reference, and the AZ91D alloy as the working electrode. A small sinusoidal perturbation signal was employed with amplitude of ±10 mV (rms) in the frequency range from 100 kHz to 1 Hz and acquisition of 10 points per frequency decade. The tests were conducted using an Autolab M101 potentiostat/galvanostat.

2.5. X-Ray Photoelectron Spectroscopy (XPS). XPS surface mapping was employed to check the distribution of Mg, Al and O on the surface of the AZ91D alloy after immersion

for different times in 3.5 wt.% NaCl solution at room temperature. The immersion times were the same employed for the immersion test described in Section 3.2. The spectra were obtained using a ThermoFisher Scientific K-alpha+ spectrometer operating with a monochromatic Al-kα radiation source. The spot size was 80 μm. The area map was defined with 16×10 points (approximately 3.0 mm^2).

3. Results and Discussion

3.1. Structural Characterization. Figure 3 shows the XRD pattern of the AZ91D alloy. The main crystalline phase is the magnesium matrix α-Mg. The presence of the intermetallic Mg$_{17}$Al$_{12}$ (β-phase) is unequivocally identified. These results are in agreement with the literature [31, 32].

The alloy microstructure is displayed in the CLSM micrographs shown in Figure 4. A general view is shown in Figure 4(a) whereas a more detailed representation is seen in Figure 4(b). The microstructure consists of a mixture of α-Mg, lamellar eutectic microconstituent along the grain boundaries of the magnesium matrix and β-Mg$_{17}$Al$_{12}$ as labeled in Figure 4(b). These features are typical of the AZ91D alloy [33, 34]. Microstructure was further characterized by SEM and EDS analysis. The SEM micrograph shown in Figure 5(a) reveals the different microconstituents of the AZ91D alloy. Their elemental composition was checked by EDS analysis. Mapping analyses for Al and Mg are displayed in Figures 5(b) and 5(c), respectively. It is seen that aluminum is enriched in the eutectic region with respect to the matrix. Yet, it is mainly present in the β-phase whereas the magnesium signal is much more intense in α-Mg. As inferred from the literature, compositional differences between each phase can give rise to regions with distinct chemical activities and, therefore, promote the formation of microgalvanic cells that trigger corrosion process of the alloy [19].

The results presented in this section were allowed to fully characterize the as-cast microstructure of the AZ91D. Next, the evolution of its corrosion process is discussed based on

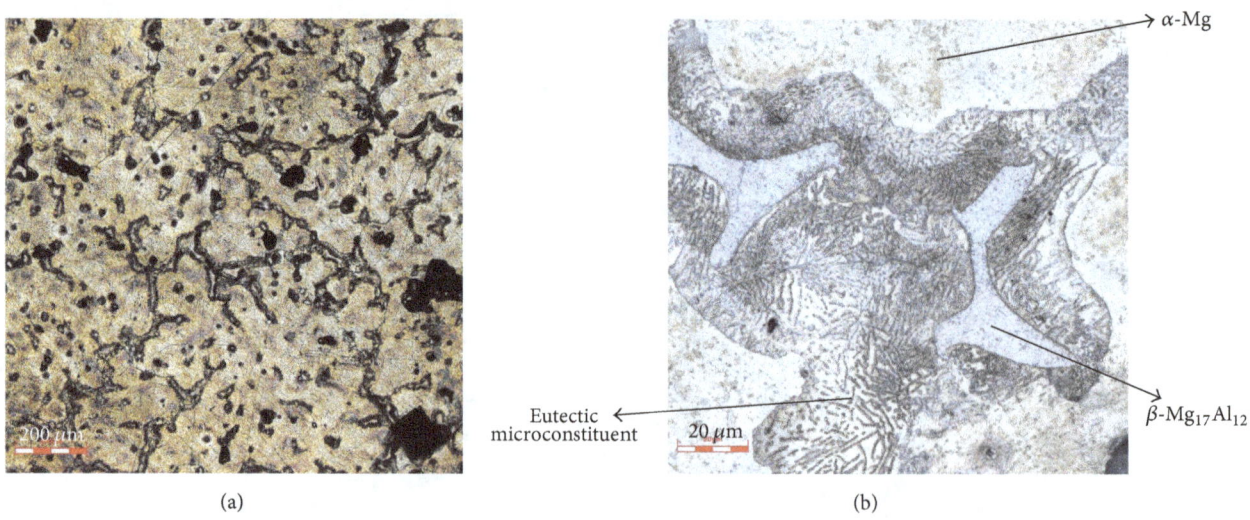

FIGURE 4: CLSM micrographs of the AZ91D alloy: (a) general view; (b) detailed view.

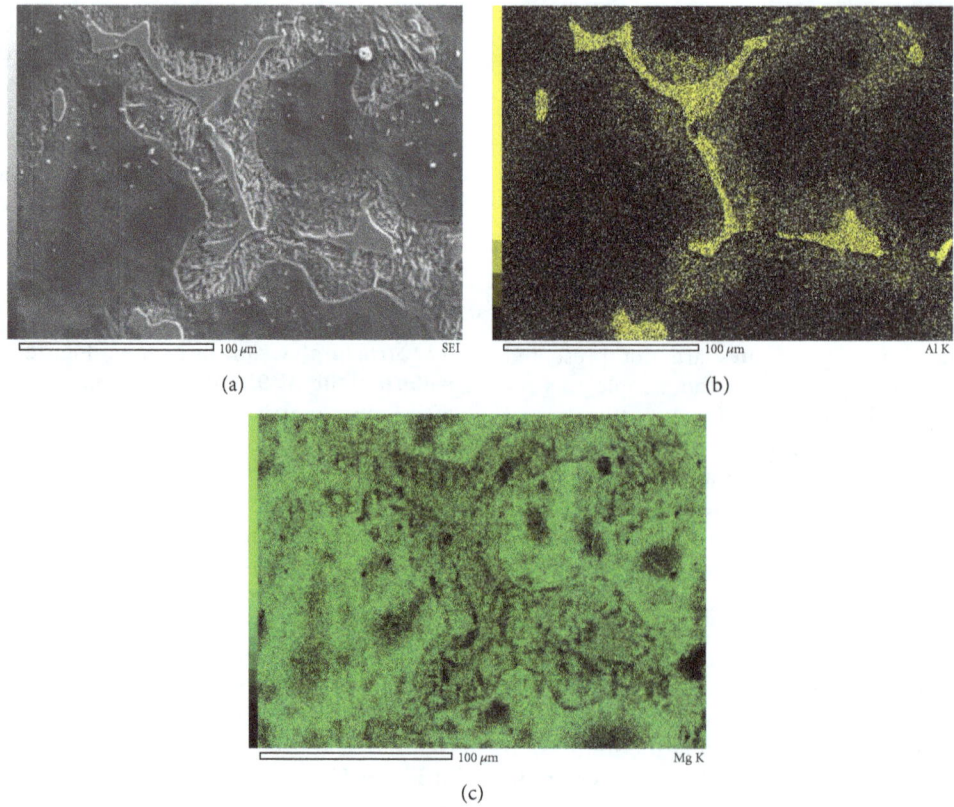

FIGURE 5: (a) SEM micrograph of the AZ91D alloy; (b) EDS mapping analysis for Al; (c) EDS mapping analysis for Mg.

CLSM analyses carried out in accordance with the immersion test described in Section 2.3.

3.2. Immersion Test. Figure 6 shows CLSM micrographs of the AZ91D alloy immersed for up to 3 h in 3.5 wt.% NaCl solution at room temperature. The Vickers microhardness impression is seen at the bottom right of each micrograph. It is somewhat attacked during the chemical etching procedure

employed to reveal the alloy microstructure. The general microstructure can be perceived in the as-received condition (Figure 6(a)). It was little affected by the NaCl solution up to 3 h of immersion. As shown in Section 3.1, the brighter regions are related to the presence of eutectic-α and β-$Mg_{17}Al_{12}$. At this time, corrosion spots were formed as indicated by the small stained region at the lower right part of the micrograph (pointed by the red circle in Figure 6(d)).

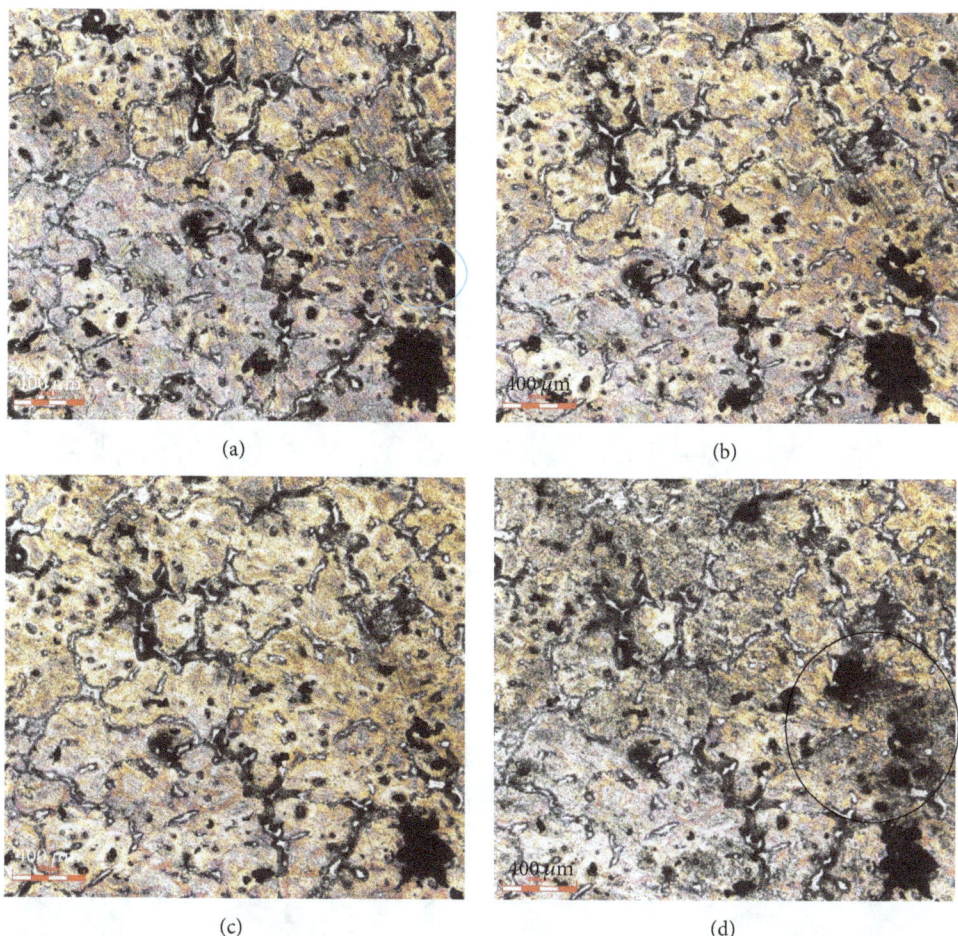

(a)

(b)

(c)

(d)

FIGURE 6: CLSM micrographs of the AZ91D alloy immersed for different times in 3.5 wt.% NaCl solution at room temperature: (a) as-received; (b) 1 min; (c) 30 min; (d) 3 h.

Interestingly, this part of the microstructure presents discontinuous distribution of the β-phase when compared to the upper left part of the micrograph. According to the literature, β-phase morphology is closely related to the onset of corrosion processes of the AZ91D alloy. When it is evenly distributed within the α-Mg matrix, it acts as a barrier against dissolution. If, in turn, it is discontinuous as observed in the region where corrosion spots started to appear, it triggers the formation of microgalvanic cells that drive the localized corrosion attack [15, 16]. Our results point that this mechanism took place for the AZ91D alloy and could be perceived after 3 h of immersion in the electrolyte.

The evolution of the corrosion process was accompanied up to 72 h. Figures 7(a)–7(d) show CLSM micrographs of the same region observed in Figure 6. The micrographs were acquired after 6 h, 15 h, 40 h, and 72 h of immersion in 3.5 wt.% NaCl solution at room temperature. Corrosion spread out slowly. New stained areas are visible after 40 h (Figure 7(c)). Intensively corroded areas were found after 72 h, propagating from right to left (Figure 7(d)). It is noteworthy, though, that some regions remained unaffected by corrosion even after 72 h of immersion. Such regions are mainly the brighter parts of the microstructure, wherein

the β-phase is more evenly distributed. Notwithstanding, corrosion seems to spread even at the β-phase after 72 h (Figure 7(d)).

The morphological features of the corrosion process are not as marked up to 15 h of immersion as they are for 40 h and 72 h. In this respect, in order to give a more clear interpretation of the corrosion spots formed at first hours of immersion, transverse profiles were obtained from the CSLM micrographs as shown in Figure 8 for the as-received condition (Figure 6(a)) and for the specimen immersed for 15 h in 3.5 wt.% NaCl solution at room temperature (Figure 7(b)). These lines were taken from the lower part of the CLSM micrographs shown in Figures 6(a) and 7(b) for the as-received and 15 h conditions, respectively. It is important to mention, though, that the first corrosion spots were observed after 3 h of immersion, as shown in Figure 6(d), and the micrograph from the 15 h condition (Figure 7(b)) was chosen only to exemplify how the transverse profile of the corroded area appears when evaluated using the CLSM measuring tool.

The profiles show the more intense variation of the height for the specimen immersed for 15 h. Moreover, it is clear that a depressive region appear at right for the 15 h condition as indicated by the black circle. Such region was not observed in

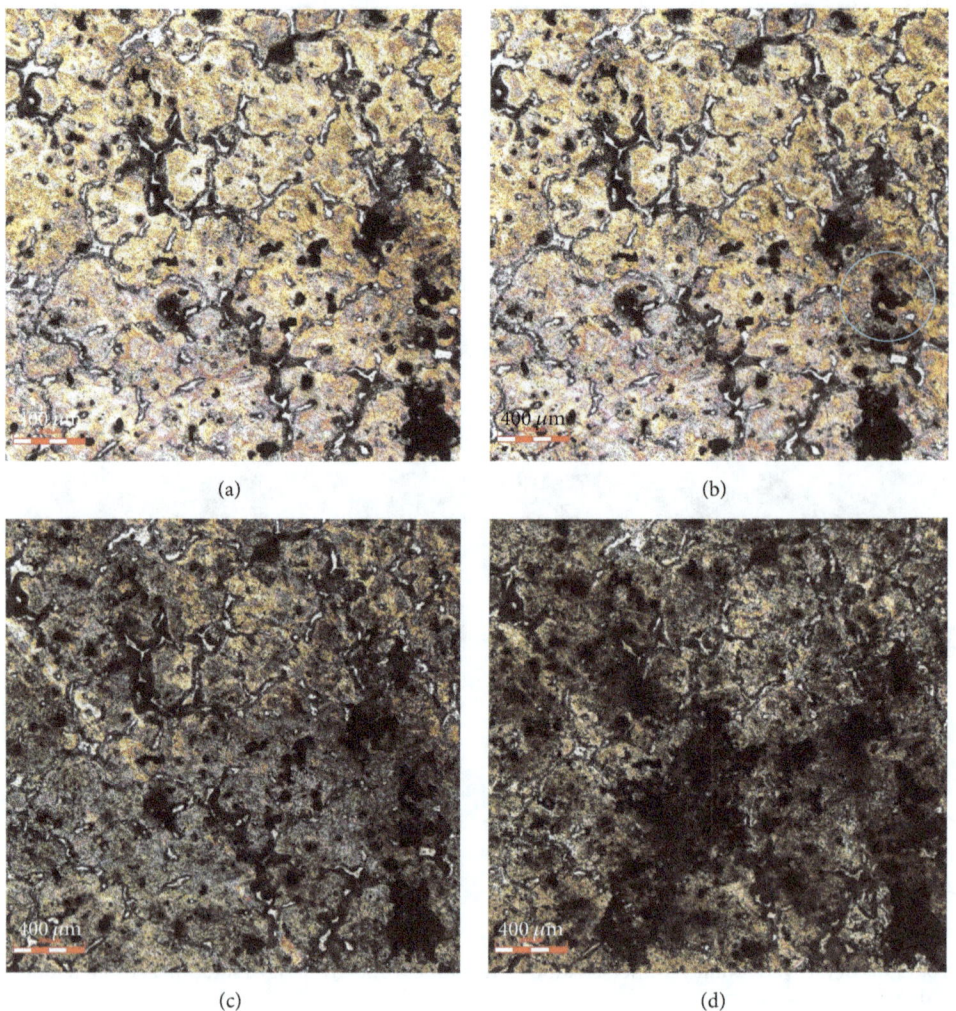

FIGURE 7: CLSM micrographs of the AZ91D alloy immersed for different times in 3.5 wt.% NaCl solution at room temperature: (a) 6 h; (b) 15 h; (c) 40 h; (d) 72 h.

the profile obtained for the as-received alloy. At this point, it is important to correlate these profiles with the corresponding microstructures so that the corroded areas can be clearly related to the height variation in the z-axis. Hence, the blue circles shown in Figures 6(a) and 7(b) were inserted to indicate their corresponding regions in the transverse profiles displayed in Figures 8(a) and 8(b), respectively.

By carefully examining the CLSM micrographs for the as-received and 15 h immersed alloy and their corresponding transverse profiles at the regions marked in Figures 6(a) and 7(b), it is noteworthy that the depressive region marked by the black circle in Figure 8(a) corresponds to the region where the corrosion spots firstly appear in the AZ91D microstructure. This corroded region was perceived after 3 h of immersion (Figure 6(d)). CLSM analysis aimed to successfully indicate the onset of corrosion spots by combining their visual identification at the micrograph with the quantitative evaluation of the surface profile.

Notwithstanding, in spite of the interpretation given above, the micrographs shown in Figures 6 and 7 did not allow

one to clearly identify the microstructural features related to the onset of corrosion. Higher magnification micrographs are needed to show the corroded regions in detail, allowing identification of the different phases of the AZ91D alloy. In this regard, higher magnification CLSM micrographs of the AZ91D alloy immersed for 6 h, 15 h, 40 h, and 72 h in 3.5 wt.% NaCl solution at room temperature are shown in Figures 9(a)–9(d). These periods were selected as the initial corrosion spots and their evolution could be promptly examined from their micrographs. Shorter immersion times did not allow an unequivocal visualization of the initial corrosion sites.

Figure 9(a) shows dark points indicating the onset of corrosion mainly at the grain boundaries of the eutectic microconstituent, along with the boundaries of the β-phase. The β-phase, in turn, is not affected. After 15 h of immersion (Figure 9(b)), it remains unaffected, but there were several dark spots indicating corrosion within the α-Mg matrix and also at the eutectic microconstituent. This scenario holds up to 40 h of immersion (Figure 9(c)). Corrosion propagated throughout the whole microstructure after 72 h (Figure 9(d)).

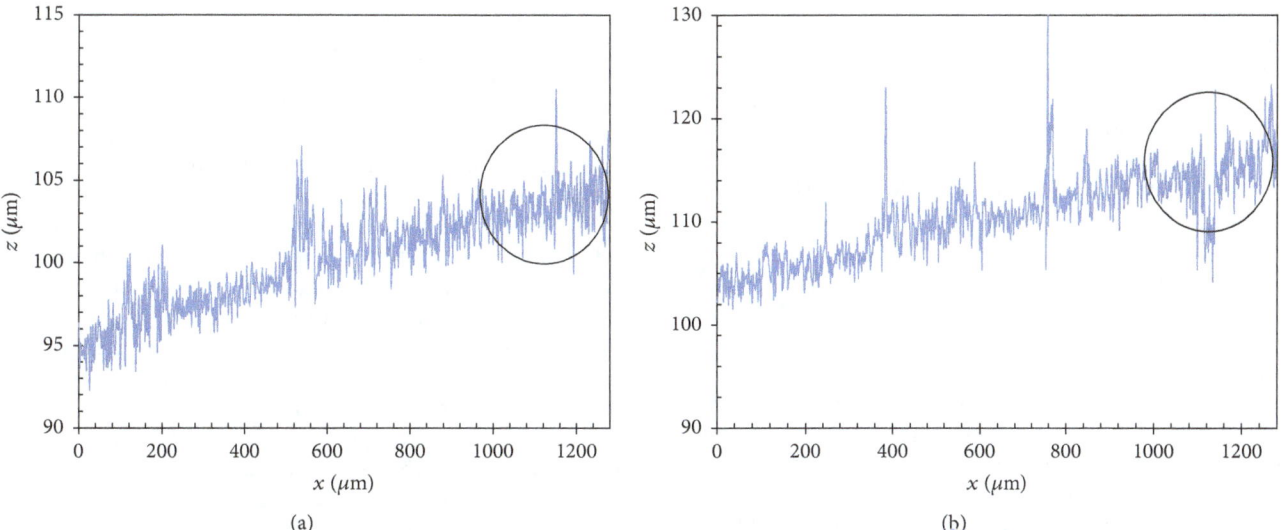

(a)

(b)

FIGURE 8: Transverse profiles obtained from the CLSM micrographs of the AZ91D alloy: (a) the as-received condition; (b) after immersion for 15 h in 3.5 wt.% NaCl solution at room temperature.

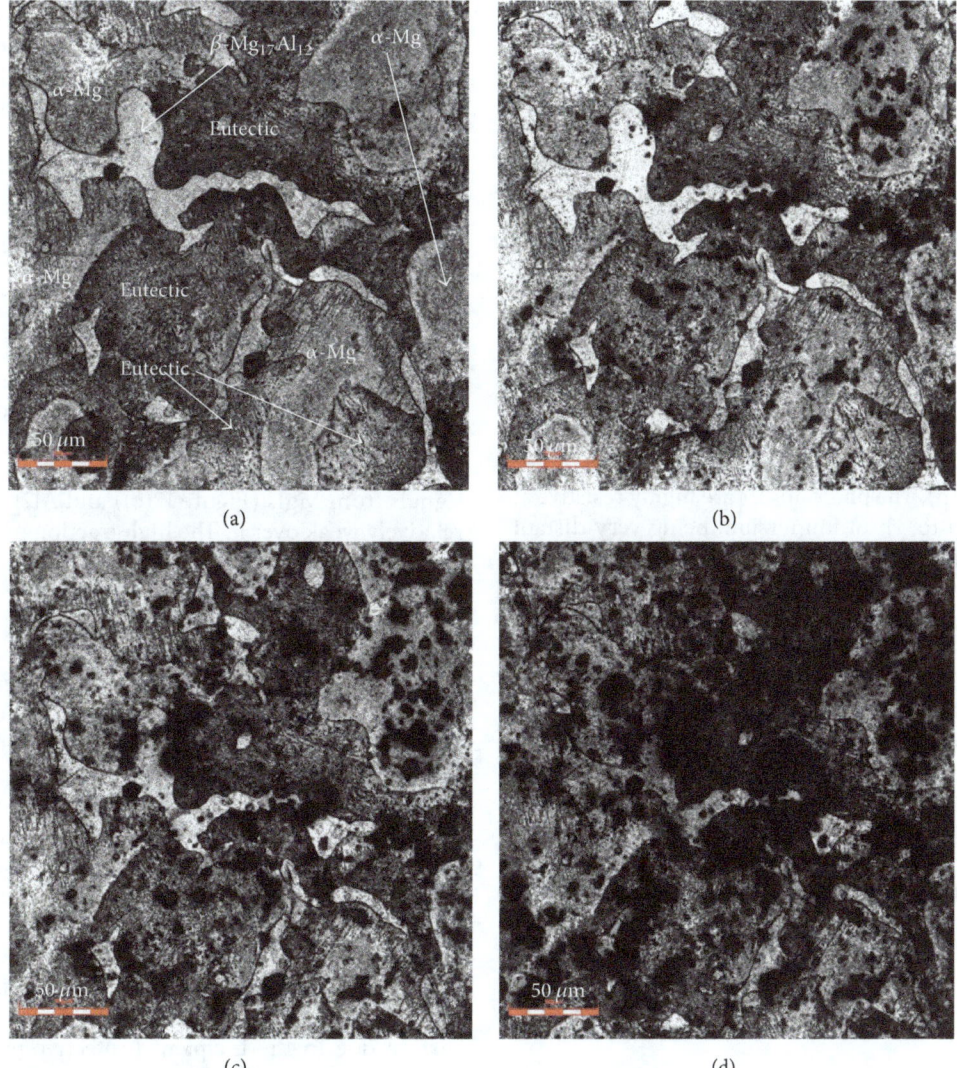

(c)

(d)

FIGURE 9: CLSM micrographs showing the evolution of the corrosion process of the AZ91D alloy in 3.5 wt.% NaCl solution at room temperature: (a) 6 h; (b) 15 h; (c) 40 h; (d) 72 h.

The β-phase was also dissolved. The eutectic microconstituent was severely attacked as well as the α-Mg matrix.

The evolution of the corrosion process can be checked by examining 3D CLSM micrographs of the AZ91D alloy and the corresponding roughness profile along specific lines. This is exemplified in Figures 10(a)–10(c) for the AZ91D alloy immersed for 6 h, 15 h, and 72 h in NaCl 3.5 wt.% at room temperature.

3.3. Electrochemical Impedance Spectroscopy (EIS). EIS diagrams of the AZ91D alloy immersed in 3.5 wt.% NaCl solution at room temperature for different times are shown in Figure 11.

Nyquist plots (Figure 11(a)) are characterized by capacitive loops whose diameters depend on the immersion time. It is well-known that the capacitive loop diameter is associated with the charge transfer resistance of the electrode surface, being, therefore, related to its corrosion resistance [35, 36]. In this respect, its evolution with the immersion can be an indication of the corrosion reactions taking place on the alloy surface. From Figure 11(a) one can observe that there is a marked decrease of the impedance values from the 1 min condition up to 3 h. The capacitive loops of the 30 min and 3 h conditions are very small compared to the 1 min loop. Thus, in order to more clearly observe them with respect to the other immersion times, the inset of Figure 11(a) shows the Nyquist plots with expanded scales. This would indicate that corrosion reactions are taking place right after immersion in the electrolyte. As a consequence, the alloy surface becomes oxidized. This, in turn, would lead to the formation of a thicker oxide layer, giving rise to an impedance rise which is confirmed by the increase of the capacitive loop diameter for the 6 h and 15 h immersion times. It is noteworthy that the onset of visual corrosion spots were observed for these times, as described in Section 3.2.

Additional evidence for the reasoning derived from the Nyquist plots can be achieved by evaluating the Bode plots shown in Figure 11(b). The phase angles are highly resistive at low frequencies up to 3 h of immersion, being very distant from the ideal capacitive response at −90° which denotes the propensity of the electrode surface to charge transfer reactions [37]. As corrosion proceeds, the phase angles rise up to −40° and −50° for the 6 h and 15 h conditions, leading to a more capacitive response with the development of a more intensively corroded layer. However, as this layer is not compact, it is not protective against corrosion and the phase angles decrease at the low frequencies for the 40 h immersion condition. A new rise toward more capacitive values is found after 72 h as a consequence of the intensified corrosion process. EIS proved to be very sensitive to the evolution of the corrosion process on the AZ91D, allowing one to correlate its results with those obtained from the immersion test (Section 3.2). Further information on the composition of the corroded layer and its correlation with the immersion time was obtained by XPS analysis, as described in the next section.

3.4. X-Ray Photoelectron Spectroscopy (XPS). XPS mapping was used to elucidate the elemental distribution of magnesium, aluminum and oxygen over the AZ91D surface at the same immersion times evaluated by CLSM and EIS. The results are shown in Figures 12–19. The data are expressed as quantified atomic percentage XPS maps for Mg1s, O1s, and Al2p. Only these signals were detected in significant quantities on the surface of the AZ91D alloy independently of the immersion time. The maps were interpolated up to an equivalent 128 × 128 pixel density using the Avantage software©.

The maps for the as-received alloy (Figure 12) show that Al2p is present up to 9.0 at% in some areas whereas small regions present a very small amount. For the most part of the area the atomic concentration of Al2p is near the highest value of the scale. Mg1s predominates across the whole surface but it is less intense where the O1s signal is more intense. The O1s signal reaches a maximum of 39.7 at.%.

This scenario promptly changes after 1 minute of immersion (Figure 13). The Al2p maximum signal decreased to 6.37 at.% whereas that of O1s increased to 56.7 at.%. The Mg1s signal became less intense over the upper part of the area where the O1s signal is the most intense. XPS mapping indicates that the surface of the AZ91D alloy became more oxidized right after 1 minute of immersion in 3.5 wt.% NaCl solution at room temperature.

This trend was intensified for the sample immersed for 30 minutes (Figure 14). The area where the Al2p signal is weak increased with respect to the as-received alloy (Figure 14(a)). The O1s signal, in turn, is strong over a bigger part of the probed area (Figure 14(b)) whereas the Mg1s atomic concentration is at the lowest level of the scale in most part of the area (Figure 14(c)).

As corrosion proceeds for longer times the O1s signal remains high over the whole area for 3 h and 6 h of immersion (Figures 15(b) and 16(b)) whereas the signal of Al2p is weak (Figures 15(a) and 16(a)). The Mg1s signal does not follow a clear trend, being less intense at 6 h (Figure 16(c)) than at 3 h (Figure 15(c)). For the sample immersed for 15 h, though, the O1s atomic concentration is relatively high and spreads over almost the whole surface (Figure 17(b)) whereas the signals from Mg1s (Figure 17(c)) and Al2p (Figure 17(a)) are relatively weak over it. The high fraction of the O1s signal for this condition allows one to hypothesize that the impedance increase observed from 6 h to 15 h of immersion (Figure 11(a)) is actually due to the formation of a more spread oxide layer on the surface of the AZ91D alloy after 15 h.

After 40 h of immersion the Al2p signal (Figure 18(a)) seems to be more spread over the whole surface than at 15 h. The Mg1s signal (Figure 18(c)) is more intense than at 15 h whereas the O1s signal presents an opposite trend. For the 72 h condition the Al2p signal decreased with respect to the 40 h condition (Figure 19(a)), the O1s signal increased (Figure 19(b)), and the Mg1s signal decreased (Figure 19(c)).

The more intense Al2p and Mg1s signals observed for the sample immersed for 40 h can be related to the lower impedance values of this condition when compared to the sample immersed for 15 h (Figure 11(a)). In this respect, the charge transfer reactions would take place at the electrode surface due to the incipient protective oxide layer formed during immersion. This layer would be deteriorated from 15 h to 40 h, exposing more aluminum and magnesium to the NaCl solution. As a consequence, the impedance values are

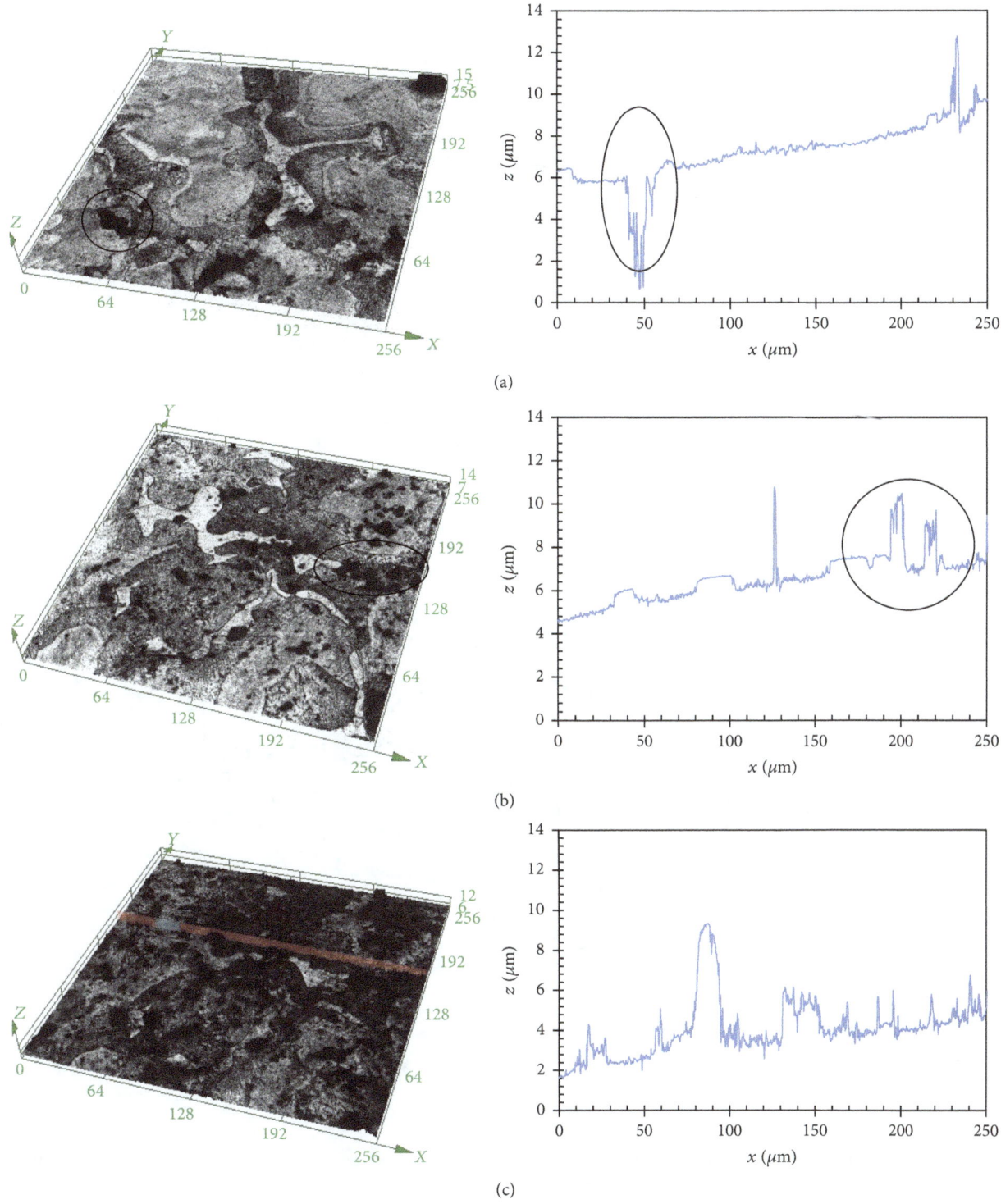

FIGURE 10: CLSM 3D images and corresponding roughness profiles for the AZ91D immersed in 3.5 wt.% NaCl solution for different times: (a) 6 h; (b) 15 h; (c) 72 h.

low at 40 h than at 15 h. Next, as the oxidation process evolves on the surface of the AZ91D alloy, the oxygen signal becomes more intense and the oxide layer formed at 72 h would lead to an increase of the impedance values with respect to the 40 h period, as observed in Figure 11(a).

XPS mapping and EIS measurements gave valuable results, regarding the chemical stability and composition of the oxide layer formed on the surface of the AZ91D alloy upon the first hours of immersion in 3.5 wt.% NaCl solution at room temperature. However, they do not allow identification

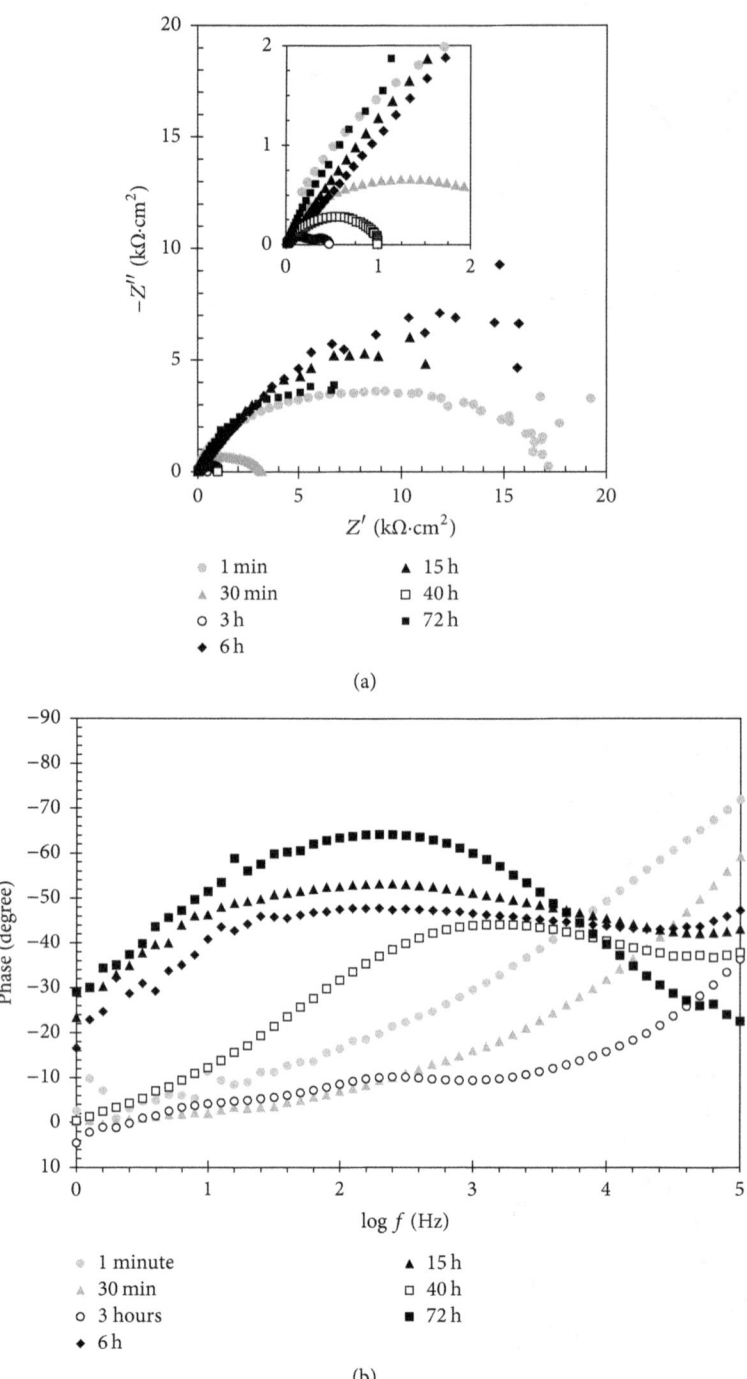

FIGURE 11: EIS diagrams of the AZ91D alloy immersed in 3.5 wt.% NaCl solution at room temperature for different times: (a) Nyquist and (b) Bode (phase angle versus log f) plots.

of the microstructural features associated with the first corrosion spots as CLSM does. In this respect, coupling these techniques is an interesting route to understand the corrosion process of the AZ91D alloy.

4. Conclusions

The evolution of the corrosion process of the AZ91D magnesium alloy was examined by CLSM during the first hours of immersion in 3.5 wt.% NaCl solution at room temperature. This technique proved to be a valuable tool for assessing the onset of corrosion spots in the alloy microstructure. Corrosion started at the interface between the eutectic microconstituent and the β-phase after 6 h of immersion in the electrolyte. Next, it propagated within the eutectic microconstituent and the magnesium matrix. The β-phase remained unaffected by the corrosion process up to 40 h of immersion. After 72 h, the microstructure was severely

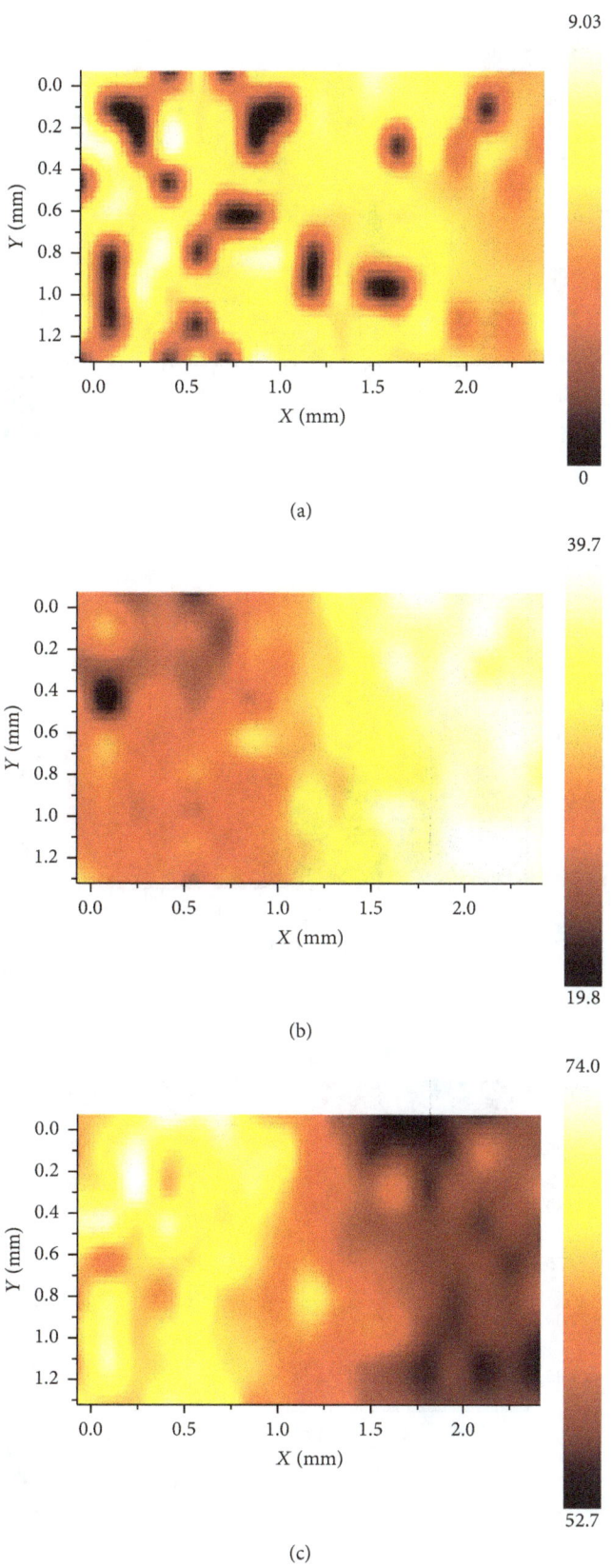

FIGURE 12: XPS maps for the atomic percentages of (a) Al2p; (b) O1s; and (c) Mg1s for the as-received AZ91D alloy.

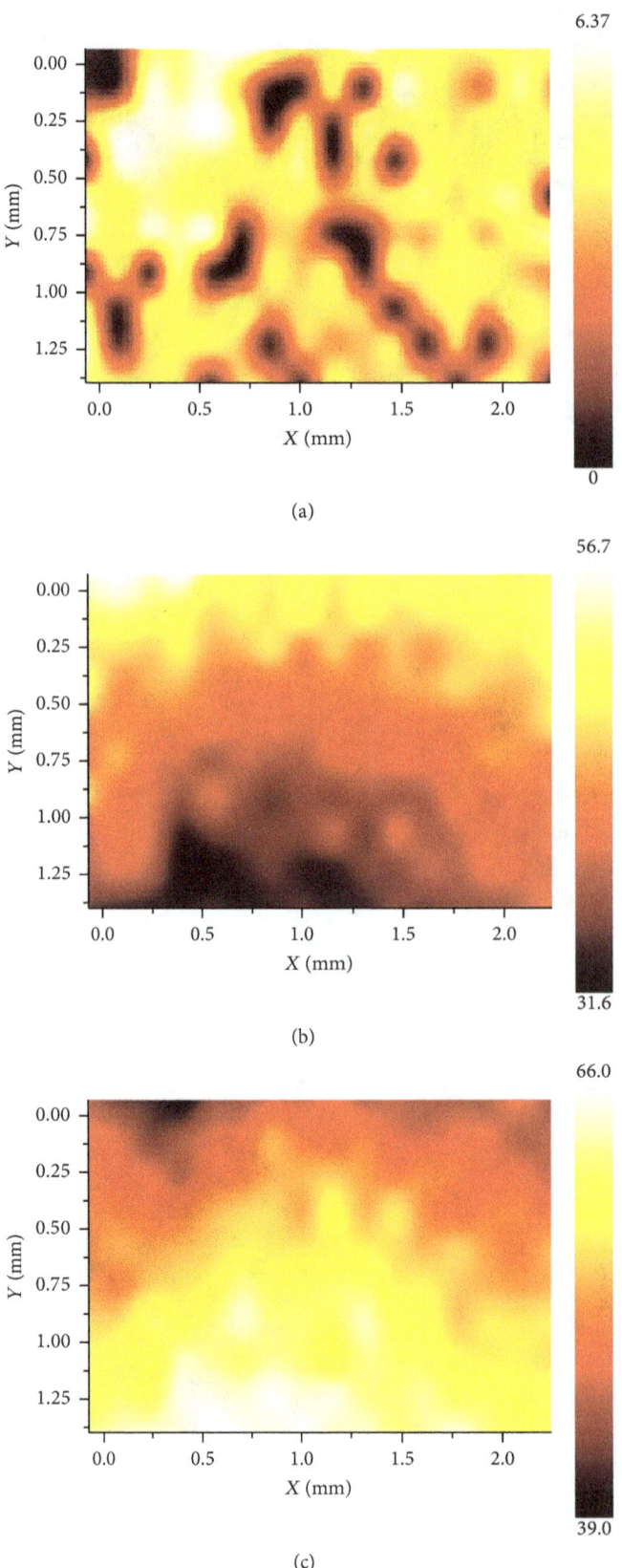

FIGURE 13: XPS maps for the atomic percentages of (a) Al2p; (b) O1s; and (c) Mg1s for the AZ91D alloy immersed for 1 minute in 3.5 wt.% NaCl solution at room temperature.

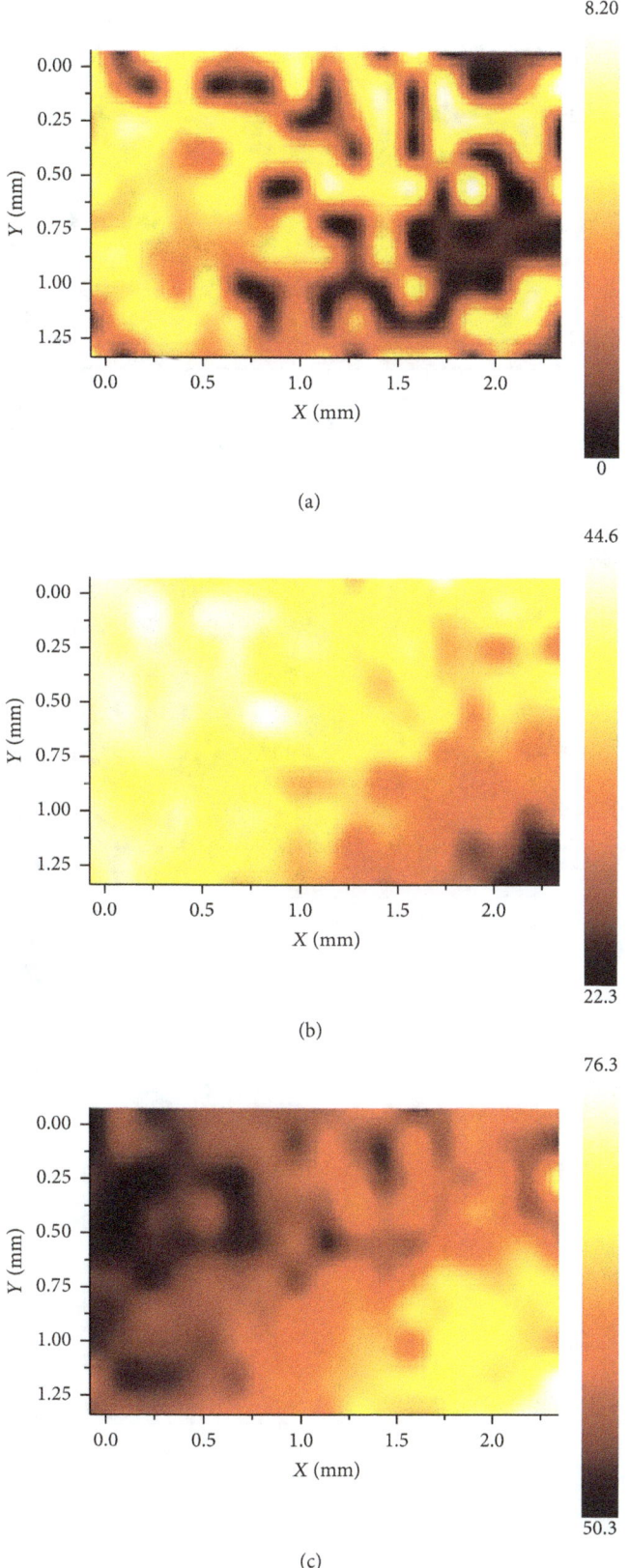

FIGURE 14: XPS maps for the atomic percentages of (a) Al2p; (b) O1s; and (c) Mg1s for the AZ91D alloy immersed for 30 minutes in 3.5 wt.% NaCl solution at room temperature.

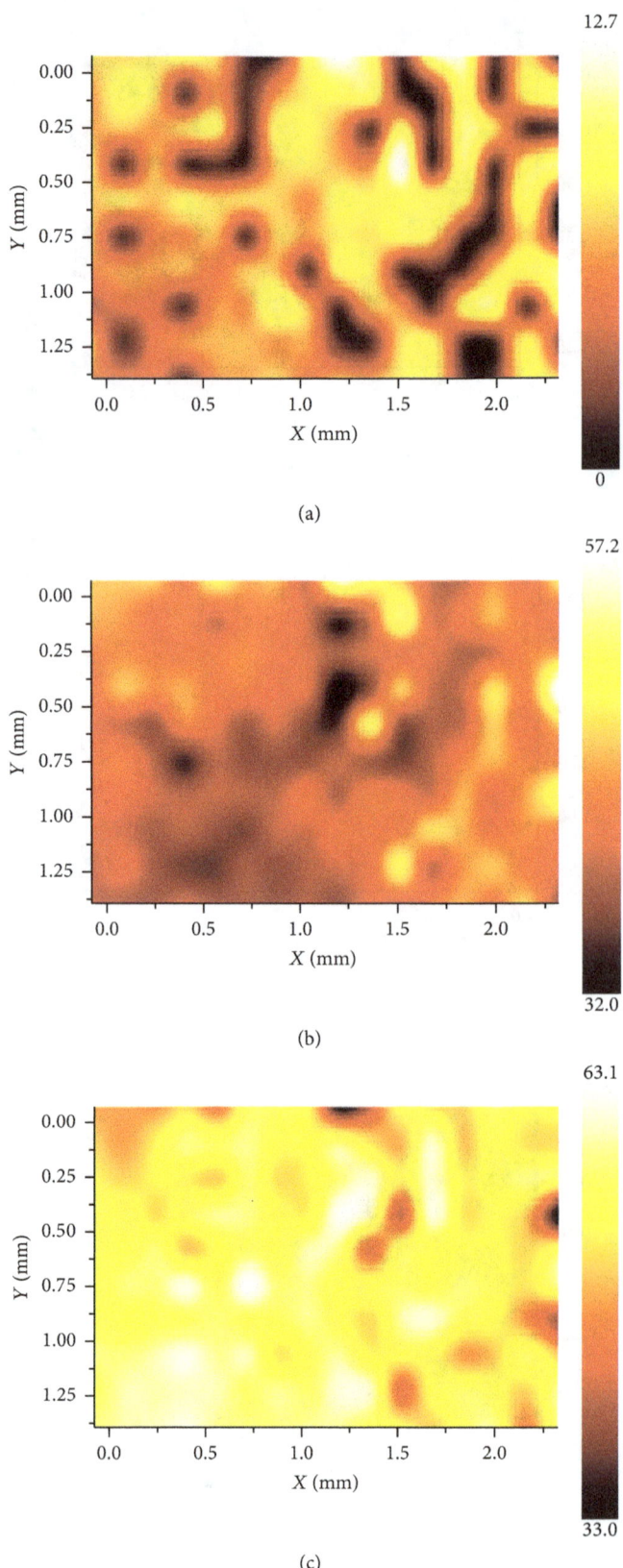

FIGURE 15: XPS maps for the atomic percentages of (a) Al2p; (b) O1s; and (c) Mg1s for the AZ91D alloy immersed for 3 h in 3.5 wt.% NaCl solution at room temperature.

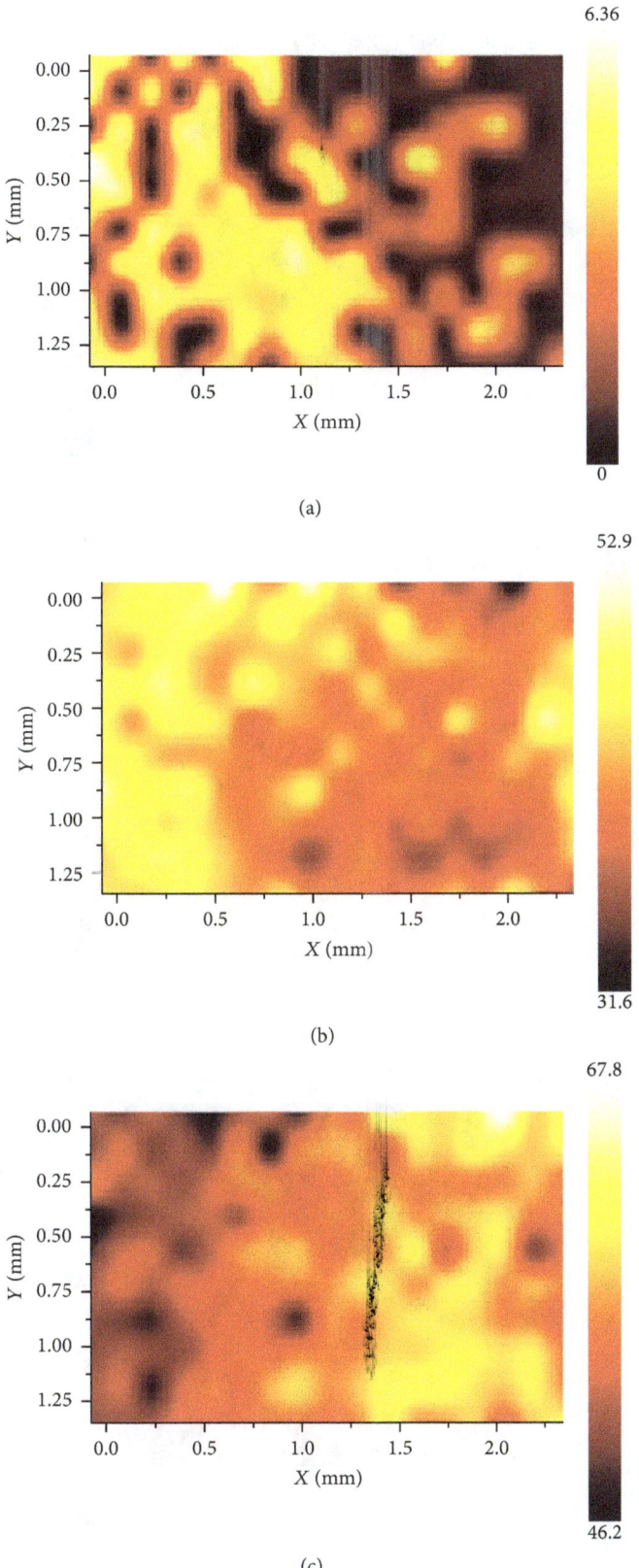

FIGURE 16: XPS maps for the atomic percentages of (a) Al2p; (b) O1s; and (c) Mg1s for the AZ91D alloy immersed for 6 h in 3.5 wt.% NaCl solution at room temperature.

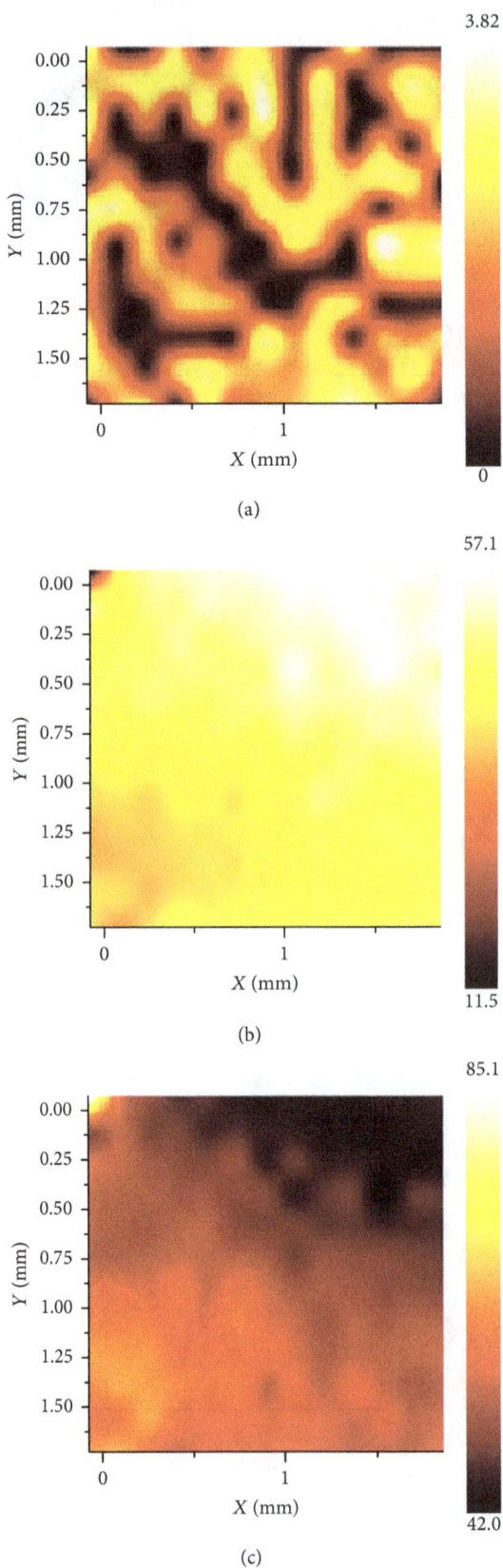

FIGURE 17: XPS maps for the atomic percentages of (a) Al2p; (b) O1s; and (c) Mg1s for the AZ91D alloy immersed for 15 h in 3.5 wt.% NaCl solution at room temperature.

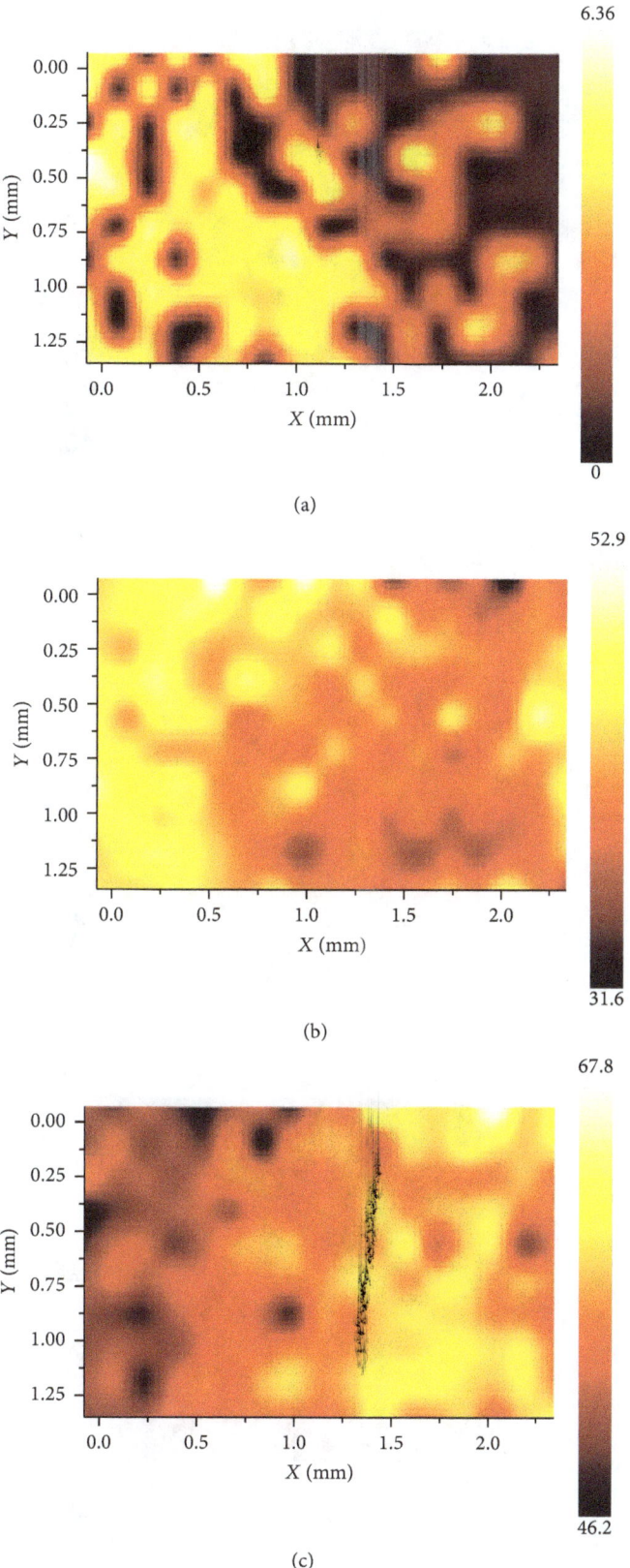

(a)

(b)

(c)

FIGURE 16: XPS maps for the atomic percentages of (a) Al2p; (b) O1s; and (c) Mg1s for the AZ91D alloy immersed for 6 h in 3.5 wt.% NaCl solution at room temperature.

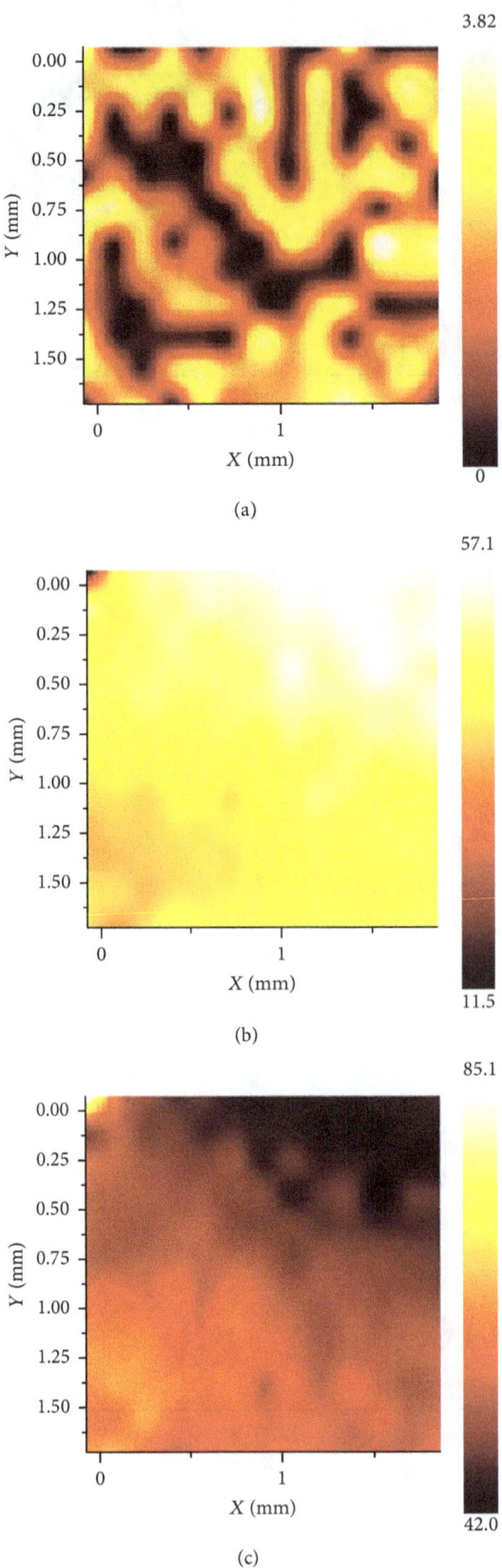

FIGURE 17: XPS maps for the atomic percentages of (a) Al2p; (b) O1s; and (c) Mg1s for the AZ91D alloy immersed for 15 h in 3.5 wt.% NaCl solution at room temperature.

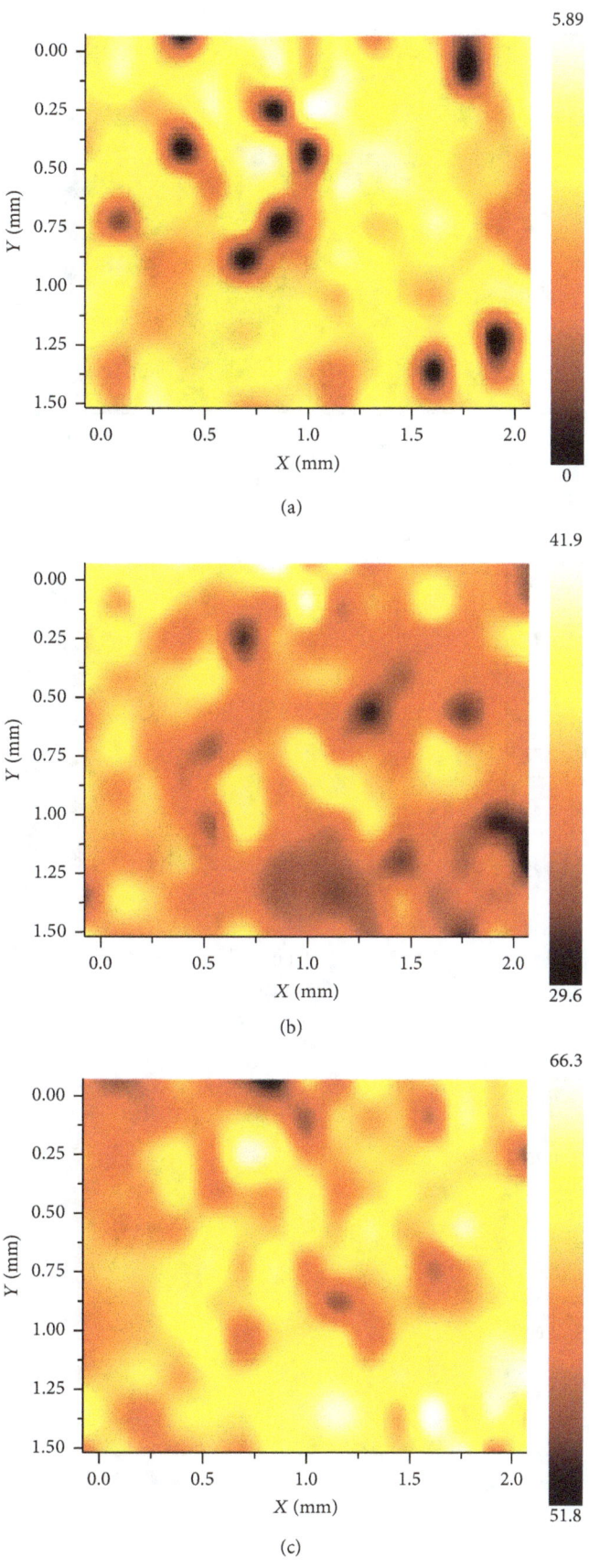

FIGURE 18: XPS maps for the atomic percentages of (a) Al2p; (b) O1s; and (c) Mg1s for the AZ91D alloy immersed for 40 h in 3.5 wt.% NaCl solution at room temperature.

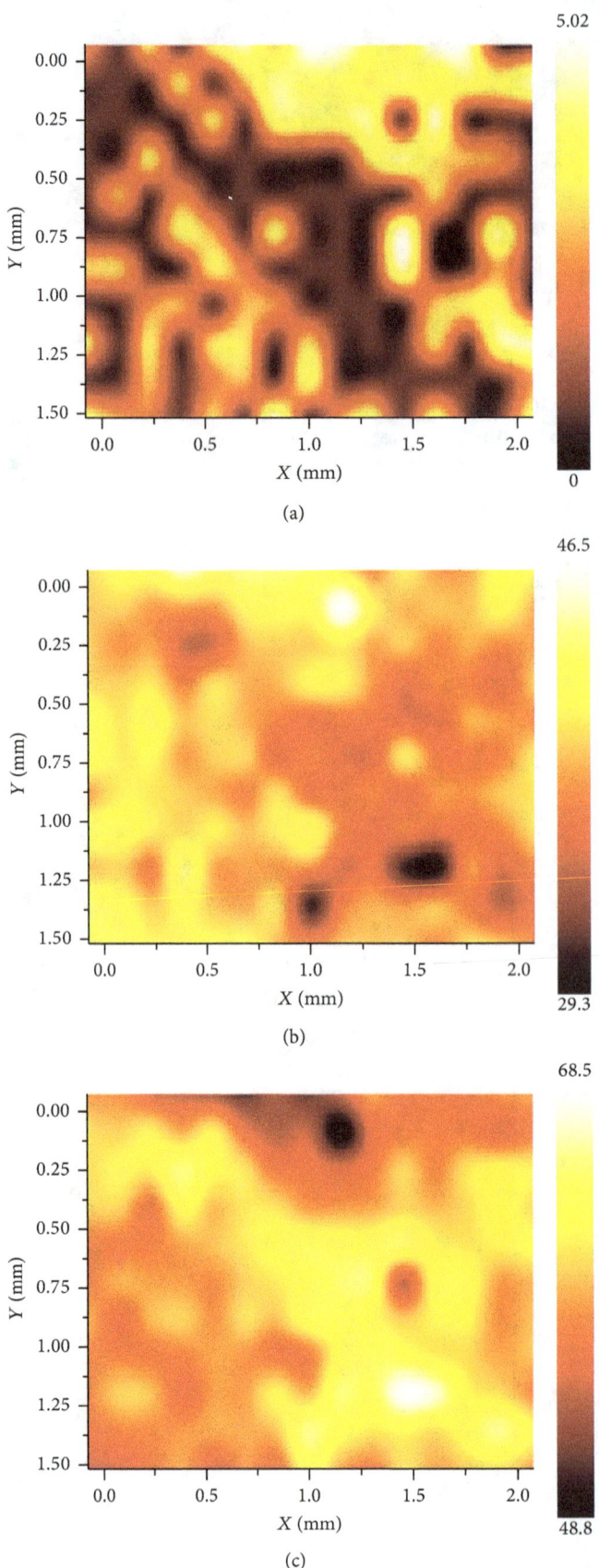

FIGURE 19: XPS maps for the atomic percentages of (a) Al2p; (b) O1s; and (c) Mg1s for the AZ91D alloy immersed for 72 h in 3.5 wt.% NaCl solution at room temperature.

corroded, even at the β-phase. Corrosion was found to preferentially begin at the regions where the β-phase was discontinuously distributed within the magnesium matrix. XPS mapping revealed that the oxygen atomic concentration was dependent on the immersion time and the composition of the oxide layer was related to the electrochemical behavior observed by EIS.

Conflicts of Interest

The authors declare that they have no conflicts of interest.

Acknowledgments

Rima Industrial Magnésio (Brazil) is kindly acknowledged for providing the AZ91D ingot studied in this work. The authors are also thankful to Dr. Nelson Batista de Lima (IPEN/CNEN-SP) for the X-ray diffraction analysis.

References

[1] Y. Tian, L.-J. Yang, Y.-F. Li et al., "Corrosion behaviour of diecast AZ91D magnesium alloys in sodium sulphate solutions with different pH values," *Transactions of Nonferrous Metals Society of China*, vol. 21, no. 4, pp. 912–920, 2011.

[2] T. Li, H. Zhang, Y. He, and X. Wang, "Comparison of corrosion behavior of Mg-1.5Zn-0.6Zr and AZ91D alloys in a NaCl solution," *Materials and Corrosion*, vol. 66, no. 1, pp. 7–15, 2015.

[3] M. Laleh and F. Kargar, "Effect of surface nanocrystallization on the microstructural and corrosion characteristics of AZ91D magnesium alloy," *Journal of Alloys and Compounds*, vol. 509, no. 37, pp. 9150–9156, 2011.

[4] A. A. Luo, "Magnesium casting technology for structural applications," *Journal of Magnesium and Alloys*, vol. 1, no. 1, pp. 2–22, 2013.

[5] M. O. Oteyaka, E. Ghali, and R. Tremblay, "Corrosion behaviour of AZ and ZA magnesium alloys in alkaline chloride media," *International Journal of Corrosion*, vol. 2012, Article ID 452631, 10 pages, 2012.

[6] D. Y. Hwang, Y. M. Kim, D.-Y. Park, B. Yoo, and D. H. Shin, "Corrosion resistance of oxide layers formed on AZ91 Mg alloy in $KMnO_4$ electrolyte by plasma electrolytic oxidation," *Electrochimica Acta*, vol. 54, no. 23, pp. 5479–5485, 2009.

[7] M. C. L. De Oliveira, V. S. M. Pereira, O. V. Correa, and R. A. Antunes, "Corrosion performance of anodized AZ91D magnesium alloy: effect of the anodizing potential on the film structure and corrosion behavior," *Journal of Materials Engineering and Performance*, vol. 23, no. 2, pp. 593–603, 2014.

[8] L. Zhang, Q. Wang, W. Liao et al., "Effect of homogenization on the microstructure and mechanical properties of the repetitive-upsetting processed AZ91D alloy," *Journal of Materials Science and Technology*, vol. 33, no. 9, pp. 935–940, 2017.

[9] Z. Li, A. A. Luo, Q. Wang, H. Zou, J. Dai, and L. Peng, "Fatigue characteristics of sand-cast AZ91D magnesium alloy," *Journal of Magnesium and Alloys*, vol. 5, no. 1, pp. 1–12, 2017.

[10] G. Ballerini, U. Bardi, R. Bignucolo, and G. Ceraolo, "About some corrosion mechanisms of AZ91D magnesium alloy," *Corrosion Science*, vol. 47, no. 9, pp. 2173–2184, 2005.

[11] M. C. L. de Oliveira, V. S. M. Pereira, O. V. Correa, N. B. de Lima, and R. A. Antunes, "Correlation between the corrosion resistance and the semiconducting properties of the oxide film formed on AZ91D alloy after solution treatment," *Corrosion Science*, vol. 69, pp. 311–321, 2013.

[12] W. Zhou, T. Shen, and N. N. Aung, "Effect of heat treatment on corrosion behaviour of magnesium alloy AZ91D in simulated body fluid," *Corrosion Science*, vol. 52, no. 3, pp. 1035–1041, 2010.

[13] D. Zander and C. Schnatterer, "The influence of manufacturing processes on the microstructure and corrosion of the AZ91D magnesium alloy evaluated using a computational image analysis," *Corrosion Science*, vol. 98, pp. 291–303, 2015.

[14] G. Song, A. Atrens, X. Wu, and B. Zhang, "Corrosion behaviour of AZ21, AZ501 and AZ91 in sodium chloride," *Corrosion Science*, vol. 40, no. 10, pp. 1769–1791, 1998.

[15] G. Song, A. Atrens, and M. Dargusch, "Influence of microstructure on the corrosion of diecast AZ91D," *Corrosion Science*, vol. 41, no. 2, pp. 249–273, 1998.

[16] R. Ambat, N. N. Aung, and W. Zhou, "Evaluation of microstructural effects on corrosion behaviour of AZ91D magnesium alloy," *Corrosion Science*, vol. 42, no. 8, pp. 1433–1455, 2000.

[17] M. Jönsson and D. Persson, "The influence of the microstructure on the atmospheric corrosion behaviour of magnesium alloys AZ91D and AM50," *Corrosion Science*, vol. 52, no. 3, pp. 1077–1085, 2010.

[18] O. Lunder, J. E. Lein, T. K. Aune, and K. Nisancioglu, "Role of Mg17Al12 phase in the corrosion of Mg alloy AZ91," *Corrosion*, vol. 45, no. 9, pp. 741–748, 1989.

[19] M. Gobara, M. Shamekh, and R. Akid, "Improving the corrosion resistance of AZ91D magnesium alloy through reinforcement with titanium carbides and borides," *Journal of Magnesium and Alloys*, vol. 3, no. 2, pp. 112–120, 2015.

[20] M. Esmaily, J. E. Svensson, S. Fajardo et al., "Fundamentals and advances in magnesium alloy corrosion," *Progress in Materials Science*, vol. 89, pp. 92–193, 2017.

[21] I. B. Singh, M. Singh, and S. Das, "A comparative corrosion behavior of Mg, AZ31 and AZ91 alloys in 3.5% NaCl solution," *Journal of Magnesium and Alloys*, vol. 3, no. 2, pp. 142–148, 2015.

[22] M. Esmaily, D. B. Blücher, J. E. Svensson, M. Halvarsson, and L. G. Johansson, "New insights into the corrosion of magnesium alloys—the role of aluminum," *Scripta Materialia*, vol. 115, pp. 91–95, 2016.

[23] R. Leiva-García, J. García-Antón, and M. J. Muñoz-Portero, *Application of Confocal Laser Scanning Microscopy to the In-situ and Ex-situ Study of Corrosion Processes, Laser Scanning, Theory and Applications*, C. C. Wang, Ed., InTech, 2011.

[24] M. Jönsson, D. Persson, and R. Gubner, "The initial steps of atmospheric corrosion on magnesium alloy AZ91D," *Journal of The Electrochemical Society*, vol. 154, no. 11, pp. C684–C691, 2007.

[25] O. Schneider, G. O. Ilevbare, J. R. Scully, and R. G. Kelly, "Confocal laser scanning microscopy as a tool for in situ monitoring of corrosion underneath organic coatings," *Electrochemical and Solid-State Letters*, vol. 4, no. 12, pp. B35–B38, 2001.

[26] P. Jakupi, J. J. Noël, and D. W. Shoesmith, "The evolution of crevice corrosion damage on the Ni-Cr-Mo-W alloy-22 determined by confocal laser scanning microscopy," *Corrosion Science*, vol. 54, no. 1, pp. 260–269, 2012.

[27] M. P. Garcia, G. L. Mantovani, R. Vasant Kumar, and R. A. Antunes, "Corrosion behavior of metal active gas welded joints

of a high-strength steel for automotive application," *Journal of Materials Engineering and Performance*, vol. 26, no. 10, pp. 4718–4731, 2017.

[28] R. M. Fernández-Domene, E. Blasco-Tamarit, D. M. García-García, and J. García-Antón, "Repassivation of the damage generated by cavitation on UNS N08031 in a LiBr solution by means of electrochemical techniques and Confocal Laser Scanning Microscopy," *Corrosion Science*, vol. 52, no. 10, pp. 3453–3464, 2010.

[29] R. Leiva-García, J. García-Antón, and M. J. Muñoz-Portero, "Contribution to the elucidation of corrosion initiation through confocal laser scanning microscopy (CLSM)," *Corrosion Science*, vol. 52, no. 6, pp. 2133–2142, 2010.

[30] C. Y. Cui, X. G. Cui, Y. K. Zhang et al., "Microstructure and corrosion behavior of the AISI 304 stainless steel after Nd:YAG pulsed laser surface melting," *Surface and Coatings Technology*, vol. 206, no. 6, pp. 1146–1154, 2011.

[31] A. H. Feng and Z. Y. Ma, "Enhanced mechanical properties of Mg-Al-Zn cast alloy via friction stir processing," *Scripta Materialia*, vol. 56, no. 5, pp. 397–400, 2007.

[32] Y. Fan, G. Wu, H. Gao, and C. Zhai, "Influence of Ca on corrosion resistance of AZ91D," *Journal of The Electrochemical Society*, vol. 153, no. 8, Article ID 002608JES, pp. B283–B288, 2006.

[33] M. Sumida, S. Jung, and T. Okane, "Solidification microstructure, thermal properties and hardness of magnesium alloy 20 mass % Gd added AZ91," *Materials Transactions*, vol. 50, no. 5, pp. 1161–1168, 2009.

[34] Y. C. Guan, W. Zhou, and H. Y. Zheng, "Effect of laser surface melting on corrosion behaviour of AZ91D Mg alloy in simulated-modified body fluid," *Journal of Applied Electrochemistry*, vol. 39, no. 9, pp. 1457–1464, 2009.

[35] A. S. Hamdy, E. El-Shenaw, and T. El-Bitar, "Electrochemical impedance spectroscopy study of the corrosion behavior of some niobium bearing stainless steels in 3.5% NaCl," *International Journal of Electrochemical Science*, vol. 1, no. 4, pp. 171–180, 2006.

[36] E.-S. M. Sherif, "A comparative study on the electrochemical corrosion behavior of iron and X-65 steel in 4.0 wt % sodium chloride solution after different exposure intervals," *Molecules*, vol. 19, no. 7, pp. 9962–9974, 2014.

[37] F. Farelas and A. Ramirez, "Carbon dioxide corrosion inhibition of carbon steels through bis-imidazoline and imidazoline compounds studied by EIS," *International Journal of Electrochemical Science*, vol. 5, no. 6, pp. 797–814, 2010.

21

Effect of Piperidin-4-ones on the Corrosion Inhibition of Mild Steel in 1 N H$_2$SO$_4$

Glory Tharial Xavier,[1] Brindha Thirumalairaj,[2] and Mallika Jaganathan[2]

[1]*Department of Science and Humanities, Sri Krishna College of Engineering and Technology, Coimbatore, Tamil Nadu 641008, India*
[2]*Department of Chemistry, PSG College of Arts and Science, Coimbatore, Tamil Nadu 641014, India*

Correspondence should be addressed to Mallika Jaganathan; jmpsgcas@gmail.com

Academic Editor: Michael I. Ojovan

The corrosion inhibition of mild steel in 1 N sulphuric acid solution by 2,6-diphenylpiperidin-4-ones with various substituents at 3- and 3,5-positions (01–06) has been tested by weight loss, potentiodynamic polarization, electrochemical impedance spectroscopic methods, and FTIR and UV absorption spectra. The surface morphology of the mild steel specimen has been analyzed by SEM. The effect of temperature (300 to 323 ± 1 K) on the corrosion behavior of mild steel in the presence of the inhibitors (01–06) was studied using weight loss techniques. The effect of anions (Cl$^-$, Br$^-$, and I$^-$) on the corrosion behavior of mild steel in the presence of the same inhibitors was also studied by weight loss method and the synergism parameters were calculated. The adsorption characteristics of the inhibitors have been determined from the results.

1. Introduction

Corrosion inhibitors find vast application in the industrial field as components in acid descaling, oil well acidizing, acid pickling, acid cleaning, and so forth. Most of the effective inhibitors used contain heteroatom such as O, N, S, and multiple bonds in their molecules through which they are adsorbed on the metal surface. It has been observed that adsorption depends mainly on certain physiochemical properties of the inhibitor group, such as functional groups, electron density at the donor atom, π-orbital character, and the electronic structure of the molecule [1–8]. Several researchers investigated the corrosion inhibiting properties of organic compounds on mild steel in acidic medium [9–12]. The aim of this work was to investigate the effect of substitution at 3- and 3,5-positions of 2,6-diphenylpiperidin-4-one on the corrosion inhibition of mild steel 1 N H$_2$SO$_4$ solution. Corrosion inhibition was investigated using weight loss, electrochemical impedance spectroscopy (EIS), and potentiodynamic polarization (Tafel) methods and their results were compared.

2. Experimental Details

The variously substituted 2,6-diphenylpiperidin-4-ones (01–06) were prepared via following reported procedures [13, 14]. Mild steel specimens of the following composition have been used all over the present investigations (Carbon: 0.07%, Sulphur: nil, Phosphorous: 0.008%, Manganese: 0.34%, remaining ferrous).

2.1. Methods. In the gravimetric experiment, mild steel specimen of the dimensions 2.5 cm × 1 cm × 0.1 cm has been used. It was polished using 1/0, 2/0, 3/0, and 4/0 emery papers, washed with double distilled water, dried, and finally degreased with the acetone. In this method previously weighed coupon was completely immersed in 100 mL 1 N sulphuric acid with and without inhibitor in an open beaker. After an hour, the corrosion product was removed by washing each coupon using double distilled water. The washed coupons were rinsed with acetone and dried in the air before reweighing. From the average weight loss (mean of three replicates analysis) results, the inhibition efficiency (IE%) of the inhibitor and the corrosion rate of mild steel (CR) were calculated using the following equations:

$$\text{Corrosion rate (mmpy)} = \frac{(87.6 \times W)}{(D \times A \times T)}, \qquad (1)$$

where W is weight loss in mg, D is density in mg, A is area of exposure in cm^2, and T is time in hours.

Inhibition efficiency has been determined by using the following relationship:

$$\text{IE (\%)} = W_o - \frac{W_t}{W_o}. \qquad (2)$$

W_o is weight loss without inhibitor. W_t is weight loss with inhibitor.

The effects of temperature on the corrosion inhibition performance for the variously substituted piperidin-4-ones (01–06) were studied in the range of (300 to 323 ± 1 K).

All the electrochemical measurements were carried out using a glass cell of 100 mL capacity. A platinum electrode and a saturated calomel electrode (SCE) were used as counter and reference electrodes, respectively. The mild steel specimens were placed in the test solution for 10 minutes before electrochemical measurements.

The electrochemical impedance measurements were carried out over the frequency range of 10 KHz to 0.01 Hz carried with AC signed amplitude of the 10 mV at the corrosion potential. The measurements were automatically controlled by Z_{view} software and the impedance diagrams were given as Nyquist plots. From the plots the electrochemical parameters such as double layer capacitance (C_{dL}) and charge transfer resistance (R_{ct}) were calculated. The potentiodynamic polarization measurements were made for a potential range of −200 mV to +200 mV with respect to open circuit potential, at a scan rate of 1 mV/sec. From the plot of E versus log I, the corrosion potential (E_{corr}), corrosion current (I_{corr}) were obtained. Tafel slopes b_a and b_c were obtained in the absence and presence of inhibitors. From the I_{corr} values the inhibition efficiencies were calculated. The corrosion rates and inhibition efficiency were obtained from the relationships:

$$\text{Corrosion rate } (C_R \text{ mmpy}) = 3.2 \times I_{\text{corr}} \left(\text{mA/cm}^2\right)$$

$$\times \frac{\text{Equivalent weight}}{\text{Density}}, \qquad (3)$$

$$\text{IE\%} = \frac{I_{\text{corr}(0)} - I_{\text{corr}(i)}}{I_{\text{corr}(0)}} \times 100,$$

where $I_{\text{corr}(0)}$ is the corrosion current in the absence of inhibitor and $I_{\text{corr}(i)}$ is the corrosion current in the presence of inhibitor.

The surfaces of the corroded and corrosion inhibited mild steel specimens were examined using scanning electron microscopy (JEOL-JSM-35-CF). A Shimadzu FT-IR 8000 Spectrophotometer in the 4000–400 cm^{-1} region using the KBr disc technique was employed to examine the interaction of inhibitors with the metal surface.

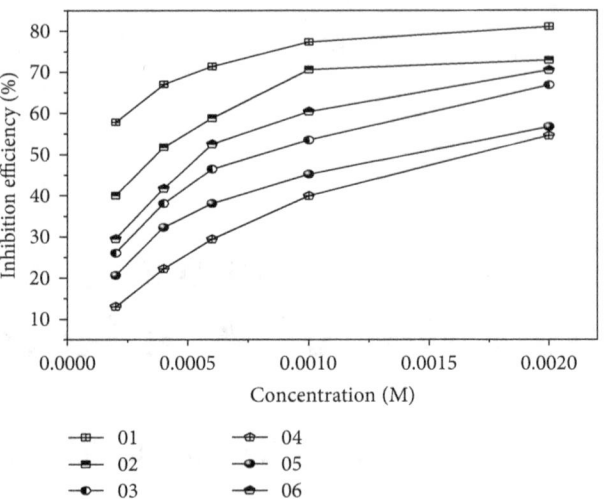

FIGURE 1: Variation of inhibition efficiency with inhibitor (01–06) concentration for mild steel in 1 N H$_2$SO$_4$ at 300 ± 1 K.

3. Results and Discussion

3.1. Effect of Inhibitor Concentration. The corrosion of mild steel in 1 N sulphuric acid solution in the absence and presence of various concentrations (0.2×10^{-3} to 2×10^{-3} M) of inhibitors (01 to 06) was investigated at 300 ± 1 K for 1-hour immersion period using weight loss measurements.

From the weight loss data, plots of inhibition efficiency versus concentration were made and shown in Figure 1. The plots confirm that the inhibition efficiency of all the studied inhibitors (01–06) increases with increase in concentration of the inhibitor.

The extent of inhibition depends on the substituents at 3-position or 3,5-positions of the 2,6-diphenylpiperidin-4-one. At lower concentrations (0.2 to 1×10^{-3} M) the inhibition efficiency increases gradually and it almost reaches saturation above 2×10^{-3} M concentration. The order of inhibition efficiencies of these inhibitors at 300 ± 1 K is 01 > 02 > 06 > 03 > 05 > 04. Analysis of the inhibition efficiencies of inhibitors (01 to 06) reveals that the inhibition efficiency decreases when it is substituted with various groups (methyl, ethyl, isopropyl) at either 3-position of the piperidin-4-one ring or at 3,5-positions of the piperidin-4-one ring. The observed trend can be explained by considering the conformations of substituted piperidin-4-ones and steric hindrance caused by the substitutes.

2,6-Diphenylpiperidin-4-one exists as an equilibrium mixture of both boat and chair conformation [15], whereas 3-alkyl substituted 2,6-diphenylpiperidin-4-one exists in chair conformation with phenyl and alkyl groups at equatorial positions [16–18] (except in compound (06) where one of the C(3)-methyl groups must necessarily be axially oriented). Piperidin-4-ones contain two potential inhibiting groups, namely, the carbonyl group and ring nitrogen [19–22]. The interaction of 2,6-diphenylpiperidin-4-one (01) with the metal surface could occur through either carbonyl group or ring nitrogen. The less electronegativity of nitrogen than oxygen favours ring nitrogen to be the inhibiting site.

Sankarapapavinasam et al. [21] compared the inhibition efficiency of piperidin-4-one and cyclohexanone under identical conditions. They reported that the observed inhibition efficiency of piperidin-4-one (65–80%) was achieved even at very low concentration of the substrate (1.0 mM) while that of cyclohexanone (40%) was achieved at a much higher concentration (100 mM). The inhibition efficiency of unsubstituted piperidine-4-one is higher than that of cyclohexanone. Further the inhibition efficiency of 2,6-diphenylpiperidin-4-one is found to be higher than that of substituted piperidine-4-one. This result reveals that the phenyl rings present in the 2,6-position which are flanking the ring nitrogen play an important role in the inhibition of corrosion. These phenyl groups are in the equatorial position. These equatorial phenyl groups of the substituted piperidin-4-ones lie parallel to the mild steel surface in chair conformation. The interaction between the π electrons of phenyl rings and the mild steel surface prevents the corrosion effectively.

The efficiency of 3-methyl-2,6-diphenylpiperidin-4-one (02) was found to be less than that of 2,6-diphenylpiperidin-4-one (01). This can be explained as follows. In the case of inhibitor (01) the two phenyl rings at 2,6 positions are lying parallel to the metal surface and cause effective interaction with the metal surface, whereas in the case of 3-methyl-2,6-diphenylpiperidin-4-one (02) the methyl group at the C-3 position creates an indirect steric effect known as buttressing effect. It pushes the phenyl rings towards the ring nitrogen thereby changing the orientation of the lone pair of electrons on the nitrogen atom. This effect also decreases the π electron interaction between the phenyl ring and metal surface. By varying the substituents at C-3 position we can study the buttressing effect in detail. Substitution by ethyl, isopropyl, 3,3-dimethyl, and 3,5-dimethyl groups increases the buttressing effect and decreases the inhibition efficiency. In compound (02) methyl, (03) ethyl and in (04) isopropyl group is present. As we move from compound (02) to compound (04), the bulkiness of the group increases, so the buttressing effect increases and, hence, the inhibition efficiency decreases. Generally, the $+I$ effect of alkyl group increases in the order methyl < ethyl < isopropyl. Isopropyl group has greater $+I$ effect than ethyl and methyl and so it is expected to give the highest inhibition efficiency. But in our current investigation inhibitor (04) with isopropyl group shows the least inhibition efficiency, because in this case buttressing effect dominates over $+I$ effect. The inhibition efficiency follows the order 02 > 03 > 04. In addition to this, increase in the degree of chain branching may also lead to lesser protection efficiency owing to higher steric hindrance [23]. Compound (06) with two methyl groups at C-3 position and compound (05) with one methyl group each at 3- and 5-position exhibit lesser inhibition efficiency than compound (02). On correlating experimental results with molecular structures it can be concluded that the increase of steric hindrance at C-3 and C-5 position of substituted piperidin-4-one ring decreased the corrosion inhibiting ability of the studied piperidin-4-ones [24, 25]. One may expect higher inhibition efficiency for inhibitors (06) and (05) when compared to inhibitor (02) because more number of electron releasing methyl groups is present in these inhibitors. On the contrary, inhibitor (02) exhibited higher

efficiency. The above observation can be explained as follows. If the number of groups in a molecule increases it leads to overcrowding which causes a strain in the molecule, which results in loss of planarity of the molecule. Since the planarity is disturbed, the phenyl rings may not have much interaction with the metal surface. That is, it may not be able to come closer with the metal surface to have direct contact, which results in lesser protection efficiency for (06) and (05) than inhibitor (02). Moreover the presence of groups which do not act as adsorption centers may not contribute much for the inhibition efficiency of the compound. In addition to this, increase in the degree of chain branching may also lead to lesser protection efficiency owing to higher steric hindrance [23].

3.2. Effect of Temperature. The effect of temperature on the corrosion inhibition performance of inhibitors (01–06) on mild steel in $1 N H_2SO_4$ is investigated by weight loss measurements in the temperature range 300 to 323 ± 1 K in the absence and presence of inhibitors (01–06) at 0.4×10^{-3} M concentration. Table 1 shows the values of corrosion rate, surface coverage, and inhibition efficiency obtained from weight loss measurements at various temperatures (300 to 323 ± 1 K). Data in Table 1 suggest that all inhibitors get adsorbed on the mild steel surface at all temperatures studied and corrosion rate increases with increase in temperature in both uninhibited and inhibited solutions. This can be attributed to the fact that the rate of corrosion reaction increases with increase in temperature. But the rate of corrosion is less in inhibited solutions compared with uninhibited solutions. Decrease in inhibition efficiency with increase in temperature is suggestive of physical adsorption mechanism [26, 27] and may be attributed to increase in the solubility of the protective films and of any reaction products precipitated on the surface of mild steel that may otherwise inhibit the corrosion process. As the temperature increases the number of adsorbed molecules decreases, leading to a decrease in the inhibition efficiency. A decrease in inhibition efficiency with the increase in temperature in this case may also be due to weakening of physical adsorption [28].

In order to calculate the activation energy (E_a) for the corrosion process, Arrhenius equation is used

$$C_R = A \exp\left(-\frac{E_a}{RT}\right), \qquad (4)$$

where C_R is the corrosion rate, A is the Arrhenius preexponential factor, E_a is the activation energy for the corrosion process, R is the universal gas constant, and T is the absolute temperature.

The apparent activation energies (E_a) at optimum concentration 0.4×10^{-3} M of the inhibitors (01–06) were determined by linear regression between $\log C_R$ and $1/T$ and are represented in Figure 2. Plots of $\log C_R$ versus $1/T$ gave straight line with slope ($-E_a/2.303R$) and intercept A. The linear regression coefficients were close to unity for all the inhibitors analysed. The calculated value of activation energies (E_a) and preexponential factors (A) for the inhibitors are presented in Table 2. Arrhenius law predicts that corrosion

TABLE 1: Corrosion parameters for mild steel in 1 N H$_2$SO$_4$ in the absence and presence of optimum concentration of variously substituted piperidin-4-ones (0.4×10^{-3} M) obtained from weight loss measurements at different temperatures.

Temperature (±1 K)	Inhibitor	Inhibitor efficiency (%)	Corrosion rate (mmpy)	Surface coverage (θ)
300	Blank	—	57.88	—
	(01)	51.76	18.25	0.5038
	(02)	84.62	08.90	0.8462
	(03)	79.62	11.80	0.7962
	(04)	47.31	30.50	0.4731
	(05)	51.15	28.27	0.5115
	(06)	80.16	09.25	0.8016
313	Blank	—	97.06	—
	(01)	33.49	64.56	0.3349
	(02)	57.57	41.18	0.5757
	(03)	49.54	48.98	0.4954
	(04)	28.21	69.68	0.2821
	(05)	30.05	67.90	0.3005
	(06)	52.40	44.65	0.5240
318	Blank	—	136.02	—
	(01)	30.93	93.94	0.3093
	(02)	48.28	70.35	0.4828
	(03)	44.52	75.47	0.4452
	(04)	26.19	100.40	0.2619
	(05)	31.59	93.05	0.3159
	(06)	46.73	73.18	0.4673
323	Blank	—	152.05	—
	(01)	19.91	121.77	0.1991
	(02)	33.82	100.62	0.3382
	(03)	26.65	111.53	0.2665
	(04)	14.93	129.34	0.1493
	(05)	20.50	120.88	0.2050
	(06)	30.15	105.84	0.3015

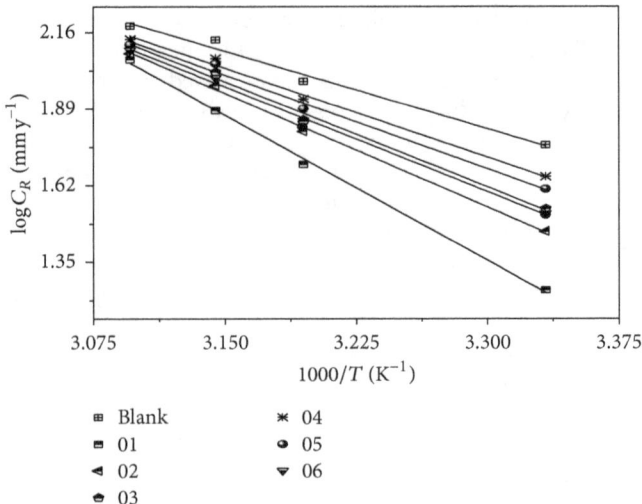

FIGURE 2: Arrhenius plots of $\log C_R$ versus $1/T$ for mild steel in 1 N H$_2$SO$_4$ in the absence and presence of inhibitors (01–06) at optimum concentration (0.4×10^{-3} M).

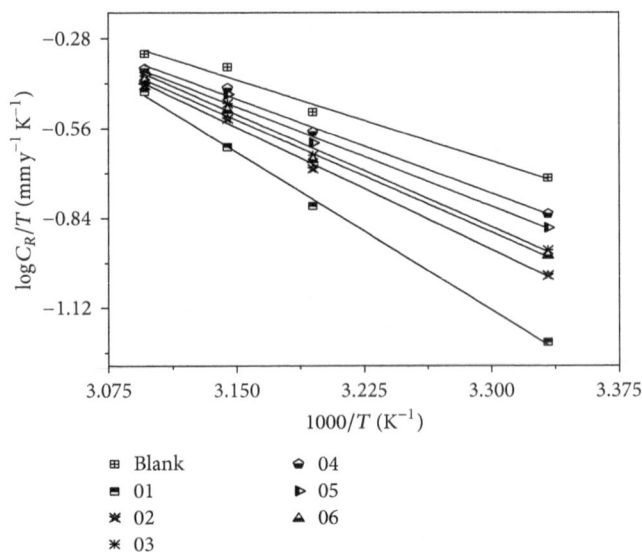

FIGURE 3: Arrhenius plots of $\log C_R/T$ versus $1/T$ for mild steel in 1 N H$_2$SO$_4$ in the absence and presence of inhibitors (01–06) at optimum concentration (0.4×10^{-3} M).

rate increases with the temperature and E_a and A may vary with temperature in accordance with (4). Inspection of Table 2 showed that the E_a value for all the inhibitors studied was greater than 20 kJ/mol in both inhibited and uninhibited solutions, which revealed that the entire process is controlled by the surface reaction [29].

The value of E_a for the uninhibited solution was found to be 34.96 kJ/mol and that obtained in the presence of inhibitors (40.07–65.31 kJ/mol) for 01–06. It is clear that activation energy is higher in inhibited solution than in uninhibited solution. The increase in the apparent activation energy in the inhibited solution suggests physical adsorption of the inhibitors on the mild steel surface. The increase in activation energy can also be attributed to an appreciable decrease in the adsorption of the inhibitor on the mild steel surface with increase in temperature. As adsorption decreases more desorption of inhibitor molecules occurs because these two

opposite processes are in equilibrium. Due to more desorption of inhibitor molecules at higher temperatures the greater surface area of mild steel comes in contact with aggressive environment resulting in increased corrosion rates with increase in temperature [30]. The very high activation energy in the presence of the inhibitor may be due to the physical adsorption of the inhibitor species on the mild steel surface. This type of inhibitor retards corrosion at ordinary temperatures but inhibition is diminished at elevated temperature [31]. Thus these inhibitors perform well at room temperature.

Figure 3 showed the plot of $\log(C_R/T)$ versus $1/T$ for the corrosion of mild steel in the absence and presence of inhibitors (01–06) in 1 N H$_2$SO$_4$. In order to calculate

TABLE 2: Activation parameters E_a, $\Delta H°$, and $\Delta S°$ for the mild steel dissolution in 1 N H_2SO_4 in the absence and presence of variously substituted piperidin-4-ones at optimum concentration (0.4×10^{-3} M).

Inhibitor	Preexponential factor ($g\,cm^{-1}\,h^{-1}$)	E_a ($kJ\,mol^{-1}$)	$E_a - \Delta H°$ ($kJ\,mol^{-1}$)	$\Delta H°$ ($kJ\,mol^{-1}$)	$\Delta S°$ ($J\,mol^{-1}$)
Blank	6.99×10^7	34.96	2.6	32.38	−45.94
(01)	4.08×10^{12}	65.31	2.6	62.72	45.30
(02)	2.24×10^{10}	52.27	2.6	49.69	05.71
(03)	5.67×10^9	47.22	2.6	44.64	−09.39
(04)	4.22×10^{10}	40.07	2.6	37.49	−30.99
(05)	8.83×10^8	42.17	2.6	39.59	−24.85
(06)	5.28×10^{10}	47.14	2.6	44.56	−09.98

TABLE 3: Activation parameters K and $\Delta G°_{ads}$ for the mild steel dissolution in 1 N H_2SO_4 and in the absence and presence of variously substituted piperidin-4-ones at optimum concentration (0.4×10^{-3} M).

Inhibitors	R^2	Slope	K (mol^{-1})	$-\Delta G°_{ads}$ ($kJ\,mol^{-1}$)
(01)	1.00	1.17	8075	32.46
(02)	0.99	1.20	3501	30.37
(03)	0.99	1.23	1788	28.70
(04)	1.00	1.17	0756	26.55
(05)	0.99	1.43	1414	28.11
(06)	0.99	1.19	2214	29.23

the activation parameters like $\Delta H°$ and $\Delta S°$ for the corrosion process, transition state equation [24] was used:

$$C_R = \left(\frac{RT}{Nh}\right) \exp\left(\frac{\Delta S°}{R}\right) \exp\left(-\frac{\Delta H°}{RT}\right), \qquad (5)$$

where h is Planck's constant, N is Avogadro's number, R is the universal gas constant, T is the absolute temperature, $\Delta S°$ is the entropy of activation, and $\Delta H°$ is the enthalpy of activation. Plot of $\log(C_R/T)$ versus $1/T$ gave straight lines with slope $(-\Delta H°/2.303R)$ and intercept $[\log(R/Nh) + (\Delta S°/2.303R)]$, from which $\Delta H°$ and $\Delta S°$ were calculated and listed in Table 2. Inspection of these data revealed that the entropy of activation for the dissolution reaction of mild steel in 1 N H_2SO_4 in the presence of all inhibitors (0.4×10^{-3} M) is higher (37.49–62.72 $kJmol^{-1}$) than that in the absence of inhibitors (32.38 $kJmol^{-1}$). Positive values of enthalpy of activation ($\Delta H°$) in the absence and presence of inhibitor reflect the endothermic nature of the mild steel dissolution process meaning that dissolution of steel is difficult. It is evident from Table 3 that the value of $\Delta H°$ increased in the presence of the inhibitor compared to the uninhibited solution indicating higher protection efficiency. This may be attributed to the presence of energy barrier for the reaction; hence the process of adsorption of inhibitor leads to rise in enthalpy of the corrosion process. The values of $\Delta H°$ and E_a are nearly the same and are higher in the presence of the inhibitor. This indicates that the energy barrier of the corrosion reaction increased in the presence of the inhibitor without changing the mechanism of dissolution. The value of E_a is found to be larger than the corresponding $\Delta H°$ value indicating that the corrosion process must involve a gaseous reaction, simply

the hydrogen evolution reaction, associated with a decrease in the total reaction volume [32, 33]. Moreover, the difference value of the $E_a - \Delta H°$ is found to be 2.6 kJ/mol, which is approximately equal to the average value of RT (2.63 kJ/mol). This indicates that the corrosion process is a unimolecular reaction as it is characterized by the following equation:

$$E_a - \Delta H° = RT. \qquad (6)$$

This result shows that the inhibitors acted equally on E_a and $\Delta H°$.

On comparing the values of the entropy of activation ($\Delta S°$) in Table 2, it is clear that the positive entropy of activation is obtained in the presence of inhibitors, while negative value (-45.94 $Jmol^{-1}$) is observed in the case of free 1 N H_2SO_4 solution. Such variation is concerned with the phenomenon of ordering and disordering of the inhibitor molecules at the electrode surface and could be explained as follows. The adsorption of organic inhibitor molecules from the aqueous solution can be regarded as a quasisubstitution process between the organic compounds in the aqueous phase and water molecules at the electrode surface. The adsorption of inhibitors on the mild steel surface is accompanied by desorption of water molecules from the surface. Thus while the adsorption process for the inhibitor is believed to be exothermic and associated with a decrease in entropy of the solute, the opposite is true for the solvent. The thermodynamic values obtained are the algebraic sum of the adsorption of organic molecules and desorption of water molecules. Hence, the gain in entropy is attributed to the increase in solvent entropy and to more positive water desorption enthalpy. The positive values of $\Delta S°$ also suggest that an increase in disordering takes place on going from reactants to the metal/solution interface [26], which is the driving force for the adsorption of inhibitors onto the mild steel surface.

3.3. Adsorption Isotherm. Organic molecules are used to inhibit corrosion as they are adsorbed on the metal solution interface. The adsorption depends on the nature of the inhibitor, its conformation in aqueous solution, chemical composition of the medium, nature of the metal surface, temperature, and electrochemical potential at the metal solution interface. Various adsorption isotherms (Henry, Frendlich, Langmuir, Frumkin, and Temkin) have been tried for the inhibitors (01–06) on the mild steel surface in acidic medium at room temperature. The correlation coefficients (R^2) were

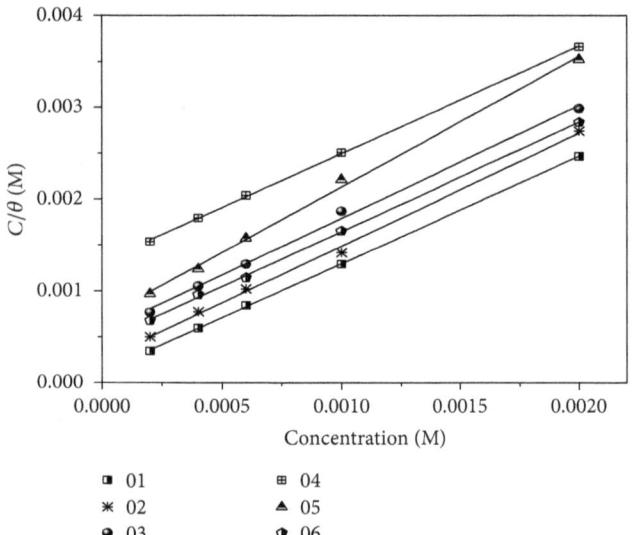

FIGURE 4: Langmuir isotherm plots for adsorption of inhibitors (01–06) on the mild steel in 1 N H_2SO_4.

used to determine the best fit. For the studied inhibitors best fit was obtained with Langmuir adsorption isotherm.

The Langmuir isotherm is given by the following equation:

$$\frac{C}{\theta} = \frac{1}{K} + C, \qquad (7)$$

where K is the equilibrium constant of the adsorption process and C is the molar concentration of inhibitor. The plot of C/θ versus C gave straight line (Figure 4). The linear regression coefficients (R^2) are almost equal to unity. Though the plot of C/θ versus C is linear, the slope obtained is found to deviate from unity. The deviation of the slope from unity might be due to interaction among the adsorbed species on the surface of the mild steel specimen [34–39] or the adsorbate occupies more than one adsorption site on the metal surface. Similar behavior was observed by several researchers [40–43] and has suggested the modified Langmuir adsorption isotherm. Langmuir's adsorption isotherm parameters are presented in Table 3.

The values of equilibrium constant K calculated using Langmuir (8) isotherm are presented in Table 3. A large value of K obtained indicates more efficient adsorption, and hence better inhibition efficiency. The free energy of adsorption (ΔG°_{ads}) is related to the adsorption constant (K) with the following equation:

$$K = \frac{1}{55.5} \exp\left(\frac{-\Delta G^\circ_{ads}}{RT}\right), \qquad (8)$$

where the value of 55.5 is the concentration of water in solution expressed in $molL^{-1}$. The ΔG°_{ads} values obtained from Langmuir were calculated at room temperature and are listed in Table 3. The large negative values of ΔG°_{ads} indicate the spontaneity of the adsorption process and the stability of the adsorbed species on the mild steel surface. In general,

the values of ΔG°_{ads} of the order of -20 kJ/mol or lower indicate a physisorption [44] while those more negative than -40 kJ/mol involve sharing or transfer of electrons from the inhibitor molecules to the metal surface to form a coordinate type of bond (chemisorption) [45]. It is clear from Table 3 that the calculated ΔG°_{ads} values are in the range between 26.55 and 32.46 kJ/mol for H_2SO_4 solution for optimum concentration (0.4×10^{-3} M) of the inhibitor. This indicates that the adsorption of these inhibitors on the mild steel surface may involve complex interactions involving both physical adsorption and chemical adsorption [46].

It is well-known that surface charge density and zero charge potential of metals play an important role in the process of the electrostatic adsorption [47]. The surface charge of the metal is due to the electric field existing at the metal/solution interface [48]. The surface charge can be defined by the position of E_{corr} with respect to the respective potential of zero charge (PZC) $E_{q=0}$. When the difference $u = (E_{corr} - E_{q=0})$ is negative, the electrode surface acquires a negative net charge and the adsorption of cations is favoured. On the contrary, the adsorption of anions is favored when you become positive. Generally mild steel surface is positively charged in 1 N H_2SO_4 solution [49, 50]. However the inhibitors under investigation, namely, the 2,6-diphenyl-piperidin-4-ones, are protonated (piperidinium cation) in acidic solutions and exist in equilibrium with the corresponding molecular form.

The inhibitors studied have ΔG°_{ads} value between 26.55 and 32.46 kJ/mol for 1N H_2SO. This indicates that the inhibitors adsorb on the mild steel surface through both modes of adsorption (physical adsorption and chemical adsorption) in both acid media. The K values increase in the order of (04) < (05) < (03) < (06) < (02) < (01) which is the same as the order followed by inhibition efficiency obtained from weight loss method.

3.4. Synergistic Effect. It is generally accepted that the presence of halide ions in acidic media synergistically increases the inhibition efficiency of most of the organic compounds. The halide ions are able to improve adsorption of the organic cations by forming intermediate bridges between positively charged metal surface and the positive end of the organic inhibitor [51]. For studying the synergistic inhibitive effect of halide ions, an optimum concentration (0.4×10^{-3} M) of variously substituted 2,6-diphenylpiperidin-4-ones is taken (Scheme 1) and to each of these inhibitors 1×10^{-3} M KCl/KBr/KI is added and the inhibition efficiencies were found using weight loss method and the observations are presented in the Table 4 for 1 N H_2SO_4. Analysis of the results indicates a further reduction in corrosion rate; that is, the inhibition efficiency of the inhibitors was enhanced by the addition of the anions (1×10^{-3} M KCl/KBr/KI).

As far as the inhibition is concerned, it is generally assumed that the adsorption of the inhibitors at the metal/aggressive solution interface is the first step in the inhibition mechanism [52]. The dissolution of mild steel is reduced by either adsorption of the inhibitor molecule on the mild steel surface or blocking the active centers available for corrosion.

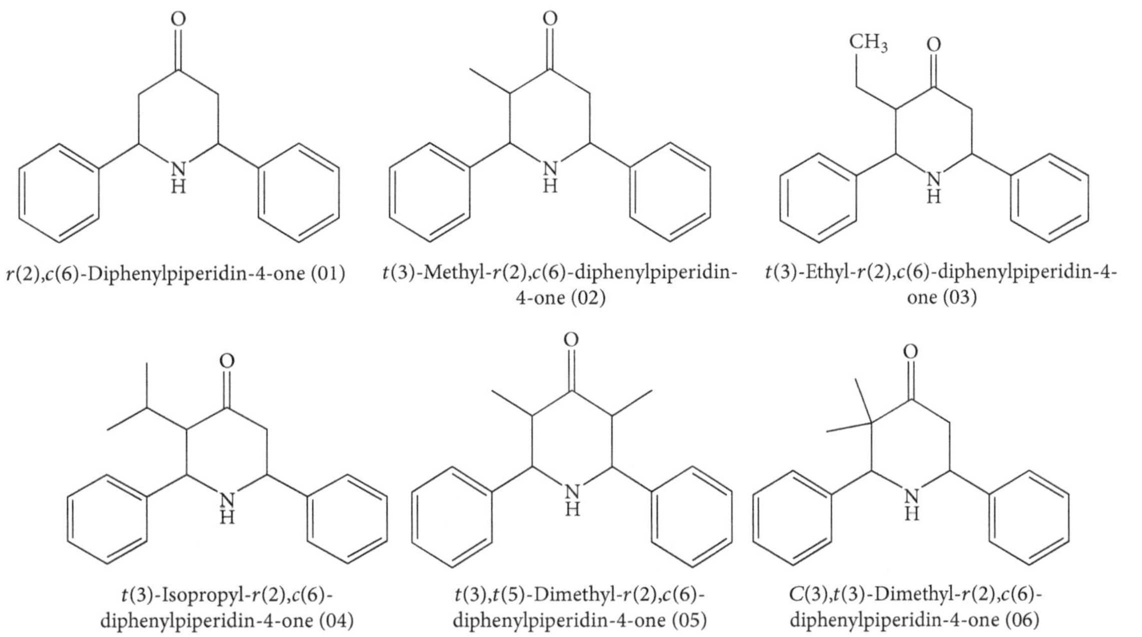

r(2),c(6)-Diphenylpiperidin-4-one (01)

t(3)-Methyl-r(2),c(6)-diphenylpiperidin-4-one (02)

t(3)-Ethyl-r(2),c(6)-diphenylpiperidin-4-one (03)

t(3)-Isopropyl-r(2),c(6)-diphenylpiperidin-4-one (04)

t(3),t(5)-Dimethyl-r(2),c(6)-diphenylpiperidin-4-one (05)

C(3),t(3)-Dimethyl-r(2),c(6)-diphenylpiperidin-4-one (06)

Scheme 1: Structures of variously substituted 2,6-diphenylpiperidin-4-ones (01–06).

Table 4: Inhibition efficiency and synergism parameter of variously substituted 2,6-diphenylpiperidin-4-ones (0.4×10^{-3} M) and KCl, KBr, and KI (1×10^{-3} M) systems in 1 N H_2SO_4 solution.

Inhibitors	Without KCl, KBr, and KI IE (%)	KCl		KBr		KI	
		IE (%)	S_θ	IE (%)	S_θ	IE (%)	S_θ
(02)	51.76	61.08	1.14	72.38	1.47	87.44	2.69
(03)	38.03	53.97	1.24	66.10	1.54	82.84	2.53
(04)	22.27	36.40	1.12	50.20	1.31	71.12	1.88
(05)	32.26	45.19	1.14	58.16	1.36	76.99	2.06

It is well-known that the adsorption process is made possible due to the presence of heteroatoms such as N and O which are regarded as active adsorption centers. As discussed earlier, the inhibitors studied (01–06) contain nitrogen, oxygen, and two phenyl rings with π electrons. The compound could be adsorbed by the interaction between the lone pair of electrons of the oxygen and nitrogen atoms or the electron rich π systems of the aromatic rings and the mild steel surface. This process as earlier reported by Umoren and Ebenso [53] may be facilitated by the presence of vacant d-orbital in steel. In addition to the molecular form, piperidin-4-ones can be present as a protonated species in acidic solutions. Generally mild steel surface is positively charged in H_2SO_4 solution, so direct adsorption of a protonated species (piperidinium cation) on the positively charged mild steel surface is difficult. In order to facilitate the adsorption of piperidinium cation on the mild steel surface, halide ions are added to the aggressive medium. The added halide ions being negatively charged are attracted to the positively charged mild steel surface and are adsorbed on metal surface; this leads to development of excess negative charge at metal solution interface, thus lending the mild steel surface negative. Now electrostatic interaction occurs between the positively charged nitrogen atom in the piperidinium cation and negatively charged mild steel surface [54] resulting in physisorption of the inhibitor on the mild steel surface thereby inhibiting the corrosion process. From the above discussion it is clear that the addition of halide ions to the aggressive medium facilitates the adsorption of the piperidin-4-ones on the metal surface.

From Table 4 it is clear that the synergistic ability of the halides increased in the order $Cl^- < Br^- < I^-$ and similar observation has been reported by several researchers [55, 56]. The inhibition efficiency values increase significantly on moving from Cl^- to Br^- to I^-. In other words the surface coverage (θ) value increases in the same order. This is similar to the findings of Oguzie et al., [41]. They have reported that the corrosion inhibitor synergism results from increased surface coverage arising from the ion pair interaction between the organic cation and the anions. The great influence of the iodide ion is often attributed to its large ionic radius, high hydrophobicity, and low electronegativity compared to the other halide ions. In addition, the more stable chemisorption of I^- ions because of the easy deformability of their electron shells also results in higher inhibition effect. This stabilization

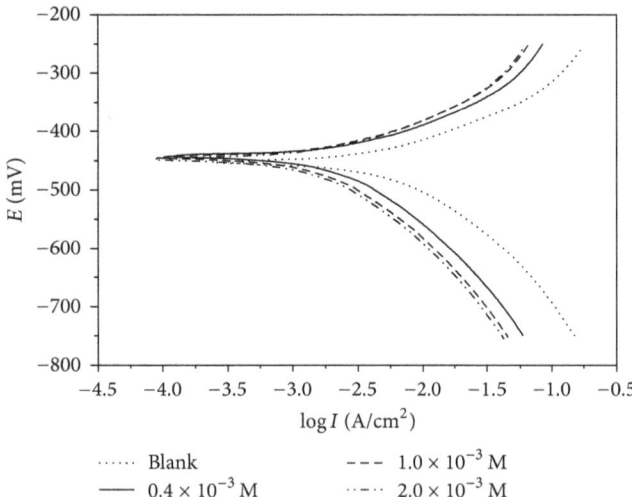

FIGURE 5: Polarization curves for mild steel in 1 N H$_2$SO$_4$ in the absence and presence of different concentration of inhibitor (01).

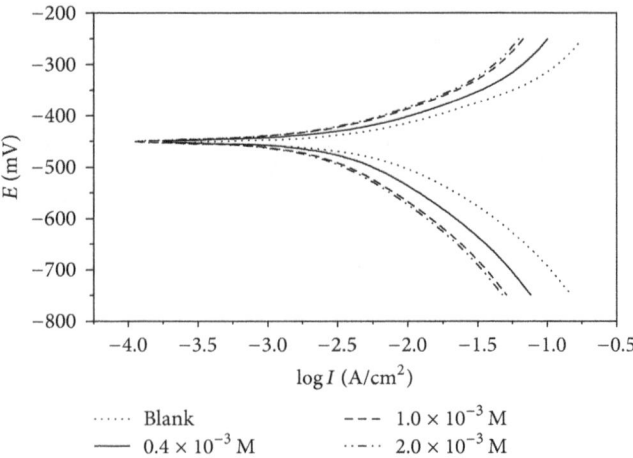

FIGURE 6: Polarization curves for mild steel in 1 N H$_2$SO$_4$ in the absence and presence of different concentration of inhibitor (02).

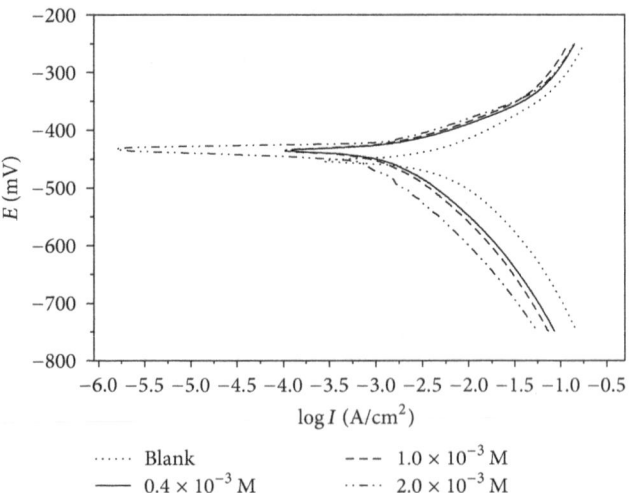

FIGURE 7: Polarization curves for mild steel in 1 N H$_2$SO$_4$ in absence and presence of different concentrations of inhibitor (03).

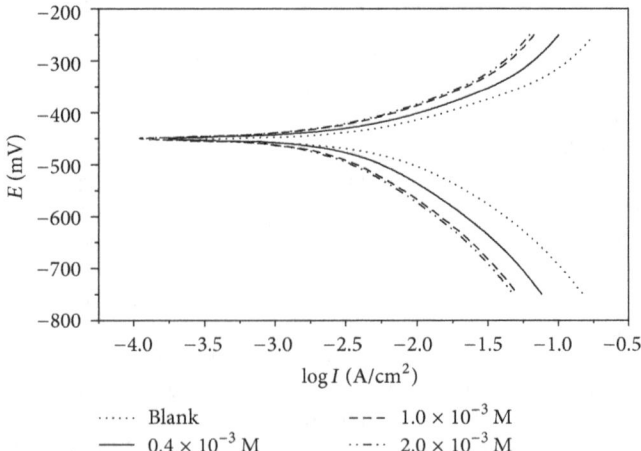

FIGURE 8: Polarization curves for mild steel in 1 N H$_2$SO$_4$ in the absence and presence of different concentration of inhibitor (04).

may be caused by the interaction between the inhibitor and I$^-$ ions. This interaction enhances the inhibition efficiency to a considerable extent due to the increase of the surface coverage in the presence of iodide ions. According to Hackerman and Makrides [57], the halide ions in general and the iodide ions in particular prevent the steel dissolution in acid solution by a strong interaction with the metal surface, possibly through chemisorption.

The synergism parameter S_θ is evaluated using the relationship given by several authors [58–60]:

$$S_\theta = \frac{1 - \theta_{1+2}}{1 - \theta'_{1+2}}, \tag{9}$$

where θ_1 is surface coverage by the halide ion. θ_2 is surface coverage by the inhibitor, $\theta_{1+2} = (\theta_1 + \theta_2) - (\theta_1\theta_2)$, and θ'_{1+2} is measured surface coverage for the inhibitor in combination with halide ion.

The synergism parameters derived from the inhibition efficiency values obtained from weight loss measurements are summarized in Table 4. It is evident from Table 4 that the S_θ values are greater than unity which indicates that the improved inhibition efficiency caused by the addition of halide ions to the inhibitors is only due to synergistic effect. The highest values of S_θ for I$^-$ ions confirm the highest synergistic influence of I$^-$ ions among halides.

3.5. Potentiodynamic Polarization Measurements. Polarization curves for mild steel in 1 N H$_2$SO$_4$ solutions at room temperature without and with addition of different concentrations of inhibitors (01–06) are shown in Figures 5–10. The anodic and cathodic current potential curves are extrapolated up to their intersection at the point where corrosion current density (I_{corr}) and corrosion potential (E_{corr}) are obtained. The electrochemical parameters I_{corr}, E_{corr}, anodic,

TABLE 5: Polarization parameters for mild steel in 1 N H_2SO_4 in the absence and presence of different concentrations of variously substituted piperidin-4-ones.

Inhibitor	Concentration (10^{-3} M)	$-E_{corr}$ (mV vs SCE)	I_{corr} (mA/cm^2)	b_a (mV/dec)	b_c (mV/dec)	Corrosion rate (mmpy)	Inhibition efficiency (%)
—	Blank	467	12.85	171	250	292.24	Nil
(01)	0.4	454	04.37	133	250	099.25	66.03
	1	462	03.03	133	225	068.82	76.44
	2	456	02.53	116	225	057.51	80.32
(02)	0.4	476	06.25	170	229	142.15	51.35
	1	471	03.83	155	221	087.04	70.21
	2	465	03.55	145	229	080.67	72.39
(03)	0.4	472	08.04	162	266	182.70	37.47
	1	467	06.03	145	250	137.00	53.11
	2	479	04.33	125	225	098.34	66.34
(04)	0.4	445	10.09	154	291	229.47	21.46
	1	476	07.80	180	266	177.31	39.31
	2	477	05.92	145	250	134.50	53.96
(05)	0.4	456	08.71	133	250	198.03	32.22
	1	458	07.05	155	266	160.23	45.16
	2	475	05.70	145	208	129.58	55.65
(06)	0.4	476	07.23	142	250	164.34	43.75
	1	463	05.12	142	250	116.34	60.18
	2	469	03.87	154	238	088.05	69.86

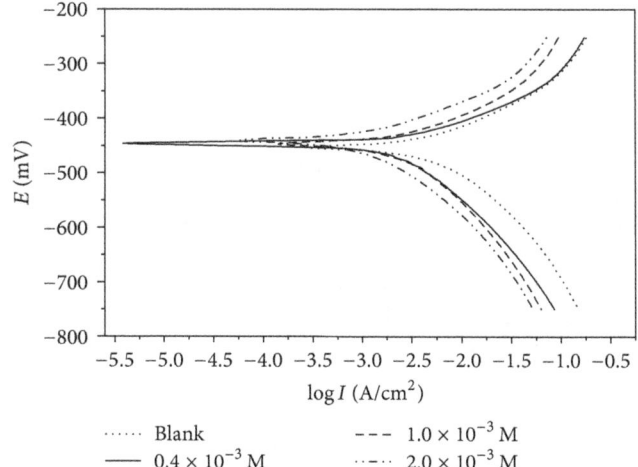

FIGURE 9: Polarization curves for mild steel in 1 N H_2SO_4 in the absence and presence of different concentration of inhibitor (05).

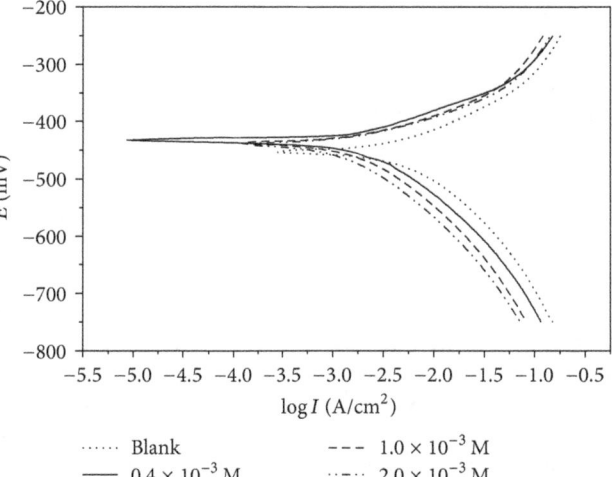

FIGURE 10: Polarization curves for mild steel in 1 N H_2SO_4 in the absence and presence of different concentration of inhibitor (06).

and cathodic Tafel slopes (b_a and b_c) obtained from polarization measurement are listed in Table 5. The inhibition efficiency was calculated from the expression:

$$IE\,(\%) = I_{corr(0)} - I_{corr(i)} \times \frac{100}{I_{corr(0)}}, \qquad (10)$$

where $I_{corr(0)}$ and $I_{corr(i)}$ are corrosion current densities obtained in the absence and presence of inhibitors, respectively. It is clear from Table 5 and Figures 5–10 that the addition of inhibitors to 1 N H_2SO_4 solution brings about a change in both the anodic and cathodic Tafel slopes. These results suggested that the addition of the studied inhibitors reduces the anodic dissolution and also retards the cathodic hydrogen

TABLE 6: Impedence parameters for mild steel in $1 \text{N H}_2\text{SO}_4$ in the absence and presence of optimum concentration (0.4×10^{-3} M) of variously substituted piperidin-4-ones.

Inhibitors	R_{ct} ($\Omega\,\text{cm}^2$)	C_{dl} ($\mu\text{F cm}^{-2}$)	Inhibition efficiency (%)
Blank	6.548	59.44	Nil
(01)	19.55	22.88	66.51
(02)	13.60	30.39	51.85
(03)	10.60	37.20	38.23
(05)	09.62	42.45	31.93
(06)	11.26	35.89	41.85

evolution reaction, indicating that these inhibitors influence both cathodic and anodic inhibition reactions. Further the addition of inhibitors decreased the corrosion current density (I_{corr}) significantly, which further decreases with increase in concentration of the inhibitors ($0.4–2 \times 10^{-3}$ M) and reaches a minimum value at 2×10^{-3} M concentration of inhibitors. According to Li et al. [61], if the displacement in E_{corr} is >85 mV with respect to $E°_{corr}$ (blank), the inhibitor can be viewed as a cathodic or anodic type. In our study the maximum displacement was found to be (12) mV, which indicates that the inhibitors could be considered as mixed type [62]. The behavior of Tafel slopes b_a and b_c also suggested mixed type of behavior for the inhibitors. Therefore, all the studied inhibitors can be classified as mixed inhibitors for mild steel in $1 \text{N H}_2\text{SO}_4$ solution. The inhibition efficiency was found to increase with increase in concentration of the inhibitors. The values of inhibition efficiency obtained from polarization measurements are almost equal to that obtained from impedance analysis and weight loss measurements. The inhibition efficiency values obtained follow the order

$$(01) > (02) > (06) > (03) > (05) > (04).$$

The above trend is the same as that obtained from weight loss measurement.

3.6. Electrochemical Impedance Measurements. The corrosion behavior of mild steel in acidic solution in the absence and presence of optimum concentration (0.4×10^{-3} M) of inhibitors (01–06) was investigated by the electrochemical impedance spectroscopy method at 300 ± 1 K and the impedance parameters R_{ct} and C_{dl} derived from these investigations are given in Table 6. Nyquist plots of mild steel in inhibiting and uninhibited acidic solutions containing an optimum concentration (0.4×10^{-3} M) of inhibitors are shown in Figure 11. The semicircular nature of impedance diagrams indicates that the corrosion of mild steel is mainly controlled by a charge transfer process, and the presence of the inhibitor does not affect the dissolution mechanism of mild steel [63]. The impedance spectra of the Nyquist plots were investigated by fitting the experimental data to a simple equivalent circuit model as shown in Figure 12, which includes the solution resistance R_s and the double layer capacitance C_{dl} which is placed in parallel to the charge transfer resistance R_{ct}.

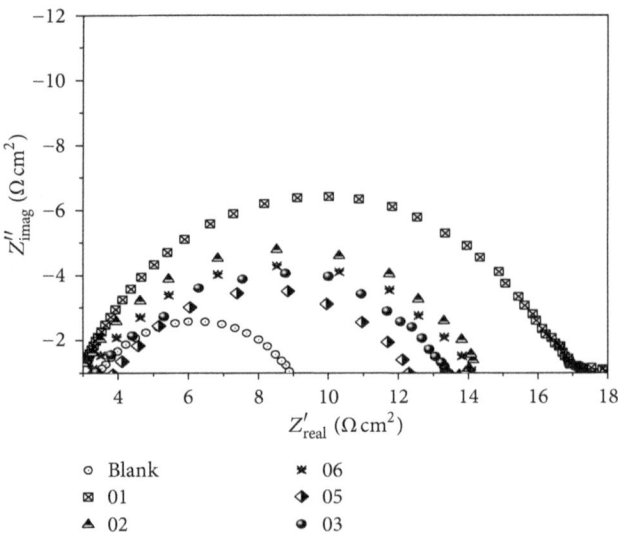

○	Blank	✳	06
⊠	01	◆	05
▲	02	●	03

FIGURE 11: Nyquist plots of mild steel in $1 \text{N H}_2\text{SO}_4$ in the absence and presence of optimum concentration of various inhibitors (01–06).

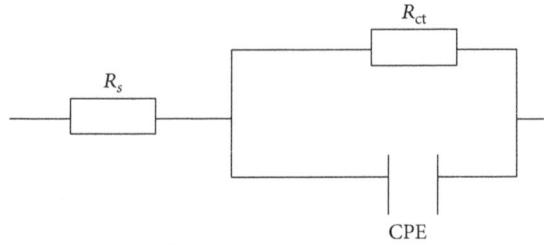

FIGURE 12: Equivalent circuit model for fitting impedance spectra.

The charge transfer resistance R_{ct} values are calculated from the difference in impedance at low and high frequencies. The R_{ct} value is a measure of electron transfer across the mild steel surface and it is inversely proportional to the corrosion rate. The double layer capacitance C_{dl} was calculated at the frequency f_{max} at which the imaginary component of the impedance is maximal using the equation [64]

$$C_{dl} = \frac{1}{2\pi f_{max} R_{ct}}. \qquad (11)$$

Analysis of the data presented in Table 6 indicates that the magnitude of R_{ct} value increased while that of C_{dl} decreased with the addition of inhibitors to $1 \text{N H}_2\text{SO}_4$ medium at optimum concentration of inhibitors. The decrease in C_{dl} values results from the adsorption of the inhibitor molecules at the metal surface. The double layer between the charged metal surface and the solution is considered as an electrical capacitor. The adsorption of inhibitors on the mild steel surface decreases its electrical capacity as they displace the water molecules and other ions originally adsorbed on the surface leading to the formation of a protective adsorption layer on the electrode surface which increases the thickness of the electrical double layer. The thickness of this protective layer

(d) is related to C_{dl} in accordance with Helmholtz model, given by the following equation [27]:

$$C_{dl} = \frac{\varepsilon \varepsilon_o A}{d}, \qquad (12)$$

where ε is the dielectric constant of the medium and ε_o is the permittivity of free space (8.854×10^{-14} F/cm) and A is the effective surface area of the electrode. From (12), it is clear that as the thickness of the protective layer, that is, the film formed by inhibitor molecules, increases, the C_{dl} should decrease. In our present studies C_{dl} value was found to be highest for uninhibited solution. Addition of optimum concentration (0.4×10^{-3} M) of inhibitors to the aggressive medium is found to decrease the C_{dl} value and also lowest value is obtained for inhibitor 01 with highest inhibition efficiency, because in this case the inhibitor exists as an equilibrium mixture of both boat and chair form and boat form of the inhibitor could adsorb through both CO and NH groups and the parallel orientation of the plain containing the C-2, C-3, C-5, and C-6 atoms of piperidone ring screens the mild steel surface from the attack of the corrosive media. The inhibition efficiency of inhibitors for the corrosion of mild steel in 1 N H$_2$SO$_4$ medium is calculated using R_{ct} values as follows:

$$IE\ (\%) = R_{ct(i)} - R_{ct(0)} \times \frac{100}{R_{ct(i)}}, \qquad (13)$$

where $R_{ct(0)}$ and $R_{ct(i)}$ are the charge-transfer resistance values in absence and presence of inhibitor, respectively. The R_{ct} value was found to be the highest for the inhibitor 01 in both acid media, indicating that the system corrodes slower in 01 when compared to others. The order of inhibition efficiency obtained from R_{ct} values is as follows:

$$(01) > (02) > (06) > (03) > (05) > (04).$$

The inhibition efficiencies obtained from R_{ct} are in good agreement with those obtained from potentiodynamic and weight loss measurements.

3.7. FTIR Spectral Studies. The FTIR spectra of 3-methyl-2,6-diphenylpiperidin-4-one (02) as well as the adsorbed film of this inhibitor over mild steel surface were recorded using Shimadzu IR Affinity-1 spectrometer. The FTIR spectra of the inhibitor and the adsorbed film are shown in Figures 13 and 14. All the variously substituted piperidin-4-ones involved in the present investigation have been shown to exist in the chair conformation with alkyl and phenyl groups in equatorial orientations. Infrared spectroscopy has been proved to be a valuable tool in the analysis of the stereochemistry of heterocyclic compounds, mainly on the basis of a series of bands in the region 2800–2600 cm^{-1} called Bohlmann bands [65]. These bands have also been used in the assessment of conformational equilibria in decahydroquinolines [66] and piperidin systems [67, 68]. Piperidin-4-ones also exhibit these bands in the region 3000–2800 cm^{-1} (Figure 13) and these bands disappear on adsorption through nitrogen (Figure 14) due to the lack of antiperiplanarity with respect to the lone pair of electrons as a result of change in conformation [69]. The sharp peak around 3296 cm^{-1} due to N–H

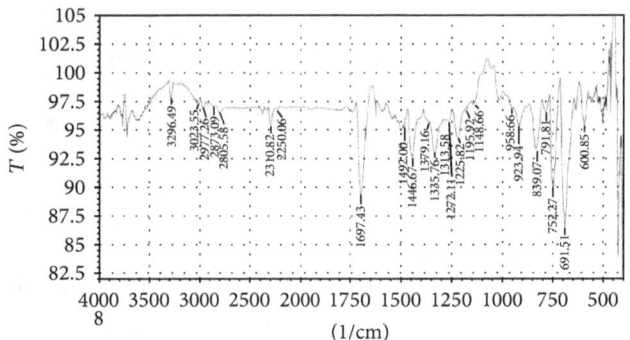

FIGURE 13: FTIR spectrum of t(3)-methyl-r(2),c(6)-diphenylpiperidin-4-one (02).

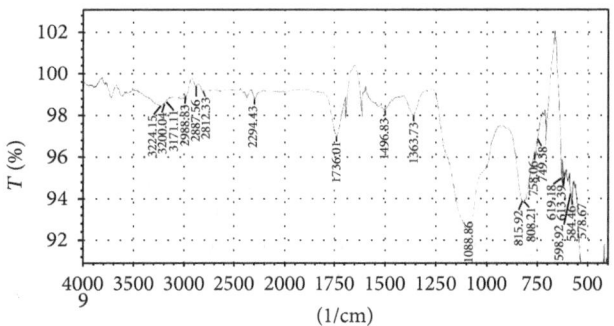

FIGURE 14: FTIR spectrum of corrosion product from mild steel surface after immersion in 1 N H$_2$SO$_4$ containing t(3)-methyl-r(2),c(6)-diphenylpiperidin-4-one (02).

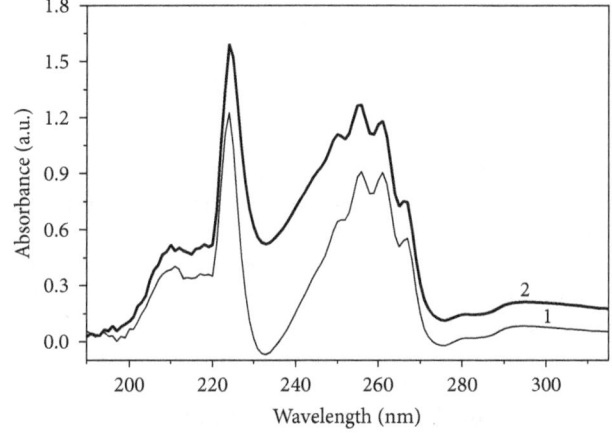

FIGURE 15: UV spectra of (2×10^{-3} M) t(3)-methyl-r(2),c(6)-diphenylpiperidin-4-one (02) used as inhibitors for mild steel corrosion in 1 N H$_2$SO$_4$ solutions [1] before and [2] after weight loss measurements.

stretching of the piperidin-4-one (02) has also been shifted to lower frequency region (3224 to 3171 cm^{-1}) and is broadened in the IR spectrum of the adsorbed film.

The carbonyl stretching frequency of the inhibitor (02) appears at 1697 cm^{-1}. This absorption peak is shifted to higher frequency (1736 cm^{-1}) in the spectrum of adsorbed film.

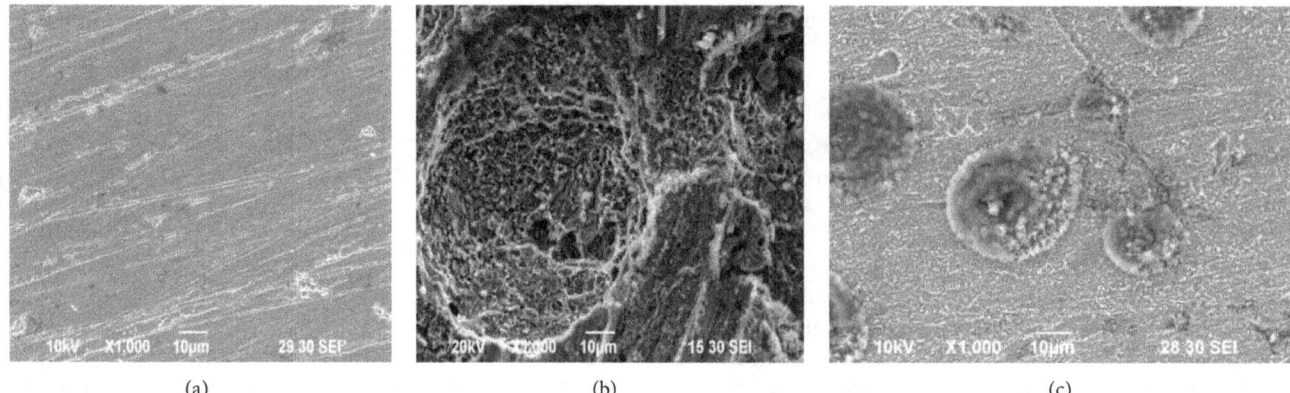

FIGURE 16: SEM photographs of mild steel sample (a) polished surface, (b) after immersion in 1 N H_2SO_4 solution, and (c) in the presence of $(2 \times 10^{-3}$ M) inhibitor (02).

Generally ring carbonyl absorption frequency will be shifted to higher frequency as the ring becomes more and more strained [70]; similar high frequency shifts of carbonyl absorption are noted in the piperidin-4-one complexes of cobalt (II), nickel (II), and copper (II) [66].

According to Rengamani et al. [1] the high frequency shift is not attributed to carbonyl participation but due to conformational change of the ring during complex formation through ring nitrogen. The IR spectrum of the free inhibitor (02) shows no appreciable change in the carbonyl shifts due to electron releasing groups. Hence in the present studies it is reasonable to assume that the high frequency shift of carbonyl group in the adsorbed film may be due to the ring strain caused by the interaction between ring nitrogen and mild steel surface. A new band is observed in the spectrum of the adsorbed film in the region 520–508 cm^{-1} and this may be attributed to metal nitrogen bonding.

3.8. UV Absorption Spectra.

The UV regions of the electronic spectra of the inhibited solutions before and after immersion of mild steel specimens were recorded using *Shimadzu UV-1800* spectrometer and are given in Figure 15. The spectra of the inhibited solution before immersion of the mild steel specimens show two bands, a high intensity band at 224 nm and a low intensity band at 256 nm. The low intensity band may be assigned to $n \rightarrow \pi^*$ transition of carbonyl group present in the inhibitors. The high intensity band may be due to $\pi \rightarrow \pi^*$ transition of the carbonyl group as well as phenyl groups. The spectra of the inhibited solution after immersion of mild steel plate show very slight shifts compared with the spectra of inhibited solution before immersion. But the absorption intensity corresponding to the above mentioned bands is high in the inhibited solution after mild steel immersion. This may be attributed to the change in the orientation of the phenyl groups of piperidin-4-ones due to adsorption on mild steel surface. The change in orientation of the phenyl groups leads to increase in ring strain and this is supported by FTIR spectrum of the inhibitor. The substitutional effect studied in this chapter also supports the presence of ring strain.

3.9. Scanning Electron Microscopy (SEM).

The discussion presented above makes it clear that piperidin-4-ones are proved to be good inhibitors for mild steel in 1 N H_2SO_4. To confirm the obtained results, electron microscope photographs of mild steel specimen were taken before and after immersion in 1 N H_2SO_4 solution and in the presence of 2.0×10^{-3} M concentration of inhibitor $t(3)$-methyl-$r(2)$, $c(6)$-diphenylpiperidin-4-one (02). The SEM photographs are shown in Figures 16(a)–16(c).

Figure 16(a) illustrates the brightly polished surface of the mild steel specimen before immersion in the test solution, while Figure 16(b) depicts the effect of 1 N H_2SO_4 solutions on the mild steel specimen after 1-hour immersion. It clearly shows large pits that are caused by the attack of 1 N H_2SO_4 solution. On comparing these microphotographs with those in Figure 16(c) it is evident that, in the presence of 2.0×10^{-3} M of inhibitor (02), the surface morphology of the mild steel specimen has been changed due to the presence of the adsorbed layer of inhibitor molecules, and the large pits shown in Figure 16(b) in the absence of inhibitor have been reduced and the depth of pits has decreased. Thus the addition of inhibitor (02) to the aggressive medium has reduced the corrosion of mild steel specimen in 1 N H_2SO_4 solution.

4. Conclusion

(1) The various substituted 2,6-diphenyl-methylpiperidine-4-ones (01–06) exhibit maximum efficiency towards the corrosion inhibition of mild steel in 1 N H_2SO_4. The optimum efficiency of these compounds was achieved even at very low concentration of the inhibitors (0.4×10^{-3} M). Thus piperidin-4-ones act as efficient inhibitors. Also, the order of inhibition efficiency is as follows: 01 > 02 > 06 > 03 > 05 > 04. At higher temperatures, the rate of corrosion is less in inhibiting solutions compared with the uninhibited solutions.

(2) Piperidin-4-ones contain two potential anchoring sites, namely, the carbonyl group and the ring

nitrogen; the results indicate the participation of only ring nitrogen.

(3) The increase in apparent activation energy in the inhibited solution suggests physical adsorption of the inhibitors on the mild steel surface.

(4) Positive values of enthalpy of activation (ΔH°) in the inhibited solution reflect the endothermic nature of mild steel dissolution. Also, the positive values of entropy of activation (ΔS°) suggest that an increase in disordering takes place in metal/solution interface.

(5) The values of ΔG°_{ads} indicate that the adsorption of inhibitors on the mild steel surface may involve complex interactions involving both physical and chemical adsorption. The FT-IR and SEM analysis supports the formation of protective film on the mild steel surface.

(6) The adsorption on piperidine-4-ones on mild steel in 1 N H_2SO_4 obeys Langmuir adsorption isotherm.

(7) The variation of Tafel constants b_a, b_c and E_{corr} values with the increase in concentration of piperidine-4-ones (01–06) suggest that these compounds act as mixed type inhibitors.

Conflict of Interests

The authors declare that there is no conflict of interests regarding the publication of this paper.

References

[1] S. Rengamani, S. Muralidharan, M. Anbu Kulandainathan, and S. Venkatakrishna Iyer, "Inhibiting and accelerating effects of aminophenols on the corrosion and permeation of hydrogen through mild steel in acidic solutions," *Journal of Applied Electrochemistry*, vol. 24, no. 4, pp. 355–360, 1994.

[2] S. S. Abd El-Rehim, M. A. M. Ibrahim, and K. F. Khaled, "4-Aminoantipyrine as an inhibitor of mild steel corrosion in HCl solution," *Journal of Applied Electrochemistry*, vol. 29, no. 5, pp. 593–599, 1999.

[3] M. A. Quraishi and D. Jamal, "Inhibition of mild steel corrosion in the presence of fatty acid triazoles," *Journal of Applied Electrochemistry*, vol. 32, no. 4, pp. 425–430, 2002.

[4] K. C. Emregül, R. Kurtaran, and O. Atakol, "An investigation of chloride-substituted Schiff bases as corrosion inhibitors for steel," *Corrosion Science*, vol. 45, no. 12, pp. 2803–2817, 2003.

[5] K. F. Khaled, K. Babic-Samardzija, and N. Hackerman, "Piperidines as corrosion inhibitors for iron in hydrochloric acid," *Journal of Applied Electrochemistry*, vol. 34, no. 7, pp. 697–704, 2004.

[6] H.-L. Wang, R.-B. Liu, and J. Xin, "Inhibiting effects of some mercapto-triazole derivatives on the corrosion of mild steel in 1.0 M HCl medium," *Corrosion Science*, vol. 46, no. 10, pp. 2455–2466, 2004.

[7] F. Bentiss, M. Lebrini, and M. Lagrenée, "Thermodynamic characterization of metal dissolution and inhibitor adsorption processes in mild steel/2,5-bis(n-thienyl)-1,3,4-thiadiazoles/hydrochloric acid system," *Corrosion Science*, vol. 47, no. 12, pp. 2915–2931, 2005.

[8] S. Zor, P. Dogan, and B. Yazici, "Inhibition of acidic corrosion of iron and aluminium by SDBS at different temperatures," *Corrosion Reviews*, vol. 23, no. 2-3, pp. 217–232, 2005.

[9] M. A. Quraishi, D. Jamal, and R. N. Singh, "Inhibition of mild steel corrosion in the presence of fatty acid thiosemicarbazides," *Corrosion*, vol. 58, no. 3, pp. 201–207, 2002.

[10] A. Popova, M. Christov, S. Raicheva, and E. Sokolova, "Adsorption and inhibitive properties of benzimidazole derivatives in acid mild steel corrosion," *Corrosion Science*, vol. 46, no. 6, pp. 1333–1350, 2004.

[11] A. Y. Musa, A. A. H. Kadhum, A. B. Mohamad, M. S. Takriff, A. R. Daud, and S. K. Kamarudin, "Adsorption isotherm mechanism of amino organic compounds as mild steel corrosion inhibitors by electrochemical measurement method," *Journal of Central South University of Technology*, vol. 17, no. 1, pp. 34–39, 2010.

[12] S. Vishwanatham and M. A. Kumar, "Corrosion inhibition of mild steel in binary acid mixture," *Corrosion Reviews*, vol. 23, no. 2-3, pp. 181–194, 2005.

[13] M. B. Balasubramanian and N. Padma, "Studies on conformation-I. Preparation and stereochemistry of some 4-piperidinols," *Tetrahedron*, vol. 19, no. 12, pp. 2135–2143, 1963.

[14] C. R. Noller and V. Baliah, "The preparation of some piperidine derivatives by the Mannich reaction," *Journal of the American Chemical Society*, vol. 70, no. 11, pp. 3853–3855, 1948.

[15] B. Vijayan, *[Ph.D. thesis]*, Annamalai University, Chidambaram, India, 1981.

[16] M. B. Balasubramanian and N. Padma, "Studies on conformation—I: preparation and stereochemistry of some 4-piperidinols," *Tetrahedron*, vol. 19, no. 12, pp. 2135–2143, 1963.

[17] T. M. Ikramuddeen, *[Ph.D. thesis]*, Bharathiar University, Coimbatore, India, 1994.

[18] K. Selvaraj, M. Narasimhan, and J. Mallika, "Cobalt(II) nitrate complexes with piperidin-4-ones," *Transition Metal Chemistry*, vol. 26, no. 1-2, pp. 224–227, 2001.

[19] S. Muralidharan, R. Chandrasekar, and S. V. K. Iyer, "Effect of piperidones on hydrogen permeation and corrosion inhibition of mild steel in acidic solutions," *Proceedings of the Indian Academy of Sciences: Chemical Sciences*, vol. 112, no. 2, pp. 127–136, 2000.

[20] A. N. Senthilkumar, K. Tharini, and M. G. Sethuraman, "Studies on a few substituted piperidin-4-one oximes as corrosion inhibitor for mild steel in HCL," *Journal of Materials Engineering and Performance*, vol. 20, no. 6, pp. 969–977, 2011.

[21] S. Sankarapapavinasam, F. Pushpanaden, and M. F. Ahmed, "Piperidine, piperidones and tetrahydrothiopyrones as inhibitors for the corrosion of copper in H_2SO_4," *Corrosion Science*, vol. 32, no. 2, pp. 193–203, 1991.

[22] A. N. Senthilkumar, K. Tharini, and M. G. Sethuraman, "Corrosion inhibitory effect of few piperidin-4-one oximes on mild steel in hydrochloric acid medium," *Surface Review and Letters*, vol. 16, no. 1, pp. 141–147, 2009.

[23] R. T. Vashi, H. M. Bhajiwala, and S. A. Desai, "Hexamine as corrosion inhibitor for zinc in (HNO_3 + H_2SO_4) binary acid mixture," *Der Pharma Chemica*, vol. 5, no. 2, pp. 237–243, 2013.

[24] K. F. Khaled, K. B. Samardzija, and N. Hackerman, "Piperidines as corrosion inhibitors for iron in hydrochloric acid," *Journal of Applied Electrochemistry*, vol. 34, no. 7, pp. 697–704, 2004.

[25] K. Babić-Samardžija, K. F. Khaled, and N. Hackerman, "N-heterocyclic amines and derivatives as corrosion inhibitors for iron in perchloric acid," *Anti-Corrosion Methods and Materials*, vol. 52, no. 1, pp. 11–21, 2005.

[26] S. K. Shukla and E. E. Ebenso, "Corrosion inhibition, adsorption behavior and thermodynamic properties of streptomycin on mild steel in hydrochloric acid medium," *International Journal of Electrochemical Science*, vol. 6, no. 8, pp. 3277–3291, 2011.

[27] I. Ahamad, R. Prasad, and M. A. Quraishi, "Thermodynamic, electrochemical and quantum chemical investigation of some Schiff bases as corrosion inhibitors for mild steel in hydrochloric acid solutions," *Corrosion Science*, vol. 52, no. 3, pp. 933–942, 2010.

[28] I. B. Obot, N. O. Obi-Egbedi, and S. A. Umoren, "Adsorption characteristics and corrosion inhibitive properties of clotrimazole for aluminium corrosion in hydrochloric acid," *International Journal of Electrochemical Science*, vol. 4, no. 6, pp. 863–877, 2009.

[29] A. Zarrouk, B. Hammouti, H. Zarrok, S. S. Al-Deyab, and M. Messali, "Temperature effect, activation energies and thermodynamic adsorption studies of L-Cysteine Methyl Ester Hydrochloride as copper corrosion inhibitor in nitric acid 2M," *International Journal of Electrochemical Science*, vol. 6, no. 12, pp. 6261–6274, 2011.

[30] C. S. Venkatachalam, S. R. Rajagopalan, and M. V. C. Sastry, "Mechanism of inhibition of electrode reactions at high surface coverages—II," *Electrochimica Acta*, vol. 26, no. 9, pp. 1219–1224, 1981.

[31] A. A. Khadom, A. S. Yaro, A. S. Altaie, and A. A. H. Kadum, "Electrochemical, activations and adsorption studies for the corrosion inhibition of low carbon steel in acidic media," *Portugaliae Electrochimica Acta*, vol. 27, no. 6, pp. 699–712, 2009.

[32] E. A. Noor, "Temperature effects on the corrosion inhibition of mild steel in acidic solutions by aqueous extract of fenugreek leaves," *International Journal of Electrochemical Science*, vol. 2, pp. 996–1017, 2007.

[33] M. Lebrini, F. Robert, and C. Roos, "Inhibition effect of alkaloids extract from *Annona squamosa* plant on the corrosion of C38 steel in normal hydrochloric acid medium," *International Journal of Electrochemical Science*, vol. 5, no. 11, pp. 1698–1712, 2010.

[34] I. B. Obot, S. A. Umoren, and N. O. Obi-Egbedi, "Corrosion inhibition and adsorption behaviour for aluminuim by extract of *Aningeria robusta* in HCl solution: synergistic effect of iodide ions," *Journal of Materials and Environmental Science*, vol. 2, no. 1, pp. 60–71, 2011.

[35] I. M. Mejeha, A. A. Uroh, K. B. Okeoma, and G. A. Alozie, "The inhibitive effect of *Solanum melongena* L. leaf extract on the corrosion of aluminium in tetraoxosulphate (VI) acid," *African Journal of Pure and Applied Chemistry*, vol. 4, pp. 158–165, 2010.

[36] D. Ben Hmamou, R. Salghi, A. Zarrouk et al., "Alizarin red: an efficient inhibitor of C38 steel corrosion in hydrochloric acid," *International Journal of Electrochemical Science*, vol. 7, no. 6, pp. 5716–5733, 2012.

[37] M. A. Migahed, H. M. Mohamed, and A. M. Al-Sabagh, "Corrosion inhibition of H-11 type carbon steel in 1 M hydrochloric acid solution by N-propyl amino lauryl amide and its ethoxylated derivatives," *Materials Chemistry and Physics*, vol. 80, no. 1, pp. 169–175, 2003.

[38] E. E. Oguzie, B. N. Okolue, E. E. Ebenso, G. N. Onuoha, and A. I. Onuchukwu, "Evaluation of the inhibitory effect of methylene blue dye on the corrosion of aluminium in hydrochloric acid," *Materials Chemistry and Physics*, vol. 87, no. 2-3, pp. 394–401, 2004.

[39] E. E. Ebenso, N. O. Eddy, and A. O. Odiongeny, "Corrosion inhibition and adsorption properties of methocarbamol on mild steel in acidic medium," *Portugaliae Electrochimica Acta*, vol. 27, no. 1, pp. 13–22, 2009.

[40] R. F. V. Villamil, P. Corio, J. C. Rubim, and S. M. L. Agostinho, "Sodium dodecylsulfate-benzotriazole synergistic effect as an inhibitor of processes on copper — chloridric acid interfaces," *Journal of Electroanalytical Chemistry*, vol. 535, no. 1-2, pp. 75–83, 2002.

[41] E. E. Oguzie, Y. Li, and F. H. Wang, "Corrosion inhibition and adsorption behavior of methionine on mild steel in sulfuric acid and synergistic effect of iodide ion," *Journal of Colloid and Interface Science*, vol. 310, no. 1, pp. 90–98, 2007.

[42] I. B. Obot and N. O. Obi-Egbedi, "Ipomoea involcrata as an ecofriendly inhibitor for aluminium in alkaline medium," *Portugaliae Electrochimica Acta*, vol. 27, no. 4, pp. 517–524, 2009.

[43] T. Umasankareswari and T. Jeyaraj, "Salicylideneaniline as inhibitor for the corrosion of mild steel in 1.0 N hydrochloric acid," *Journal of Chemical and Pharmaceutical Research*, vol. 4, no. 7, pp. 3414–3419, 2012.

[44] A. K. Singh and M. A. Quraishi, "Investigation of the effect of disulfiram on corrosion of mild steel in hydrochloric acid solution," *Corrosion Science*, vol. 53, no. 4, pp. 1288–1297, 2011.

[45] M. Bouklah, B. Hammouti, M. Lagrenée, and F. Bentiss, "Thermodynamic properties of 2,5-bis(4-methoxyphenyl)-1,3,4-oxadiazole as a corrosion inhibitor for mild steel in normal sulfuric acid medium," *Corrosion Science*, vol. 48, no. 9, pp. 2831–2842, 2006.

[46] E. E. Ebenso, I. B. Obot, and L. C. Murulana, "Quinoline and its derivatives as effective corrosion inhibitors for mild steel in acidic medium," *International Journal of Electrochemical Science*, vol. 5, no. 11, pp. 1574–1586, 2010.

[47] M. A. Amin, K. F. Khaled, and S. A. Fadl-Allah, "Testing validity of the tafel extrapolation method for monitoring corrosion of cold rolled steel in HCl solutions—experimental and theoretical studies," *Corrosion Science*, vol. 52, no. 1, pp. 140–151, 2010.

[48] A. Łukomska and J. Sobkowski, "Potential of zero charge of monocrystalline copper electrodes in perchlorate solutions," *Journal of Electroanalytical Chemistry*, vol. 567, no. 1, pp. 95–102, 2004.

[49] H. H. Hassan, E. Abdelghani, and M. A. Amin, "Inhibition of mild steel corrosion in hydrochloric acid solution by triazole derivatives. Part I. Polarization and EIS studies," *Electrochimica Acta*, vol. 52, no. 22, pp. 6359–6366, 2007.

[50] M. A. Amin, S. S. Abd El-Rehim, E. E. F. El-Sherbini, and R. S. Bayoumi, "The inhibition of low carbon steel corrosion in hydrochloric acid solutions by succinic acid. Part I. Weight loss, polarization, EIS, PZC, EDX and SEM studies," *Electrochimica Acta*, vol. 52, no. 11, pp. 3588–3600, 2007.

[51] E. E. Ebenso, H. Alemu, S. A. Umoren, and I. B. Obot, "Inhibition of mild steel corrosion in sulphuric acid using alizarin yellow GG dye and synergistic iodide additive," *International Journal of Electrochemical Science*, vol. 3, no. 12, pp. 1325–1339, 2008.

[52] L. Niu, H. Zhang, F. Wei, S. Wu, X. Cao, and P. Liu, "Corrosion inhibition of iron in acidic solutions by alkyl quaternary ammonium halides: correlation between inhibition efficiency and molecular structure," *Applied Surface Science*, vol. 252, no. 5, pp. 1634–1642, 2005.

[53] S. A. Umoren and E. E. Ebenso, "The synergistic effect of polyacrylamide and iodide ions on the corrosion inhibition of mild steel in H_2SO_4," *Materials Chemistry and Physics*, vol. 106, no. 2-3, pp. 387–393, 2007.

[54] Y. Feng, K. S. Siow, W. K. Teo, and A. K. Hsich, "The synergistic effects of propargyl alcohol and potassium iodide on the inhibition of mild steel in 0.5 M sulfuric acid solution," *Corrosion Science*, vol. 41, no. 5, pp. 829–852, 1999.

[55] K. Parameswari, *[Ph.D. thesis]*, Bharathiar University, Coimbatore, India, 2006.

[56] S. A. Umoren, M. M. Solomon, I. I. Udosoro, and A. P. Udoh, "Synergistic and antagonistic effects between halide ions and carboxymethyl cellulose for the corrosion inhibition of mild steel in sulphuric acid solution," *Cellulose*, vol. 17, no. 3, pp. 635–648, 2010.

[57] N. Hackerman and A. C. Makrides, "Inhibition of acid dissolution of metals .1. Some general observations," *Journal of Physics and Chemistry*, vol. 56, p. 707, 1955.

[58] K. Aramaki, M. Hagiwara, and H. Nishihara, "The synergistic effect of anions and the ammonium cation on the inhibition of iron corrosion in acid solution," *Corrosion Science*, vol. 27, no. 5, pp. 487–497, 1987.

[59] B. Sherine, A. J. Abdul Nasser, and S. Rajendran, "Inhibitive action of hydroquinone: Zn^{2+} system in controlling the corrosion of carbon steel in well water," *International Journal of Engineering Science and Technology*, vol. 2, pp. 341–357, 2010.

[60] G. Y. Elewady, A. H. Elaskalany, and A. F. Molouk, "Some β-aminoketone derivatives as corrosion inhibitors for nickel in hydrochloric acid solution," *Portugaliae Electrochimica Acta*, vol. 26, p. 503, 2008.

[61] W. H. Li, Q. He, S. T. Zhang, C. L. Pei, and B. Hou, "Some new triazole derivatives as inhibitors for mild steel corrosion in acidic medium," *Journal of Applied Electrochemistry*, vol. 38, no. 3, pp. 289–295, 2008.

[62] N. Goudarzi, M. Peikari, M. Reza Zahiri, and H. Reza Mousavi, "Adsorption and corrosion inhibition behavior of stainless steel 316 by aliphatic amine compounds in acidic solution," *Archives of Metallurgy and Materials*, vol. 57, no. 3, pp. 845–851, 2012.

[63] S. T. Arab and A. M. Al-Turkustani, "Corrosion inhibition of steel in phosphoric acid by phenacyldimethyl sulfonium bromide and some of its p-substituted derivatives," *Portugaliae Electrochimica Acta*, vol. 24, pp. 53–69, 2006.

[64] G. Y. Elewady, "Pyrimidine derivatives as corrosion inhibitors for carbon-steel in 2M hydrochloric acid solution," *International Journal of Electrochemical Science*, vol. 3, no. 10, pp. 1149–1161, 2008.

[65] F. W. Vierhapper and E. L. Eliel, "Conformational analysis. 38. 8-tert-Butyl-trans-decahydroquinolines: carbon-13 and proton nuclear magnetic resonance and infrared spectra. The nitrogen-hydrogen conformational equilibrium," *The Journal of Organic Chemistry*, vol. 44, no. 7, pp. 1081–1087, 1979.

[66] V. Venkatachalam, K. Ramalingam, D. Natrajan, and N. Bhavani, "*bis*(2-Aryldecahydroquinolin-4-onedithiocarbamato)-metal(II) complexes: a new preparative method and charac-

terization along with ^{13}C and ^1H NMR decoupling studies," *Synthesis and Reactivity in Inorganic and Metal-Organic Chemistry*, vol. 26, no. 5, pp. 735–759, 1996.

[67] M. Golfier, *Dermination of Configuration by Infrared Spectroscopy*, G. T. Publication, Stuttgart, Germany, 1st edition, 1977.

[68] T. Masamune, M. Tagasugi, and M. Matsuki, "Infrared-spectral studies on the orientation of the lone pairs in piperidine derivatives," *Bulletin of the Chemical Society of Japan*, vol. 41, no. 10, pp. 2466–2472, 1968.

[69] N. Manjula, *[Ph.D. thesis]*, Bharathiar University, Coimbatore, India, 2001.

[70] R. M. Silverstein, G. C. Bassler, and T. C. Morill, *Spectrometric Identification of Organic Compounds*, Wiley, New York, NY, USA, 1978.

Permissions

All chapters in this book were first published in IJC, by Hindawi Publishing Corporation; hereby published with permission under the Creative Commons Attribution License or equivalent. Every chapter published in this book has been scrutinized by our experts. Their significance has been extensively debated. The topics covered herein carry significant findings which will fuel the growth of the discipline. They may even be implemented as practical applications or may be referred to as a beginning point for another development.

The contributors of this book come from diverse backgrounds, making this book a truly international effort. This book will bring forth new frontiers with its revolutionizing research information and detailed analysis of the nascent developments around the world.

We would like to thank all the contributing authors for lending their expertise to make the book truly unique. They have played a crucial role in the development of this book. Without their invaluable contributions this book wouldn't have been possible. They have made vital efforts to compile up to date information on the varied aspects of this subject to make this book a valuable addition to the collection of many professionals and students.

This book was conceptualized with the vision of imparting up-to-date information and advanced data in this field. To ensure the same, a matchless editorial board was set up. Every individual on the board went through rigorous rounds of assessment to prove their worth. After which they invested a large part of their time researching and compiling the most relevant data for our readers.

The editorial board has been involved in producing this book since its inception. They have spent rigorous hours researching and exploring the diverse topics which have resulted in the successful publishing of this book. They have passed on their knowledge of decades through this book. To expedite this challenging task, the publisher supported the team at every step. A small team of assistant editors was also appointed to further simplify the editing procedure and attain best results for the readers.

Apart from the editorial board, the designing team has also invested a significant amount of their time in understanding the subject and creating the most relevant covers. They scrutinized every image to scout for the most suitable representation of the subject and create an appropriate cover for the book.

The publishing team has been an ardent support to the editorial, designing and production team. Their endless efforts to recruit the best for this project, has resulted in the accomplishment of this book. They are a veteran in the field of academics and their pool of knowledge is as vast as their experience in printing. Their expertise and guidance has proved useful at every step. Their uncompromising quality standards have made this book an exceptional effort. Their encouragement from time to time has been an inspiration for everyone.

The publisher and the editorial board hope that this book will prove to be a valuable piece of knowledge for researchers, students, practitioners and scholars across the globe.

List of Contributors

A. V. Afanasyev and A. A. Mel'nikov
Samara National Research University, 34 Moskovskoye Shosse, Samara 443086, Russia

S. V. Konovalov
Samara National Research University, 34 Moskovskoye Shosse, Samara 443086, Russia
Wuhan Textile University, 1 Fang Zhi Road, Wuhan 430073, China

M. I. Vaskov
"Gazprom Transgaz Samara" LLC, 106a Novo-Sadovaya Str., Samara 443086, Russia

Yingbo Hou, Deqing Lei, Shujin Li and Wei Yang
School of Civil Engineering and Architecture, Wuhan University of Technology, Wuhan 430070, China

Chun-Qing Li
School of Engineering, RMIT University, Melbourne, VIC 3001, Australia

Gino Ebell, Andreas Burkert and Jürgen Mietz
Federal Institute for Materials Research and Testing (BAM), Berlin, Germany

S. Malarvizhi
Department of Science and Humanities, Faculty of Engineering, Avinashilingam Institute for Home Science and Higher Education for Women, Coimbatore 641 043, India

Shyamala R. Krishnamurthy
Department of Chemistry, Avinashilingam Institute for Home Science and Higher Education for Women, Coimbatore 641 043, India

Muhammad Rifai, Motohiro Yuasa and Hiroyuki Miyamoto
Department of Mechanical Engineering, Doshisha University, Kyoto, Japan

Agus Maryoto and Nor Intang Setyo Hermanto
Department of Civil Engineering, Engineering Faculty, Jenderal Soedirman University, Jl. Mayjen Sungkono KM 5, Blater, Purbalingga, Jawa Tengah, Indonesia

Buntara Sthenly Gan
Department of Architecture, College of Engineering, Nihon University, 1 Nakagawara, Koriyama, Fukushima, Japan

Rachmad Setijadi
Department of Geology Engineering, Engineering Faculty, Jenderal Soedirman University, Jl. Mayjen Sungkono KM 5, Blater, Purbalingga, Jawa Tengah, Indonesia

Rachid Radouani, Younes Echcharqy and Mohamed Essahli
Laboratory of Applied Chemistry and Environment, Faculty of Science and Technology, University of Hassan 1, Settat, Morocco

A. Nikitin, L. Schleuss, R. Ossenbrink and V. Michailov
Department of Joining and Welding Technology, Brandenburg University of Technology Cottbus-Senftenberg, Konrad-Wachsmann-Allee 17, 03046 Cottbus, Germany

Jan-Ervin C. Guerrero and Drexel H. Camacho
Chemistry Department, De La Salle University, 2401 Taft Avenue, 0922 Manila, Philippines

Omid Mokhtari and Hiroshi Nishikawa
Joining and Welding Research Institute, Osaka University, 11-1 Mihogaoka, Ibaraki, Osaka 567 0047, Japan

M. Cabrini, S. Lorenzi and T. Pastore
Department of Engineering and Applied Sciences, University of Bergamo, Viale Marconi 5, Dalmine, 24044 Bergamo, Italy

T. Monetta, A. Acquesta, A. Carangelo and F. Bellucci
Department of Chemical Engineering, Materials and Industrial Production, University of Napoli Federico II, Piazzale Tecchio 80, 80125 Napoli, Italy

Jundi Geng, Junzhe Liu, Jiali Yan, Mingfang Ba, Zhimin He and Yushun Li
Faculty of Architectural, Civil Engineering and Environment, Ningbo University, Ningbo 315211, China

Qinghua Lu and Huiping Hu
Foshan University, Foshan, China

Kai Wang
Foshan University, Foshan, China
Guangdong Welding Institute (China-Ukraine E.
O. Paton Institute of Welding), Guangzhou, China

Yaoyong Yi, Jianglong Yi and Ben Niu
Guangdong Welding Institute (China-Ukraine E.
O. Paton Institute of Welding), Guangzhou, China

Zexin Jiang and Jinjun Ma
Guangzhou Shipyard International Co., Ltd,
Guangzhou, China

Cuixia Yan
Faculty of Materials Science and Engineering,
Kunming University of Science and Technology,
68Wenchang Road, Kunming, Yunnan 650093,
China

Peng Liu
Faculty of Materials Science and Engineering,
Kunming University of Science and Technology,
68 Wenchang Road, Kunming, Yunnan 650093,
China
Guangdong Welding Institute (China-Ukraine E.
O. Paton Institute of Welding), Guangdong
Key Laboratory of Modern Welding
Technology, 363 Changxing Road, Guangzhou,
Guangdong 510650, China

Shanguo Han and Yaoyong Yi
Guangdong Welding Institute (China-Ukraine E.
O. Paton Institute of Welding), Guangdong Key
Laboratory of Modern Welding Technology, 363
Changxing Road, Guangzhou, Guangdong 510650,
China

Johan Tidblad, Bror Sederholm and Simon Leijonmarck
Swerea KIMAB, Isafiordsgatan 28 A, Kista 16440
Sweden

Johan Ahlström
Swerea KIMAB, Isafiordsgatan 28 A, Kista 16440
Sweden

Division of Building Technology, Chalmers
University of Technology, Gothenburg 41296
Sweden

Luping Tang
Division of Building Technology, Chalmers
University of Technology, Gothenburg 41296
Sweden

Aisha H. Al-Moubaraki and Hind H. Al-Rushud
Chemistry Department, Faculty of Sciences, King
Abdulaziz University, Al Faisaliah Campus, Jeddah,
Saudi Arabia

Ximing Li
Chemical and Biomolecular Engineering, The
University of Akron, Akron, OH 44325, USA

Cheng Sun
State Key Laboratory for Corrosion and Protection,
Institute of Metals Research, Chinese Academy of
Sciences, Shenyang 110016, China

Qingmiao Ding, Yanyu Cui and Yujun Wang
Airport School, Civil Aviation University of China,
Tianjin, China

Liping Fang
Guangxi Colleges and Universities Key Laboratory
of Beibu Gulf Oil and Natural Gas Resource
Effective Utilization, Qinzhou University, Qinzhou
535011, China

Bian Li Quan, Jun Qi Li and Chao Yi Chen
College of Materials and Metallurgy, Guizhou
University, Guiyang 550025, China
Guizhou Province Key Laboratory of Metallurgical
Engineering and Process Energy Saving, Guiyang
550025, China

Vanessa Mandarano Pinela, Leandro Antônio de Oliveira, Mara Cristina Lopes de Oliveira and Renato Altobelli Antunes
Universidade Federal do ABC (UFABC), Centro de
Engenharia, Modelagem e Cíncias Sociais Aplicadas
(CECS), 09210-580 Santo André, SP, Brazil

Index